Industrial Air Quality and Ventilation

Controlling Dust Emissions

Industrial Air Quality and Ventilation

Controlling Dust Emissions

Ivan Nikolayevich Logachev
Konstantin Ivanovich Logachev

CRC Press
Taylor & Francis Group
Boca Raton London New York

CRC Press is an imprint of the
Taylor & Francis Group, an **informa** business

CRC Press
Taylor & Francis Group
6000 Broken Sound Parkway NW, Suite 300
Boca Raton, FL 33487-2742

First issued in paperback 2017

Version Date: 20130923

ISBN 13: 978-1-138-07619-8 (pbk)
ISBN 13: 978-1-4822-2216-6 (hbk)

Library of Congress Cataloging-in-Publication Data

Logachev, Ivan Nikolaevich.
Industrial air quality and ventilation : controlling dust emissions / by Ivan Nikolayevich Logachev, Konstantin Ivanovich Logachev.
pages cm
Includes bibliographical references and index.
ISBN 978-1-4822-2216-6 (hardback)
1. Factories--Heating and ventilation. 2. Dust control. I. Logachev, Konstantin Ivanovich, 1968- II. Title.

TH7692.L64 2014
697.9'324--dc23 2013036690

Visit the Taylor & Francis Web site at
http://www.taylorandfrancis.com

and the CRC Press Web site at
http://www.crcpress.com

Contents

Preface xi
List of Symbols xiii

Chapter 1 Dust and Air Mechanics of Bulk Material Transfer 1

1.1 Transfer Groups as Air Pollution Sources 1
1.1.1 Intensity of Dust Emissions 1
1.1.2 Primary Means of Dust Emission Control 8
1.2 Theoretical Models of Air Suction with a Gravitational Solid Stream 12
1.2.1 Butakov–Hemeon Model and Its Development 14
1.2.2 Semiempirical Models 22
1.2.3 Dynamic Theory and Research Methodology for Injection Properties of a Particle Stream 24
1.2.3.1 Mathematical Modeling 25
1.2.3.2 Experimental Studies 26
1.2.3.3 Industrial Evaluation 28
1.3 Classification of Bulk Material Streams 29

Chapter 2 Aerodynamic Properties of Particles in the Gravitational Flow of a Chuted Bulk Material 33

2.1 Peculiarities of Bulk Material Motion in Chutes 33
2.1.1 Modes of Motion 35
2.1.2 Particle Distribution 38
2.1.3 Motion Speed 41
2.2 Aerodynamic Characteristic of a Single Particle 45
2.2.1 Geometric Shape 48
2.2.2 Dynamic Shape of Particles 50
2.2.3 Resistance Coefficient 53
2.3 Sedimentation of Particles 55
2.3.1 Particle Motion in the Air Stream 55
2.3.2 Aerodynamic Drag of a Particle Moving at an Increasing Rate 57
2.4 Method for Evaluating the Aerodynamic Characteristic of Particle Gravitational Flow 65
2.4.1 Channel Pressure Variation 66
2.4.2 Experimental Evaluation of the Method for Determining the Particle Drag Factor 72

Chapter 3 Air Injection in Chutes 75

3.1 Isothermal Flow 75
3.1.1 Averaged Aerodynamic Characteristic of Particles 78
3.1.1.1 Monofractional Stream 78
3.1.1.2 Polyfractional Stream 81
3.1.2 Air Injection with a Stream of Particles in a Prismatic Chute 84
3.1.2.1 Pressure Distribution 86
3.1.2.2 Induced Air Velocity 95
3.1.3 Peculiarities of the Dynamic Interaction of Air and a Bulk Material Stream in Laminar Flow in a Chute 101
3.1.4 Air Injection in a Bin Chute with a Uniform Distribution of Particles 104
3.1.5 Air Mechanics of a Stream of Particles with High Bulk Concentrations 107
3.2 Effect of Heat and Mass Exchange 110
3.2.1 Inter-Component Heat Exchange in an Inclined Chute 111
3.2.2 Thermal Head 113
3.2.3 Air Velocity in the Chute 116
3.2.4 Influence of Mass Exchange on the Volumes of Induced Air 118
3.3 Aerodynamics of an Unsteady Particle Flow in the Chute 120
3.3.1 Sudden Change in the Material Flow 120
3.3.2 Smooth Change in the Material Flow 129

Chapter 4 The Aerodynamics of Solid-Particle Jets 133

4.1 Air Injection in a Jet of Freely Falling Particles 134
4.1.1 Initial Equations 134
4.1.1.1 Changes in Volumetric Particle Concentration in a Jet of Material 134
4.1.1.2 Volumetric Forces of Interaction between Components 137
4.1.1.3 Fluid Dynamics Equations 139
4.1.2 Structure of Air Streams in a Flat Jet of Loose Matter 144
4.1.2.1 Self-Similar Motion Equations 144
4.1.2.2 Approximate Solution of Self-Similar Motion Equation 152
4.1.2.3 Uniformly Distributed Particles 154
4.1.2.4 One-Dimensional Problem 157
4.1.2.5 Exponentially Distributed Particles 169
4.1.2.6 Effect of Pressure Gradient 175

4.1.3 Injection of Air in an Axially Symmetric Jet of Freely Falling Particles 184
4.1.3.1 Self-Similar Motion Equations 185
4.1.3.2 Solving the Self-Similar Equation 190
4.2 The Aerodynamics of a Jet of Particles in a Channel 196
4.2.1 Plane-Parallel Flow 197
4.2.2 One-Dimensional Flow 200

Chapter 5 Engineering Solutions for Dust Release Containment and Air Dedusting 213
5.1 Basic Premises for Calculating Local Suction Capacity 214
5.1.1 Initial Equations 214
5.1.2 Determining the Minimum Negative Pressure 215
5.1.2.1 Interaction between an Injected Air Jet and Suction Spectrum of a Local Suction Unit 215
5.1.2.2 Compressive Effect 219
5.1.2.3 Thermal Pressure in the Cowl 221
5.1.2.4 Optimizing the Choice of Negative Pressure 221
5.1.3 Choosing an Aspiration Layout and Calculating the Performance of Local Suction Units at Handling Facilities 222
5.1.3.1 Conveyor-to-Conveyor Transfers 224
5.1.3.2 Conveyor (Feeder)–Crusher–Conveyor 229
5.1.3.3 Conveyor–Screen–Conveyor 232
5.1.3.4 Dry Magnetic Separation Assembly 234
5.1.3.5 Cascade Installations 236
5.1.3.6 Specific Issues of Injection-Driven Air Discharge Calculations for Complex Configurations of Chutes 237
5.1.4 Calculations of Air Replacement in High-Speed Machinery 242
5.1.4.1 Hammer Breakers as Fans 242
5.1.4.2 Aspiration Volumes 246
5.2 Dust Release Intensity and Mitigation of Initial Dust Concentration in Aspirated Air 251
5.2.1 Overview and Primary Features of Dust Release Sources 251
5.2.1.1 Dust Carryover from Aspiration Cowls 251
5.2.1.2 Concentration and Particle Size Composition of Dust in Aspirated Air 253
5.2.2 Decrease in Dust Release Intensity 257
5.2.2.1 Changes in Total Dust Releases Depending on Structural and Process Parameters of Load-Handling Facilities 260

5.2.3 Techniques for Intensifying Inertial Dust Precipitation in Aspirating Cowls ... 271
5.2.3.1 Inertial Trap Using a Plate Grid inside Cowl ... 271
5.2.4 Reduction of Dust Concentration in Aspiration Funnels ... 279
5.2.4.1 Initial Dust Concentration as a Function of Air Velocity in Aspiration Funnels ... 279
5.2.4.2 Dust Precipitation in Dust Receivers/ Separators ... 281
5.2.4.3 Dust Precipitation in a Local Cyclone-Type Suction Unit/Dust Trap ... 291
5.2.4.4 Dust Precipitation in a Local Dust-Separating Suction Unit with a Filter Element ... 295
5.3 Sources of Fugitive Atmospheric Dust Releases in Outdoor Storage of Iron Ore Pellets ... 298
5.3.1 Outdoor Storage Locations as Atmospheric Emission Sources at Ore Beneficiation Plant Industrial Sites ... 298
5.3.1.1 Storage Process Layout ... 299
5.3.1.2 Primary Dust Emission Sources ... 301
5.3.2 Examining Dust Release Intensity at Iron Ore Pellet Storage Sites ... 303
5.3.2.1 Distribution of Dust Released by a Ground-Based Source ... 303
5.3.2.2 Field Surveys of Dust Plume Structure ... 309
5.3.2.3 Intensity of Primary Dust Emission Sources ... 315
5.3.2.4 Procedure for Determining Losses of Powdered Material during Stockpiling of Fired Pellets ... 318
5.3.3 Dust Release Containment in Fired Pellet Storage ... 321
5.3.3.1 Fencing Off the Flowing Material When Pellets Are Dumped into the Stockpile ... 321
5.3.3.2 Local Suction Units and Aspiration Systems of Storage Facilities ... 329
5.4 Fugitive Emissions and Containment of Dust during Loading of Iron-Ore Pellets in Railway Cars ... 339
5.4.1 Dust Emission Containment Designs for Loading Bins ... 339

5.4.2 Performance Calculations for Local Suction Units of Pellet Loading Bins ... 344
5.4.3 Improving Aspiration Efficiency for Pellet Handling in Transfer Bin Housings ... 351

Conclusion ... 359

Appendix Initial Aerodynamic Equations for a Bulk Material Stream ... 363
A.1 Phenomenological Method of Dynamic Equation Development for Two-Component Stream ... 363
A.1.1 Inter-Component Interaction ... 364
A.1.2 Accounting Equations ... 367
A.2 Space–Time Method of Averaging Accounting Equations ... 373
A.2.1 Mass Transfer Equation ... 374
A.2.2 Pulse Transfer Equation ... 376
A.2.3 Energy Transfer Equation ... 378
References ... 379

References ... 383

Index ... 393

Preface

There has always been interest in the most precise answer to the question of suction hood capacity. The lack of an in-depth analysis of aerodynamic processes and properly equipped computer facilities has meant that specialists had to be content with the simplest proportions. Typically used was an empirical approach based on crude models (if not on one's intuition) or on such vague notions as "practical data" or "countertypes." Therefore, an answer was quite often approximate: dust exhaustive plant capacity was either assumed to be within a great margin, which contributed to lower service quality and higher power consumption, or was much lower than the required values, which decreased the sanitary and hygienic effect.

This volume is devoted to studying air injection into granular material streams and to defining the closed hood capacity widely used in mechanical reprocessing of minerals. An air injection mechanism used with a solid stream has been discovered for two typical cases of bulk material flow: when transferring in closed chutes and in gravity bulking, which allowed for detailing accurate methods of aspiration volume calculation for transfer groups featuring diverse chute configurations in view of the aerodynamic connections of extract hoods.

The authors did not integrate published study findings for this subject but took a chance on familiarizing the reader primarily with findings from their own studies conducted during several years of work in the All-Union Research and Development Institute of Occupational Safety in Metal Mining Industry (VNIIBTG, Krivoy Rog) and in the Belgorod Shukhov State Technology University (BSTU), from which the members' support, assistance, and positive help are sincerely appreciated.

We also credit our teachers V. V. Nedin, O. D. Neikov, and A. V. Sheleketin, and our colleagues V. A. Minko, R. N. Shumilov, A. M. Golyshev, S. I. Zadorozhny, V. V. Kachanov, V. I. Stukanova, L. M. Chernenko, and all workers at the VNIIBTG Industrial Ventilation Laboratory and at the BSTU Department of Heat, Gas Supply, and Ventilation, whose attention and direct cooperation, creative debates, and discussions of findings enabled the authors to practically demonstrate their ideas.

The reported study was partially supported by the Council for Grants of the President of the Russian Federation (projects NSH-588.2012.8), RFBR (project number 12-08-97500-p_center_a) and Strategic Development Program of Belgorod State Technological University named after V. G. Shukhov (project number A-10/12).

Symbols

a_T	acceleration of a stream of particles in a chute, m²/s
$B(b)$	half-width of a plane jet of particles, m
c	airborne speed of particles, m/s
c_y	conventional airborne speed, m/s
c_1	heat capacity of material particles, J/(kg·K)
c_2	air heat capacity (with p = const), J/(kg·K)
D	hydraulic diameter of a chute (channel), m
d, d_E, d_e	particle diameter (sphere diameter equivalent to a particle in terms of volume), m
E	specific energy, J/kg
e	specific enthalpy, J/kg
F_{21}	interacting force between air and stream volume unit particles, N/m³
F	leakage area (F_b, upper hood; F_H, lower hood), m²
f_M, f_P	particle frontal area, m²
G	mass flow (G_1, particles; G_2, air; G_B, dry air), kg/s
g	gravity factor (g_x, chute x-direction gravity factor), m/s²
H	drop height of particles, m
$h = x = x/l_\infty$	dimensionless drop height of particles
I	intensity of interphase transformations, kg/(s·m³)
k	particle drag coefficient (k_g, k_f, k_s, geometric; k_∂, dynamic)
k_m	particle frontal area/volume ratio, 1/m
L_E, Q_E	induced airflow in a chute, m³/s
l	chute length, m
l_∞	characteristic length (inertial course length), m
M	mass force (M_1, particles; M_2, air), N/kg
m, m_P	particle mass, kg
n_P, n_1	particle count, 1/m³
n	relation of the initial particle speed in a chute to the particle speed in the chute channel
P	pressure (p_E, p_e chute injection pressure; P_T, chute thermal pressure; P_α, P_0, outside chute; P_j, chute interphase pressure), Pa
$\mathbf{P} = P/(\rho_2 c^2)$	dimensionless pressure
P_p	particle weight, N

Q_{ch}	chute airflow, m^3/s
Q_{21}	air-to-particles heat exchange rate, W/m^3
q	heat flow, W/m^2
R	aerodynamic drag of bombarding particles, N
R_{21}	air impact on solid particle, N
P_{Π}	aerodynamic force of stream particle, N
R, R_0	aerodynamic force of single particle, N
R_{ch}	chute hydraulic characteristic, kg/m^7; $Pa/(m^3/s)^2$
S	area of particles flow section, m^2
S, S_{ch}	cross sectional area of a chute (channel), m^2
s	surface (s_P, particles; s_L, sphere), m^2
T	temperature,°K
T_{2mean}	mean air temperature in a chute, °K
T_0	average air temperature outside a chute, °K
t, τ	time (τ_∞, relaxation time), s
V	volume (V_P, particle volume), m^3
υ, v, ϑ	velocity (υ, υ_1, particles; υ_{1k}, υ_k, particles at the chute outlet; υ_{10}, υ_{1H}, particles at the chute inlet; υ_2, u, air), m/s
u_{BX}	exhaust pipe entry section air velocity, m/s
$w = \upsilon - u$	relative particle velocity, m/s
w	material humidity,%
x	path of particles over a chute, m
α	interelement exchange ratio (α_m, mass, kg/(s·m²·K); α_T, α, heat, W/(m²·K)
β	volume concentration (β_1, particles; β_2, air), m^3/m^3
β_T	air thermal expansion coefficient, 1/°K
ε	air-to-particles density ratio
ζ	local drag factor (LDF)
η, μ	absolute viscosity coefficient, Pa·s
θ	horizontal chute angle
λ	hydraulic resistance coefficient
λ_g	air thermal conductivity, W/(m·K)
ν	air kinematic viscosity coefficient, m^2/s
$\vec{\Pi}$	surface force vector, N/m^2
Π_C	material particle constraint ratio, without unit of measurement
Π_d	dynamical interference activity factor, without unit of measurement

ρ	density (ρ_1, ρ_m, particle material; ρ_2, ρ, particle stream air; ρ_0, air outside a chute; ρ_{2H}, ρ_{2K}, air at the chute inlet and outlet), kg/m^3
τ	time, s
τ	tangential stress, Pa
φ, φ_k	component slip ratio (relation of the induced air speed to the particle speed at the chute outlet), without unit of measurement
ψ	particle resistance coefficient (ψ_0, particles in the area of self-similarity; ψ_{0L}, sphere in the area of self-similarity; ψ_c, airborne particles; ψ^*, stream particles), without unit of measurement

CRITERIA

$Re = wd\rho/\eta$	Reynolds number
$Fr = gh / \upsilon_1^2$	Froude number
$Fr^* = G_1 g / \left(\upsilon_1^3 b \rho_1\right)$	modified Froude number
$Bu = \psi^* k_m G_1 \upsilon_{1k} / \left(\sum \zeta a_T S_{ch} \rho_1\right)$	Butakov–Neikov number
$Eu = S_{ch} \frac{c_y}{2} \rho_0 / (G_1 v_{1k}), Eu^* = \Delta p / \left(0,5 \sum \zeta \upsilon_{1k}^2 \rho_2\right)$	Euler number
$Gr = \beta_T \frac{gH^3}{\nu^2} (T_{2av} - T_0)$	Grashof number
$Nu = \alpha d / \lambda_g$	Nusselt number

1 Dust and Air Mechanics of Bulk Material Transfer

Bulk material transfer (gravity transportation by chutes) is the most widespread operation for reprocessing mineral raw materials: mining and beneficiation of ore and coal, sintering of concentrates, stock preparation in ferrous and nonferrous metallurgy, and production of building materials. Bulk material flow results in considerable dust emission. With the great volume of mineral raw materials that are reprocessed, such dust emissions significantly impact the overall balance of airborne atmospheric pollution. Dust emissions are dangerous not only from the standpoint of toxicity and occupational disease but also because of the negative impact on the environment.

Ore preparation plants that serve major iron ore deposits are primary sources of dust emissions in terms of capacity and diversity. Highly intensive bulk material transfer operations at plants such as the Northern, Novo-Krivorozhskiy, Southern, and Inguletskiy mining and concentration complexes of Krivbass; the Lebedinskiy, Mikhailovskiy, and Stoylenskiy mining and concentration complexes of the Kursk Magnetic Anomaly (KMA) basin; the Kostomukshskiy, Olenegorskiy, and Kovdorskiy mining and concentration complexes of the southwestern district of Russia; and the Kachkanarskiy (Ural) and Sokolovsko-Sarbayskiy (Kazakhstan) mining and concentration complexes feature any number of dust-producing sources: crushed ore, iron ore concentrate, agglomerate, pellets, bentonite, limestone, and charred coal. The most environmentally unfriendly are agglomerate and pellets generated from sintering of fine-grained concentrate. Transferring such materials produces a major dust release (e.g., when loading and unloading rail cars or stacking unused materials in storage).

The main cause of dust discharge is ejection, that is, directional air flow formation within a stream of a bulk material resulting from interaction between bombarding particles and air. Studying regularities in induced air flow occurrence enables forecasting air pollution levels and aerosol emission, thereby making it possible to select the optimum engineering solutions for air containment and dedusting. This can be demonstrated using bulk material transfer technology in ore preparation plants as an example, including the diversity of materials, the material handling processes, and the process equipment.

1.1 TRANSFER GROUPS AS AIR POLLUTION SOURCES

1.1.1 Intensity of Dust Emissions

In terms of atmospheric pollution, dust transfer groups are conventionally divided into external and internal types. The dust emissions from outdoor (external) transfer groups pollute the ground level air of mine sites. Internal transfer groups are located in production areas and pollute the intrashop air. The dust generation mechanism is the same for both and differs only in dust cloud propagation. Although dust particles in a shop are transported exclusively by means of diffusion and convective air flows when transferring hot materials, the outdoor process is supplemented by wind force.

An immediate dusting of the ground level air occurs:

- When conveying, grating, or breaking the ore mass (typical of the cyclical and continuous method of ore delivery in open-cut mines)
- When feeding receiving funnels of dressing plant primary crushers
- When discharging agglomerate raw materials from indurating and sintering machines
- When loading rail cars with agglomerate and burnt pellets
- With outdoor storage and the blending of bulk mining materials
- With open-cut mines
- In mining and concentration complex plants

The intensity of dust emissions depends on the type of process operations and the physical and mechanical properties of the reprocessed material as well as the availability of dust control arrangements (Table 1.1).

The transfer of agglomerate and pellets results in the highest dust emission intensity. This can be demonstrated by analyzing the specific gross dust emissions by iron ore integrated works and by reprocessing operations in general (Figure 1.1). Gross dust emissions from all transfer groups at sintering plants (such as the sintering plants of YuGOK and NKGOK and the pellet plants of SevGOK) are greater than dust emissions at crushing and dressing plants (InGOK). This excess is noticeable in specific dust emissions in terms of mass (q, kg per ton of reprocessed material) and volume (Q, thous. m^3/t; i.e., aspirated dusty air volume per ton of reprocessed material). Sintering and pelletizing processes are much "dustier" than dressing and crushing processes. This is also noticeable when feeding conveyers (Figure 1.2): due to high strength and apparent humidity, natural minerals (e.g., iron ore) feature much poorer dust-making properties than artificial materials resulting from thermal treatment (agglomerate, pellets). The greatest amount of dust-making results from loading agglomerate and pellets in rail cars (hopper cars, pellet cars, dump cars) and from stocking operations (Figure 1.3).

Dust generation, when transferring bulk materials, is mainly caused by dust fractions that have been suspended for a time. Dust fractions result from mechanical reduction of minerals in crushers and mills, as well as from the impact of bombarding particles with each other and with chute walls.

Strong minerals are reprocessed in the metal mining industry; therefore, dust could be formed mainly out of fine fractions present in transferred materials. More fractions are found with artificial materials such as iron-ore pellets and agglomerate. Fraction content, in this case, is also determined by the quality of charging material and the evenness of its sintering in indurating machines. For instance, pellet firing in pipe furnaces (Poltavskiy mining and concentration

TABLE 1.1
Intensity of Bulk Material Transfer Dust Emissions

Equipment or Process Description	Intensity of Dust Emissions	
	Absolute, g/s	Specific, g/t
1. Iron ore conveying in an open-cut mine		
(a) w/o dust control arrangements	0.4–3.0	3–22
(b) w/suction devices	0.03–0.3	0.02–2.2
2. Iron ore conveyer		
(a) w/o dust control arrangements	0.1–0.4	0.7–3
(b) w/iron ore sprinkling devices	0.05–0.2	0.3–1.5
3. Rumbling when screening ore at the CPT site		
(a) w/o means of containment	0.8–1.0	4–5
(b) w/ventilated hoods	0.07–0.09	0.3–0.5
4. Transferring iron ore from the conveyer to the CPT site storage stockpile		
(a) w/iron ore sprinkling	0.1–0.12	0.5–0.6
(b) w/containment of dust emissions	0.015–0.03	0.03–0.06
5. When breaking ore using self-propelled crushers		
SDA-300 (a) w/o means of containment	0.5–0.7	2.5–3.5
(b) w/suction devices	0.1–0.12	0.5–0.6
SDA-1000 (a) w/o means of containment	0.8–1.7	1.6–3.6
(b) w/suction devices	0.5–0.7	1.0–1.4
SDA-2000 (a) w/o means of containment	7–11	8–12
(b) w/suction and hydraulic dust control devices	1.8–2.3	2–2.5
6. Storing of chalky marl stones using ZP-5500 stocker (w/o dust control arrangements)	8–12	3–4
7. Transferring iron ore from a dump car into a short-shaft crusher receiving funnel		
(a) w/o dust control arrangements	16–30	1.6–3
(b) w/suction devices	2.5–5	0.3–0.5
8. Discharging agglomerate from sintering machine into a hopper		
(a) w/o ventilated tunnel	20	500
(b) w/suction devices	4	100
9. Discharging burnt pellets from bins into a hopper		
(a) w/o ventilated tunnel	15	300
(b) when loading via a telescopic chute	7	140
(c) w/ventilated tunnel	3	60
10. Transferring iron-ore pellets from UK-550 stacker to a stock pile		
(a) w/o dust control arrangements	15	30
(b) w/water sprinkling	8	16
(c) w/ventilated hoods	2	4

continued

TABLE 1.1 (Continued)
Intensity of Bulk Material Transfer Dust Emissions

Equipment or Process Description	Intensity of Dust Emissions: Absolute, g/s	Intensity of Dust Emissions: Specific, g/t
11. Transferring iron-ore pellets from a conveyer to UK-550 stacker beam conveyer		
(a) w/o dust control arrangements	3–7	6–14
(b) w/ventilated hoods	0.3–0.8	0.6–1.6
12. Transferring from 2R-550 rotary reclaimer wheel buckets when delivering pellets from a stock pile		
(a) w/o dust control arrangements	20	40
(b) w/hydraulic dust control devices	12	25

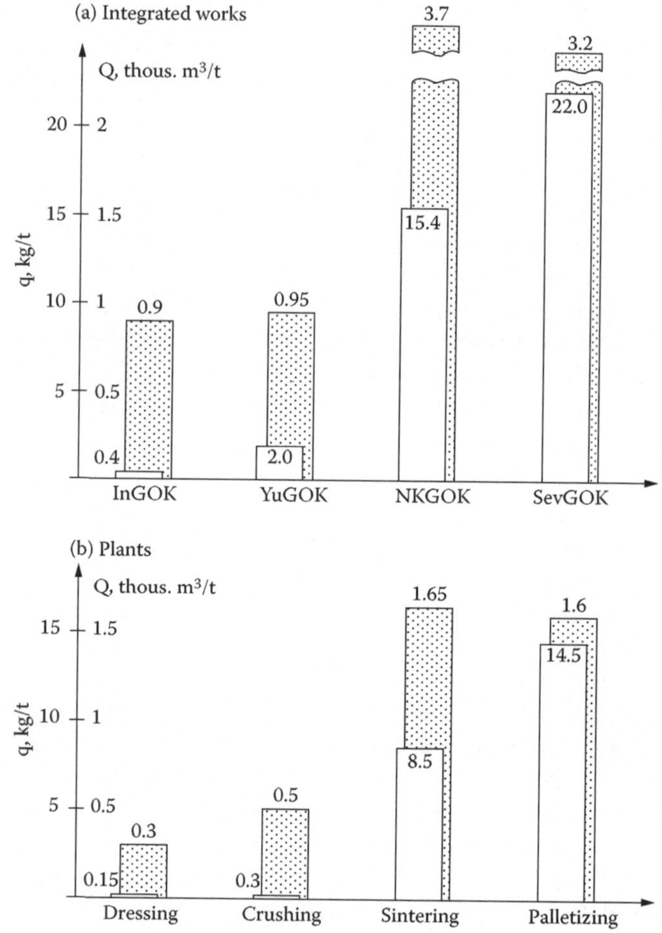

FIGURE 1.1 Specific dust emissions at Krivbass mining and concentration complexes.

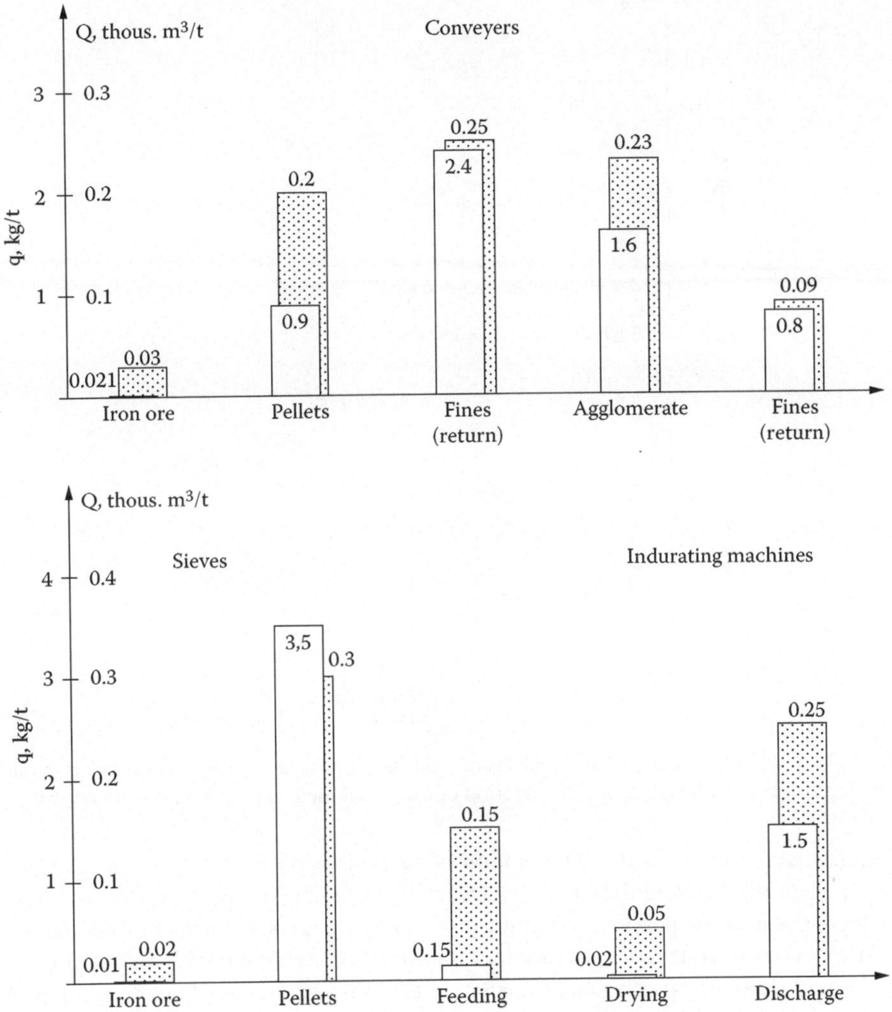

FIGURE 1.2 Specific dust emissions of ore preparation plants' processing equipment.

complex), where even heat treatment conditions are more favorable, yields stronger products with less dust content, especially compared to firing using conveyer-type machines.

The three successively alternating stages of dust emissions in bulk material transfer are:

- Free-falling material stream aeration
- Dynamic interaction of a particle stream bombarding at an increasing rate with air in transfer chutes
- Bleeding of induced dusty air from the stream when stacking particles on the conveyer belt

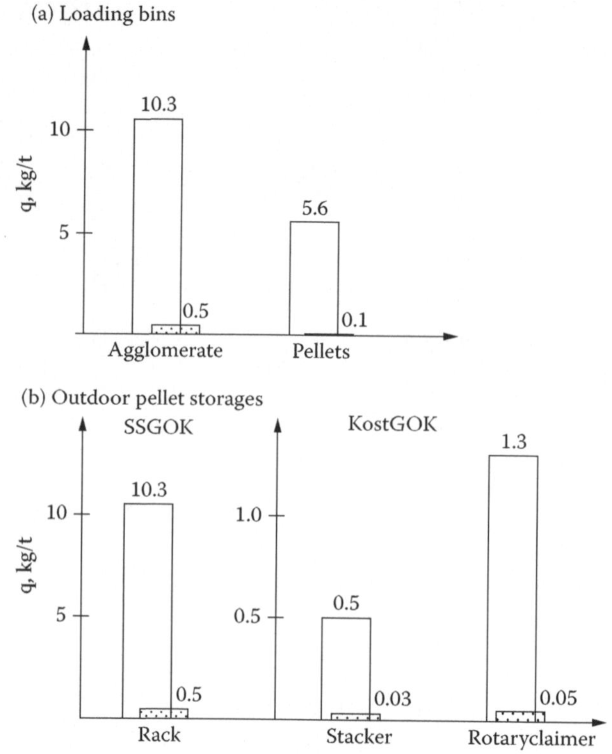

FIGURE 1.3 Specific dust emissions from land-based sources of iron ore sintering plants (the lower level results from the introduction of technical means described in Chapter 5).

The first stage features the interruption of self-adhesion forces between dust particles when discharging the material stream from the upper conveyer driving drum or feeder. An air dispersion system or dust aerosol begins to form. In free fall, the particle conglomerate discontinuity increases due to interaction with air and the collision with coarser particles and transfer chute walls. The induced air flow intensely fills with dust particles and forms an adhering jet of dusty air when bulk material is stacked on the lower conveyer.

Two facets of this stage are an inertial separation of particles and their sedimentation on the piled material surface and a blow-off of settled particles into the atmosphere. Therefore, the intensity of dust emissions is significantly influenced by the transferred material's humidity (which enhances the self-adhesion of fine particles) and by the material's pouring height (which determines the stream falling rate and the intensity of the dynamic interaction between particles and air).

Multiple experiments (see Chapter 5) showed that the primary factors determining the intensity of dust emissions are (Figure 1.4):

(a) Process parameters and physical and mechanical properties of bulk material:
- Material humidity (W, %)

- Particle-size distribution—mean particle size (d_{mean}) and dust fraction weight content (a_d, %)
- Material flow rate (G_m, kg/s)
- Temperature (T_m,°K)
- Density of particles (ρ_m, kg/m^3)

(b) Design parameters of transfer chutes and hoods:

- Transfer height, or pouring height (H, m)
- Shape of chutes—inclination of straight portions of chutes (α_i, deg.), height of the same (H_i, m). and cross-sectional area (S_i, m^2)
- Hood type, which defines the optimum vacuum-gauge pressure (P_{opt}, Pa) and air injection resistance ($\Sigma\zeta$)
- Hood pressurization degree, which defines the leakage area (F_l, m^2)

Most of the parameters influence induced air volume, which defines dust discharge from hoods immediately in terms of lack of suction due to so-called unorganized sources of dust emissions and through suction volumes when such sources become unorganized. Air injection defines induced emission volume and has a significant effect on exhaust air dust concentration.

The quantitative interrelation among these parameters was first determined by V. A. Minko [61] and his students. He determined that the dust concentration depends on the weight content of dust fractions in the transferred material, a_d (particles finer than d_{max}, that is, the maximum diameter of dust particles blown out from

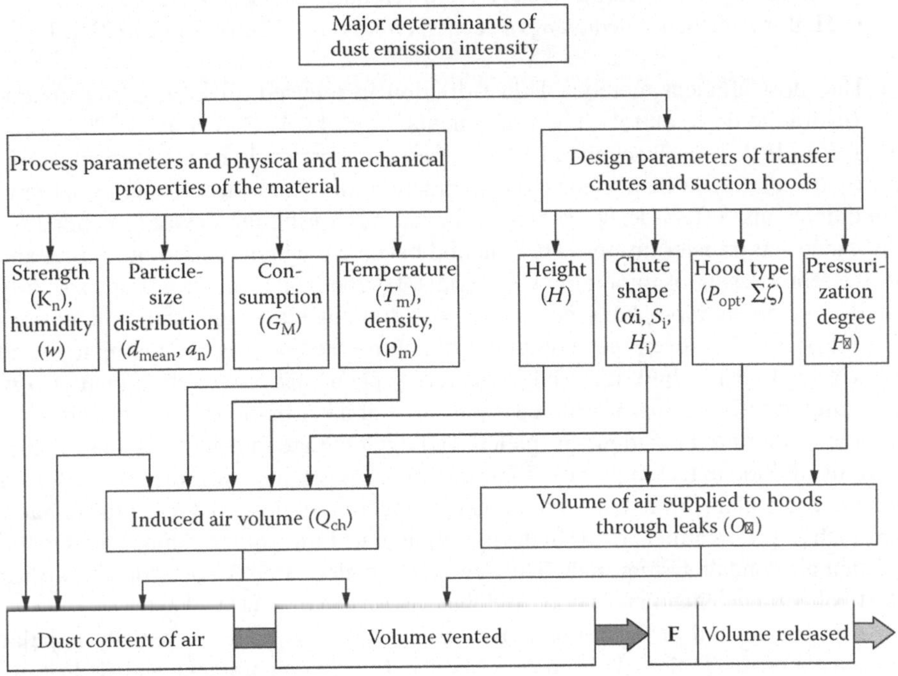

FIGURE 1.4 Major determinants of gross dust emissions in the transfer of bulk materials.

the hood). The maximum diameter value, in its turn, depends on the induced air flow, Q_{ch}, m³/s; on the suction volume, Q_a, m³/s; and on the geometric dimensions of the hood [204,205]:

$$d_{max} = 5780 \cdot \sqrt{\frac{Q_a}{\rho_M \cdot S_{\Pi} \cdot \left(1 + 0{,}08 \cdot \frac{Q_a \cdot S_{ch} \cdot L}{Q_{ch} \cdot S_{\Pi} \cdot H}\right)}},$$

where ρ_M is a density of particles, kg/m³; S_{ch}, S_{Π} are cross-sectional areas of the chute and dust-collecting bag, m²; H is the hood height, m; and L is the distance between the chute and the dust-collecting bag, m.

Dust discharge from a ventilated hood is similar to dust particle gravity sedimentation in a dust chamber: the bigger the hood and the lower the induced air volume, the lower the maximum size of particles blown out with the exhaust air, thereby resulting in lower dust content at the hood outlet.

1.1.2 Primary Means of Dust Emission Control

An analysis of current industrial ecology applications at ore preparation plants highlights three main trends for dust emission control in the transfer of bulk materials (Figure 1.5):

- Dust dilution in induced air [150,154,155,167,206]
- Reduction of air volume exhaust from ventilated hoods [130,150, 207,208]
- High-performance dedusting of suction emissions [123,164–166,209,210,211]

The most efficient method of dust dilution in induced air is watering materials (hydraulic dust control). The fundamental work by V. P. Zhuravlev [29], A. A. Tsytsura [212], I. G. Ishchuk [213], and their students explains the mechanism of dust particulate and dispersed liquid interaction, discloses the optimum operating conditions, and offers design solutions for various sprinkling devices intended for bulk material transfer groups. This method became commonly used in mining and in the reprocessing of mineral raw materials. Hydraulic dust control is successfully used at ore preparation plants, at crushing and dressing plants, and with iron ore conveyer systems. However, the hydraulic dust control method was not commonly used in heat treatment of bulk materials at sintering plants because of additional energy consumption (for drying of watered material) and deterioration of production quality due to thermal breakdown of pellets and agglomerate in drip irrigation. That is why, in addition to techniques used for forming an indiscrete mass of the transferred material, the plants utilize dry methods for reduction of dust content in the exhaust air, such as pre-treatment of air in the direction of its flow from the chute outlet to the suction air conduit system inlet. This method is widely used for developing various dust-collecting elements for hoods and dust-collection bags (see Chapter 5).

The dry method of dust emission control (suction) is more universally popular and, as seen in Table 1.1, is more effective for air containment and dedusting. Therefore, of the three trends in dust emission control, the second is the most

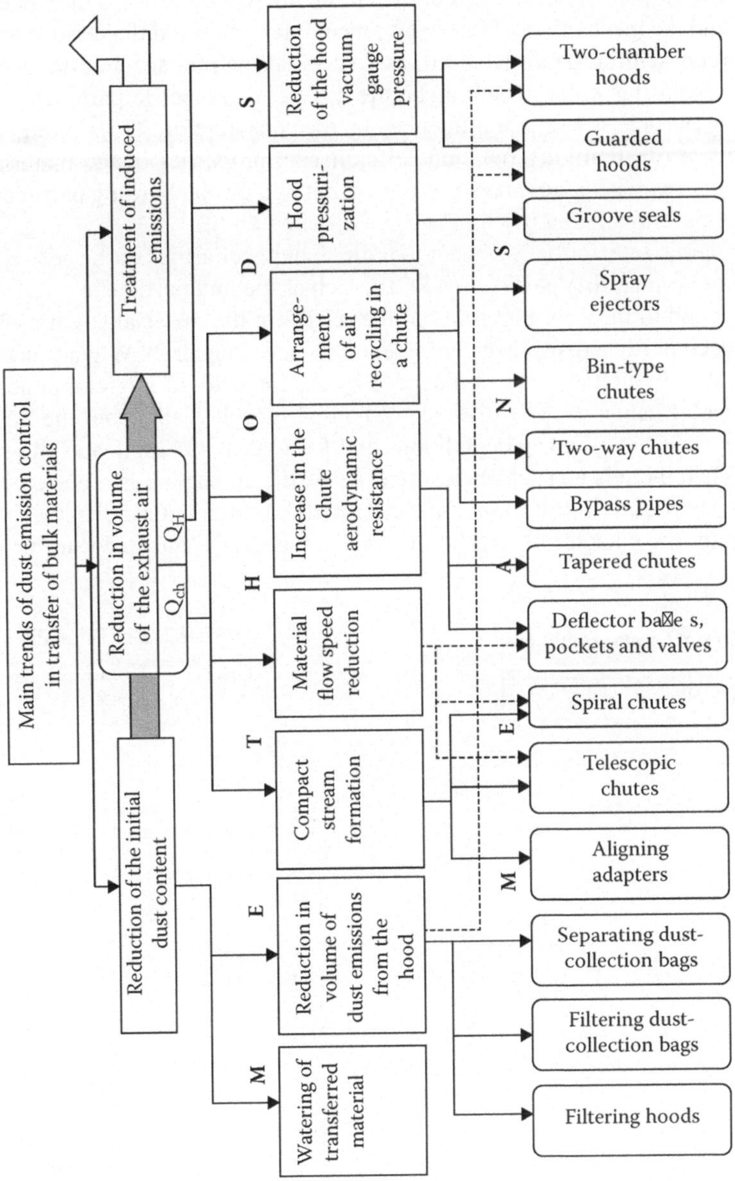

FIGURE 1.5 Primary methods and means of dust emission control in the transfer of bulk materials.

significant: reducing the induced air volume by controlling the air ventilation processes and sealing the hoods. By minimizing the output of suction hoods, it is possible to decrease the suction emission volume and significantly reduce the power consumption of ventilation units.

In order to implement effective control of the air suction process, it is necessary to understand the mechanism of intercomponent interaction and the regulation of the particle stream within the directed air, as well as taking into account the peculiarities of the enclosing walls' location (Figure 1.6). The geometric parameters of the bombarding particle stream are influenced by the consumption (G_M), initial velocity (υ_{init}), fineness (d), humidity (w), and self-adhesion properties of the material particles (σ_{self}). Stream behavior and structure are defined by bombarding particle velocity (υ), cross-sectional area (R), and particle distribution (β).

This dynamic interaction is subject to individual peculiarities of the aerodynamic resistance of bombarding particles (ARBP), such as the unit particle resistance coefficient (ψ_0), and to the common traits of the ARBP in the material stream—known as the reduced particle resistance coefficient (ψ^*) (see Chapter 2). When transferring hot materials, air suction is also influenced by the intensity of intercomponent heat exchange (see Chapter 3). The distance of non-permeable walls from the flow axis (r_0) creates various air leakage conditions and facilitates or complicates the suction process. When there is no such enclosure ($r_0 \rightarrow \infty$), the air suction is represented by a free flow of particles. In this case, an accelerated flow stream of induced air occurs in the stream (see Chapter 4). As the material stream nears the enclosure walls, air

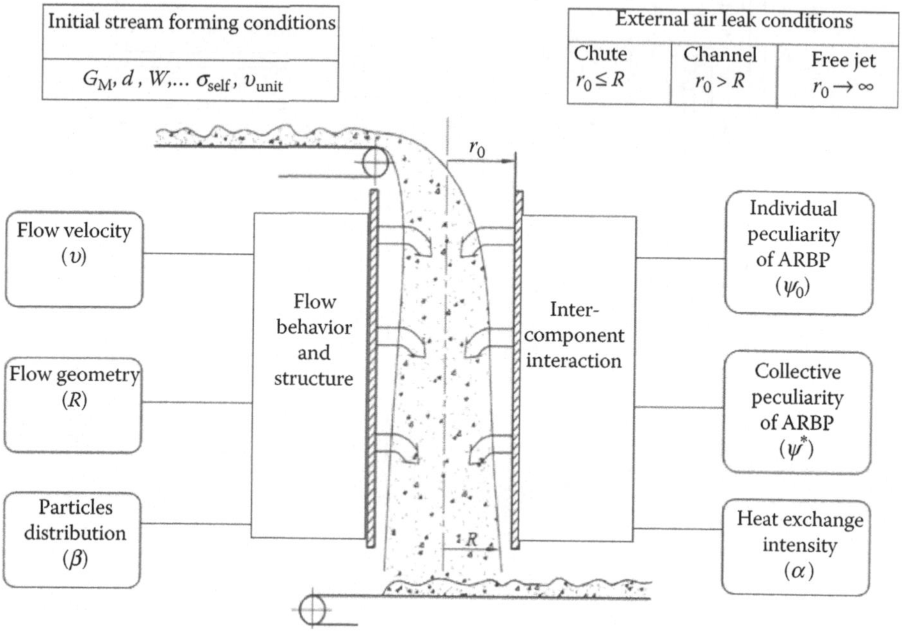

FIGURE 1.6 Qualitative structure and key factors that define the process of air suction with a bombarding particle stream.

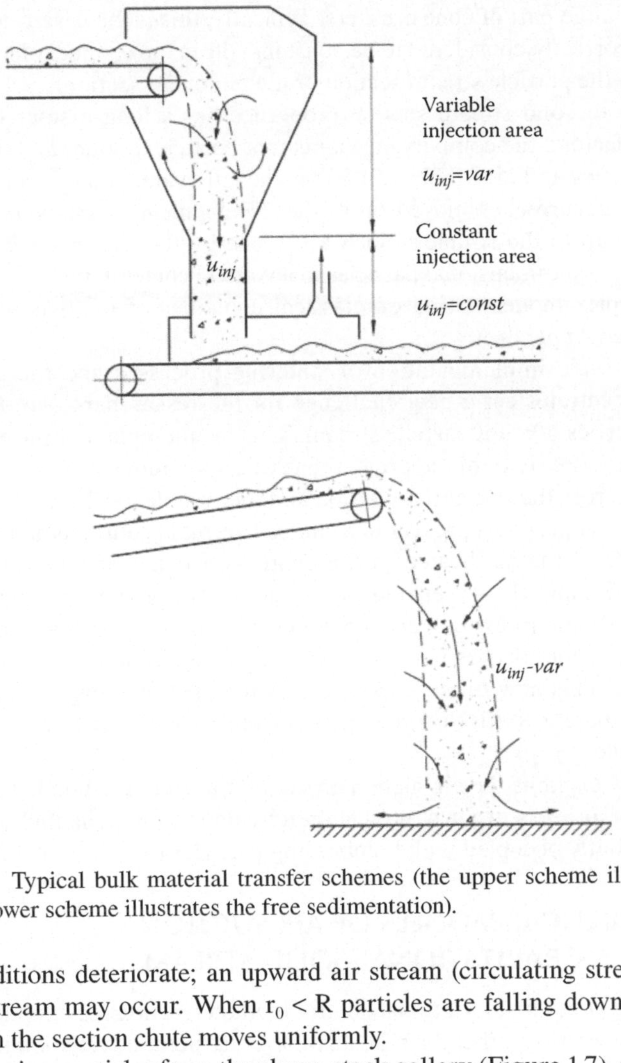

FIGURE 1.7 Typical bulk material transfer schemes (the upper scheme illustrates chute transfer; the lower scheme illustrates the free sedimentation).

leakage conditions deteriorate; an upward air stream (circulating stream) and/or a downward stream may occur. When $r_0 < R$ particles are falling down, the induced air formed in the section chute moves uniformly.

When pouring particles from the above-stack gallery (Figure 1.7), a free jet may be observed. In general, the most common chute transfer has combined leakage conditions. The most favorable air leakage conditions form at the receiving funnel inlet. First, the induced air jet is formed (accelerated suction area), then a uniform flow of induced air occurs (constant suction area), where particles enter into the straight portion of the small section chute ($r_0 < R$). This correlation between areas may be different in practice, however. In a receiving funnel, the chute height usually is much greater than the drop height, which impacts suction at the particle inlet stream.*

In bunker-type chutes, where the initial portion is much greater than the height of straight portions, transfers occur regularly—such as in chutes adjacent to sieves

* Nearly all design method guidelines skip the accelerated suction area except for OST 14-17-98-83 [73].

or to the discharge part of cone crushers. Typically, this is the case when the suction process is incorrectly considered to be constant within the channel of a phantom section (equal to the particle stream section or the bin outlet section).

The study of solid stream suction properties has a long history detailing suction process factors, the complex mechanism of particle motion, and the interaction between particles and air (Table 1.2) Experimental evaluation of suction properties in individual occurrences moved on to the development of mathematical models. These ranged from the simplest, such as an energy theory for uniformly accelerated stream of equidimensional particles in a vertical chute of uniform cross-section, to more complex models based on classical equations of multicomponent stream mechanics (see Appendix).

The large-scale implementation of sintering processes and the pelletizing of iron ore concentrates set a new challenge for the researchers—to determine the suction properties of a hot particle stream. This meant replacing the energy theory model with a more dynamic approach that treats air movement in a chute that is the result of forces that we call induction and thermal heads. Induction is the aerodynamic force of particles present in a chute. Thermal heads account for buoyancy forces that affect the air heated in the chute as a result of intercomponent heat exchange. This new theory enabled us to solve the problem of air suction and heated particles and to explain certain experimental facts, such as why reverse air flow (or anti-suction) occurs in a chute when unheated sand is poured into it (A. S. Serenko [85]). This new theory also explains the pressure surges that result when bulk material begins to fill (or stops filling) a pressurized vessel with a bulk material (see Chapter 2).

This theory explains the air suction process for a stream of bombarding particles and a complex process of air stratification (circulation) in a channel when a cross-section is partially occupied with bombarding particles (see Section 4.2).

1.2 THEORETICAL MODELS OF AIR SUCTION WITH A GRAVITATIONAL SOLID STREAM

When looking at the history of dust control method (suction) development from the quantitative (scientific) rather than the qualitative (structural) viewpoint, two periods of study should be considered.

The first period (1941–1949) is marked by experimental study of the suction process as a technical means of dust emission source containment. The most well-known studies are those conducted by Altmark, Rekk, Stakhorskiy, and Naumov in the Soviet Union and by Pring in the United States. These studies focused primarily on the problem of quantitatively assessing the phenomenon of air injection into a bulk material stream.

The second period of study, focusing on air injection research, may, in turn, be divided into two stages. The first stage involves suction property assessment in terms of energy. The fundamental efforts in this field were a study by S. E. Butakov (1949) of uniformly accelerated and distributed particle streams in a chute and the experimental study injecting air into a stream of water drops that was conducted in Utah by Pring, Knudsen and Dennis (1949). This field of study was further advanced in

TABLE 1.2
Studies of Solid Stream Suction Properties

Effects, Regularities	Methods, Notions	Authors
	Experimental Estimates	
Air movement in a vertical pipe when pouring sand (suction).	Inclined velocity of particles considered $u_{inj} = 0,48 v_k$.	M. K. Altmark 1941
Reverse air flow when sand is moving in a chute.	Velocity and flow rate of particles as well as the chute cross-section considered.	A. S. Serenko 1953 [85]
	The same.	M. T. Kamyshenko 1955 [37]
	The same.	A. V. Sheleketin 1959 [102]
	All key factors considered.	E. N. Boshnyakov 1965 [11]
	The same.	Degner and Futterer 1969 [107]
	Mathematical Models	
A. Energy theory (based on the equation of the law of variation of kinetic energy of a stream of particles)		
	Subject to the analysis of the variation of kinetic energy of the uniformly accelerated stream of particles, there was an analytical relation obtained with the aim of determining the induced air flow rate.	S. E. Butakov 1949 [15]
	The same, the induction ratio notion was introduced.	O. D. Neykov 1965 [66]
Reduction in volume of the induced air with increase in the material flow rate.	The same as for powder material, "particle packet" and "nominal diameter" notions were introduced.	V. A. Minko 1969 [60]
B. Dynamic theory (based on the equation of variation of momentum of "solid particles-air" double speed continuum)		
Inhibiting effect on the volume of induced air of a stream of particles at the chute inlet. Reverse air flow in a chute when transferring particles at a high temperature (induction inversion).	There was the dynamic equation of the uniform air flow in a chute accounting of bulk forces of the dynamic and thermal interaction of components. The induction head notion was introduced.	I. N. Logachev 1969 [49]
Pressure surge when starting and ending to fill a sealed bin with bulk material.	There was an experimental method of determining the aerodynamic resistance of a group of bombarding particles in a chute (pressure measuring method).	1969 [52]
	Analytic studies of transient processes for an unsteady heated solid stream.	1974 [68]

continued

TABLE 1.2 (Continued)
Studies of Solid Stream Suction Properties

Effects, Regularities	Methods, Notions	Authors
	Analytic studies of the boundary-layer equation for a jet of air induced by a stream of bombarding particles.	1981 [69]
	There was a possibility of air circulation in a chute analytically demonstrated when the chute was partially filled with bombarding particles.	1987 [42]

the Soviet Union by O. D. Neykov (1965), E. N. Boshnyakov (1965), and V. A. Minko (1969), and in the United States by Hatch (1954), Hemeon (1955), Anderson (1964), and Cruise and Bianconi (1966). The second stage will be discussed in Section 1.2.3 of this chapter.

1.2.1 Butakov–Hemeon Model and Its Development

The Butakov–Hemeon model was built on the assumption that part of the momentum energy of a stream of particles E_1 is lost when surmounting environmental resistance. These losses are determined through the material particles' air drag R_0:

$$dE_1 = N_k \cdot R_0 \cdot dx = N_k \cdot R_0 \cdot v_1 \cdot d\tau, \tag{1.1}$$

where N_k is the number of bombarding particles per second. This energy is transmitted to the air, thereby moving it in order to surmount the chute drag.

The quantity of air energy (power) E_2 can be expressed through air flow rate and drag as

$$dE_2 = L_E dp. \tag{1.2}$$

If dE_1 and dE_2 are equal, the integration results in the following:

$$L_E \cdot p = \int_0^l N_k \cdot R_0 \cdot dx. \tag{1.3}$$

It should be noted that some degree of inaccuracy is assumed in this case. When comparing Equations 1.1 and 1.2, it is assumed that the lost energy of bombarding particles is fully applied to the translational motion of air in a chute. However, only a portion of the lost energy is actually applied to accomplish this "useful" work while the rest of the energy goes to "mix" the induced air with a penetrating stream of particles. Introducing the energy transfer coefficient η_T to account for the portion of the bombarding particles' energy that is consumed to create a directional air flow, we obtain a more accurate result:

$$L_E \cdot p = \eta_T \cdot \int_0^l N_k \cdot R_0 \cdot dx. \tag{1.4}$$

The pressure difference is expressed by the sum of local drag factors:

$$p = \sum \zeta \cdot \frac{v_2^2}{2} \cdot \rho_2. \tag{1.5}$$

Then

$$R_{ch} \cdot L_E^3 = \eta_T \cdot \int_0^l N_k \cdot R_0 \cdot dx, \tag{1.6a}$$

where

$$R_{ch} = \sum \zeta \cdot \frac{\rho_2}{2 \cdot S_{ch}^2}. \tag{1.6b}$$

Expanding the integral value on the right side of Equation 1.6 with

$$R_0 = \psi_0 \cdot \frac{\pi \cdot d^2}{4} \cdot \frac{(v_1 - v_2)^2}{2} \cdot \rho_2, \tag{1.7}$$

$$v_1 = \sqrt{2 \cdot g \cdot x}, \tag{1.8}$$

for $\eta_T = 1$, S. E. Butakov obtained [15] the following:

$$Q^3 + a \cdot Q^2 + b \cdot Q + c = 0, \tag{1.9}$$

where $a = -A \cdot h/K \cdot F^2$, $b = 0{,}6 \cdot A \cdot h^{1{,}5}/K \cdot F$, $c = -A \cdot h^2/K$,

$$A = n \cdot k \cdot \frac{\pi \cdot d^2}{4} \cdot \frac{\gamma_b}{2} = 0{,}392 \cdot n \cdot k \cdot \gamma_b \cdot d^2,$$

$$K = \sum \zeta \cdot \frac{\gamma_b}{2 \cdot g \cdot F^2}, \quad n = \frac{6 \cdot G_M}{\pi \cdot d^3 \cdot \gamma_M};$$

where h = material drop height, m; K = chute hydraulic characteristic; F = chute cross-sectional area, m^2; γ_b = specific air weight, kG/m^3; d = diameter of particles, m; k = head drag coefficient of particles; n = number of particles per 1 sec; G_M = material weight flow rate, kG/s; γ_M = specific material weight, kG/m^3; and Q = induced air flow rate, m^3/s. Therefore, Equation 1.9 may be rewritten as:

$$\frac{\varphi^3}{6 \cdot \varphi^2 - 8 \cdot \varphi + 3} = \frac{Bu}{12}, \tag{1.10}$$

where the number

$$Bu = \frac{2 \cdot G_1 \cdot v_{1k}}{\Sigma\zeta \cdot c^2 \cdot \rho \cdot S_{ch}} \tag{1.11}$$

is hereinafter referred to as the Butakov–Neykov criterion (an inverse value of a modified Euler criterion)

$$Bu \equiv 1/Eu_m. \tag{1.12}$$

The initial equation (1.9) was first reduced to a dimensionless equation (1.10) by O. D. Neykov [66], who had analyzed the quantitative results of S. E. Butakov's model. In particular, multiple values were noted with respect to functions $\varphi = f(Bu)$ in the range $8.7 < Bu < 13.92$. It is therefore assumed that only the ranges $0 < \varphi < 0.807$ corresponds to the physics of the phenomenon in question and resultsing in the acceptance of $\varphi = 0.807$ and $Bu > 13.92$ without further proof.

It is important to bear in mind that Equation 1.7 does not account for the reversed direction of particle drag force at different levels in a chute (the second inaccuracy found in S. E. Butakov's model). A more accurate form of this force is represented as follows:

$$R = \psi \cdot f_M \cdot \frac{|v_1 - v_2| \cdot (v_1 - v_2)}{2} \cdot \rho_2. \tag{1.13}$$

At the chute inlet, the induced air speed may exceed the material movement speed when the latter is at its maximum, drag force $R < 0$ (i.e., particles at the chute inlet may cause additional flow resistance to the air suction).

Because of this, Equation 1.10 yields a slightly conservative value for induced air volumes.

Considering this same phenomenon, N. F. Grashchenkov, V. S. Kharkovskiy, and B. Tsoy developed the following formula for the induced air quantity [27]:

$$Q = 0{,}63 \cdot k \cdot \sqrt[3]{c \cdot \rho \cdot G \cdot t \cdot (\omega_k^3 - \omega_0^3)/(R \cdot d)}, \tag{1.14}$$

where G is material flow rate, m^3/s; ρ is air density, kg/m^3; c is a head drag coefficient; d is an equivalent sphere diameter, m; R is an aerodynamic drag of the chute, kg·s^2/m^8; k is a correction factor (k = 0,18 for vertical chutes); ω_0, ω_k are relative velocities of material particles at the chute inlet and outlet, respectively, m/s; and t is a time period during which particles are in a chute, s.

Considering Equation 1.8, Equation 1.14 can be easily reduced to the following form:

$$\frac{\varphi^3}{(1-\varphi)^3 + \varphi^3} = \frac{k^3}{3}.Bu. \tag{1.15}$$

Looking at S. E. Butakov's model for a situation where drag force is proportional to relative velocity squared and is in a different direction based on the relative velocity sign, P. Ch. Chulakov, N. N. Korabekov, and K. S. Salimzhanov [101] obtained:

$$\frac{K}{N} = \frac{\lambda^3}{6 \cdot \lambda^2 - 2 \cdot \lambda^4 - 8 \cdot \lambda + 3}, \quad (1.16)$$

where

$$\frac{K}{N} = \frac{G \cdot c \cdot \rho \cdot h}{8 \cdot v_k \cdot d \cdot \gamma_M \cdot R \cdot F_T^3}, \lambda = \frac{v_2}{v_k}; \quad (1.17)$$

G is the material weight flow rate, N/s; γ_M is the material specific weight, N/m^3; d is a mean equivalent diameter of pieces, m; c is a head drag coefficient; h is a chute height, m; F_T is a chute cross-sectional area, m^2; v_k is the bounded bombarding velocity of particles, m/s; R is an aerodynamic drag of the chute, N·s^2/m^8; and ρ is air density, kg/m^3.

Using these symbols, Equation 1.16 will appear as:

$$\frac{\varphi^3}{6 \cdot \varphi^2 - 2 \cdot \varphi^4 - 8 \cdot \varphi + 3} = \frac{\text{Bu}}{12} \quad (1.18)$$

When integrating dynamic equations for a particle and converting Equation 1.3, V. A. Minko [60] assumed that

$$\psi = 4{,}1 \cdot \chi \cdot \text{Re}^{-0{,}3}. \quad (1.19)$$

To obtain the following design ratio for particles of 0,2 mm < d < 2,5 mm and $v_1 < c$:

$$\frac{\lambda^3}{1 - 2{,}28\lambda + 1{,}28\lambda^2} = 2{,}8 \cdot 10^{-2} \cdot \frac{H \cdot v_1^{0{,}7}}{d^{1{,}3}}, \quad (1.20)$$

where

$$H = \frac{0{,}135 \cdot \chi \cdot G}{\rho_M \cdot R \cdot F^3},$$

$$R = \sum \zeta \cdot \frac{\rho_b}{2 \cdot F^2}, \quad (1.21)$$

and λ is the relation of the induced air speed to the material particles' bombarding speed; v_1 is the particles' bombarding speed in a stationary phase, m/s; G is the material flow rate, kg/s; ρ_M is the material density, kg/m^3; χ is an impact factor of particle shape; F is a chute cross-sectional area, m^2; Σζ is a sum of local drag factors for a chute; ρ_b is air density, kg/m^3; and d is a diameter of particles, m.

Inserting these symbols into Equation 1.20, we obtain:

$$\frac{\varphi^3}{1 - 2{,}28 \cdot \varphi + 1{,}28 \cdot \varphi^2} = \frac{\text{Bu}}{3{,}7}, \quad (1.22)$$

where

$$\mathrm{Bu} = \frac{\psi \cdot k_m \cdot G_1 \cdot v_{1k}}{g \cdot \rho_1 \cdot S_{ch} \cdot \Sigma\zeta}, \tag{1.23}$$

and ψ is the coefficient determined from Equation 1.19.

O. A. Bogaevskiy and U. H. Bakirov [8] considered a stream of particles with the initial velocity v_{1H} and the acceleration equal to:

$$a_m = 0{,}5(g+a_k), \tag{1.24}$$

where a_k is a particle's acceleration at the end of its fall in still air, m/s^2. They assumed that the process of air induction with such a particle stream is similar to S. E. Butakov's model and obtained:

$$Q = 3\varepsilon G \cdot h\rho/(8 \cdot \gamma_M \cdot r), \tag{1.25}$$

where Q is the induced air volume, m^3/s; G is the material weight flow rate, kg/s; γ_M is the specific weight of material particles, kg/m^3; r is a radius of particles, m; h is a drop height, m; ε is an aerodynamic drag factor; and ρ is a correction factor (for iron ore of normal humidity $\rho = 0{,}3$).

Converting to these symbols, we obtain:

$$L_E = 6 \cdot \psi \cdot G_1 \cdot v_k^2 \cdot \rho_2 / (16 \cdot g \cdot \rho_1 \cdot d) \tag{1.26}$$

or

$$\varphi_k = \frac{1}{4} \rho \cdot \mathrm{Bu} \cdot \Sigma\zeta. \tag{1.27}$$

P. I. Kilin [39,40], having replaced the integral of the right side of Equation 1.3 with the sum of averaged values, studied S. E. Butakov's model with respect to chutes with a random number of straight sections. In particular, he suggested the following equation for a chute with a straight section:

$$\lambda = (\sqrt{9 + N \cdot M} - 3) / M, \tag{1.28}$$

where

$$N = 3 + 2 \cdot \frac{k \cdot d}{c_x \cdot l} \cdot \frac{v_k^2 - v_H^2}{v_M^2}; \quad M = \frac{d}{c_x \cdot l} \cdot \frac{\Sigma\zeta_{ch}}{S} \cdot \frac{v_k + v_H}{v_M} - 3; \tag{1.29}$$

$$S = \frac{G_M}{F \cdot \rho_M \cdot v_M}; \quad v_M = \frac{2}{3} \cdot \frac{v_k^3 - v_H^3}{v_k^2 - v_H^2}; \tag{1.30}$$

v_H, v_k are material velocities at the chute inlet and outlet, m/s; $\lambda = v_B/v_M$; v_B is air velocity in a chute, m/s; G is the material flow rate, kg/s; F is a chute cross-sectional area, m²; ρ_M is a density of material particles, kg/m³; c_x is a head drag coefficient; d is the mean diameter of material particles, m; l is a chute length, m; $\sum\zeta_{ch}$ is a sum of local drag factors of a chute; and k is a factor of apparent mass (assumed to be equal to 0,5).

For a vertical chute with $v_{1H} = 0$, $N \approx 3$, Equation 1.28 becomes:

$$\varphi_k = \frac{3}{2} \cdot \frac{\sqrt{Eu_m} - 1}{Eu_m - 1} \cdot \qquad (1.31)$$

V. A. Popov [77,78] theoretically analyzed S. E. Butakov's model for a bulk material considering the impact of environmental resistance on falling velocity. Assuming that iron ore concentrate and apatite move as a stream of blocks (10–60 mm in size with the conveyer belt width reaching 1000 mm), he proposed the following equation for induced air velocity (v_B) when transferring these materials:

$$v_B^3 - \frac{K \cdot h}{N} \cdot v_B^2 + 2 \cdot A \cdot \frac{K}{N} \cdot v_B - B \cdot \frac{K}{N} = 0, \qquad (1.32)$$

where

$$K = ncf\rho; \quad N = 2RF^3; \qquad (1.33)$$

h is a material drop height; A and B are coefficients accounting for variations in the velocity of blocks of material particles that are due to environmental resistance; n is the number of blocks per 1 sec; $c = 1.15$ is a drag coefficient of blocks; f is a block cross-sectional area; ρ is air density; R is a chute hydraulic characteristic; and F is a chute cross-sectional area.

Incorporating these factors, Equation 1.32 will appear as:

$$\frac{\varphi_k^3}{6 \cdot \varphi_k^2 - 8 \cdot k_1 \cdot \varphi_k + 3 \cdot k_2} = \frac{\text{Bu}}{12}, \qquad (1.34)$$

where

$$k_1 = \frac{1,5 \cdot A}{h \cdot \sqrt{2 \cdot g \cdot h}}; \quad k_2 = \frac{B}{g \cdot h^2}. \qquad (1.35)$$

To determine aerodynamic force (assuming still air), Hemeon solved Equation 1.3 as follows [109]:

$$L_E^3 \frac{\rho_2}{2S_{ch}^2} = \int_{v_{1H}}^{v_{1K}} \bar{\beta}_1 l S_{ch} k_m \psi \frac{v_1^2}{2} \rho_2 dv_1, \qquad (1.36)$$

where $\bar{\beta}_1$ is a bulk concentration averaged along the chute length

$$\bar{\beta}_1 = \frac{1}{l} \int_0^l \frac{G_1}{\rho_1 S_{ch} v_1} dx = \frac{2G_1}{\rho_1 S_{ch}(v_{1H} + v_{1K})}, \qquad (1.37)$$

$$\bar{\beta}_1 S_{ch} l = \frac{2lG_1}{\rho_1 (v_{1H} + v_K)}. \tag{1.38}$$

(The last formula is simply a chute volume filled with material.)

Hemeon expanded the right-hand side of Equation 1.36 for three cases: (a) for the self-similarity area

$$\psi = \psi_0 \text{ at Re} > 500; \tag{1.39}$$

(b) for the transition area

$$\psi = a/\text{Re}^{0,6}; \tag{1.40}$$

and (c) for the airborne area when

$$v_1 = c - const.$$

In the latter case, the drag force in Equation 1.3 was replaced with the gravity force. Thus, the hydraulic resistance of the chute and the air motion within the chute were not considered. It was assumed that the count concentration (and, hence, the bulk concentration) is constant throughout the chute length.

For the self-similarity area ($\psi_0 = 0.44$) with $v_{1H} = 0$, $v_1 = \sqrt{2gh}$ Hemeon obtained:

$$Q = \sqrt[3]{7 \cdot \frac{R \cdot S^2}{\gamma_3 d} \cdot A^2 \cdot 1200}, \tag{1.41}$$

where Q is the induced air flow, m^3/s; S is the total drop height, m; R is the material flow rate, kg/s; A is a flow area of particles, m^2; γ_3 is the material density, kg/m^3; d is a diameter of particles, m; and h is the present bombarding height of particles, m.

Equation 1.41 will then appear as follows:

$$L_E = 20,3 \cdot \sqrt[3]{G_1 \cdot H^2 \cdot S_{ch} / (\rho_1 \cdot d)} \tag{1.42}$$

or

$$\frac{\varphi_k^3}{(1-n)\cdot(1-n^3)} = \frac{\Sigma\zeta}{3 \cdot Eu_m}. \tag{1.43}$$

Hatch [108], having noticed excessive results from Equation 1.41, introduced the efficiency factor:

$$Q = 0,78 \cdot \sqrt[3]{E \cdot T \cdot A^2 \cdot h^2 / (z \cdot d)}, \tag{1.44}$$

where Q is the induced air quantity, ft^3/min; T is the material flow rate, t/hr; h is a drop height, ft; A is a flow area of particles, ft^2; d is the mean mass diameter of particles, inches; z is the material density, g/cm^3; and E is the efficiency factor.

Equation 1.44 then becomes:

$$L_E = 17{,}4 \cdot \sqrt[3]{EG_1 H^2 S_{ch}^2 / (\rho_1 \cdot d)} \tag{1.45}$$

or

$$\frac{\varphi_k^3}{(1-n)\cdot(1-n^3)} = \frac{\Sigma\zeta \cdot E_E}{3 \cdot Eu_m}\,. \tag{1.46}$$

Morrison [112] introduced a correction factor into Hemeon's equation for transfers of polyfractional material:

$$Q = 110 \cdot \sqrt[3]{T \cdot H^2 \cdot A^2 / (G \cdot D)}\,, \tag{1.47}$$

where Q is the induced air quantity, ft^3/min; S is the material flow rate, t/hr; H is a drop height, ft; A is a flow area of particles, ft^2; G is the material density, pound/ft^3; and D is the mean diameter of material particles, inches. Therefore:

$$L_E = 6{,}3 \cdot \sqrt[3]{G_1 \cdot H^2 \cdot S_{ch}^2 / (\rho_1 \cdot d)}\,. \tag{1.48}$$

Considering the hydraulic resistance of chute walls to induced air movement, Anderson and Dennis [106] replaced the chute cross-section in Hemeon's equation with the upper hood leakage area to correspond with $F_b \leq 0.15$B (where F_b is the upper hood leakage area, m^2; B is the feed conveyer belt width, m), resulting in:

$$Q_1 = 10 \cdot A_u \cdot \sqrt[3]{R \cdot S^2 / D}, \tag{1.49}$$

where Q_1 is the induced air quantity, ft^3/min; A_u is the upper hood leakage area, ft^2; R is the material flow rate, t/hr; S is a drop height, ft; and D is the mean diameter of particles, ft. Using symbols, this becomes:

$$L_E = 1{,}5 \cdot F_b \cdot \sqrt[3]{G_1 \cdot H^2 / d},\ \text{m}^3/\text{s}. \tag{1.50}$$

Cruise and Bianconi [110] took the material flow area for an initial parameter and did not relate it to the chute cross-section (introduced in Hemeon's formula as the chute cross-sectional area):

$$F_{con} = k \cdot G_1 / (\rho_n \cdot v_1), \tag{1.51}$$

where ρ_n is material mass in a stream volume unit determined by the empirical function:

$$\rho_n = 5{,}4 \cdot \rho_1 \cdot d^{\,0,3}; \tag{1.52}$$

ρ_1 is a density of particles, g/cm^3; d is a diameter of particles, inches; and k is a trial coefficient.

This data satisfactorily matched experimental data from studies of coal transfers with $S_{ch} = 0.56 \div 1.12$ m², $G_1 \geq 0.28$ kg/s, d ≥ 1.27 mm, $\rho_1 = 1300$ kg/m³, and $H \sim 2$ m obtained with calculations according to the formula:

$$Q = 10,5 \cdot T \cdot \sqrt[3]{h} \cdot d^{-0,5} \cdot z^{-1} \cdot \exp(-6,5 \cdot k), \quad (1.53)$$

where Q is the induced air quantity, ft³/min; T is the material quantity transferred, t/hr; h is a drop height, ft; d is the mean diameter of material, inches; z is the material density, g/cm³; k is the efficiency factor equal to $k = N{\cdot}90/(h{\cdot}\theta)$; N is the number of chute revolutions; and θ is the chute inclination angle, deg. Using symbols:

$$L_E = 132 \cdot G_1 \cdot H^{1/3} \cdot d^{-0,5} \cdot \rho_1^{-1} \cdot \exp(-1,98 \cdot k). \quad (1.54)$$

1.2.2 Semiempirical Models

Now we will focus on some empirical formulas widely used to assess the injective capacity of a stream of bulk material.

Using a stream of crushed granite ($\rho_1 = 2630 \div 2660$ kg/m³, d = 22.6 mm and 11.2 mm), M. T. Kamyshenko [37] obtained the following empirical equation (with $G_1 = 1.4 \div 18.1$ kg/s, $H = 1.315$; 1.755; 2.275 m in a vertical pipe of $D = 260$ mm):

$$Q_B = \frac{G_M}{1,2} \cdot \frac{F_T}{f_M} \cdot \mathrm{tg}\beta, \quad (1.55)$$

where

$$f_M = G_M / (\gamma_M \cdot v_{BK} \cdot 3600);$$

Q_B is the induced air volume, m³/hr; F_T is the chute cross-sectional area, m²; G_M is the material flow rate, t/hr; f_M is the chute section area filled with falling material; γ_M is the material bulk weight, t/m³; v_{BK} is the bombarding velocity of particles at the chute inlet assumed to be equal to the upper conveyer speed, m/s; and tgβ is slope ratio of linear dependence.

Having assumed that $\mathrm{tg}\beta = 0.0038{\cdot}v_K^2$, A. M. Gervasiev [21] converted Kamyshenko's equation ($F_B/S_{ch} \leq 0.3$) to determine induced air quantity (Q_E) by using the following formula:

$$Q_E = 0,04 \cdot k_y \cdot Q_M \cdot v_K^2, \quad (1.56)$$

where Q_M is the material volume flow rate, m³/hr; k_y is hood structure-dependent factor ($k_y = 1.35 \div 3.0$); and v_K is the material flow rate at the chute outlet, m/s.

For transfers of quartzite particles (with a fineness of 0.5 ÷ 1 mm and 3 ÷ 5 mm with $F_{ch} = 0.075$; 0.06; 0.035 m²; $H = 1, 2, 3$ m; $\Theta = 45, 50, 70$ deg.), A. V. Shelektin [102] found:

$$Q_{ch} = 1,16 \cdot k \cdot G_M^{0,2} \cdot F_{ch}^{0,8}, \quad (1.57)$$

where Q_{ch} is the induced air volume, m³/hr; G_M is the material quantity transferred, kg/hr; F_{ch} is the chute cross-sectional area, m²; and k is a factor accounting for a drop height H and the chute inclination Θ.

For coal transfers (in $2.5 < v_K < 11.5$ m/s; $5 < Q_y < 170$ dm³/s; $0.14 < F_{ch} < 1.25$ m²; $40° < \Theta < 90°$), A. P. Lyubimova [56] found:

$$Q_E = \frac{0,29 \cdot k_\alpha \cdot v_k \cdot Q_y^{0,3} \cdot F_{ch}^{0,7}}{d^{0,34 \cdot \varphi} \cdot c_H^{0,87}}, \tag{1.58}$$

where Q_E is the induced air volume flow rate, m³/s; Q_y is coal volume flow rate, m³/s; v_K is a bounded coal falling velocity, m/s; F_{ch} is the chute cross-sectional area, m²; d is the particle diameter, m; k_α, φ are factors accounting for the influence of nonuniformity of an in-depth distribution of a solid ingredient feed concentration and the quantity of a surface of an active interaction of particles with air based on the chute inclination; and c_H is the relation of the chute cross-sectional area to the leakage area.

Having analyzed the air induction with a stream of steel spheres, V. D. Olifer [71] obtained the design ratios for the dynamic interaction force:

$$P_E = \rho \cdot (\overline{v}_M - \overline{v}_B)^2 \cdot \left(\frac{h}{h_{e\partial}}\right)^{1,75} \cdot \overline{k}^{1,25} \cdot \left[0,81 + \frac{1,68 \cdot 10^6 \cdot \nu^2}{(\overline{v}_M - \overline{v}_B)^2 \cdot d_{av}^2}\right] \tag{1.59}$$

as well as for velocity of the induced air in a chute for transfers of spherical particles and irregularly shaped particles:

$$\overline{v}_B^2 \cdot \left(\frac{\zeta_n}{1,265} - 1\right) + 2 \cdot \overline{v}_M \overline{v}_B - \overline{v}_M^2 - \frac{482}{d_{av}^2} - 1,32 \cdot \frac{P}{S} = 0, \tag{1.60}$$

where $\overline{v}_B$ is an average air velocity in a chute, m/s; $\overline{v}_M = 0,7 v_M^H + 0,3 v_M^K$ is an average material velocity in a chute, m/s; v_M^H, v_M^K are material velocities at the chute inlet and outlet, respectively, m/s; d_{av} is the mean diameter of particles, mm; P is the lower hood vacuum-gauge pressure, Pa; and S is the sum total:

$$S = m \cdot N \cdot \overline{k} \cdot (h / h_{e\partial})^{1,5}, \tag{1.61}$$

where $m = 1.3f_p - 0.3$; f_p is a particle geometric form factor; N is a coefficient (with $d_{av} > 3.5$ mm, $N = 1$); $\overline{k} = 100 \cdot W / (F_{ch} \cdot \overline{v}_M)$; W is the material volume flow rate, m³/s; F_{ch} is the chute cross-sectional area, m²; ζ_n is a sum of local drag factors of the chute and the upper hood; h is the chute height, m; and $h_{e\partial}$ is the chute unit height (equal to 3 m).

When $N = 1$, $v_{1H} = 0$, $P = 0$, $h = h_{e\partial} = 3$ m, Equation 1.60 may be rewritten as follows:

$$\varphi_k^2 \cdot \left[\frac{10,8}{d_{av}} \cdot Eu_m - 1\right] + 0,6 \cdot \varphi_k - 0,09 - \frac{8}{d_{av}^2} = 0, \tag{1.62}$$

where d_{av} is the mean diameter of particles, mm.

Following experimental studies, E. N. Boshnyakov [10,11] obtained this design ratio for induced air volume (Q_e):

$$Q_e = 3{,}165 \cdot k_H \cdot k_G \cdot k_{v0} \cdot k_F \cdot k_{\sum\zeta} \cdot k_c \cdot k_d \cdot k_\gamma \cdot k_h, \tag{1.63}$$

where k_H is the material transfer height (the material velocity at the chute outlet), k_G is the material flow rate, k_{v0} is the initial velocity, k_F is the chute cross-sectional area, $k_{\Sigma\zeta}$ are local drag factors, k_c is a head drag of the material particles, k_d is the fineness of particles, k_γ is the material density, and k_h is the hood vacuum-gauge pressure.

After reducing the experimental data, Degner and Futterer [107] obtained the following equation for transfers at coal preparation plants:

$$Q = \frac{k_1 \cdot M^\alpha \cdot F_{E0}^\beta \cdot (F_{su}^\nu + k_2) \cdot H^\delta \cdot v_B^\vartheta \cdot F_s}{d^\varepsilon \cdot \rho^\zeta \cdot z^\eta}, \tag{1.64}$$

where M is the material mass flow rate; F_{E0} is the leakage area in the chute receiver portion hood; F_{su} is the leakage area in the chute discharge outlet hood; F_s is the chute cross-sectional area; H is a material transfer height; v_B is the feed conveyer belt speed; d is the mean diameter of material grains; ρ is the material density; z is the number of seal covers; and k_1, k_2, α, β, ν, δ, ϑ, ε, ζ, and η are trial coefficients.

After having analyzed the air mechanics of a stream of steel spheres, particles of coal, millet, peas, rice, wheat, and lentils in a vertical pipe, V. P. Pavlov [74] obtained the following empirical equation (with $D_0/d_e = 9 \div 27$; $l/d_e = 75 \div 614$; and $v_f/v_{Bum} = 0 \div 1.34$) for the air velocity along the material stream axis (v_f^0):

$$v_f^0 \big/ v_{Bum} = 0{,}0174 \cdot \left(l \big/ d_e \right)^{0{,}51} \cdot \left(v_f \big/ v_{Bum} + 1 \right)^{1{,}82} \cdot \left(D_0 \big/ d_e \right)^{-0{,}2}, \tag{1.65}$$

where v_f is the velocity of undisturbed air flow outside the jet; v_{Bum} is the airborne velocity of particles; l is the jet length; D_0 is the initial diameter of a jet; and d_e is a diameter of a sphere equivalent to a particle in its volume.

Experimental studies by M. T. Kamyshenko (1951), A. S. Serenko (1953), and A. V. Sheleketin (1959) built empirical relations for determining L_E and for discovering new effects (reverse air flows and pressure surges in closed chutes) that had been unexplainable for a considerable time period.

1.2.3 Dynamic Theory and Research Methodology for Injection Properties of a Particle Stream

The second stage, the study of aerodynamic processes in a bulk material stream in terms of two-component flow dynamics, was initiated by one of the authors in 1964 at the Krivoy Rog branch of the Institute of Mining Affairs of the Academy of Sciences of the Ukrainian Soviet Socialist Republic (presently NIIBTG) following the solution of hot material transfer suction problems [36,51,52].

Initial theories introduced thermal and induction pressure forces that explained chute air flow dynamics in terms of a one-dimensional problem described by the hydraulic equation

$$P_2 - P_1 + P_E - P_T = \sum \zeta \frac{\upsilon_2^2}{2} \rho_2, \tag{1.66}$$

where P_1, P_2 are vacuum-gauge pressure values supported by suction hoods in the upper and lower hood, respectively, Pa; P_E is the induction pressure in a chute, Pa; P_T is the thermal pressure in a chute, Pa; and $\Sigma\zeta$ is the sum of local drag factors in a chute. The simplicity of this easy-to-demonstrate equation contributed to its rapid spread in design practice in Russia and elsewhere [2,3,61,72].

This study explains the fundamental provisions (based on classical laws of motion) of bulk material gravitational flow aerodynamics in dedusting ventilation.

We first built a mathematical model depicting the interaction between solid particles and air with regard to a stream of bulk material. We then determined bulk material particle characteristics and then formulated the fundamental provisions of air mechanics for a material stream in closed chutes (thereby solving the one-dimensional problem). We explained the regularities when air streams are induced by a bulk material stream (thereby analyzing the two-dimensional problem). We followed the classic comprehensive method of analysis through our mathematical modeling, experimental correction of theoretical models, and industrial evaluation of findings.

1.2.3.1 Mathematical Modeling

The theoretical description of the interaction mechanism of a bulk material stream and air was made using general dynamic equations of heterogeneous media (see Appendix). Fundamental studies concerning these media mechanics describe this interaction for a number of practical tasks with a carrying continuous medium (liquid and gas) and with moving or stationary discrete medium (solid particles, liquid drops, gas bubbles). These are primarily flows of aerosols and suspensions, air-dispersed mixtures and liquid gas mixtures, fluidization and filtration processes, pneumo- and hydrotransport, and drifts and snowstorms. A stream of bulk material and the air it drags should be considered as a separate subclass of two-component flows in which the carrying medium is a discrete medium of solid particles and the carried medium is a pseudo-continuous dispersion medium (air). Streams of particles affected by the Earth's gravitational field are moving at a growing rate even though the aerodynamic processes are occurring at a low-intensive rate (air stream velocity is typically less than particles velocity), which makes them significantly different from thoroughly studied dispersion through flows common for pneumo- and hydrotransport.

There are two methodological approaches used to describe the mechanics of two-component streams: the phenomenological approach (in which a heterogeneous medium stream is regarded as a motion of interpenetrating multispeed continua) and the method of averaging classical accounting equations in spatial and time microscales.

As noted, the first approach is based on the same provisions as the mechanics of a homogenous continuous medium. It is assumed that the elementary volume of a mixture as well as the elementary volume of components contains a sufficiently great number of particles in spite of the smallness of such volumes. The components' dynamical interaction represents a bulk force resulting from particles' aerodynamic resistance due to the relative velocity of components. These forces are included in accounting equations of momentum and energy conservation. By analyzing energy conservation equations separately for each component and for the mixture in general, it is possible to demonstrate that the mixture's momentum energy is increased due to the work of intercomponent forces and interphase transformations. This was not considered in the studies of Butakov, Hemeon, and others who looked at the phenomenon of air injection with a bulk material stream in terms of the work-kinetic energy theorem.

1.2.3.2 Experimental Studies

Experimental studies organically supplemented and updated mathematical models of processes analyzed, evaluated the results of theoretical studies, and, finally, provided answers to those practical questions where the theory was of no help. Therefore, experiments were conducted in several directions.

First, the studies uncovered the basic regularities of interaction between particles and air, and they quantified the aerodynamic properties of individual and collective particles and heat exchange between the components with respect to an accelerated stream of particles. These studies were preceded with analyses of bulk material flow patterns: variation in bulk concentration of particles in a stream and modes of motion based on structural dimensions of chutes. This trend was analyzed using experimental arrangements, such as noting which structural components showed the most explicitly studied process or served as indicators. For instance, the main component in the study of the dynamic characteristics of a stream of particles, particle air mechanics, and heat exchange was a chute with a variable inclination angle and cross-section. Aerodynamic properties of individual particles were determined by measuring the airborne velocity in a tapered tube that also served as the velocity meter.

Second, in the experimental specification of the physical model of accelerated particle stream and air interaction, the maximum possible approximation of simplifying assumptions that underlie the theoretical provisions was determined. The requirement for particle uniformity in size and shape and in flow area distribution, and for material flow rate stability led to the necessity of using uniformly sized water drops (produced by a slow fluid discharge from orifices of equal sizes located at the reservoir bottom). The mathematical models of air suction with a solid stream were evaluated in straight chutes with a variable cross-section and an inclination angle.

Methods were developed for calculating the optimum output of suction hoods; the efficiency of suction system elements was primarily evaluated using a multipurpose semi-commercial plant at the All-Russia Institute for Occupational Safety in Ore Mining (VNIIBTG) in Krivoy Rog. Such plant development was undertaken because interaction between material particles affected by gravity and air is difficult to study; it is impossible to simulate Earth's gravitational field. However, industrial experiments virtually exclude any parameter variability affecting the process in question even within close limits. It is absolutely impossible to analyze processes using different materials under identical conditions. Experimental plants that use a chute that connects two bins (an upper feed bin and a lower receiving bin) create difficulties due to material discharge transcience that, in turn, limits material flow and increases an experiment's labor intensity. Because of transience, making a set of dust measurements is virtually impossible. Therefore, for experiments to replicate natural conditions as much as possible, we developed and assembled a semi-commercial plant with serially produced process and ventilation equipment.

The assembled plant used (Figure 1.8) a closed-loop conveying system where the horizontal transportation of bulk material was carried out by two conveyers with a belt width of 650 mm; vertical transportation was provided using an EPG-200 chain-and-bucket elevator and two gravity chutes. The upper conveyer (11.0 m long) was installed on the laboratory facility's first floor, and the lower conveyer (14.5 m long) was installed on the ground floor sloping bench. The elevator (3, in Figure 1.8) lifted material from the lower to the upper conveyer. In order to ensure the material was uniformly fed, the maximum possible bucket pitch was 170 mm. The lower conveyer connected with the elevator via a capacious collecting bin (6) with 5m^3 of usable volume. The material flow rate was controlled with a rack-and-pinion gate (7) mounted on the bin discharge chute. The upper conveyer featured LTM-1M automatic belt scales (8) appropriate for the material flow measurements. The maximum output of the plant was 90 t/hr (for iron-ore pellets).

In the process of bulk transportation, the material was naturally milled to reproduce the transfer quality of closed-loop plants. In order to conduct an experiment of high quality, the assembled plant had to be capable of continuous (1–2 hours with

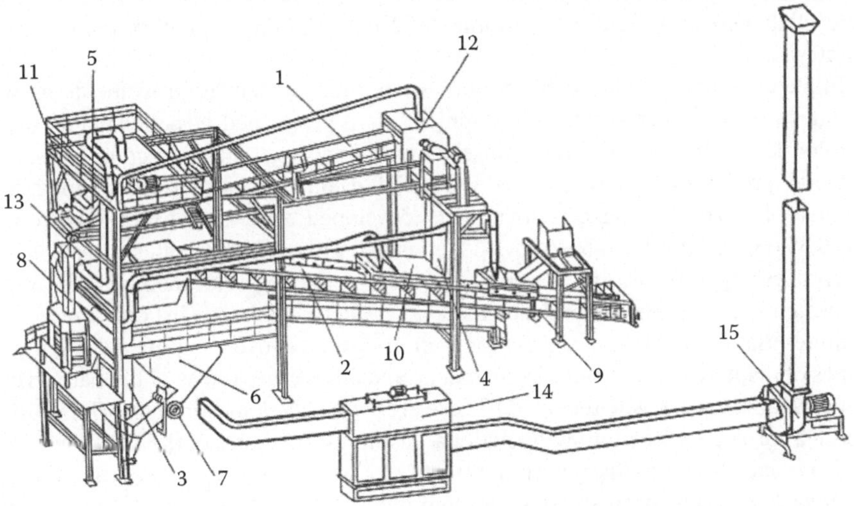

FIGURE 1.8 Semi-commercial suction plant: 1 = upper conveyer; 2 = lower conveyer; 3 = elevator; 4, 8 = chutes; 5 = suction manifold; 6 = bin; 7 = rack-and-pinion gate; 9–13 = hoods; 14 = sleeve filter; 15 = fan.

low material flow rates) operation involving consistently fine material. Constant supply was ensured by filling the collecting bin with a considerable quantity of material.

Dedusting of process equipment was performed using a suction manifold system made up of suction hoods of various types (9–13, in Figure 1.8), branched ductwork, a vertical prism manifold, a cloth filter (14), and a fan (15) VVD No. 11. Suction hoods were also provided for the bin, the elevator feed, and discharge sections. For the purpose of suction volume control, all suction hoods were equipped with electrically driven single-leaf dampers. Remote control of process and ventilation equipment and registration of parameters was performed from the control panel equipped with the appropriate instruments.

1.2.3.3 Industrial Evaluation

Industrial evaluation was the final stage of dust exhaust system design basis development. Suction volumes were specified and efficient layouts for suction hoods were determined. The efficiency of structural suction system components was established and engineering means and methods of optimizing the dust exhaustion plants were defined under industrial conditions. Such evaluation was necessary due to the inevitable laboratory simplifications in determining the theoretical basis for dust exhaustion and the ideal conditions of experimental analyses. The practical tasks of dedusting some process equipment (indurating machines, cone crushers, sieves, etc.) are associated with specific structural and layout solutions that could not be reproduced under laboratory conditions.

Iron ore preparation and iron ore concentrate pelletization were the most dust-forming productions selected for industrial evaluation of developed dust exhaustion means. Moreover, iron ore concentrate pelletization features application of diverse process equipment and emits harmful impurities such as dust, heat, and moisture.

The capacity calculation methods for individual suction hood production units were evaluated at virtually all mining and concentration complex plants in the country.

Implementation of dust exhaustion system enhancement improvements as well as industrial tests and system development were performed at pellet plants in the Sokolovsko-Sarbayskiy and Lebedinskiy mining and concentration complexes.

Good professional practices for industrial evaluation have been defined in a number of normative design documents developed with our direct involvement [18,19,35,63,73,79,81,93]. These are widely used for the design of enterprises engaged in reprocessing dust-forming materials. It is impossible to list all facilities that utilize the design basis developed for dust exhaustion systems. Here, we list only the leading institutes that replied to our request for their use of the above mentioned design. (The names of institutes, ministries, departments, and cities are given as of the date written replies to our request were received.) These are primarily domestic enterprises built or rebuilt as projects of the following national institutes: Institutes of the USSR Ministry of Iron and Steel Industry: Hypromez (Moscow), Uralhypromez (Sverdlovsk), Ukrhypromez (Dnepropetrovsk), Sibhypromez (Novokuznetsk), Gruzhypromez (Rustavi), Hyproruda (Leningrad), Centrohyproruda (Belgorod), Yuzhhyproruda (Kharkov), Syberian branch of Hyproruda (Novokuznetsk), Mechanobr (Leningrad), Mechanobrchermet (Krivoy Rog), Krivbassproject (Krivoy Rog), and All-Union

Refractory Institute (Leningrad), Hyprocokes (Kharkov); USSR institutes of non-ferrous metals: Hypronickel (Leningrad), Kavkazhyprotsvetmet (Ordzhonikidze), and Kazhyprotsvetmet (Ust-Kamenogorsk); institutes of the USSR Ministry of Construction Materials Producing Industry: Soyzhypronerud (Leningrad), Yuzhhyprocement (Kharkov), NIPIotstrom (Novorossiysk), and NIIstromproject (Tashkent); USSR Gostroy institutes: Kharkov Santekhproject, Alma-Ata branch of GPI Santekhproject, Ural branch of GPI Santekhproject (Sverdlovsk), Leningrad PromstroyNIIproject, Kharkov PromstroyNIIproject, Chelyabinsk PromstroyNIIproject, and Kazakhstan PromstroyNIIproject (Alma-Ata).

Normative materials [79] were also used in the design of foreign facilities. For instance, the Ural PromstroyNIIproject designed dust exhaust plants for iron and steel works located in Arna Mehr (Iran) and Helwan (Egypt).

1.3 CLASSIFICATION OF BULK MATERIAL STREAMS

Referring to the source of motion, a bulk material stream and the air that it drags will be analyzed as a separate subclass of two-component streams featuring a discrete dispersion medium of solid particles as the carrying medium and a pseudo-continuous dispersion medium (air) as the carried medium. In the streams in question, the carrying medium (a stream of particles) is moving at an increasing rate influenced by the Earth's gravitational field; the aerodynamic processes that are taking place are low-intensive, which makes them significantly different from thoroughly studied dispersion through flows common in case of pneumo- and hydrotransport.

Bulk material streams (Figure 1.9) should be classified by the geometry of channels in which the material stream is moving (I); the stream kinematics (II); the intensity of the dynamic interaction of components (III); the fineness and composition of particles (IV); the distribution of particle bulk concentration in the flow area (V); and the material's temperature and humidity (VI).

Classifying streams by the first criterion due to a difference in the air stream structure results in the stream being constrained with no-flow boundaries. Quantitatively, the constraint may be evaluated using the relation of the flow area (S_Π) to the chute clear area (S_{ch}):

$$\Pi_C = S_\Pi / S_{ch}. \tag{1.67}$$

Prism chutes or tubes for which $\Pi_c = 1$ typically feature a rod-like motion of the induced air and a lack of longitudinal and transverse velocity gradients.

The picture will significantly change if the walls enclosing the stream are moved away for a considerable distance ($\Pi_c \le 0.1$). The induced air stream has clear velocity gradients in both directions that differ from a free jet only by increases in the quantity of motion due to the intercomponent interaction forces. With unconstrained inflow of ambient air, an external closed-loop air circulation is practically absent. With a capacious chute ($0.1 < \Pi_c < 1$), unlike with a free jet, ascensional closed circulations occur and feed the induced air jet.

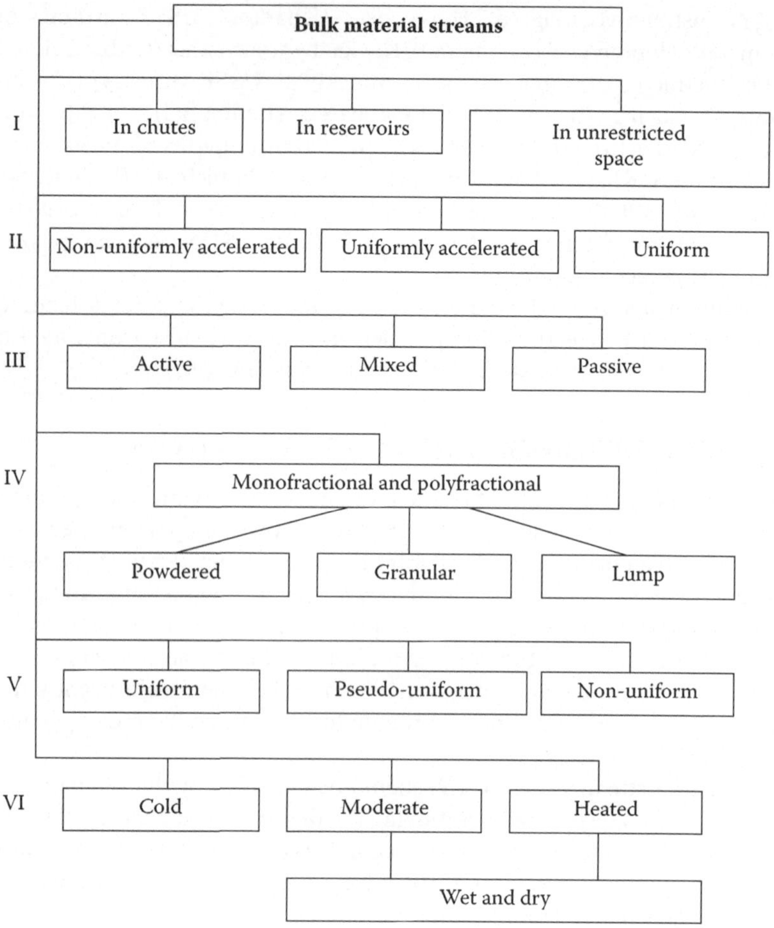

FIGURE 1.9 Classification of bulk material streams.

The second classification criterion reflects stream differences in kinematic terms. Because the environment resists free motion of particles, the stream may be non-uniformly accelerated:

$$\frac{d\upsilon}{dt} = g - f(\upsilon, u). \tag{1.68}$$

However, with particles of a large mass and low drop height, the drag force may be neglected. When evaluating the dynamic interaction in

$$2\bar{x}Fr < 1,\ \bar{x} = x / l_{\infty}, \tag{1.69}$$

the stream of particles may accelerate uniformly. Another extreme case may be observed with a stream of fine particles when they reach a steady rate

$$\frac{dv}{dt} \approx 0, \tag{1.70}$$

and the stream is virtually in uniform motion.

The third criterion defines component force interaction, which is essential in evaluating aerodynamic effects in a bulk material stream. As a criterion for the dynamic interaction intensity, we use the relation of particle's aerodynamic force in a stream (R_{Π}) to the aerodynamic force of a single particle (R_0), with the same average relative velocity of components:

$$\Pi_{\partial} = \left(R_n / R_0\right)\Big|_{\upsilon_1-\upsilon_2-idem}. \tag{1.71}$$

Let us consider two extreme cases. Assume, for instance, that a stream of uniformly distributed particles in the channel section is dispersed to the same degree that the mutual influence of particles on the aerodynamic flow environment is virtually absent ($\beta_1 \le 0{,}001$). In this case, $\Pi_{\partial} \approx 1$, so the stream is aerodynamically active because the dynamic interaction forces are equitable or higher than the aerodynamic forces of a single particle. The average relative velocity of a cloud of particles falling in an unlimited space will be equal to the falling velocity. However, the actual relative velocity for most of the particles within the cloud will be less than the cloud falling velocity (in an extreme case, with a sufficient particle–particle packing $\beta_1 > 0{,}4$, the actual relative velocity will be nearly equal to zero). Therefore, $\Pi_{\partial} << 1$, and with respect to the air induction, such a stream will be aerodynamically passive. In borderline cases, streams will be dynamically mixed (i.e., one portion of a stream can be active and another can be passive). This may be demonstrated by a wide array of bulk material transfers by chutes. A stream portion at the chute bottom is passive due to a larger, cloud-like particle–particle packing while another portion of the stream (above the "layer") actively interacts with air, engaging it in motion.

Separation of materials by particle size is primarily due to specific suction hood design requirements and to a difference in stream structure based on particle size. For powder materials, more than 50% of contained particles are less than 0.5 mm in size, with a maximum particle size not exceeding 1–2 mm. For granular materials, more than 50% of contained particles are less than 3 mm in size, with a maximum size not exceeding 10 mm. For lump materials, more than 50% of contained particles are larger than 3 mm.

The fifth classification criterion is based on differences in the bulk material stream structure, namely the distribution of particles in the cross-section area. A uniform distribution of particles may be observed in a wide range of bulk concentrations—from a densely packed layer (for example, in a chute or a tube completely filled with a material) to a dispersed layer (where there is no mutual influence of particles on their flow). Such streams show active dynamic interaction of components. In another extreme case, a bulk concentration may have a noticeable transverse gradient. The aerodynamic activity of particles is too different (such as bulk material streams moved in a bound mode by capacious chutes). A mixed case may occur when practically all particles are dynamically active (a noticeable concentration gradient, notwithstanding). We call such streams pseudo-uniform. A quantitative criterion for pseudo-uniformity may be an average bulk concentration. Studies have demonstrated that chute streams are pseudo-uniform at $\beta_1 \le 0,01$.

The material temperature and humidity determines the nature of intercomponent heat and moisture exchange. Ascensional forces of emitted gaseous components quantitatively and qualitatively alter the mechanism of air induction by gravity flows.

In spite of certain conditions, the classification system presented here should help provide readers with a more detailed explanation of air induction aerodynamic process regularity in various bulk material streams about which additional details will be provided in the following chapters of this book.

2 Aerodynamic Properties of Particles in the Gravitational Flow of a Chuted Bulk Material

Chutes are linking elements used for transfers of reprocessed materials from one transportation line group or from one type of equipment to another. Transfer groups can be divided into four sub-groups (Figure 2.1): conveyer to conveyer loading of material; conveyer to equipment loading of material; equipment to conveyer loading of material; equipment to equipment loading of material.

In all cases, the material being transferred is first supplied to the funnel adjacent to the process equipment or mounted at the belt conveyer pulley, and then the material is chuted by gravity to the lower transport conveyer or to the process equipment.

The type of chuted material motion and associated aerodynamic processes are determined by the aggregate physical and mechanical properties of the material being transferred and by the structural design of the chute.

Structurally, chutes are subdivided by shape into prismatic, cylindrical, and pyramid-shaped (bin) and into vertical, tip, and kinked chutes by the bottom slope angle.

The most common structures are tip chutes of a prismatic or a pyramid shape.

2.1 PECULIARITIES OF BULK MATERIAL MOTION IN CHUTES

A granular material model was selected to study chute mechanical properties—in this case, a flow of crushed granite (ρ_1 = 2750 kg/m^3). The granite particles were 1.25–2.5 mm (d_e = 1.56 mm) and 0.625–1.25 mm (d_e = 0.74 mm) in size. Granite was selected because its shape and aerodynamic properties resemble those of granular materials widely used in the ore preparation industry (crushed iron ore, chalkstone, agglomerate, fine agglomerated iron ore concentrate, etc.)

The study of crushed granite particle motion and distribution in the tip chute cross-section, as well as the measurement of the particles' stream velocity, was conducted on a test bench. The bench's main component was a 3-m-long rectangular cross-sectional chute installed at various angles to the horizontal plane. The granite particles were supplied to the chute from the upper bin through pre-tared diaphragms (Figure 2.2).

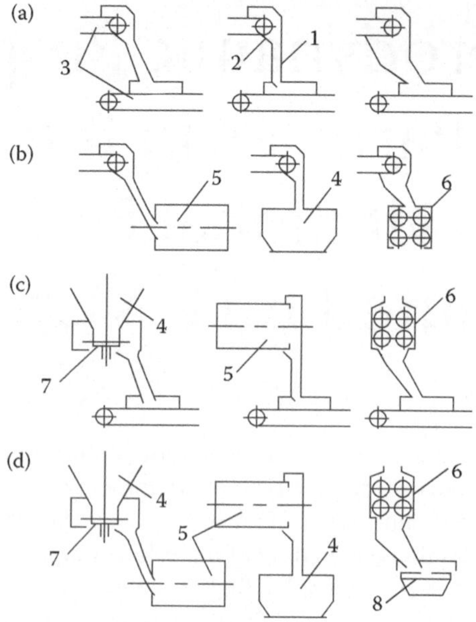

FIGURE 2 .1 (a) Conveyer to conveyer loading pattern, (b) conveyer to equipment loading pattern, (c) equipment to conveyer loading pattern, and (d) equipment to equipment loading pattern; 1 = chute; 2 = funnel; 3 = conveyers; 4 = bins; 5 = drums (for material cooling, mixing etc.); 6 = crushers; 7 = disc feeds; and 8 = sieve.

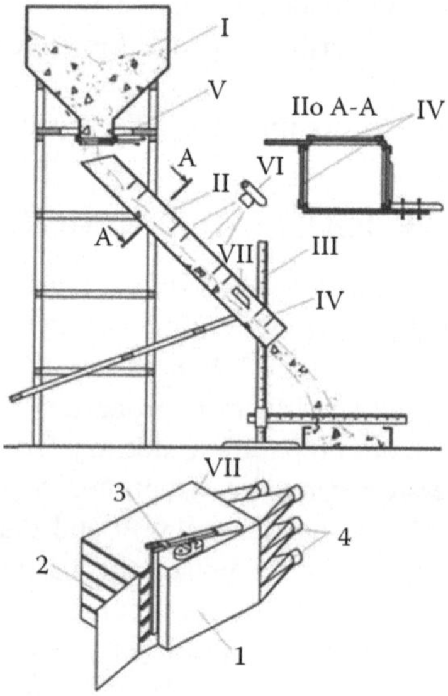

FIGURE 2.2 Diagram of the experimental arrangement for the study of physical and mechanical properties of a bulk material stream: I = upper bin; II = chute; III = coordinate spacer; IV = windows; V = diaphragm; VI = photo camera; and VII = flow divider (1 = body, 2 = shelves, 3 = valve controller, and 4 = bins).

2.1.1 Modes of Motion

When chuting crushed granite (such as grain flows in inclined pipes, first studied by P. N. Platonov [75,76]), three modes of motion may be observed: constrained motion, intermediate motion, and unconstrained motion.

In the constrained mode, material is moved as an indiscrete mass with no noticeable discontinuity of contact among the particles. There is no bulk concentration gradient. The intermediate mode features local discontinuities in the indiscrete mass of particles. The unconstrained mode features total decomposition of the indiscrete mass into particles or jets completely separate from each other.

Analyzing the motion of grain under high specific loads, Platonov suggested using a chute inclination angle, Θ, as the parameter defining the nature of bulk material motion. For instance, he noticed that the mode of constrained motion takes place if

$$\Theta_H < \Theta < \Theta_B, \tag{2.1}$$

although the mode of unconstrained motion occurs when

$$\Theta > \Theta_B, \tag{2.2}$$

where Θ_H, Θ_B are external and internal friction angles, respectively.

When studying the stream of crushed granite particles in the tip chute, we observed the mode of unconstrained motion at the chute inclination angle was less than at the internal friction angle. Therefore, we used the Froude number* to describe the stream kinetic movement (instead of the chute inclination angle) as a criterion for a change in the motion modes:

$$Fr = gh / v_1^2,$$

where h is a stream depth, m.

In order to clarify the physical meaning of this criterion for a bulk material motion in a chute under small specific loads, we compared this motion with water flow in inclined drop structures. Let us evaluate particle energy in a cross-section of the stream.

The flow strength of a material moving through surface ds (Figure 2.3) per unit time with respect to datum 0–0 drawn through the lower point of the cross-section in question is:

$$dE = \beta_1 \rho_1 v_1 \frac{v_1^2}{2} dS + \beta_1 \rho_1 v_1 g dS \cdot y \cos\Theta. \tag{2.3}$$

---

* The practice of applying this criterion in cases where a bulk material flow with vertical chutes fully filled has been used before. For instance, when studying the motion of crushed graphite in a vertical pipe, Z. R. Gorbis [26] defined the area of critical values of Froude numbers as $1.65 < Fr_{cr} < 5$, at which one mode is changed to another.

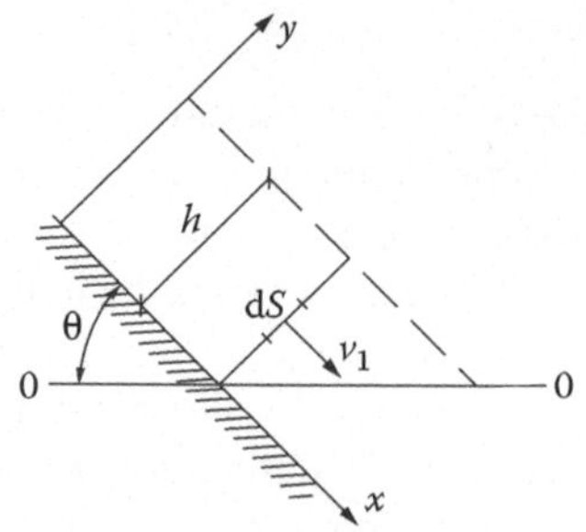

FIGURE 2.3 Bulk material flow energy in a chute.

Then, the total material flow strength is

$$E = \int_S \beta_1 \rho_1 v_1 \frac{v_1^2}{2} dS + \int_S \beta_1 \rho_1 g v_1 y \cos\Theta \cdot dS. \quad (2.4)$$

Dividing the flow strength value by the bulk material weight flow rate

$$G_T = \beta_1 \rho_1 g \bar{v}_1 \cdot S \quad (2.5)$$

and assuming that there is no bulk concentration gradient throughout the chute cross-section, we obtain the energy per weight unit of the material passing through the cross-section area per time unit;

$$E = \int_S v_1^3 dS / (2gS\bar{v}_1) + \left(\int_S y v_1 dS \right) \cdot \cos\Theta / (S\bar{v}_1), \quad (2.6)$$

where $\bar{v}_1$ is a medium flow speed of the bulk material motion.

For a rectangular chute, when

$$v_1 = f(x, y),\ G_T = \beta_1 g \rho_1 B h \bar{v}_1, \quad (2.7)$$

we have:

$$E = \frac{\alpha_0 \bar{v}_1^2}{2g} + \frac{k_0 h}{2} \cos\Theta, \quad (2.8)$$

where B is the chute width, m; h is a bulk material flow depth, m; α_0 is the flow momentum correction factor equal to:

$$\alpha_0 = \left(B \int_0^h v_1^3 dh \right) \Big/ \left(B \bar{v}_1^3 h \right); \quad (2.9)$$

and k_0 is the flow potential energy correction factor equal to:

$$k_0 = \left(B \int_0^h v_1 y\, dy \right) \Big/ \left(\bar{v}_1 B \frac{h}{2} \right). \quad (2.10)$$

Therefore, it is obvious that, in cases with gradientless motion of bulk material, $\alpha_0 = 1$ and $k_0 = 1$.

The extreme value of specific energy E in view of Equation 2.7 occurs in our case when the flow depth $h = h_{cr}$:

$$\frac{dE}{dh} = -\frac{\alpha_0 G_T^2}{(\rho_1 g \beta_1 B h_{cr})^2 \cdot g h_{cr}} + \frac{1}{2} k_0 \cos\Theta = 0 \tag{2.11}$$

or in the criterion form:

$$Fr_{cr} = \frac{2\alpha_0}{k_0 \cos\Theta}, \tag{2.12}$$

where Fr_{cr} is the critical Froude number value equal to

$$Fr_{cr} = g h_{cr} / \overline{v}_1^2. \tag{2.13}$$

When studying inclined drop structures [100], we noticed that the critical Froude number value describes the transition of a subcritical fluid flow into a rapid flow. The latter is characterized by fluid jet discontinuity, especially at the jet's free surface, and by an abundant aeration of the flow. The transition of constrained motion of bulk material into unconstrained motion is accompanied by a similar phenomenon: jet discontinuity and galloping particle motion (saltation). Returning to condition (2.1), we can note that by choosing Fr number as the transition criterion it is possible to consider the material flow rate in addition to the chute inclination angle. In other words, the Fr criterion provides a great deal of information on a bulk material stream.

The research data from our experiment concerning the motion of a crushed granite particle stream in a chute is shown in Figure 2.4.

As in the case with vertical motion of material [26], the dependence diagram $\beta_1 = f(Fr)$ is clearly divided into three areas that correspond to three modes of stream motion. In area Fr > 1.7 (c), the bulk concentration is constant and is virtually equal to the material concentration at rest. This area corresponds to the mode of constrained motion. In the interval 0.8 < Fr < 1.7 (area (b), the bulk concentration

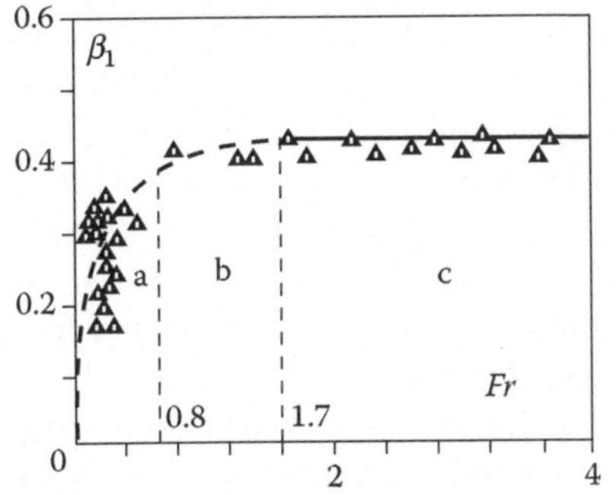

FIGURE 2.4 Dependence between bulk concentration of crushed granite particles and Froude number.

is sharply decreased, the material motion is characterized by local discontinuities among particle groups, and there is a velocity gradient in the flow depth. This is called the transition area, and the numbers within it are the critical numbers. At Froude numbers below 0.8 (area a), dependence $\beta_1 = f(\text{Fr})$ is curvilinear, and the lowest Froude number corresponds to the lowest bulk concentration (unconstrained motion mode). Note that according to Equation 2.12, the area of the critical Froude numbers at the inclination angles in question is

$$2 < Fr_{cr} \leq 2{,}5\,, \tag{2.14}$$

which is in line with the experimental data.

Various modes of bulk material motion complicate the analysis of bombarding particle aerodynamic interaction. Particles are moved by air and by heat and mass exchange between material and air change.

2.1.2 Particle Distribution

In the mode of unconstrained motion, particle distribution over a chute cross-section is static. Theoretical prerequisites to the study of static regularities were developed by L. Boltsman [9], who studied a stream with a large quantity of small elastic balls. He demonstrated that the particles' concentration and speed are determined by the F distribution function. This function pattern is defined by the differential equation:

$$\frac{\partial F}{\partial t} + u_x \frac{\partial F}{\partial x} + u_y \frac{\partial F}{\partial y} + u_z \frac{\partial F}{\partial z} + X \frac{\partial F}{\partial u_x} + Y \frac{\partial F}{\partial u_x} + Z \frac{\partial F}{\partial z_x} = \Delta_c F,$$

where X, Y, and Z are external force components; u_x, u_y, and u_z are projections of particle velocities on coordinate axes; and $\Delta_c F$ is the rate of change of the fixed point distribution function due to particle collision.

In general, however, the reduced equation defies solutions. Approximations have been made [103], but with no regard to the external forces that are determinative in our case. Therefore, in order to determine particle concentration, we conducted experimental tests using a flow divider with five synchronized valves installed on the material motion path. Particles caught in the divider during a fixed interval of time were discharged from the divider bins and weighed. The experiments were conducted with R. N. Shumilov [50] and enabled us to clarify the following pattern of motion for 0.625–1.25-mm crushed granite particles. A substantial portion of the particles is moving at the chute bottom. Moreover, the number of "bottom" particles increases with a rise in material flow rate (Figure 2.5a) and with a decrease in the distance to the stream falling point at the chute bottom (*l*). This is explained by a superposition of two processes occurring in a stream of airborne particles. The first is a saltation process (a galloping motion of particles resulting from their periodic impact with the chute bottom), and the second process is the intercollision of particles.

At small flow rates or at a great distance *l*, when the concentration of particles is low, motion produces virtually no intercollision of particles. The transverse gradient of particle concentration is comparatively low.

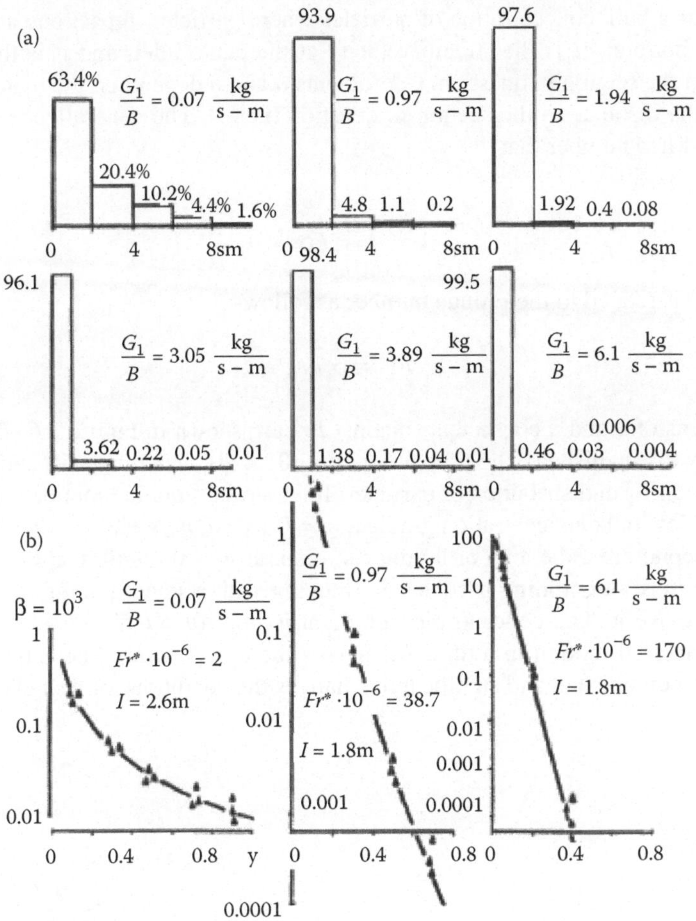

FIGURE 2.5 Distribution of crushed granite particles and chute section height.

At high flow rates, particle concentration reaches such values when their intercollision becomes so apparent that few saltating particles are able to break through the intercolliding particle mass and leave the stream. Therefore, the quantity of particles drawn out from the stream and moving above it is small and the concentration gradient is high (Figure 2.5b). The following quantitative characteristics were established.

Distribution of particles with channel height is clearly exponential

$$\beta = \beta_0 \exp(-ay^n), \tag{2.15}$$

and the bulk concentration of particles at the chute bottom is subject to the flow continuity law

$$\beta_0 = v_{1H}\beta_H / v_1,$$

where β_H is a bulk concentration of particles where particles impact one another on the chute bottom; v_{1H} is the stream velocity at the chute inlet; and v_1 is the stream velocity in the section in question. Coefficients a and n depend on the material flow rate and the distance to the section in question (i.e., l). The generalized parameter was a modified number that

$$Fr^* = \frac{G_1 g}{v_1^3 B \rho_1}, \tag{2.16}$$

(if $v_1 = \bar{v}_1$) is related to the Froude number as follows:

$$Fr^* = \beta_1 Fr \tag{2.17}$$

The plots of a and n coefficients against Fr^* are shown in Figure 2.6. Two areas are clearly distinguished. The first area is $Fr^* \cdot 10^6 < 40$, which we will call the area of pseudo-uniform distribution of particles. It features a saltating motion and a comparatively low bulk concentration gradient ($n = 0.1$-0.67; $n_{av} \approx 0.3$).

The second area, the area of laminar motion, at $Fr^* \cdot 10^6 > 40$, features a bed in which the most of colliding particles are moving with a small quantity of particles saltating above it. The concentration gradient is high ($n = 0.67$-1.2; $n_{av} \approx 1$). This unusual material motion in a tip chute makes the aerodynamic interaction pattern even more complicated and significantly changes the conditions of heat exchange.

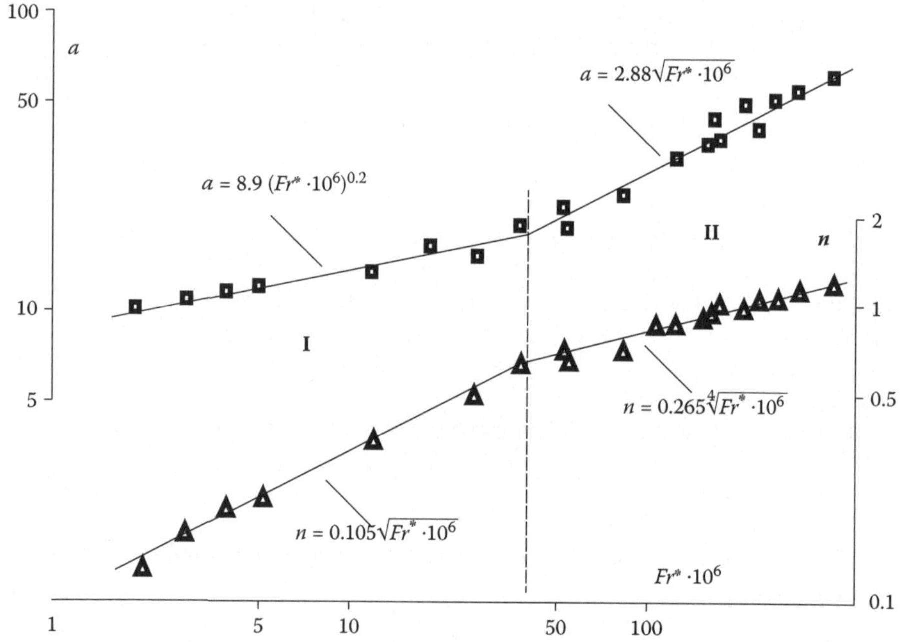

FIGURE 2.6 Variation of a and II coefficients with an increase in Fr^*.

2.1.3 Motion Speed

In industrial conditions, a bulk material is typically chuted as a non-dense bed. Here, the effect of solid particle friction is substituted with the effect of air drag forces, the frictional force resulting from particle contact with chute walls and from gravity forces.

Unlike gravity and air drag forces, the frictional force of particle contact is transient and is very hard to determine. In order to obtain design data concerning the bulk material motion velocities, we conducted experimental tests. Particle stream velocity was measured in various motion modes in a tip chute in the experimental arrangement (see Figure 2.2). The velocity value was determined using two methods: photographic and ballistic. The photographic method measured a particle's travel path within the time frame of a photographic shutter opening (exposure). By knowing the exposure time $\Delta\tau$ and measuring the particle's path sections obtained on photographic prints, it is possible to determine the mean projection of particle velocity on the chute axis:

$$v_1 = \frac{1}{N}\sum_{i=1}^{N}\frac{\Delta x_i}{\Delta\tau}.$$

Section projections Δx_i were determined by using a ruler photographed together with airborne particles. Therefore, it was not necessary to consider the scale in photographing and photocopying. This method was used at low flow rates of bulk materials when the probability of superposition of paths was not a factor.

At low material flow rates, velocity was determined using the ballistic method, which consisted of measuring the material stream path at the chute outlet. Knowing the chute inclination angle and the coordinates of the stream centerline, the material's final velocity was calculated using a reduced dynamic equation of free settling particles. The centerline coordinates were determined by using a coordinate spacer, the horizontal axis of which was positioned in the material stream for more accurate measurements.

Solving the dynamic equation for a free-falling body

$$\dot{\vec{v}}_1 = \vec{g} \tag{2.18}$$

in XOY coordinate system (Figure 2.7) with some transformations, we obtained the formula

$$v_k = \frac{x_k}{\cos\Theta}\sqrt{\frac{g}{2(y_k - x_k tg\Theta)}}, \tag{2.19}$$

which was used to calculate the particle velocity at the chute outlet. Because Equation 2.18 does not account for air drag forces (2.19), it gives rather excessive results.

Experimental tests showed that a chuted stream of bulk material particles is uniformly accelerated (Figure 2.8a).

The acceleration rate is:

$$a_T = g\sin\Theta\cdot(1 - f_{\text{Tp}} ctg\Theta). \tag{2.20}$$

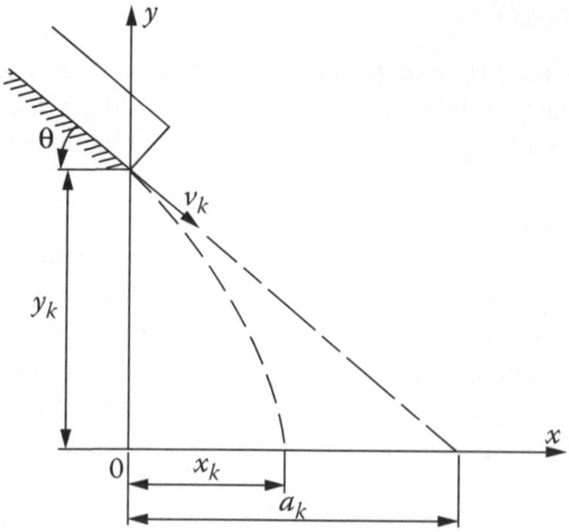

FIGURE 2.7 Defining the path of bulk material particles poured from a tip chute.

The conditional coefficient of the chute walls' resistance to particle motion f_{Tp} depends on the mode of motion (Figure 2.8 b, c).

In case of unconstrained motion, this coefficient is lower than the coefficient of sliding friction f_{CK}; the relation $\delta = f_{\text{Tp}}/f_{\text{CK}}$ is: $\delta = 0.5$ for unconstrained motion and $\delta = 1$ for the constrained motion.

Considering the fact that f_{CK} varies over a wide range and depends on many factors (such as physical and mechanical properties of the transferred material, chute wall surface condition, etc.), it is recommended that calculations of local transfer group exhausts are based on the assumption that $f_{\text{Tp}} = 0.5$.

Now, let us consider the peculiarities of material motion in kinked chutes. First, we will calculate the path and velocity of the conveyed bulk material stream (Figure 2.9).

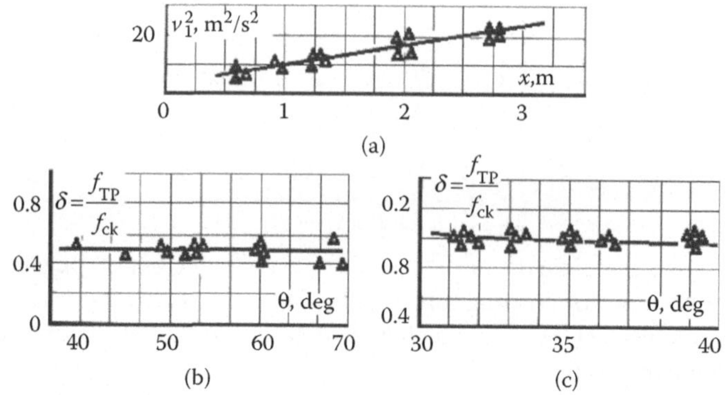

FIGURE 2.8 Chute length variation in (a) particle velocity and the relation of the friction coefficient to the chute inclination angle in (b) unconstrained and (c) constrained modes of bulk material motion.

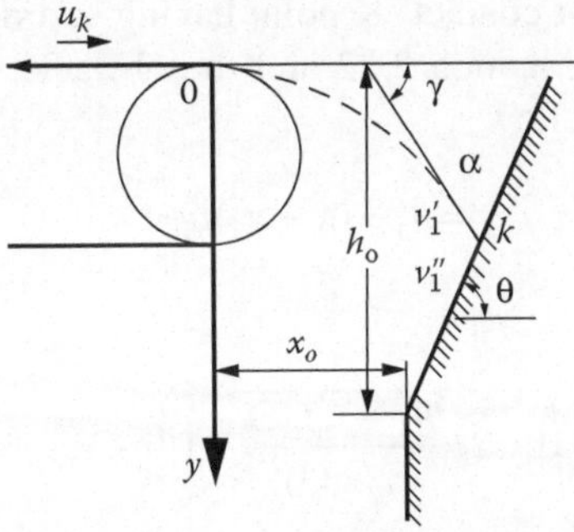

FIGURE 2.9 Defining particle settling velocity in a chute receiving funnel.

We use Equation 2.18 to do this. By integrating this equation at the initial conditions

$$v_1\big|_{t=0} = u_k, \text{ and}$$

$$x\big|_{t=0} = 0,\ y\big|_{t=0} = 0,$$

we obtain:

(a) for a jet trajectory (jet axis equation)

$$x = u_K t,\ y = gt^2/2 \tag{2.21}$$

or

$$y = 0{,}5g(x/u_k)^2; \text{ and} \tag{2.22}$$

(b) for velocity

$$v_1 = \sqrt{u_k^2 + 2gy} \tag{2.23}$$

or, in view of Equation 2.22,

$$v_1 = \sqrt{u_k^2 + (gx/u_k)^2}, \tag{2.24}$$

where u_k is the forward velocity of particles that is equal to the velocity of the horizontal conveyer belt, m/s.

Based on the chute wall position and the conveyer speed, the discharged material flow may either have contact with the belt wall or not. Contact with the belt wall leads to a sharp change in the jet trajectory and in the speed. The flow-to-wall contact condition (from Equation 2.22) is determined by the following inequation:

$$h_0 > (x_0/u_k)^2 g/2. \tag{2.25}$$

In order to find a point of contact (K point having coordinates x_k, y_k), it is necessary to jointly solve path Equation 2.22 and the obstacle surface equation. In our case, the latter appears as follows:

$$y = h_0 - (x - x_0)tg\Theta. \tag{2.26}$$

Then

$$x_k = \frac{u_k^2}{g}\left(\sqrt{1 + 2g\frac{h_0 + x_0 tg\Theta}{(u_k \mathrm{tg}\Theta)^2}} - 1\right)tg\Theta, y_k = \frac{gx_k^2}{2u_k^2}. \tag{2.27}$$

If the wall is vertical ($\Theta \to \pi/2$), it is easy to obtain

$$x_k = x_0, \ y_k = g(x_0/u_k)^2/2. \tag{2.28}$$

As soon as the coordinates are available, Equation 2.24 can be used to determine the bulk material stream velocity at the moment of contact.

Elastic forces and the wall drag forces make the stream change its direction. The impact of an irregularly shaped particle stream is not an elastic impact in the strictest sense; therefore, for a stream in general, or for single particles within the stream, the angle of reflection is not equal to the angle of incidence. R. L. Zenkov's studies [33] show that the angle of reflection for a bulk material stream is virtually equal to $\pi/2$. The stream velocity after the impact is:

$$v_1'' = Kv_1', \tag{2.29}$$

where v_1' is the stream velocity at the wall contact, m/s; v_1'' is the stream velocity after the wall contact, m/s; and K is a correction factor accounting for the reduction in speed at the chute turn.

Chute turn α, deg	0	10	20	30	40	45	50	60
K	1.0	0.97	0.93	0.85	0.75	0.69	0.63	0.45

In our case, angle α is an acute angle between the jet path tangent in the contact point and the wall plane.

The tangent slope is determined after differentiating Equation 2.22:

$$tg\gamma = gx_k / u_k^2. \tag{2.30}$$

Then it is obvious that

$$\alpha = 180 - \left(\Theta + arctg\frac{gx_k}{u_k^2}\right). \tag{2.31}$$

Further calculation of the bulk material stream velocity is based on the formula:

$$v_1 = \sqrt{2a_T l + (v_1'')^2}. \tag{2.32}$$

In cases where there are significant drops of fine material (when $h > 0.5$), the medium drag force must be considered.

2.2 AERODYNAMIC CHARACTERISTIC OF A SINGLE PARTICLE

Many theoretical and experimental efforts have dealt with the study of a particle's aerodynamic characteristic. Theoretical studies by Stokes, Oseen, and Goldstein discussed the viscous flow-around of spherical particles. L .I. Sedov [82] and G. Schlichting [104] compiled theoretical and experimental papers on the aerodynamic interaction of gas and sphere. Studies by Z. R. Gorbis dealt with the air mechanics of irregularly shaped particles [26].

The medium impact on a particle is determined by forces $\overleftarrow{\Pi}_S$ continuously distributed across the particle surface *S*; these surface forces can be expressed through direct and tangential stresses $\vec{p}$ and $\vec{\tau}$ in each point of the particle surface;

$$\vec{R} = \int_s \vec{p}\,ds + \int_s \vec{\tau}\,ds \tag{2.33}$$

is the resultant vector of the system of elementary forces distributed across the particle surface. This is called the aerodynamic force or the medium drag force.

Generally, aerodynamic force is directed at an angle to the particle gravity center relative velocity vector $\vec{w}$. Vector $\vec{R}$ is usually substituted in air mechanics, and its components in the rectangular coordinate system are related to the particle relative velocity vector $\vec{w}$. The force applied in the direction opposite to the particle's relative motion direction is called air drag $\vec{X}$ or motion drag. The force that is perpendicular to the drag force and that lies in the vertical plane is buoyancy force $\vec{Y}$. The force that is perpendicular to drag and buoyancy is lateral force $\vec{Z}$. Magnitudes of these vectors are determined by projecting the vector equation (2.33) on the selected coordinate system axes:

$$\left.\begin{aligned} X &= \Psi_x \frac{\rho w^2}{2} f_M, \\ Y &= \Psi_y \frac{\rho w^2}{2} f_M, \\ Z &= \Psi_z \frac{\rho w^2}{2} f_M, \end{aligned}\right\} \tag{2.34}$$

where

$$\Psi_x = \frac{1}{0,5\rho w^2 f_M} \int_s \left[(p - p_\infty)\cos(p,x) + \tau \sin(\tau,x) \right] ds, \tag{2.35}$$

$$\Psi_y = \frac{1}{0,5\rho w^2 f_M} \int_s \left[(p - p_\infty)\cos(p,y) + \tau \sin(\tau,y) \right] ds, \tag{2.36}$$

$$\Psi_z = \frac{1}{0,5\rho w^2 f_M} \int_s \left[(p - p_\infty)\cos(p,z) + \tau \sin(\tau,z) \right] ds, \tag{2.37}$$

are a head drag coefficient, buoyancy force coefficient, and lateral force coefficient, respectively.

Thus, aerodynamic force R is proportional to the dynamic pressure and to the specific body cross-section area f_M and depends on a dimensionless drag coefficient Ψ that is based on the body shape and flow-around conditions:

$$R = \Psi \frac{\rho w^2}{2} f_M, \tag{2.38}$$

where

$$\Psi = \sqrt{\Psi_x^2 + \Psi_y^2 + \Psi_z^2}. \tag{2.39}$$

The translational uniform motion of sphere integrals (2.36) and (2.37) are equal to zero, and the aerodynamic force is

$$\vec{R} = \vec{X} = -\Psi \rho w \vec{w} f_M / 2. \tag{2.40}$$

At low Reynolds numbers (Re < 1), the vector of stress forces in a translational sphere's motion has the same value of $3\eta \vec{w} / d$ [14] in all parts of the sphere; the aerodynamic force is determined by the Stokes law:

$$\vec{R} = -3\pi\eta\vec{w}d \tag{2.41}$$

and the drag coefficient

$$\Psi = 24 / Re. \tag{2.42}$$

Generally, the regime of flow Ψ coefficient depends on the particle motion tightness and the particle rotation around the center of gravity. The effect of the proximity of tube walls or of single particles on the regime of flow and drag force is accounted for by correction factor E. The drag factor of a particle moving in a constrained environment Ψ_{CT} is determined through the drag factor Ψ_{∞} of a sphere moving at the same relative velocity in an unconstrained environment

$$\Psi_{CT} = \Psi_{\infty} E^{-2}. \tag{2.43}$$

There are efforts [12,13,26,57] that provide a quantitative estimation of E correction. P. V. Lyashchenko's [57,62,91] formula became widely used for a stream of uniformly distributed particles moving at a constant speed in a tube

$$E = (1 - \beta_1)^n, \tag{2.44}$$

where n is a trial coefficient varying from 2.5 to 3.8 and is equal to 3 (on average).

Determination of the aerodynamic force for isometric particles—even in the laminar mode of motion (Re < 0.2)—presents mathematical difficulties. Therefore, in practice, particle drag force is compared with the drag force of a sphere equivalent to the particle in volume.

The necessity of studying aerodynamic interaction of components is driven by theoretically unstudied aerodynamic properties of irregularly shaped particles and by the specific dynamics of the class of two-component streams in question.

The gravitational motion of bulk material particles is characterized by micro- and macro-non-uniformity. As an average static collective, the stream of particles is generally accelerated by the Earth's gravitational field. Due to collisions with the channel walls or with each other, particles make complex movements accompanied by micropulsations. Typically, particles move translationally. Because air viscosity forces are small, the rotational motion of particles remains virtually unchanged. An airborne particle offers various portions of its surface to the air-flown stream. Therefore, it is equally probable that a midsection can be any projection of a particle, unlike when a particle is moving in a more viscous medium (e.g., in water) where a settling particle is oriented with most of its projected area. That is why the extensive results of hydrodynamic characteristics of various mineral grains must be used very carefully. In addition, accelerated motion does not enable a direct transfer of experimental data concerning steady state flows in pneumatic transportation of solid particles.

We know that aerodynamic interaction is determined by the mode of particle motion and by particle tightness as well as by the geometric shape of particles. Geometric shape can present difficulties; particle geometry differs even within the same bulk material. The only thing that can be noted in advance is that a stream does not contain a single pair of particles that are absolutely identical in shape. This is because any disintegration of natural minerals is clearly spontaneous. Therefore, statistical methods are used to estimate particle shape.

The list of stream peculiarities would be incomplete unless we mention the irregularity of particle concentration in a stream. We do not know of any studies devoted to assessing the aerodynamic characteristics of particles at a concentration gradient. The gravitational flow in tip chutes, for example, is characterized by reduction in particle concentration along the flow path and by a noticeable non-uniformity of particle distribution in the cross-section.

2.2.1 Geometric Shape

Quantitative and qualitative assessment of particle shape is driven by the nature of the processes being studied. For instance, a quantitative assessment of rock shapes widely used for engineering evaluation of mining techniques and for implementation of the same is based on the relation among the greatest linear dimensions [4,5,6]: length (D), width (L), and thickness (T). Rock coarseness (also called rock diameter) is determined by the geometric mean value

$$d_0 = \sqrt[3]{\mathrm{D} \cdot \mathrm{L} \cdot \mathrm{T}}. \tag{2.45}$$

The following equation was developed from numerous measurements

$$V_P = \mathrm{D} \cdot \mathrm{L} \cdot \mathrm{T}/2{,}2, \tag{2.46}$$

in order to find the interrelation between rock coarseness and equivalent diameter

$$dE = \frac{d_0}{1{,}3} \cdot \sqrt[3]{\frac{6}{\pi}} \approx 0{,}95 d_0. \tag{2.47}$$

In the study of heat and mass transfer processes, particle shape is assessed using the coefficient

$$k_s = (s_p / s_L)\big|_{V=idem}, \tag{2.48}$$

which accounts for particle surface s_p.

This is explained by the fact that particle surface is among the principal factors in the engineering design of exchange processes. The same coefficient is also used in studies of the aerodynamic interaction of particles [98]; for our purposes here (2.38), the basic parameter is the particle's midsection area rather than its surface. This determined our decision to use the particle projected area [57] for a quantitative assessment of particle shape. A similar point of view was expressed by a team of authors [41] who studied the drag of bulk material particles as applied to hydrotransportation of bulk material particles.

In the ore mining industry, bulk material particles are classified as sharp-grained (the exception is iron-ore pellets, which are round), featuring a variety of non-isometric shapes. Because these particles are quite large, the degree of their non-isometric shape may be determined by direct measurement of geometric particle dimensions. Total particle size is measured by the equivalent diameter value, while the non-isometric shape degree will be estimated using the coefficient of variation of measured projected areas:

$$r_f = \sqrt{\frac{\sum_{i=1}^{N} (f_{pi} - \bar{f}_p)^2}{N \cdot \bar{f}_p^2}}, \tag{2.49}$$

where $\overline{f}_p$ is the arithmetic mean value of the particle projected based on N measurements and f_{pi} is the area of projections in the i-position of the particle.

The projected area of a particle in a particular position was determined, using an ocular micrometer (with a mesh mounted inside), by placing the particle in question on MBS-2 stereoscopic microscope glass. The relation of the arithmetic mean projected area to a circular area of d_e diameter was used as a quantitative criterion of the geometric shape:

$$k_f = \overline{f}_p \Big/ \frac{\pi d_e^2}{4}. \tag{2.50}$$

This study used particles of granite, iron ore, agglomerate, and pellets $d_e \le 20$ mm (Figure 2.10).

After processing the results, the coefficient of variation of measured areas was found to be directly proportional to form factor k_f (see dashed curve in Figure 2.11):

$$k_f = (1 - \sqrt{r_f})^{-0{,}5}. \tag{2.51}$$

Let us compare these results with regular-shaped bodies. First, we will assume that it is equally possible for all projected areas of regular-shaped bodies to be continuously situated in the interval between the minimum area (f_{min}) and the maximum

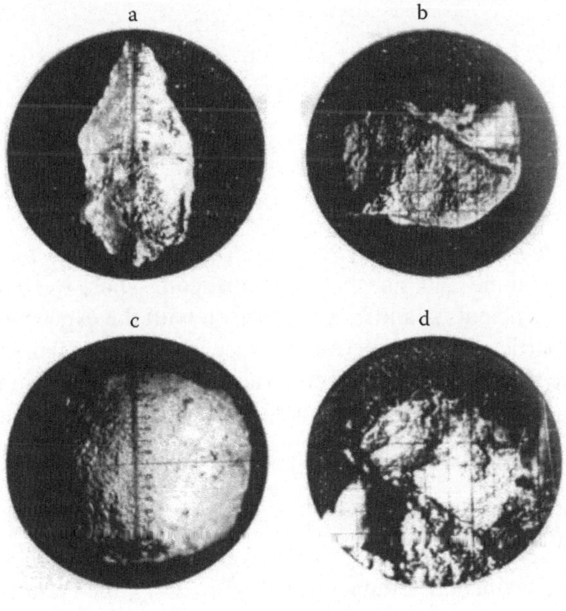

FIGURE 2.10 General view of particles of (a) granite, (b) iron ore, (c) iron-ore pellets, and (d) agglomerate.

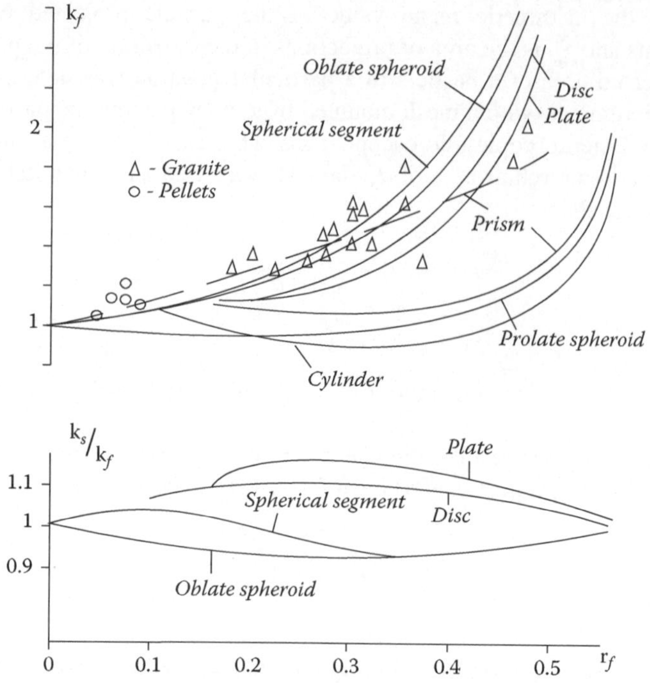

FIGURE 2.11 Variation in the geometric form factor of particles and regular-shaped bodies with an increase in the coefficient of variation of projected areas.

area (f_{max}). Such a condition can be met for regular-shaped bodies by appropriate positioning on the projection plane. In this case,

$$\bar{f}_p = 0,5(f_{min} + f_{max}); r_f = \frac{1}{\sqrt{3}} \frac{f_{max} - f_{min}}{f_{max} + f_{min}}. \quad (2.52)$$

Determining the minimum and maximum projected areas for specific bodies as well as their volume and surface is not difficult. Thus, we find the equations for calculating coefficients k_f and k_s. Comparison with the experimental data shows that the studied particles are geometrically close to oblate regular-shaped bodies for which an approximate equation of geometric form factors is specific: $k_f \approx k_s \approx = k$. The latter is important for comparison and generalization of experimental data.

2.2.2 Dynamic Shape of Particles

The aerodynamic characteristic of particles was determined by measuring airborne velocity in the experimental arrangement (Figure 2.12), the main operating part of which was a tapered tube (5° taper angle) made of organic glass.

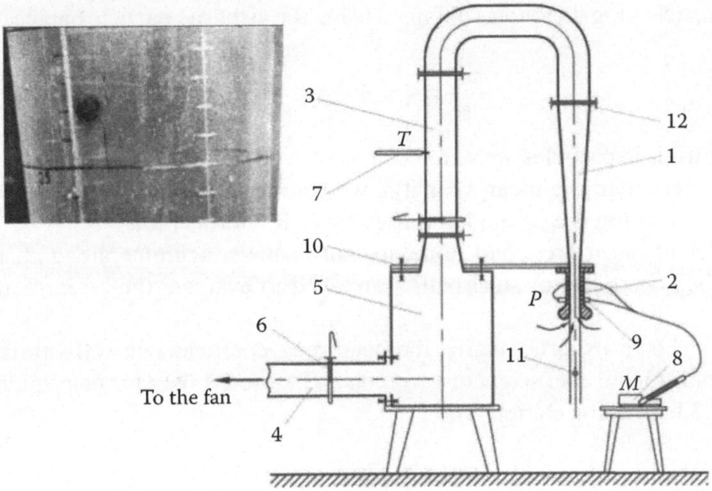

FIGURE 2.12 Diagram of the experimental arrangement for the study of airborne solid particles: 1 = tapered tube; 2 = measuring manifold; 3, 4 = air ducts; 5 = chamber; 6 = damper; 7 = thermometer; 8 = micropressure gauge; 9 = fittings; 10 = brackets; 11 = plumb; and 12 = grill.

Air was supplied to the tube through the manifold, the upstream section of which was made on the lemniscate through the ductwork and compensation chamber to the fan.

Airborne velocity was determined in the following manner. The particle in question was placed in the manifold and the particle "hanging" positions were fixed against the tapered tube inlet section (x_i). There were x_i distances measured N times considering the pulsations for the purpose of accuracy, and the design value adopted was the mean arithmetic value $\bar{x}$. The count was determined using common metrology techniques. Then, given the known $\bar{x}$, there was a relation (k) of the mean air velocity in the section wherein the particle was hanging up to the mean manifold flow velocity u_{man} by using a rating curve. The latter value was determined given the measured manifold vacuum (P_{man}) from the formula

$$u_{man} = \sqrt{\frac{2P_{man}}{(1+\zeta_{man})\rho}}, \tag{2.53}$$

where ζ_{man} is the manifold resistance coefficient that in our case was rated equal to $\zeta_{man} = 0.0018$.

The airborne velocity was determined from the formula

$$c = k \cdot u_{man},$$

and the particle drag factor was obtained using the airborne particle balance equation

$$\frac{\pi d_e^3}{6}\rho_1 g = \Psi \frac{\pi d_e^2}{4} \cdot \frac{c^2}{2}\rho_2. \tag{2.54}$$

Because airborne particles were in the area adjacent to the wall (where the true flow velocity is less than the mean velocity), we converted the coefficient into the local velocity value using the method proposed by Z. R. Gorbis [26].

Particles of burnt ore, chalkstone, iron-ore pellets, agglomerate, and steel balls were studied. Particles of steel balls were used to evaluate the research technique error.

As was shown experimentally, the head drag coefficient in self-similarity area (Ψ_0) depends on the coefficient of variation r_f (Figure 2.13) and is determined within $0 < r_f < 0.3$ by the correlation ratio

$$\Psi_0 = 2,24\sqrt{r_f} + 0,43, \tag{2.55}$$

given Equation 2.51

$$\Psi_0 = 2,24(1 - k_f^{-2}) + 0,43 \tag{2.56}$$

or

$$\Psi_0/\Psi_{0L} = 6,67 - 5,67k_f^{-2}. \tag{2.57}$$

The results allow us to determine the particle drag factor without conducting any experiments. Having determined the coefficient of variation r_f with an MBS microscope, we use Equations 2.55 and 2.51 to obtain the Ψ_0 and k_f coefficients.

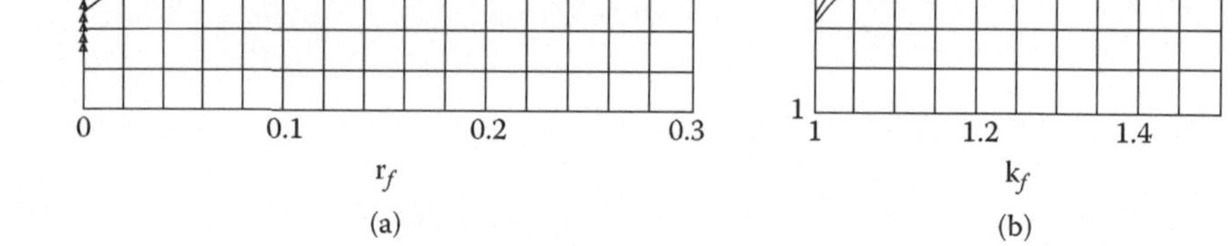

FIGURE 2.13 Aerodynamic characteristics of particles: (a) Ψ_0 function of the coefficient of variation; (b) k_∂ function of the geometric form factor (1 – according to E. Pettyjohn and E. Christiansen [98]; 2 – according to our data).

In this case, it is not necessary to use a dynamic form factor to compare the particle drag force with the aerodynamic force of a sphere equivalent to the particle in its volume:

$$k_\partial = \frac{R}{R_L}\bigg|_{v=\text{idem},\text{Re}=\text{idem}} = \frac{\Psi}{\Psi_L}\bigg|_{v=\text{idem},\text{Re}=\text{idem}}. \tag{2.58}$$

The following relations [98,99] are known for this coefficient in the Stokes area

$$k_{\partial_{st}} = (1+0{,}86\lg k_s^{-1})^{-1} = (1-0{,}373\ln k_s)^{-1} \tag{2.59}$$

and in the self-similarity area

$$k_\partial = 12{,}4 - 11{,}4k_s^{-1}. \tag{2.60}$$

It is clear from the diagrams (Figure 2.13) that the resulting particle drag factors within $k_s < 1.3$ and $r_f < 0.3$ satisfactorily match the experimental data obtained by E. Pettyjohn and E. Christiansen [98] for isometric-shaped particles.

2.2.3 Resistance Coefficient

In order to obtain design ratios for the drag factor in a wide range of Re numbers, our results were compared with those of other authors who studied the aerodynamic properties of other material particles (Figure 2.14)*. The experimental data analysis and comparison led us to the following conclusions.

(1) Experimental data for steel balls satisfactorily match the Rayleigh curve, which indicates the accuracy of our research technique.
(2) The drag factor of particles of the same material widely varies even in the self-similarity area; this is explained by differences in geometric shapes. The self-similarity area for irregularly shaped particles occurs earlier than for a ball when $\text{Re}_E \geq 400$ (for a ball $\text{Re} \geq 2 \cdot 10^3$).
(3) Bulk materials can be subdivided into two groups by drag factor value (Table 2.1). The first group of materials is composed of sharp-grained particles featuring a wide variety of geometric form factors and drag factors: $k_\Gamma = 1.3 - 2$, $\Psi_0 = 1.2 - 2$; the second group includes rounded bulk materials for which $k_\Gamma = 1 - 1.5$, $\Psi_0 = 1 - 1.1$.

By compiling the experimental data, we obtained the following Oseen-type relation [51,52,70]

$$\Psi = \Psi_0\left(\frac{a}{\text{Re}} + 1\right) = \frac{A}{\text{Re}} + B \tag{2.61}$$

* Because most authors measured k_s instead of k_f, coefficients k_s and k_f are hereinafter identified with k_r.

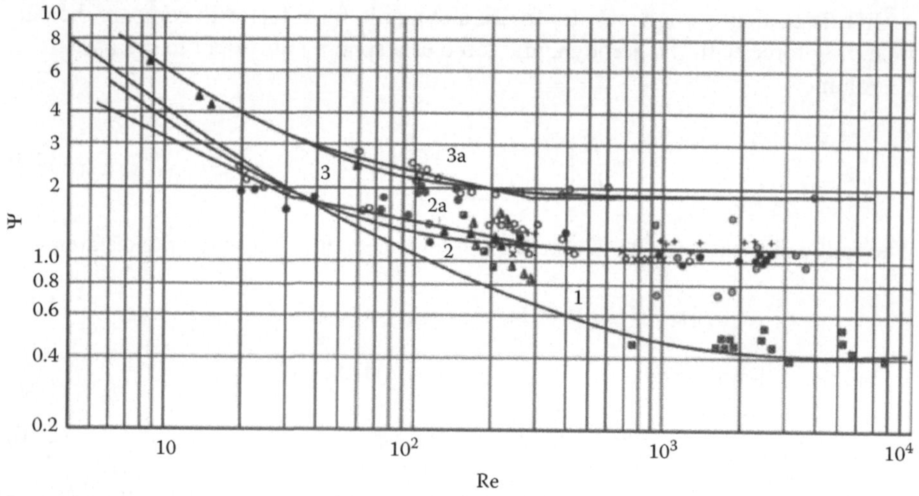

FIGURE 2.14 Dependence of the drag factor of solid particles on Reynolds number: 1 = for a sphere (Rayleigh curve); 2 = for rounded particles (by Equation 2.61 at $\Psi_0 = 1.1$); 2a = Z. R. Gorbis' approximation; 3 = for sharp-grained particles (by Equation 2.61 at $\Psi_0 = 1.8$); 3a = Z. R. Gorbis' approximation. Experimental data: G. N. Khudyakova [99] – • quartz $d_e = 0.07–0.845$ mm; Z. R. Gorbis [25,26] – ∘ quartz $d_e = 0.4–1.12$ mm, + graphite $d_e = 0.18–2.5$ mm, x – graphite $d_e = 0.36–6.74$ mm; I. A. Vakhrusheva and A. I. Skoblo [16] – ▲ – charred coal $d_e = 0.256–1.08$ mm; authors – ◪ – burnt iron ore $d_e = 2.5–4.3$ mm; ⊕ – chalkstone $d_e = 1.9–3.5$ mm; ⊙ – iron-ore pellets $d_e = 12–24$ mm; ▲ – granite $d_e = 1.16–3.2$ mm; ⊠ – steel balls $d_e = 2–4$ mm.

TABLE 2.1
Aerodynamic Characteristic of Solid Particles

				Design Values				
Particle Group	**Material Description**	k_r	Ψ_0	k_z	Ψ_0	k_∂	A	a
Sharp-grained particles	Iron ore	1.1–1.4	0.9–1.5	1.3	1.2	1.11	26.6	22.2
	Iron ore pellet fines	1.2–1.4	1.1–1.5	1.3	1.3	1.11	26.6	20.5
	Quartz	1.1–1.7	1.2–2.0	1.4	1.6	1.14	27.4	17.1
	Chalkstone	1.3–1.7	1.5–2.0	1.5	1.8	1.18	28.3	16.7
	Artificial graphite	1.4–1.9	1–1.6	1.6	1.3	1.21	29.0	22.3
	Granite	1.3–2.0	1.4–1.9	1.7	1.7	1.25	30.0	17.6
	Anthracite	1.5–2.0	1.4–2.2	1.7	1.8	1.25	30.0	16.7
	Sharp-grained sand	1.5–1.9	1.8–2.2*	1.7	2.0	1.25	30.0	15.0
	Coal dust	1.6–2.6	1.9–2.5*	2.2	2.2	1.42	34.1	15.5
Rounded particles	Iron-ore pellets	1.1–1.2	0.8–1.2	1.1	1.0	1.04	25.0	25.0
	Round sea sand	1.15–1.2	1–1.2ˣ	1.15	1.1	1.05	25.2	22.9

*The values have been determined from Equation 2.56.

that allows calculation of the particle drag factor throughout the flow-around areas (Stokes area, transition area, and self-similarity area) with sufficient accuracy. The resulting relation accurately correlates with the experimental data and the known compilations by Z. R. Gorbis and enables a comparatively simple integration of the particle dynamic equations. The latter is especially important for dispersed particles passing through flow-around areas in accelerated sedimentation.

2.3. SEDIMENTATION OF PARTICLES

2.3.1 Particle Motion in the Air Stream

The particle dynamic equation

$$m\dot{\vec{v}}_1 = \vec{R} + \vec{P}_\rho \tag{2.62}$$

in a uniform air stream ($u = const$) appears as

$$V_\rho \rho_m \dot{\vec{v}}_1 = V_\rho (\rho_m - \rho)\vec{g} - \psi f_M \vec{w} w \rho / 2. \tag{2.63}$$

Given that

$$\dot{\vec{v}}_1 = 0 \,\text{at}\, w = c$$

and

$$V_\rho (\rho_m - \rho)\vec{g} = \psi f_M c^2 \rho / 2, \tag{2.64}$$

the particle dynamic equation may be written in a way that makes it easier to further analyze

$$\dot{\vec{v}}_1 = (1 - \varepsilon) g \, [\vec{g} / g - \psi w \vec{w} / (\psi_c c^2)]. \tag{2.65}$$

Let us consider a particle's vertical motion ($g_x/g = 1$). Convert this equation into one where airborne speed is taken for the specific velocity, the relaxation time is taken for the specific time

$$t_\infty = c/(g(1\text{-}\varepsilon)), \tag{2.66}$$

and the specific length is

$$l_\infty = ct_\infty = c^2/(g(1 - \varepsilon)). \tag{2.67}$$

On the assumption that

$$v_1 = vc;\ v_2 = uc;\ v_1 - v_2 = \omega c; t = \tau t_\infty; x = hl_\infty \tag{2.68}$$

Equation 2.65 will then appear as follows:

$$\frac{d\omega}{d\tau} = 1 - \frac{\psi}{\psi_c}|\omega|\,\omega \tag{2.69}$$

or

$$(\omega + u)\frac{d\omega}{dh} = 1 - \frac{\psi}{\psi_c}|\omega|\,\omega. \tag{2.70}$$

Inserting the general expression for the drag factor (2.61) in the earlier equations, we obtain

$$\frac{d\omega}{d\tau} \equiv (\omega + u)\frac{d\omega}{dh} = 1 - \frac{r_c\omega + B\omega|\omega|}{r_c + B}, \tag{2.71}$$

where, for the sake of convenience, it is assumed that

$$r_c = A / \mathrm{Re}_c,\ \mathrm{Re}_c = d_e c / v. \tag{2.72}$$

The second term of the right-hand side of Equation 2.71 represents the dimensionless drag force

$$\bar{R} = -(r_c\omega + B\omega|\omega|) / (r_c + B) \tag{2.73}$$

that displays positive values with a negative relative velocity of particles.

An integration of Equation 2.71 presents no special difficulties. The integration results for various initial conditions are shown in Table 2.2, which contains general and specific solutions. For the sake of convenience, it is assumed in the table that

$$b = \frac{B}{r_c + B}. \tag{2.74}$$

Figure 2.15 shows the particle relative velocity curves plotted with $b = 0.9$ (the maximum value for the class of problems in question) by Equations 2.25, 2.26, 2.31, and 2.33 (Table 2.2 for negative relative velocity values and by Equations 2.7, 2.4, 2.19, and 2.16 for positive relative velocity values. Relevant curves are provided for reference with respect to Stokes flow-around of particles (see Equations 2.13 and 2.10 and Table 2.2) and free sedimentation of particles when drag forces are absent. In the last case, the right-hand side of Equation 2.71 was equal to unit and the relative velocity was determined from the equation

$$\omega^2 - \omega_0^2 + 2u(\omega - \omega_0) = 2(h - h_0). \tag{2.75}$$

Figure 2.15 also shows the curves of acceleration and drag force. The latter was calculated for free sedimentation by Equation 2.73, subject to preliminary determination of relative velocity from Equation 2.75.

After analyzing these curves, it is possible to adopt some provisions that solve a number of problems of bulk material gravitational flow aerodynamics. First,

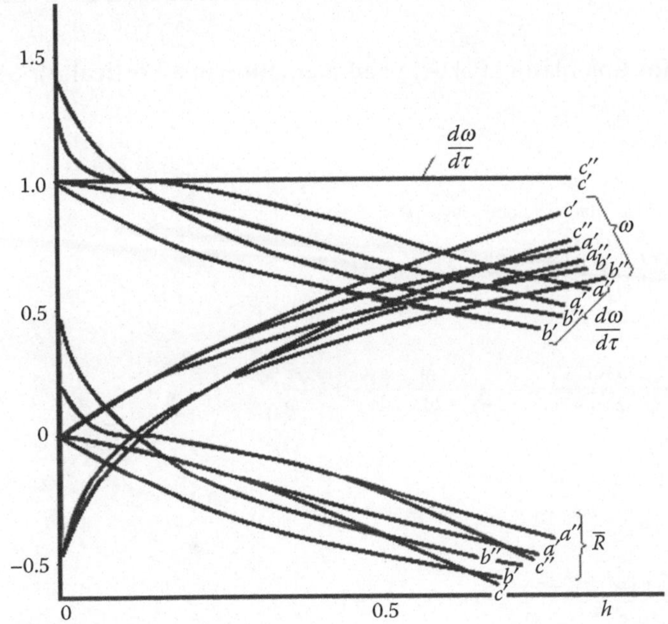

FIGURE 2.15 Variation in relative velocity (ω), acceleration (dω/dτ), and drag force (R) of a particle settling in a uniform air stream (u = 0.5; B = 0.9; case $\omega_0 = 0$ is designated with a single upper hyphen; $\omega_0 = -0.5$ is designated with two upper hyphens): a is the general law of resistance; *b* is Stokes law of resistance; *c* is sedimentation of particles without regard to the environmental resistance.

the relative velocity of particles can be accurately determined (up to 15%) within $h \le 0.5$, $u \le 0.5$, $B \le 0.9$ by reducing Equation 2.75 to a simple expression to find the dimensionless velocity of a particle:

$$v = \sqrt{v_H^2 + 2(h - h_0)}. \tag{2.76}$$

The drag force within the same area can be accurately established (up to 20%) by Equation 2.73 by determining the relative velocity for a free-settling particle.

Second, the particle relative velocity and drag force may be calculated within $b \le 0.9$ from the self-similarity area formulas. The segment of $\omega \approx 0$ (where Stokes law is valid) is small enough to assume that the particle is only moving in the self-similarity area. We extensively used these calculations to describe particle stream air mechanics within $h \ge 0.5$. These results were obtained without considering the inertial components of drag forces occurring in accelerated particle motion.

2.3.2 Aerodynamic Drag of a Particle Moving at an Increasing Rate

The aerodynamic force of a particle moving at an increasing rate differs from the aerodynamics of steady-state particle and air interaction by a value of added mass drag force and by a value of "hereditary" or Basset force resulting from an unsteady flow-around.

TABLE 2.2
Formulas for Calculating Particle Sedimentation in a Vertical Air Stream

Design Formulas	Formula No.
1	2
A. With a positive relative velocity ($\omega_0 \geq 0$)	
$$\frac{d\omega}{d\tau} \equiv (\omega + u)\frac{d\omega}{dh} = 1-(1-b)\omega - b\omega^2,$$	(1)
$$\frac{d\tau}{d\omega} = \frac{1}{1-(1-b)\omega - b\omega^2},$$	(2)
$$\frac{dh}{d\omega} = \frac{\omega + u}{1-(1-b)\omega - b\omega^2} = \frac{\omega}{1-(1-b)\omega - b\omega^2} + u\frac{d\tau}{d\omega}.$$	(3)
Integrating (2):	
$$\frac{1}{b+1}\ln\left(\frac{b\omega+1}{1-\omega}\cdot\frac{1-\omega_0}{b\omega_0+1}\right) = \tau - \tau_0$$	(4)
or	
$$\omega - \frac{(b\omega_0+1)e^{\Theta} - (1-\omega_0)}{(b\omega_0+1)e^{\Theta} + b(1-\omega_0)},\ \Theta = (\tau - \tau_0)(b+1).$$	(5)
Then	
$$\frac{d\omega}{d\tau} = (b+1)^2\frac{(b\omega_0+1)(1-\omega_0)e^{\Theta}}{[(b\omega_0+1)e^{\Theta} + b(1-\omega_0)]^2}$$	(6)
Integrating (3):	
$$h = h_0 + u(\tau - \tau_0) - \frac{1}{1+b}\left(\ln\frac{1-\omega}{1-\omega_0} + \frac{1}{b}\ln\frac{b\omega+1}{b\omega_0+1}\right)$$	(7)
or	
$$h - h_0 = \frac{1}{b+1}\left[(1+u)\ln\frac{1-\omega_0}{1-\omega} + \left(u_2 - \frac{1}{B}\right)\ln\frac{b\omega+1}{b\omega_0+1}\right].$$	(8)
Specific cases:	
(a) In Stokes area ($b = 0$)	
$$\frac{d\omega}{d\tau} = 1-\omega;\ \frac{d\omega}{dh} = \frac{1-\omega}{\omega+u}.$$	(9)
By integrating the first system equation (Equation 2.9), we obtain	
$$\ln\frac{1-\omega}{1-\omega_0} = \tau_0 - \tau$$	(10)
or	
$$\omega = 1-(1-\omega_0)e^{-\tau+\tau_0}.$$	(11)
Then	
$$\frac{d\omega}{d\tau} = (1-\omega_0)e^{-(\tau-\tau_0)}.$$	(12)

TABLE 2.2 (Continued)
Formulas for Calculating Particle Sedimentation in a Vertical Air Stream

Design Formulas	Formula No.
1	2
By integrating the second system equation (Equation 2.9), we obtain	
$$h-h_0=u(\tau-\tau_0)-(\omega-\omega_0)-\ln\frac{1-\omega}{1-\omega_0}$$	(13)
or	
$$h-h_0=(1+u)\ln\frac{1-\omega_0}{1-\omega}-(\omega-\omega_0)=(1+u)(\tau-\tau_0)-(\omega-\omega_0).$$	(14)
(b) In the self-similarity area (B = 1)	
$$\frac{d\omega}{d\tau}=1-\omega^2;\ \frac{d\omega}{dh}=\frac{1-\omega^2}{\omega+u}.$$	(15)
By integrating the first system equation (15), we obtain	
$$\tau-\tau_0=\frac{1}{2}\ln\frac{(\omega+1)(1-\omega_0)}{(1-\omega)(1+\omega_0)}$$	(16)
or	
$$\omega=\text{th}(\tau-\tau_0+\Theta_0),\Theta_0=\frac{1}{2}\ln\frac{1+\omega_0}{1-\omega_0}.$$	(17)
Then	
$$\frac{d\omega}{d\tau}=4\frac{(1-\omega_0^2)e^{2(\tau-\tau_0)}}{\left[(1+\omega_0)e^{2(\tau-\tau_0)}+(1+\omega_0)\right]^2}=ch^{-2}(\tau-\tau_0+\Theta_0)$$	(18)
By integrating the second system equation (Equation 2.9), we obtain	
$$h-h_0=u(\tau-\tau_0)-\frac{1}{2}\ln\frac{1-\omega^2}{1-\omega_0^2},$$ $$v^2=1-(1-v_H^2)e^{-2(h-h_0)}\ \text{with}\ u=0$$	(19)
or	
$$h-h_0=\frac{u}{2}\ln\frac{(1+\omega)(1-\omega_0)}{(1-\omega)(1+\omega_0)}-\frac{1}{2}\ln\frac{1-\omega^2}{1-\omega_0^2}.$$	(20)
B. With a negative relative velocity ($\omega\le 0,\ u\ge v$)	
$$\frac{d\omega}{d\tau}=1-(1-b)\omega+b\omega^2,$$	(21)
$$\frac{d\omega}{dh}=\left[1-(1-b)\omega+b\omega^2\right]/(\omega+u),$$	(22)
$$\frac{dh}{d\omega}=\frac{\omega+u}{1-(1-b)\omega+b\omega^2}=\left(u+\frac{1-b}{2b}\right)\frac{d\tau}{d\omega}+\frac{1}{2b}\cdot\frac{2b\omega-(1-b)}{1-(1-b)\omega+b\omega^2},$$	(23)
$$\frac{d\tau}{d\omega}=\frac{1}{1-(1-b)\omega+b\omega^2}.$$	(24)

continued

TABLE 2.2 (Continued)
Formulas for Calculating Particle Sedimentation in a Vertical Air Stream

Design Formulas	Formula No.
1	2
By integrating Equations 23 and 24, we obtain	
$h-h_0=\left(\frac{1-b}{2b}+u\right)(\tau-\tau_0)+\frac{1}{2b}\ln\frac{1-(1-b)\omega+b\omega^2}{1-(1-b)\omega_0+b\omega_0^2},$	(25)
$\tau-\tau_0=\frac{2}{\sqrt{4b-(1-b)^2}}\left(arctg\frac{2b\omega+b-1}{\sqrt{4b-(1-b)^2}}-arctg\frac{2b\omega_0+b-1}{\sqrt{4b-(1-b)^2}}\right)$ at $1\geq b>3-2\sqrt{2}$	(26)
$\tau-\tau_0=2\left(\frac{1}{2b\omega+b-1}-\frac{1}{2b\omega_0+b-1}\right)$ at $b=3-2\sqrt{2}$,	(27)
$\tau-\tau_0=\frac{1}{c}\ln\frac{2b\omega+b-1-c}{2b\omega_0+b-1-c}\cdot\frac{2b\omega_0+b-1+c}{2b\omega+b-1+c}$, $c=\sqrt{(1-b)^2-4b}$ at $0\leq b<3-2\sqrt{2}$,	(28)
Specific cases:	
(a) In Stokes area (with b = 0), we have the same result as in the first case (see Equations 9–14).	
(b) In the self-similarity area (with b = 1),	
$\frac{d\omega}{d\tau}=1+\omega^2;\ \frac{d\omega}{dh}=\frac{1+\omega^2}{\omega+u}.$	(29)
By integrating the first system equation (29), we obtain	
$\tau-\tau_0=\text{arctg}\omega-\text{arctg}\omega_0,$	(30)
$\omega=tg(\tau-\tau_0+arctg\omega_0).$	(31)
Then	
$\frac{d\omega}{d\tau}=\cos^{-2}(\tau-\tau_0+\text{arctg}\omega_0).$	(32)
By integrating the second system equation (29), we obtain	
$h-h_0=u(\tau-\tau_0)+\frac{1}{2}\ln\frac{1+\omega^2}{1+\omega_0^2}.$	(33)

The following equation was thoroughly analyzed by A. Fortie [96] and is true for a spherical particle at low Reynolds numbers:

$$\vec{R}=-3\pi\eta\vec{w}d-0,5\frac{\pi d^3}{6}\rho\frac{d\vec{w}}{dt}-1,5d^2\rho\sqrt{\pi v}\int_0^t\left.\frac{d\vec{w}}{dt}\right|_{t=\tau}\cdot(t-\tau)^{-0,5}d\tau. \quad (2.77)$$

There are other forms of the Basset–Boussinesq–Oseen equation [12,17,97,114,17], and Chen converted it for an unsteady medium stream. The Chen equation has been analyzed in studies by domestic [12,97] and foreign scientists [14,89,90].

Odar suggested an equation [14] for Re < 200

$$\vec{R} = -\Psi f_m \frac{\vec{w} \cdot w}{2} \rho - c_A \frac{\pi d^3}{6} \rho \frac{d\vec{w}}{dt} - c_H \frac{d^2}{4} \rho \sqrt{\pi \nu} \int_0^t \left. \frac{d\vec{\omega}}{dt} \right|_{t=\tau} \cdot (t-\tau)^{-0,5} d\tau, \quad (2.78)$$

where

$$c_A = 1,05 - \frac{0,066}{N_A^2 + 0,12}; \; c_H = 2,88 + \frac{3,12}{(N_A + 1)^3}; N_A = \frac{u^2}{\frac{dw}{dt} . d}. \quad (2.79)$$

Studies [14,90,96,97] prove that when determining the settling velocity of airborne solid particles ($\rho << \rho_m$) the second and the third terms of Equations 2.77 and 2.78 may be neglected, and the particle dynamics is described by Equation 2.65.

Let us evaluate the influence these additional factors have on the aerodynamic force value*). If the influence on the velocity value is insignificant, inertial corrections for dynamic interaction force should not be ignored. It was clear from our analysis of the particle settling velocity that drag force is noticeable even when it is within h < 0.5.

Even straight particle motion in a uniform downward air stream at zero relative velocity at the particle settling beginning should be evaluated:

$$V_p \rho_m \frac{dw}{dt} = V_p(\rho_m - \rho) g - 3\pi\eta w d - 0,5 V_p \rho \frac{dw}{dt} - 1,5 d^2 \rho \sqrt{\pi \nu} \int_0^t \left. \frac{dw}{dt} \right|_{t=z} \cdot \frac{dz}{\sqrt{t-z}}. \quad (2.80)$$

And for the transition are

$$V_p \rho_m \frac{dw}{dt} = V_p(\rho_m - \rho) g - \psi f_M \frac{w^2}{2} \rho - c_A V_p \rho \frac{dw}{dt} - c_H \frac{d^2}{4} \rho \sqrt{\pi \nu} \int_0^t \left. \frac{dw}{dt} \right|_{t=z} \cdot \frac{dz}{\sqrt{t-z}}. \quad (2.81)$$

Using dimensionless values h, τ, and ω determined by Equations 2.66, 2.67, and 2.68, Equations 2.80 and 2.81 may be presented as follows:

$$\frac{d\omega}{d\tau} = 1 - \omega - 0,5\varepsilon \frac{d\omega}{d\tau} - \frac{9}{\sqrt{\pi}} \varepsilon \sqrt{\frac{\nu t_\infty}{d^2}} \int_0^\tau \left. \frac{d\omega}{d\tau} \right|_{\tau=\xi} \cdot \frac{d\xi}{\sqrt{\tau - \xi}}, \quad (2.82)$$

$$\frac{d\omega}{d\tau} = 1 - \frac{\psi}{\psi_c} \omega^2 - c_A \varepsilon \frac{d\omega}{d\tau} - c_H \frac{3}{2\sqrt{\pi}} \varepsilon \sqrt{\frac{\nu t_\infty}{d^2}} \int_0^\tau \left. \frac{d\omega}{d\tau} \right|_{\tau=\xi} \cdot \frac{d\xi}{\sqrt{\tau - \xi}}, \quad (2.83)$$

being easily convertible at $\varepsilon \to 0$ into the earlier mentioned Equation 2.9 and Equation 1 in Table 2.2.

Given that

$$\frac{\nu t_\infty}{d^2} = \frac{1}{\text{Re} \cdot \text{St}}, \quad (2.84)$$

* Aerodynamic force was evaluated for the simplest case of uniformly accelerated motion by A. Fortie [96].

where St is the so-called Strouhal number describing the non-stationarity through kinematic parameters [65], that is

$$\mathrm{St} = \frac{d}{ct_\infty}, \quad \mathrm{Re} = \frac{cd}{\nu}. \tag{2.85}$$

In view of Equation 2.61, the last equation will be rewritten as follows:

$$\frac{d\omega}{d\tau} = 1 - \frac{r_c\omega + B\omega^2}{r_c + B} - c_A\varepsilon\frac{d\omega}{d\tau} - c_H\frac{3}{2\sqrt{\pi}}\frac{\varepsilon}{\sqrt{\mathrm{Re}\cdot St}}\int_0^\tau \left.\frac{d\omega}{d\tau}\right|_{\tau=\xi} \cdot \frac{d\xi}{\sqrt{\tau-\xi}} \tag{2.86}$$

in the Stokes flow-round area

$$\mathrm{Re}\cdot\mathrm{St} = 18\varepsilon, \tag{2.87}$$

and Equation 2.82 will appear as

$$\omega'(\tau) = 1 - \omega(\tau) - 0,5\varepsilon\omega'(\tau) - \frac{3\sqrt{\varepsilon}}{\sqrt{2\pi}}\int_0^\tau \frac{\omega'(\xi)d\xi}{\sqrt{\tau-\xi}}. \tag{2.88}$$

Here, for the sake of convenience, the upper hyphen denotes a derivative with respect to τ and the parentheses denote an argument with the required function ω.

For the purpose of further analysis, let us convert the integro-differential Equation 2.88 into a differential one. To this effect, we divide both sides of the equation by $\sqrt{x-\tau}$, where x is a random argument, and integrate over τ within the range from 0 to x:

$$(1+0,5\varepsilon)J_1(x) = J_2(x) - \frac{3\sqrt{\varepsilon}}{\sqrt{2\pi}}\cdot J_3(x), \tag{2.89}$$

then the result is differentiated with respect to x

$$(1+0,5\varepsilon)\frac{dJ_1(x)}{dx} = \frac{dJ_2(x)}{dx} - \frac{3\sqrt{\varepsilon}}{\sqrt{2\pi}}\frac{dJ_3(x)}{dx}, \tag{2.90}$$

where for the sake of convenience we have

$$J_1(x) = \int_0^x \frac{\omega'(\tau)d\tau}{\sqrt{x-\tau}}; \; J_2(x) = \int_0^x \frac{1-\omega(\tau)}{\sqrt{x-\tau}}d\tau; \; J_3(x) = \int_0^x\left(\int_0^\tau \frac{\omega'(\xi)d\xi}{\sqrt{\tau-\xi}}\right)\frac{d\tau}{\sqrt{x-\tau}}. \tag{2.91}$$

Let us consider each integral and determine the integral derivatives with respect to the upper limit. Due to the initial Equation 2.88:

$$\frac{\sqrt{2\pi}}{3\sqrt{\varepsilon}}\{[1-\omega(\tau)] - (1+0,5\varepsilon)\omega'(\tau)\} = \int_0^\tau \frac{\omega'(\xi)d\xi}{\sqrt{\tau-\xi}}. \tag{2.92}$$

Successively performing two argument substitutions in this equation (at first assuming that $\tau = x$, then $\xi = \tau$) results in the following expression for the first integral

$$\frac{\sqrt{2\pi}}{3\sqrt{\varepsilon}}\{[1-\omega(x)]-(1+0,5\varepsilon)\omega'(x)\} = \int_0^x \frac{\omega'(\tau)d\tau}{\sqrt{x-\tau}} \equiv J_1(x), \tag{2.93}$$

from which the derivative is obtained

$$\frac{dJ_1(x)}{dx} = -\frac{\sqrt{2\pi}}{3\sqrt{\varepsilon}}\left[\frac{d\omega(x)}{dx} + (1+0,5\varepsilon)\frac{d\omega'(x)}{dx}\right]. \tag{2.94}$$

The second integral can be easily converted if integrated by parts. Given that $\omega(0) = 0$, we have

$$J_2(x) = 2\sqrt{x} - 2\int_0^x \omega'(\tau)\sqrt{x-\tau}\,d\tau. \tag{2.95}$$

The derivative will appear as

$$\frac{dJ_2(x)}{dx} = \frac{1}{\sqrt{x}} - \int_0^x \frac{\omega'(\tau)}{\sqrt{x-\tau}}d\tau. \tag{2.96}$$

Or, given Equation 2.93,

$$\frac{dJ_2(x)}{dx} = \frac{1}{\sqrt{x}} - \frac{\sqrt{2\pi}}{2\sqrt{\varepsilon}}\{[1-\omega(x)]-(1+0,5\varepsilon)\cdot\omega'(x)\}. \tag{2.97}$$

The third integral is determined by the Dirichlet formula for a double integral [88]:

$$J_3(x) = \int_0^x\left(\int_\xi^x \frac{d\tau}{\sqrt{(\tau-\xi)(x-\tau)}}\right)\omega'(\xi)d\xi = \pi\int_0^x \omega'(\xi)d\xi, \tag{2.98}$$

then

$$\frac{dJ_3(x)}{dx} = \pi\omega'(x). \tag{2.99}$$

Inserting the resulting derivatives in Equation 2.90 and considering the fact that x variable is chosen on a random basis (and, therefore, possible to assume that $x = \tau$), simple algebraic transformations result in the following linear differential equation of second order:

$$(1+0,5\varepsilon)^2\omega'' + (2-3,5\varepsilon)\omega' + \omega = 1 - \frac{3\sqrt{\varepsilon}}{\sqrt{2\pi}}\cdot\frac{1}{\sqrt{\tau}}. \tag{2.100}$$

One study [96] gives a similar equation with no derivation for the case of particles settling in still air. Therefore, we studied the solution for $\varepsilon > \frac{5}{8}$, $\varepsilon = \frac{5}{8}$ and $\varepsilon < \frac{5}{8}$, for the problem of airborne motion of particles in question $\varepsilon \ll 1$.

The solution of Equation 2.100, at the initial conditions, that $\omega'(0) = \omega(0) = 0$ is

$$\omega = \frac{1}{a}e^{-B\tau}\cdot\sin a\tau + c\int_0^\tau\left(1-\frac{3\sqrt{\varepsilon}}{\sqrt{2\pi\xi}}\right)e^{B(\xi-\tau)}\cdot\sin a(\tau-\xi)d\xi \qquad (2.101)$$

or

$$\omega = 1-e^{-B\tau}(\cos a\tau - \frac{1-B}{a}\sin a\tau) - c\frac{3\sqrt{\varepsilon}}{\sqrt{2\pi}}\int_0^\tau \frac{e^{B(\xi-\tau)}}{\sqrt{\xi}}\sin a(\tau-\xi)d\xi, \qquad (2.102)$$

where a, B, and c constants are related to the relative density of the medium by:

$$a = \frac{3\sqrt{\varepsilon(8-5\varepsilon)}}{(2+\varepsilon)^2} \approx \frac{3\sqrt{\varepsilon}}{\sqrt{2}}\left(1-\frac{21}{16}\varepsilon\right) \approx \frac{3\sqrt{\varepsilon}}{\sqrt{2}}; \qquad (2.103)$$

$$B = \frac{4-7\varepsilon}{(2+\varepsilon)^2} \approx 1-2,75\varepsilon \approx 1; c = \frac{4}{3\sqrt{\varepsilon(8-5\varepsilon)}} \approx \frac{\sqrt{2}}{3\sqrt{\varepsilon}}\left(1+\frac{5}{16}\varepsilon\right) \approx \frac{\sqrt{2}}{3\sqrt{\varepsilon}}. \qquad (2.104)$$

In view of a smallness (with $\sin a(\tau-\xi) \approx a(\tau-\xi)$), Equation 2.101 can be presented in the following way:

$$\omega = 1-e^{-B\tau}\left(\cos a\tau - \frac{1-B}{a}\sin a\tau\right) - \frac{ac}{\sqrt{B^3}}\frac{3\sqrt{\varepsilon}}{\sqrt{2\pi}}\left[(2B\tau+1)W - \sqrt{B\tau}\right], \qquad (2.105)$$

where W is tabulated inner function $\sqrt{B\tau}$ [38]

$$W = e^{-B\tau}\int_0^{\sqrt{B\tau}} e^{x^2}dx. \qquad (2.106)$$

For the relative motion acceleration, we have

$$\frac{d\omega}{d\tau} = e^{-B\tau}\left[\cos a\tau + \frac{a^2+B^2-B}{a}\sin a\tau\right] + \frac{ac}{\sqrt{B}}\frac{3\sqrt{\varepsilon}}{\sqrt{2\pi}}\left[(2B\tau-1)W - \sqrt{B\tau}\right]. \qquad (2.107)$$

Figure 2.16 shows charts of aerodynamic force behavior at the time when particles are settling in the Stokes flow-around area.

There are also ratings for general particle motion according to Equation 2.86. The charts for this equation were plotted using the approximate approach: the right-hand side velocity and acceleration were taken from Equations 5 and 6 in Table 2.2;

$$\left.\frac{d\omega}{d\tau}\right|_{\tau=\xi} = (b+1)^2\left.\frac{e^{\tau(b+1)}}{\left[e^{\tau(b+1)}+b\right]^2}\right|_{\tau=\xi} = n^2\frac{e^{n\xi}}{(e^{n\xi}+n-1)^2}; n = b+1 \qquad (2.108)$$

and then

$$J(\tau) \equiv \int_0^{\tau} \frac{d\omega}{d\tau}\bigg|_{\tau=\xi} \frac{d\xi}{\sqrt{\tau-\xi}} = 2n\sqrt{n}\int_0^{\sqrt{n\tau}} \frac{e^{n\tau}e^{-x^2}}{\left[e^{n\tau}e^{-x^2}+n-1\right]^2}dx. \quad (2.109)$$

Hence, there are specific cases—for the Stokes area ($b = 0, n = 1$)

$$J(\tau) = 2e^{-\tau}\int_0^{\sqrt{\tau}} e^{x^2}dx = 2W(\sqrt{\tau}), \quad (2.110)$$

and for the self-similarity area ($b = 1, n = 2$)

$$J(\tau) = 4\sqrt{2}\int_0^{\sqrt{2\tau}} \frac{e^{2\tau}\cdot e^{-x^2}dx}{\left[1+e^{2\tau}\cdot e^{-x^2}\right]^2} \approx 4\sqrt{2}\left[W(\sqrt{2\tau})-\sqrt{2}W(2\sqrt{\tau})\right]. \quad (2.111)$$

The following conclusions can be made based on the experimental data analysis:

1. The approximate method satisfactorily describes a particle settling process throughout its stages.
2. Inertial components are only essential at the initial moment, the particle start-up moment $\tau < 0{,}1$ ($h \leq 0.1$), that is, at a small settling path interval when the drag force is negligible due to the small velocity.

The last condition allows us to neglect inertial components of the aerodynamic force in a quantitative description of particle flow mechanics.

2.4 METHOD FOR EVALUATING THE AERODYNAMIC CHARACTERISTIC OF PARTICLE GRAVITATIONAL FLOW

Two methods are commonly used to evaluate the aerodynamic force of particles. The first method is based on the steady settling velocity measurement and is typically applied in the study of gravitational mineral dressing processes. The second method is usually used in the study of air drags of various bodies in a channel and is based on the measurement of drag values for stationary "grilles" of particles. In order to determine the air drag of an unsteady stream of particles, we used a new approach that is based on the measurement of the channel pressure during sedimentation of particles. For this purpose, features of known methods were applied—namely "non-interference" with the natural particle settling process, which is specific to the first method, and simplicity of measurements (from the second method). Thus, using simple instruments, it is possible to evaluate the process of dynamic particle and air interaction without interfering in a complex mechanism of single particle motion, thereby considering a number of factors that resist a theoretical description (such as concentration fluctuations, rotational motion of particles, velocity pulsation, collision of particles, etc.).

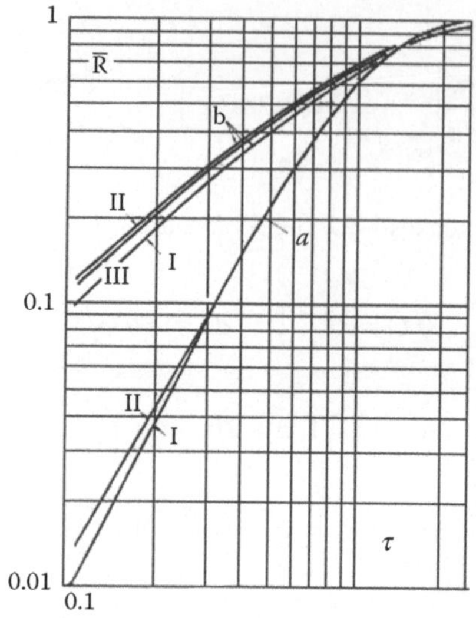

FIGURE 2.16 Variation in the aerodynamic force of a settling particle (a is for the self-similarity area; b is for the Stokes area): I is without regard to inertial components; II is with an approximate regard to inertial components; III is with a precise regard to inertial components.

2.4.1 Channel Pressure Variation

Let us analyze the pressure variation in a vertical channel with settling isomeric articles uniformly distributed across the channel section.

Assume that:

- There is no directed air flow in the channel.
- Velocity pulsations are negligible.
- The stream of particles is stationary and one-dimensional.
- There is no heat and mass exchange between particles and air.
- The *ox* axis zero is aligned with the channel head while its positive direction corresponds to the particle settling direction.
- The concentration of particles is low ($\beta_1 << 1$; $\beta_2 \approx 1$).
- Settling particles do not collide.

Ignoring pulsation moments dynamic equations for such a stream (in view of Equations 78 and 90 in the Appendix) will appear as follows for a stream of particles:

$$\frac{d\beta_1\rho_1 v_1}{dx} = 0, \qquad (2.112)$$

$$\frac{d}{dx}\beta_1\rho_1 v_1^2 = \beta_1\rho_1 g + \frac{\beta_1}{V_p}R_{21}. \qquad (2.113)$$

Or, in view of Equation 2.112,

$$\beta_1\rho_1 v_1 \frac{dv_1}{dx} = \beta_1\rho_1 g + \frac{\beta_1}{V_p} R_{21}; \tag{2.114}$$

and for still air in a channel,

$$\frac{dP}{dx} = \frac{\beta_1}{V_p} R_{12}. \tag{2.115}$$

Equation 2.112 yields the obvious relation for the particle flow rate

$$G_1 = \beta_1\rho_1 v_1 S. \tag{2.116}$$

Given the tightness, the air impact on a particle will appear as follows:

$$R_{21} = -\frac{\psi}{E^2} f_M \frac{v_1^2}{2}\rho, \tag{2.117}$$

where, in consideration of a low bulk concentration of particles, $1/E^2$ will be expressed as the second order polynomial

$$1/E^2 = 1/(1-\beta_1)^{2n} \approx 1 + a\beta_1 + b\beta_1^2, \tag{2.118}$$

that gives a fair approximation of P. V. Lyaschenko's formula within. $\beta_1 < 0, 15$. (The error does not exceed 10% with $n = 3$, $a = 6$, and $b = 21$; with the same values of n, a, and b, the error within $\beta_1 < 0,1$ does not exceed 4%.)

To reduce the equations, simplify solutions for the same, and to facilitate further comparison with the experimental data, the equations shall be converted into dimensionless equations using slightly simplified expressions for specific length and velocity (unlike Equation 2.67)

$$l_\infty = \frac{c^2}{g}; \; c = \sqrt{\frac{2V_p\rho_1 g}{\psi f_M \rho}}; \tag{2.119}$$

$$x = hl_\infty; \; v_1 = vc. \tag{2.120}$$

Assuming that the specific pressure is

$$P_\infty = \rho_1 c^2, \tag{2.121}$$

then

$$P = \bar{P}P_\infty. \tag{2.122}$$

Equations 2.114 and 2.115 will become

$$v\frac{dv}{dh} = 1 - v^2(1 + a\beta_1 + b\beta_1^2), \tag{2.123}$$

$$\frac{d\bar{P}}{dh} = \beta_1 v^2 (1 + a\beta_1 + b\beta_1^2), \tag{2.124}$$

and will conclude with the following (subject to Equation 2.116)

$$v\beta_1 = A \equiv \frac{G_1}{\rho_1 c S}. \tag{2.125}$$

At the initial conditions

$$\bar{P} = \bar{P}_a; v = v_0 \quad \text{if} \quad h = 0, \tag{2.126}$$

the solution of the system of Equations 2.123 and 2.124 will appear as

$$h = -\frac{1}{2}\ln\frac{(1 - bA^2) - aAv - v^2}{(1 - bA^2) - aAv_0 - v_0^2} - \frac{1}{2}aAJ, \tag{2.127}$$

$$\mathbf{P} = v_0 - v + J, \tag{2.128}$$

where

$$J = \int_{v_0}^{v} \frac{dv}{(1 - bA^2) - aAv - v^2} = -\Delta^{-0,5} \ln\frac{2v + aA - \sqrt{\Delta}}{2v + aA + \sqrt{\Delta}} \cdot \frac{2v_0 + aA + \sqrt{\Delta}}{2v_0 + aA - \sqrt{\Delta}}, \tag{2.129}$$

$$\Delta = 4 + A^2(a^2 - 4b), \tag{2.130}$$

$$\mathbf{P} = \frac{P - P_a}{P_\infty A} = \frac{P - P_a}{Gc / S}, \tag{2.131}$$

where P_a is the atmospheric pressure (pressure outside the tube), Pa. From which, particularly with no regard to constraints ($a = 0$; $b = 0$)

$$v = \sqrt{1 - (1 - v_0^2)e^{-2h}}, \tag{2.132}$$

$$\mathbf{P} = v_0 - v - \frac{1}{2}\ln\frac{1 - v}{1 + v} \cdot \frac{1 + v_0}{1 - v_0}. \tag{2.133}$$

A simpler solution could be obtained by ignoring the medium resistance effect on the particle velocity, on the assumption that

$$v\frac{dv}{dh} = 1. \tag{2.134}$$

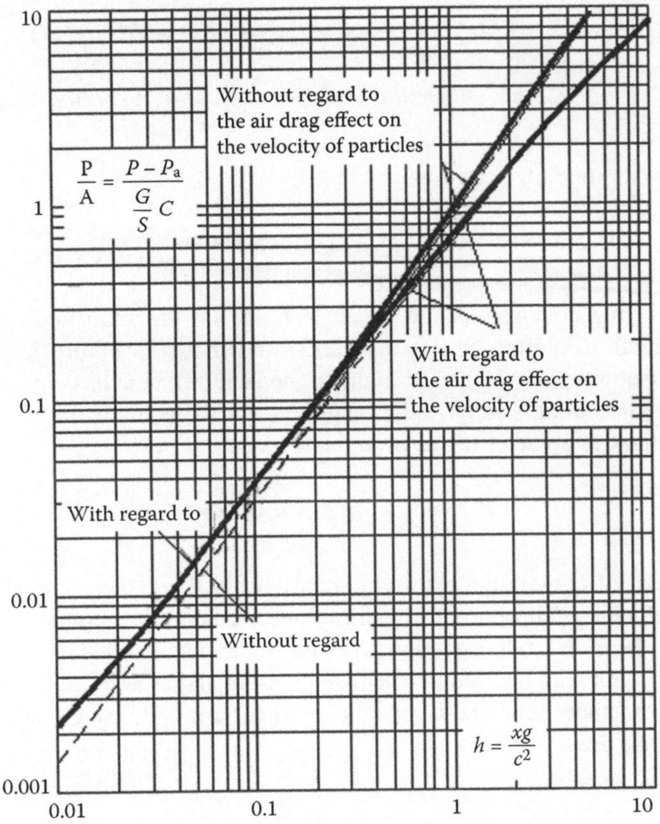

FIGURE 2.17 Pressure variation across the vertical channel length.

In this case, integrating Equation 2.124 in view of Equation 2.134, we have

$$\mathbf{P} = \frac{v^3 - v_0^3}{3} + aAh + bA^2(v - v_0), v = \sqrt{2h + v_0^2}. \quad (2.135)$$

As is clear from the results obtained (Figure 2.17), the constrained environment (at low initial flow velocities) has a significant effect on the flow acceleration area only where $h < 0.1$. In addition, the pressure distribution in $h < 0.5$ (where the medium resistance effect on particle velocity is negligible) can be quite accurately described by Equation 2.135. However, the constrained environment's influence can also be used in areas of higher bulk concentrations. Indeed, in the case of Equation 2.134, it is not necessary to substitute Lyaschenko's correction factor with a polynomial because the equation (even if $\beta_2 = 1 - \beta_1$)

$$(1 - \beta_1)\frac{d\mathbf{P}}{dh} = \frac{v}{(1 - \beta_1)^6} \quad (2.136)$$

is easily integrated:

$$\mathbf{P} = A^3\left[(t^3 - t_0^3)/3 + 4,5(t^2 - t_0^2) + 36(t - t_0) + 84\ln\frac{t}{t_0} - 126(u - u_0) - 63(u^2 - u_0^2) - 28(u^3 - u_0^3) - 9(u^4 - u_0^4) - 9/5(u^5 - u_0^5) - 1/6(u^6 - u_0^6)\right], \tag{2.137}$$

where it was assumed for convenience that

$$1/u = t = (v - A)/A,\ 1/u_0 = t_0 = (v_0 - A)/A. \tag{2.138}$$

The following conclusions can be made based on the analysis of the pressure distribution curves plotted by Equation 2.137 and shown in Figure 2.18. (1) At low initial flow velocities ($v_0 < 0,01$), the constrained environment has virtually no effect on pressure distribution across tube length except for a small initial section. The pressure value can be determined from the formula

$$\mathbf{P} = (v^3 - v_0^3)/3,\ v = \sqrt{2h + v_0^2}. \tag{2.139}$$

(2) The effect of constraints can be ignored within $\beta_0 < 0.01$, even at higher initial flow velocities. The pressure distribution calculations become much simpler: Equation 2.139 for $h < 0.5$; Equation 2.135 for higher h values.

These formulas can be used to calculate pressure and other initial conditions. For instance, if the upper tube end is airtight and the lower end is open, the initial pressure conditions are

$$P = 0 \quad \text{if} \quad h = h_k = \frac{gl}{c^2}, \tag{2.140}$$

and the pressure distribution

$$\mathbf{P} = \left[\left(\sqrt{2h + v_0^2}\right)^3 - \left(\sqrt{2h_k + v_0^2}\right)^3\right]\Big/3 \tag{2.141}$$

that is, the whole tube is under a vacuum that reaches its peak value at the tube inlet

$$\mathbf{P}_0 = -\left[\left(\sqrt{2h_k + v_0^2}\right)^3 - v_0^3\right]\Big/3 \tag{2.142}$$

(at $v_0 = 0,1$; $\beta_0 = 0,1$). A situation where both ends are airtight is also of interest:

$$\mathbf{P} = \left[\left(\sqrt{2h + v_0^2}\right)^3 - \left(\sqrt{2h_m + v_0^2}\right)^3\right]\Big/3. \tag{2.143}$$

In such an instance, the upper part of the tube is under a vacuum while the lower part is under excessive pressure.

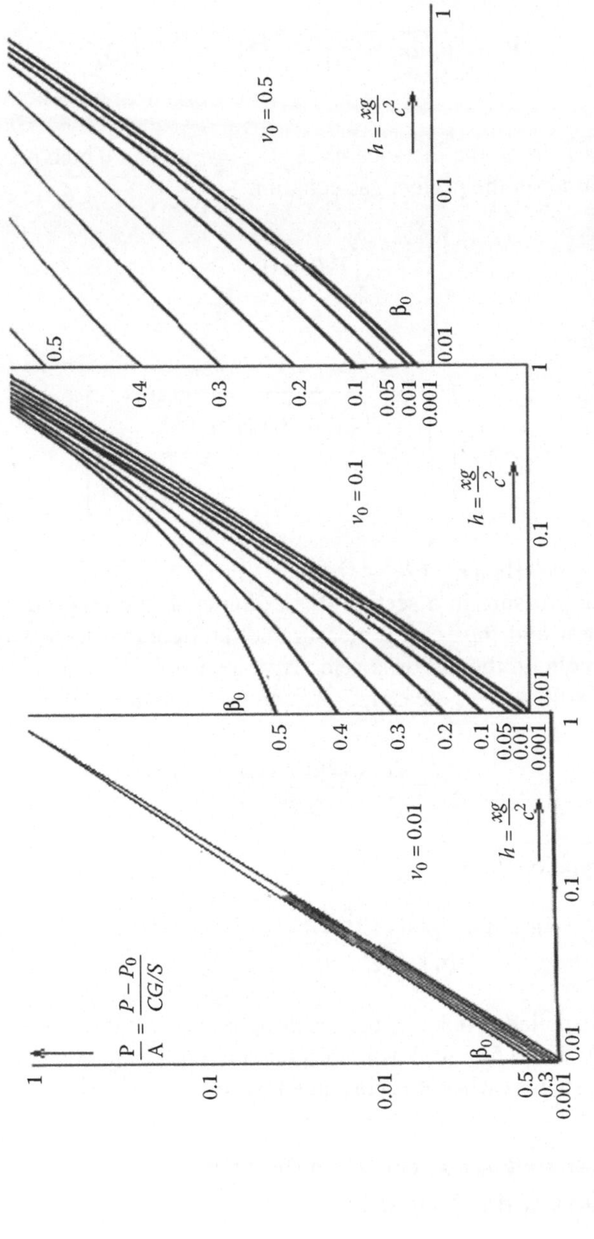

FIGURE 2.18 Pressure variation in a vertical tube at various initial velocities and bulk concentrations of particles.

With extreme pressure values at the tube inlet

$$\mathbf{P}_H = -\left[\left(\sqrt{2h_m + v_0^2}\right)^3 - v_0^3\right]\Big/3 \tag{2.144}$$

and outlet

$$\mathbf{P}_K = \left[\left(\sqrt{2h_k + v_0^2}\right)^3 - \left(\sqrt{2h_m + v_0^2}\right)^3\right]\Big/3, \tag{2.145}$$

we have zero pressure at the distance $h = h_m$ $(h_m = l_m g / c^2)$. Thus, pressure redistribution occurs based on the perfect gas equation

$$\int_0^h \mathbf{P} dh = 0. \tag{2.146}$$

In view of the above,

$$h_m = 0,5\left[\sqrt[3]{0,04\frac{\left(\sqrt{(2h_k + v_0^2)^5} - v_0^5\right)^2}{h_k^2}} - v_0^2\right]. \tag{2.147}$$

In particular, if $v_0 = 0$ then $h_m / h_k = 0,54$.

By measuring pressure in a section (for example, at the tube outlet—where the pressure is highest and thus easily measurable) and comparing it with the design pressure, we can obtain the airborne velocity or a particle drag factor for a stream.

For instance, with $p_k - p_a$ measured, we can use Equation 2.139 to obtain

$$c^2 = \frac{G}{S} \cdot \frac{\sqrt{(2gl + v_{10}^2)^3} - v_{10}^3}{3(P_k - P_a)} \tag{2.148}$$

or, in view of Equation 2.119,

$$\Psi = 4\frac{d_e \rho_1 g}{G\rho\left[\left(\sqrt{2gl + v_{10}^2}\right)^3 - v_{10}^3\right]}(P_k - P_a) \cdot S. \tag{2.149}$$

In solving some applied problems, the resulting relations can be used to determine Ψ coefficient at $l / l_\infty > 0,5$ as well as in the constrained flow area (at $\beta_0 > 0,01$). Some error is possible, such as when it is correlated by Ψ.

2.4.2 Experimental Evaluation of the Method for Determining the Particle Drag Factor

We evaluated the previously described method for determining the aerodynamic properties of stream particles when studying the air mechanics of a water drop flow in a vertical tube. Using water drops increases the accuracy of results and considerably facilitates the experiments due to uniform particle distribution across the flow area (which is in line with the theoretical model) and to a constant and steady material flow rate. A prismatic vessel with a wooden bottom was used as a drop generator. Holes were drilled in the bottom around a 0.3-m diameter circle in order to insert restrictors with a 0.4-mm internal radius. Flow uniformity was assured by maintaining a certain level of water in the vessel. The experiments were conducted at a flow rate between 0.05 and 0.18 kg/s, which was accompanied by steady drop formation at the restrictor tips. The drop diameter was 3 mm (the airborne velocity at $\Psi_0 = 0.5$ constitutes 7.8 m/s), Weber number $We = \rho_2 dc^2 / 2\sigma = 1{,}5$ (with water surface tension $\sigma = 0.0728$ N/m), that is, below the critical value with no drops broken in tests.

The drop generator was placed above the vertical tube. The tube's lower end was put in a water pan (in order to seal it). Two series of experiments were conducted. In the first series, the pipe was 2 m high and 285 mm in diameter; in the second series, the pipe was 6.3 m high and 300 mm in diameter.

Excessive pressure $P_k - P_a$ was measured at the tube end at a steady-state water flow rate, and the airborne velocity and the drops' air drag factor were determined

TABLE 2.3
Drag Factor of Spherical Particles in a Stream

Experimental Data			Design Values				
G_1, kg/s	*H, m*	$P_k–P_a$, Pa	v_{1H}, m/s	v_{1k}, m/s	$\beta \cdot 10^3$	$Re \cdot 10^{-3}$	Ψ
Water drop flow (d_e = *3 mm*) in a vertical tube							
0.095	2.0	2.0	1.06	5.38	0.46	0.62	0.57
0.116	2.0	2.4	1.06	5.40	0.56	0.62	0.56
0.125	2.0	2.6	1.06	5.39	0.61	0.62	0.57
0.135	2.0	2.7	1.06	5.42	0.65	0.62	0.54
0.180	2.0	3.7	1.06	5.40	0.87	0.62	0.55
0.058	6.3	4.4	1.08	7.55	0.19	0.82	0.46
0.089	6.3	7.1	1.08	7.39	0.29	0.82	0.49
0.101	6.3	7.7	1.08	7.53	0.33	0.82	0.46
0.132	6.3	10.0	1.08	7.55	0.43	0.82	0.45
0.45	6.3	12.0	1.08	7.28	0.49	0.80	0.51
Steel ball flow (d_e = *12.8 mm*) in a vertical chute							
1.51	3.8	6.3	2.21	8.85	1.79	4.51	0.39
1.86	4.15	9.0	2.14	9.27	2.13	4.65	0.39
2.25	3.8	9.7	2.21	8.85	2.66	4.51	0.40
2.68	4.15	12.7	2.14	9.27	3.07	4.65	0.38
2.75	3.8	11.8	2.21	8.85	3.25	3.51	0.40
3.25	3.8	14.2	2.21	8.85	3.84	4.51	0.41
3.25	3.3	11.5	2.21	8.28	4.05	4.28	0.41
2.48	3.8	14.9	2.21	8.85	4.12	4.51	0.40

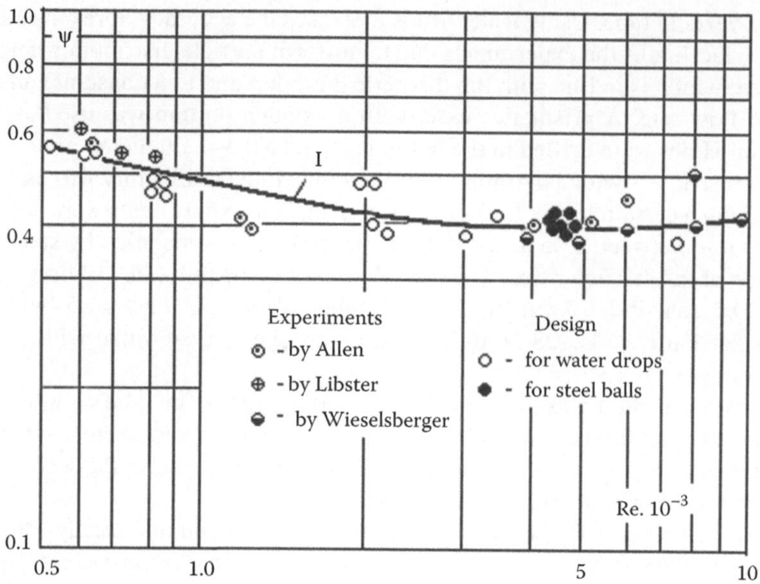

FIGURE 2.19 Variation in the air drag factor of settling spherical particles with an increase in Reynolds number (I is Rayleigh curve).

using Equations 2.148 and 2.149. The experimental data and design values are given in Table 2.3. There are also similar results for steel balls*.

It is clear from the given results that the pressure measuring method can be successfully used to evaluate aerodynamic properties of settling particles. Figure 2.19 shows design values of the Ψ coefficient. Comparing our findings with the Rayleigh curve, and with experimental data for single balls taken from G. Schlichting's monograph [104], the results correlate very accurately with the known experimental data compilations.

* The experimental data was kindly provided by V. D. Olifer, who made many measurements of pressure in a 0.14 m × 0.14 m cross-sectional vertical chute when pouring in 12.8-mm diameter steel balls.

3 Air Injection in Chutes

3.1 ISOTHERMAL FLOW

Let us consider the steady isothermal motion of a stream of material particles and air in a straight chute of the uniform cross-section area S_{ch}. To this effect, at x distance to the chute inlet (Figure 3.1), we select a unit prism of Δx length, the lateral faces of which are the chute walls. The coordinates' origin is placed in the entry section; the X-axis is directed along the chute centerline toward the bulk material particle motion. Without pulsation, the equations for component mass flow rates will appear as:

$$G_1 = \int_{S_{ch}} \beta_1 \rho_1 v_1 dS, \tag{3.1}$$

$$G_2 = \int_{S_{ch}} \beta_2 \rho_2 v_2 dS. \tag{3.2}$$

The momentum conservation equation for the material and air confined in the selected element $\Delta x \cdot S_{ch} = \Delta V_{ch}$ projected on the chute centerline will appear as follows:

$$\Delta\left(\int_{S_{ch}} \beta_1 \rho_1 v_1 v_1 dS \right) = \int_{\Delta V_{ch}} M_1 \beta_1 \rho_1 dV - \int_{\Delta V_{ch}} \frac{\beta_1}{V_p} R dV, \tag{3.3}$$

$$\Delta\left(\int_{S_{ch}} \beta_2 \rho_2 v_2 v_2 dS \right) = \int_{\Delta V_{ch}} M_2 \beta_2 \rho_2 dV + \int_{S_{\Delta V ch}} \beta_2 \Pi_2 \, dS + \int_{\Delta V_{ch}} \frac{\beta_1}{V_p} R dV, \tag{3.4}$$

where $S_{\Delta V_{ch}}$ is the selected element surface ΔV_{ch}; and Π_2 is OX-projection of surface forces.

A one-dimensional problem is formulated by substituting the current velocities, bulk concentrations, and aerodynamic interaction forces in Equations 3.3 and 3.4 for the corresponding averaged values.

In view of Equations 9 and 10 (in the Appendix), the projection of bulk forces on the chute centerline is

$$\beta_1 \rho_1 M_1 - \beta_1 \rho_1 a_T, \beta_2 \rho_2 M_2 - \beta_2 (\rho_2 - \rho_0) g_x. \tag{3.5}$$

Now, let us find the projection of surface forces. For descriptive reasons, the particle flow is positioned at the chute walls as an indiscrete mass (Figure 3.1a).

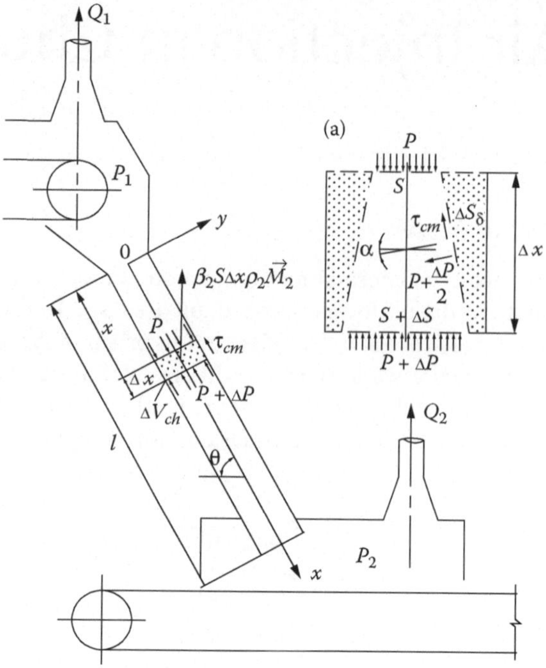

FIGURE 3.1 Derivation of dynamic equations for a one-dimensional particle stream in a tip chute; (a) represents a conventional diagram of a bulk material flow and vectors of surface forces.

The integral of surface forces for air shall be represented by the obvious relation

$$\int_{S_{\Delta V_{ch}}} \beta_2 \Pi_2 \, dS = P \cdot S - (P + \Delta P) \cdot (S + \Delta S) + (P + \Delta P / 2) \cdot \Delta S_\delta \cdot \sin\alpha - \tau_{cm} \cdot \Delta S_\delta \sin\alpha$$

$$\approx S\Delta P - \tau_{cm} \cdot \Pi \cdot \Delta x, \tag{3.6}$$

where ΔS_δ is the lateral surface of an elementary volume; Π is the chute perimeter; τ_{cm}, P is the stress of friction and pressure forces; and $S = \beta_2 S_{ch}$ is the chute cross-section area free for air passing. The momentum conservation equations will become differential as follows:

$$G_1 \frac{d\overline{v}_1}{dx} = \overline{\beta}_1 \rho_1 a_T S_{ch} - \frac{\overline{\beta}_1}{V_p} \cdot \overline{R} \cdot S_{ch}, \tag{3.7}$$

$$G_2 \frac{d\overline{v}_2}{dx} = \overline{\beta}_2 (\rho_2 - \rho_0) g_x S_{ch} - \overline{\beta}_2 S_{ch} \frac{dP}{dx} - \overline{\beta}_2 \frac{\lambda}{D} \cdot \frac{(\overline{v}_2)^2}{2} \rho_2 S_{ch} + \frac{\overline{\beta}_1}{V_p} \cdot \overline{R} \cdot S_{ch}, \tag{3.8}$$

where $\overline{v}_1, \overline{v}_2$ are the chute section-averaged material and air velocities

$$\overline{v}_1 = \frac{1}{S_{ch}} \int_{S_{ch}} v_1 dS; \; \overline{v}_2 = \frac{1}{S_{ch}} \int_{S_{ch}} v_2 dS; \tag{3.9}$$

$\bar{\beta}_1, \bar{\beta}_2$ are the chute section-averaged material and air bulk concentrations

$$\bar{\beta}_1 = \frac{1}{S_{ch}} \int_{S_{ch}} \beta_1 dS; \bar{\beta}_2 = 1 - \bar{\beta}_1; \tag{3.10}$$

and $\bar{R}$ is an averaged aerodynamic interaction force.

Henceforth, the averaging sign (a hyphen above a value) shall be omitted for the sake of convenience and unless otherwise indicated v_1, v_2, β_1, R are purportedly averaged values.

An averaged velocity and bulk concentration are easily determined from flow Equations 3.1 and 3.2:

$$\beta_1 = \frac{G_1}{\rho_1 v_1 S_{ch}}; \tag{3.11}$$

$$v_2 = \frac{G_2}{(1 - \beta_1)\rho_2 S_{ch}}. \tag{3.12}$$

The third term in the right-hand side of Equation 3.8 was entered on the assumption that τ_{cm} = const, and then

$$\beta_2 \tau_{cm} \Pi \, dx = \beta_2 \frac{\lambda}{D} \frac{v_2^2}{2} \rho_2 \cdot S_{ch}, \tag{3.13}$$

where λ is an aerodynamic drag factor for the chute walls; and D is the chute hydraulic diameter:

$$D = 4S_{ch} / \Pi. \tag{3.14}$$

In physical terms, a one-dimensional problem thus formulated corresponds to the case of a uniform distribution of particles throughout the section. It will be further demonstrated that the solution of one-dimensional Equation 3.8 also well describes the air injection process for the pseudo-uniform distribution. An experimental evaluation of a one-dimensional stream and clarification of some of its parameters was performed on the experimental arrangement intended for determination of injective bulk material properties (Figure 3.2). The principal element of that bench is a chute with a suspended "ceiling" that allows for altering the chute cross-section. The upper bin has a diaphragm that ensured the material flow at the specified rate. A sealed bin with a discharge damper was used to receive the supplied material.

The arrangement structure allowed for altering the cross-section height and the chute inclination angle as well as the transfer height. For this purpose, the structure was mounted on a sectional metal frame. The test section of the arrangement consisted of Venturi tubes installed on air ducts. Air intake or injection to the lower bin was performed with the fan.

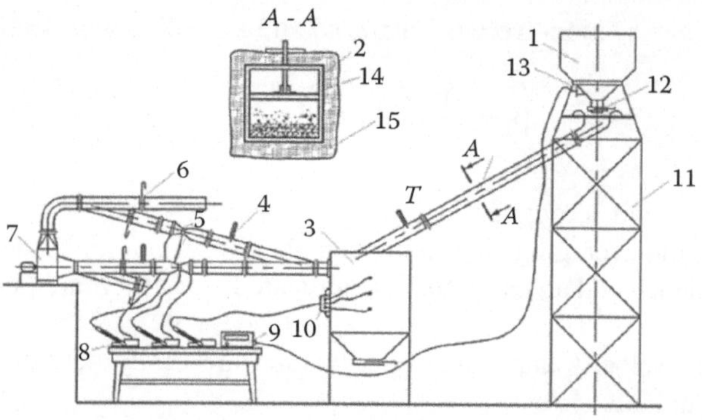

FIGURE 3.2 Diagram of the experimental arrangement for the study of bulk material injective properties: 1 = upper bin; 2 = chute; 3 = lower bin; 4 = thermometer; 5 = Venturi tube; 6 = damper; 7 = fan; 8 = micropressure gauge; 9 = galvanometer; 10 = blending chamber; 11 = metal frame; 12 = diaphragm; 13 = thermocouple; 14 = chute upper wall; 15 = heat insulation layer.

3.1.1 Averaged Aerodynamic Characteristic of Particles

The average aerodynamic characteristic of particles is determined from the equation

$$\frac{\beta_1 \Delta V_{ch}}{V_p}\psi^* \cdot f_M \frac{|v_1 - v_2|(v_1 - v_2)}{2}\rho_2 = \int_{\Delta V_{ch}} \frac{\beta_1}{V_p} R dV, \tag{3.15}$$

which defined the meaning of the averaging operation as the substitution of a sum of the aerodynamic forces of particles within the selected chute section for the product of the number of particles and the averaged aerodynamic force. The aerodynamic drag factor ψ^* is determined using the pressure measuring method described here.

Let us consider two specific cases: a stream of isometric particles and a stream of polyfractional material particles.

3.1.1.1 Monofractional Stream

Analyzing a one-dimensional stream at the minimum bulk concentration of particles ($\beta_1 \ll 1$) in absence of directional air motion in the chute $v_2 = 0$, Equation 3.8 will appear as

$$\frac{dP}{dx} = \frac{\beta_1}{V_p}\psi^* f_M \frac{v_1^2}{2}\rho_2. \tag{3.16}$$

On the assumption that the stream of particles is uniformly accelerated (a_T acceleration)

$$v_1 = \sqrt{2a_T x + v_{1H}^2}; \; v_1 dv_1 = a_T dx, \tag{3.17}$$

we integrate Equation 3.16 over the chute length to obtain

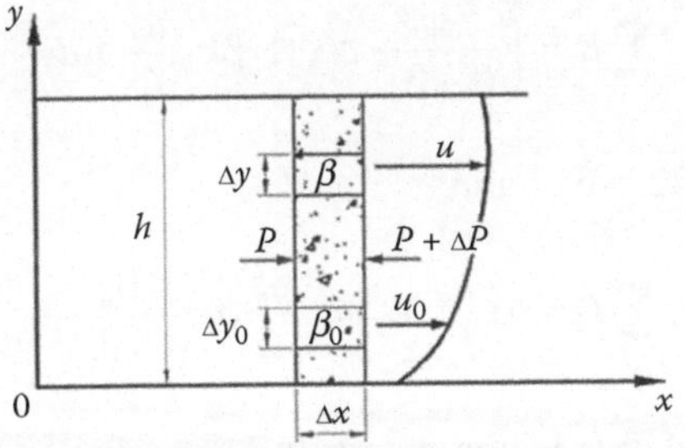

FIGURE 3.3 The air drag of particles in a bulk material stream.

$$P_\kappa = \psi^* \kappa_m \varepsilon \frac{G_1}{S_{ch} a_T} \cdot \frac{v_{1\kappa}^3 - v_{1H}^3}{3},\ K_m = f_M / V_p,\ \varepsilon = \rho_2 / \rho_1 \tag{3.18}$$

from which

$$\psi^* = P_\kappa \bigg/ \left(\kappa_m \varepsilon \frac{G_1}{S_{ch} a_T} \cdot \frac{v_{1\kappa}^3 - v_{1H}^3}{3} \right), \tag{3.19}$$

where P_K is an excessive pressure at the chute outlet, Pa.

Now, we determine ψ^* coefficient behavior with an increase in the bulk concentration of particles in the chute. To this effect, we consider the following simplified model of the aerodynamic flow interaction. Let us assume that (subject to the exponential law) particles are fixed in the rectangular chute (Figure 3.3). Air is induced into the chute at velocity u_{av}. Due to the uniform distribution of particles, the chute bottom air velocity will be less than the upper section velocity.

Select two flow tubes of $\Delta y_0 \cdot 1$ and $\Delta y \cdot 1$ section and then express the pressure drop equation. (Due to the smallness of Δy_0 and Δy, the constraint effect can be considered using Equation 2.43, 2.44.) For the first flow tube, we obtain:

$$\Delta P = \psi \frac{\Delta x \cdot \Delta y_0 \cdot 1 \cdot \beta_0}{(1-\beta_0)^6} K_m \frac{u_0^2}{2} \rho_2; \tag{3.20}$$

for the second one,

$$\Delta P = \psi \frac{\Delta x \cdot \Delta y \cdot 1 \cdot \beta}{(1-\beta)^6} K_m \frac{u^2}{2} \rho_2. \tag{3.21}$$

By solving these equations jointly, we obtain:

$$u = u_0 \left[\sqrt{\beta_0} (1-\beta)^3 \right] \Big/ \left[\sqrt{\beta} (1-\beta_0)^3 \right]. \tag{3.22}$$

Determine a sum of particle air drag forces within $\Delta x \cdot h \cdot 1$:

$$\sum R = \int_0^h \frac{\psi}{(1-\beta)^6} \Delta x \cdot 1 \cdot \beta k_m \frac{u^2}{2} \rho_2 dy. \tag{3.23}$$

In view of Equation 3.22, we obtain

$$\sum R = \Delta x \cdot h \cdot 1 \cdot \psi k_m \frac{u_0^2}{2} \rho_2 \frac{\beta_0}{(1-\beta_0)^6}. \tag{3.24}$$

Let us find the averaged drag factor ψ^* to fulfill the equation

$$\psi^* \cdot \Delta x \cdot h \cdot 1 \cdot \bar{\beta} k_m \frac{u_{av}^2}{2} \rho_2 = \sum R, \tag{3.25}$$

where $\bar{\beta}$ is h-averaged bulk concentration.
Given that (according to Equation 3.22) the mean air velocity u_{av} is

$$u_{av} = u_0 \frac{\sqrt{\beta_0}}{(1-\beta_0)^3} \cdot \frac{1}{h} \int_0^h \frac{(1-\beta)^3}{\sqrt{\beta}} dy, \tag{3.26}$$

we obtain

$$\psi^* \cdot \Delta x \cdot h \cdot 1 \cdot k_m \frac{u_0^2}{2} \rho_2 \frac{\bar{\beta} \cdot \beta_0}{(1-\beta_0)^6} \left[\frac{1}{h} \int_0^h \frac{(1-\beta)^3}{\sqrt{\beta}} dy \right]^2 = \sum R. \tag{3.27}$$

By solving Equations 3.24 and 3.27 jointly, we obtain

$$\psi^* = \frac{\psi}{\bar{\beta}} \left[\frac{1}{h} \int_0^h \frac{(1-\beta)^3}{\sqrt{\beta}} dy \right]^{-2}. \tag{3.28}$$

Therefore, it is clear that the drag factor is decreased with the increase of the bulk concentration $\bar{\beta}$; the result is different than in the case of a uniform distribution of particles when ψ^* is increased proportionally to $\bar{\beta}$.

Let us refer to the experiment. With the chute, outlet pressure measured ψ^* coefficient can be easily determined from Formula 3.19. The experiments were conducted with an open entry section of the chute ($P_0 = 0$) and the sealed lower bin. Because no air was removed from the bin ($u_{av} = 0$), the pressure in the chute outlet section is equal to the bin pressure. An averaged value of the latter was taken for the design value. As was demonstrated by numerous experiments with various materials and transfer parameters (Table 3.1), the drag factor is inversely related to the bulk concentration (Figure 3.4). The following relation was obtained after processing the experimental data

$$\frac{1}{E^2} = \frac{\psi^*}{\psi_0} = \exp\left[-1{,}8\sqrt{\beta \cdot 10^3}/(d_e \cdot 10^3)\right], \tag{3.29}$$

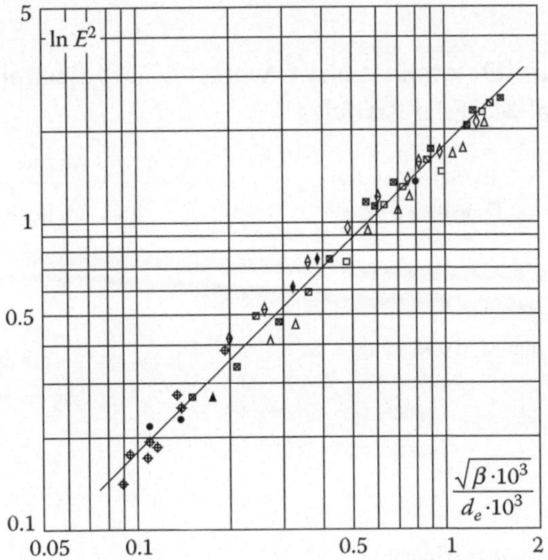

FIGURE 3.4 Variation of aerodynamic characteristic with an increase in bulk concentration of monofractional material particles (the solid line is the graph of Equation 3.29; for the symbols, see Table 3.1).

$$\beta = \frac{2G_1}{S_{ch}\rho_1 v_{1k}(1+n)}, \; n = v_{1H} / v_{1k}, \tag{3.30}$$

that allows for calculating the averaged drag factor of monofractional material particles within $0.5 < d_e < 20$ mm; $10^{-4} < \beta < 10^{-2}$.

3.1.1.2 Polyfractional Stream

Polyfractional material stream conditions make it necessary to determine the mean diameter of particles d_{av}. This is calculated using the assumption that, when substituting the actual stream for the idealized one (consisting of particles of d_{av} diameter), a particular quantitative characteristic of the stream is kept unchanged. Because our case deals with the dynamic interaction of particles and air, when substituting a polyfractional stream for a monofractional stream, it is necessary to ensure the equality of the air drag forces:

$$\sum_{i=1}^{N} \psi_i \left(\frac{w_i d_i}{v} \right) \cdot f_{Mi} \frac{w_i^2}{2} \rho_2 = N \cdot \psi \left(\frac{wd}{v} \right) \cdot f_M \frac{w^2}{2} \rho_2 . \tag{3.31}$$

Here, N is the quantity of particles in the actual stream; values with lower i index describe the i-particle; and values without the index describe a particle of the medium diameter d_{av}. Expression $\psi\left(\frac{wd}{v}\right)$ denotes the functional relation of ψ to the Reynolds number. When a stream consists of M fractions

TABLE 3.1
Test Parameters for Determination of Averaged Aerodynamic Characteristics of Monofractional Material Particles

Material Description	Particle Material Density ρ_1, kg/m^3	Chute Geometrical Parameters		Symbols for Figure 3.4
		Height, m	Incl. Angle, deg.	
Particle size 0.315–0.63 mm (d_e = 0.45 mm)				
Granite	2750	3.3	57	□
Particle size 0.63–1.25 mm (d_e = 0.88 mm)				
Granite	2750	3.3	57	⧄
Chalkstone	2600	3.3	57	◇
Burnt chalkstone	2680	3.3	57	⟠
Iron ore	3400	3.3	57	△
Particle size 1.25–2.5 mm (d_e = 1.76 mm)				
Granite	2750	1.3	75	⧅
		2.3	75	⊠
		3.3	75	⊠
		2.3	60	⊠
		3.3	60	⊠
		3.3	57	⊠
		2.0	45	⊠
Chalkstone	2600	3.3	57	⟡
Burnt chalkstone	2680	3.3	57	◆
Iron ore	3400	2.3	60	◬
Particle size 2.5–5.0 mm (d_e = 3.53 mm)				
Granite	2750	3.3	57	■
Agglomerate	3800	3.3	57	⦶
Iron ore	3400	2.3	57	▲
Particle size 5–10 mm (d_e = 7.1 mm)				
Agglomerate	3800	3.3	57	⊕
Iron ore	3400	2.3	57	▲
Particle size 10–20 mm (d_e = 14.1 mm)				
Agglomerate	3800	3.3	57	⊕
Pellets	4000	3.3	57	●

$$\sum_{j=1}^{M} \frac{G_1 m_j}{P_j} \psi_j \left(\frac{w_j d_j}{v} \right) \cdot f_{Mj} \frac{w_j^2}{2} \rho_2 = \left(\sum_{j=1}^{M} \frac{G_1 m_j}{P_j} \right) \psi \left(\frac{wd}{v} \right) f_M \frac{w^2}{2} \rho_2, \tag{3.32}$$

where the lower j index denotes values that describe the j-fraction particles, m_j is a part (by weight) of the j-fraction particles; P_j is the weight of a single particle of the j-fraction; and G_1 is the particle flow rate.

If we assume that all stream particles are settling at a steady velocity (for example, when fine particles are settling at the relative velocity equal to the airborne velocity), and there is no cross-impact of particles on the regime of flow, then in view of Equation 2.54, Equation 3.32 results in

$$1/d = \sqrt[3]{\sum_{j=1}^{M} m_j / d_j^3}, \tag{3.33}$$

that is, the mean diameter is displaced toward fine particles. When a stream contains coarse particles for which motion velocity is not very affected by the aerodynamic force, the drag factors and relative velocity for all particles are the same. In such case

$$d = \sqrt{\sum_{j=1}^{M} m_j / d_j} \bigg/ \sqrt{\sum_{j=1}^{M} m_j / d_j^3}. \tag{3.34}$$

The obtained results may be somewhat different if the condition is not met that the idealized stream contains the same quantity of particles as the actual one. For instance, the number of particles in the idealized stream is G_1/P where P is a mass of the particle of the diameter d. Then Equation 3.34 yields the mean harmonic quantity formulas

$$d = 1 \bigg/ \sum_{j=1}^{M} m_j / d_j. \tag{3.35}$$

In all analyzed cases, the mean diameter is significantly dependent on the quantity of fractions. However, in actual conditions, fine fractions are moving along the chute bottom as a layer and contribution of such fractions to the resulting aerodynamic interaction force will be much less than it was expected in theory. Particles with a mass significantly larger than that of the fine particles with which it is colliding penetrate the entire chute section in a galloping motion and thus determine the active interaction of the material flow and air. Therefore, it is preferable to use the formulas that displace the mean diameter to coarser particles. The simplest of them is the mean mass diameter formula:

$$d = \sum_{j=1}^{M} m_j d_j. \tag{3.36}$$

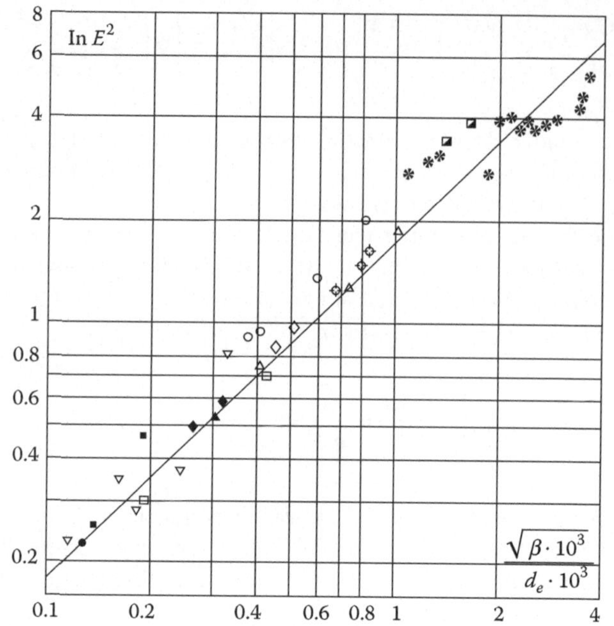

FIGURE 3.5 Variation of aerodynamic characteristic with an increase in bulk concentration of polyfractional material particles (for symbols, see Table 3.2).

Its evaluation included the tests with the polyfractional materials most common to the ore preparation industry (Table 3.2).* As it was demonstrated in the experiments outlined in Figure 3.5, the averaged drag factor determined for the monofraction of *d* size correlates very accurately with the value calculated from Equation 3.29. (The solid line in Figure 3.5 denotes the graph of Equation 3.29 at d_e equal to the mean mass diameter [3.36].)

A satisfactory agreement with the design data is evidenced by the experimental results obtained by V. A. Minko. A deviation of these results slightly greater than from the estimations is due to a less accurate method of determining of ψ^* than the chute pressure measuring method with no air motion.

3.1.2 Air Injection with a Stream of Particles in a Prismatic Chute

The resulting value of averaged coefficient ψ^* allows for estimating the chute power characteristic in full and for analyzing the aerodynamic effects occurring in bulk material motion in closed straight tubes (chutes). For this purpose, we use Equations 3.7 and 3.8, which (given that $S_{ch} = const$) will be rewritten as follows:

* The experiments with sand, coal, copper-nickel pellets, blast-furnace slag, and undersized iron ore were conducted by V. A. Minko [61] on a semi-commercial transfer group with a 500-mm-wide conveyer belt and a 1.5-m-high vertical chute of 190- × 280-mm section. (The total conveyer-to-conveyer transfer height for the material was 2850 mm.) When transferring these materials, the atmospheric pressure was maintained in the lower hood by the local exhaust operation. The induced air volume was measured as equal to the exhaust air flow.

TABLE 3.2
Particle-Size Distribution of Polyfractional Materials, %

Material Description	Symbols for Figure 3.5	Class Dimensions, mm										Mean Mass Diameter, mm
		+40	−40 −20	−20 +10	−10 +5	−5 +2.5	−2.5 +1.25	−1.25 +0.63	−0.63 +0.315	−0.315 +0.14	−0.14	
Crushed charred coal	⊠	—	—	0.8	19.9	6.8	11.8	20.4	15.9	16.8	7.6	2.6
Crushed chalkstone	◇	—	—	2.8	16.3	5.6	12.8	27.4	14.3	17.4	3.4	2.45
Iron ore	△	—	—	5.5	6.5	25.8	6.4	7.5	22.2	15.5	10.6	2.5
Agglomerated ore	⚠	—	0.3	12.8	13.1	7.2	11.4	14.7	11.3	20.4	7.8	4.5
Burnt ore	▲	1.6	—	23.6	23.1	11.7	13.3	8.5	3.5	4.2	10.5	6.7
Agglomerated fines	⌖	—	0.8	10.4	16.6	10.1	12.8	10.5	6.5	22.0	10.3	3.85
Burnt chalkstone	◆	—	1.6	28.7	26.5	9.6	13.7	9.5	5.7	3.9	0.8	7.5
Crushed charred coal	□	—	16.9	21.8	23.6	7.0	3.3	10.3	7.7	7.5	1.9	10.6
Iron-ore pellets	●	—	2.1	97.3	0.6	—	—	—	—	—	—	15.2
Sand (ρ_1 = 2600 kg/m^3)	✳	—	—	—	4.0	5.4	13.5	14.5	16.6	43.6	2.4	1.2
Coal concentrate (ρ_1 = 1400 kg/m^3)	◪	—	—	12.4	13.3	10.8	19.6	35.3	5.1	3.3	0.5	4.0
Copper-nickel concentrate pellets (ρ_1 = 3500 kg/m^3)	○	3.3	4.1	7.2	4.8	16.4	7.8	10.8	5.3	30.8	9.5	5.2
Blast-furnace slag (ρ_1 = 2300 kg/m^3)	■	13.3	16.6	19.2	10.8	8.3	7.5	16.7	5.8	1.4	0.4	14.7
Undersized iron ore (ρ_1 = 4000 kg/m^3)	▽	—	—	30	70	—	—	—	—	—	—	9.8

$$\beta_1\rho_1 v_1 \frac{dv_1}{dx} = \beta_1\rho_1 a_T - \psi^* \beta_1 k_m \frac{|v_1 - v_2|(v_1 - v_2)}{2}\rho_2, \tag{3.37}$$

$$\begin{aligned}(1-\beta_1)\rho_2 v_2 = \frac{dv_2}{dx} &= (1-\beta_1)(\rho_2 - \rho_0)g_x - (1-\beta_1)\frac{dP}{dx} - \\ &\lambda\frac{1-\beta_1}{D}\frac{v_2^2}{2}\rho_2 + \psi^*\beta_1 k_m \frac{|v_1 - v_2|(v_1 - v_2)}{2}\rho_2.\end{aligned} \tag{3.38}$$

In view of the material bulk concentration smallness

$$\beta_1 << 1\,; 1-\beta_1 \approx 1;\ v_2 \approx const;\ dv_2 / dx \approx 0, \tag{3.39}$$

the latter equation will appear as

$$\frac{dP}{dx} = (\rho_2 - \rho_0)g_x - \frac{\lambda}{D}\frac{v_2^2}{2}\rho_2 + \psi^*\beta_1 k_m \frac{|v_1 - v_2|(v_1 - v_2)}{2}\rho_2. \tag{3.40}$$

Let us assume that the process is isothermal ($\rho_2 = \rho_0$), evaluate the distribution of forces along the chute length, and determine the amount of air moved in the chute by these forces.

3.1.2.1 Pressure Distribution

The analysis of Equation 3.40 shows that the presence of materials in a tube changes the pressure gradient value and direction. Thus, in the absence of any directional air motion in the tube (for example, when a material is transferred to a pressurized vessel), the material motion results in a positive gradient equal to

$$\frac{dP}{dx} = \psi^*\beta_1 k_m \frac{v_1^2}{2}\rho_2 = \psi^* k_m \varepsilon \frac{G_1}{2S_{ch}} v_1. \tag{3.41}$$

Using this equation, we determine the pressure distribution along the tube length. In actual conditions, the absence of a directional air flow in a tube is possible in three cases: when the upper tube end is closed or open (e.g., discharge from a bin filled with a material), when the lower tube end is closed (e.g., when filling a pressurized bin), or, finally, when both ends are closed (bin-to-bin transfer of a material). In all cases, as is indicated by Equation 3.41, a positive pressure gradient occurs in the tube. The absolute pressure value is increasing along the tube length toward the material motion. However, we are interested in the excessive (against the atmosphere) pressure distribution.

On the assumption that the material motion is uniformly accelerated, in titrating, we obtain:

$$P = \psi^* k_m \varepsilon \frac{G_1}{2S_{ch}} (2a_T x + v_{1H}^2)^{1,5} / (3a_T) + C. \tag{3.42}$$

The value of C (integration constant) varies based on the tube loading and discharge pattern.

When only the upper tube end is closed, it is obvious that:

$$P\big|_{x=l} = P_a, \tag{3.43}$$

where P_a is the absolute indoor pressure. Then,

$$P = P_a - \psi^* k_m \varepsilon \frac{G_1}{2S_{ch}}\left[(2a_T l + v_{1H}^2)^{1,5} - (2a_T x + v_{1H}^2)^{1,5}\right] / (3a_T), \tag{3.44}$$

that is, with respect to the room, the entire tube is under the vacuum-gauge pressure.

The vacuum-gauge pressure is increasing from the lower tube end to the upper end where it reaches the maximum value:

$$P - P_a = -\psi^* k_M \varepsilon \frac{G_1}{2S_{ch}} \cdot \frac{v_{1k}^3 - v_{1H}^3}{3a_T}. \tag{3.45}$$

The vacuum-gauge pressure may reach a significant value—for instance, according to Voegeli [115] and A. N. Dobromyslov [28, 214], who observed the mode of water and induced air motion in high soil pipes registered vacuum-gauge pressure above 600 Pa.

Filling a pressurized vessel (only the lower tube end is virtually airtight) when

$$P\big|_{x=0} = P_a, \tag{3.46}$$

based on Equation 3.42, we obtain:

$$P = P_a + \psi^* k_M \varepsilon \frac{G_1}{2S_{ch}}\left[(2a_T x + v_{1H}^2)^{1,5} - v_{1H}^3)\right] / (3a_T). \tag{3.47}$$

The entire pipe is under excessive pressure that rises in the material motion direction to reach its maximum value at the lower tube end, the absolute magnitude of which is determined from Equation 3.45. Experimental studies indicated that the actual distinct distribution of pressure along the tube length agrees satisfactorily with the estimated value (see Figure 3.6 for results; the solid line is a graph of Equation 3.47). It is somewhat more difficult to determine constant C when both ends of the tube are airtight because there is vacuum-gauge pressure at the tube inlet and excessive pressure at the tube outlet. Then, at some distance x_a, the tube pressure is $P = P_a$.

The lengthwise pressure distribution is determined by the following equation:

$$P = P_a + \psi^* k_M \varepsilon \frac{G_1}{6a_T S_{ch}}\left[(2a_T x + v_{1H}^2)^{1,5} - (2a_T x_a + v_{1H}^2)^{1,5}\right]. \tag{3.48}$$

There is vacuum-gauge pressure in the upper end of the tube, with its maximum value at the tube inlet:

$$P - P_a = -\psi^* k_M \varepsilon \frac{G_1}{6a_T S_{ch}}\left[(2a_T x_a + v_{1H}^2)^{1,5} - v_{1H}^3)\right]. \tag{3.49}$$

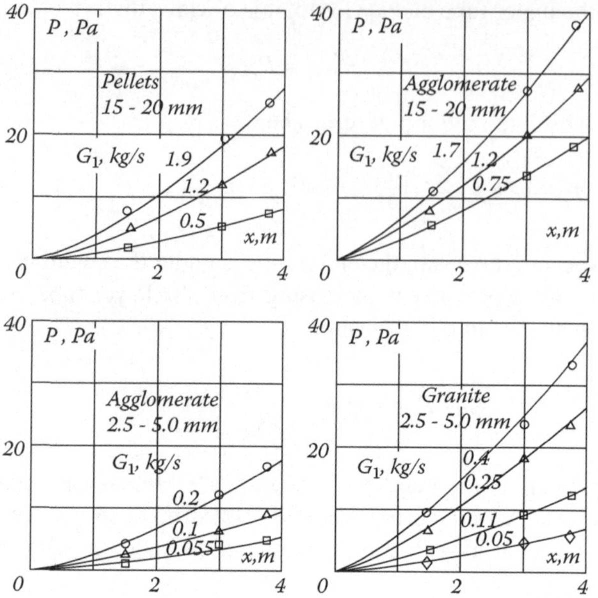

FIGURE 3.6 Chute pressure variation ($\Theta = 57°$, $S_{ch} = 0.0225$ m^2) in transfer of various bulk materials ($v_2 = 0$). The solid lines are graphs of Equation 3.47.

There is an excessive pressure in the lower end of the tube that reaches

$$P - P_a = \psi^* k_m \varepsilon \frac{G_1}{6a_T S_{ch}} \left[v_{1K}^3 - (2a_T x_a + v_{1H}^2)^{1,5} \right] \quad (3.50)$$

at the tube outlet.

Thus, pressure redistribution occurs. On the assumption that the tube volume is enclosed, we use the equation for an ideal gas condition to express

$$\int_0^l S_{ch} P dx = S_{ch} l P_a \quad (3.51)$$

or, substituting P for its value (from Equation 3.48), we will obtain after cancellations:

$$\int_0^l (2a_T x + v_{1H}^2)^{1,5} dx = \int_0^l (2a_T x_a + v_{1H}^2)^{1,5} dx, \quad (3.52)$$

from which we determine x_a. In particular, at $v_{1H} = 0$ we have $x_a = 0,543l$.

The pressure distribution along the tube length significantly changes if there is a directional air flow in the tube. In this case, there are two possible scenarios: forward flow—when the bulk material settling flow direction is the same as the air flow direction, and reverse flow—air is moving upward to the falling material.

For the forward flow, Equation 3.40 will appear as:

$$dP = -\lambda \frac{dx}{D}\frac{v_2^2}{2}\rho_2 + \psi^* \beta_1 k_m \frac{|v_1 - v_2|(v_1 - v_2)}{2}\rho_2 dx; \tag{3.53}$$

for the reverse flow:

$$dP = -\left[\lambda \frac{v_2^2}{2D}\rho_2 + \psi^* \beta_1 k_m \frac{(v_1 + v_2)^2}{2}\rho_2\right] dx. \tag{3.54}$$

By integrating Equations 3.53 and 3.54 along the tube length, we obtain, respectively:

$$P - P_0 = -\lambda \frac{x}{D}\frac{v_2^2}{2}\rho_2 + \int_0^x \psi^* \beta_1 k_m \frac{|v_1 - v_2|(v_1 - v_2)}{2}\rho_2 dx, \tag{3.55}$$

$$P - P_0 = -\lambda \frac{x}{D}\frac{v_2^2}{2}\rho_2 + \int_0^x \psi^* \beta_1 k_m \frac{(v_1 + v_2)^2}{2}\rho_2 dx. \tag{3.56}$$

For the first part, the integral represents the air pressure drop at a tube section of x length that results from the fall of material. We refer to this pressure drop value as the induction pressure [49, 70]. If there is no directional air flow in the tube, the induction pressure is equal to the excessive pressure in the tube.

On the other hand, according to Equation 3.37, we have:

$$\int_0^x \psi^* \beta_1 k_m \frac{|v_1 - v_2|(v_1 - v_2)}{2}\rho_2 dx = \int_0^x \beta_1 \rho_1 a_T dx - \beta_1 v_1 \rho_1 (v_1 - v_{1H}), \tag{3.57}$$

from which it is clear that P_E is determined by the difference in material flow velocities. Thus, the induction pressure can be determined using two methods. The first is associated with the tube air pressure measurements and the second one with the particle stream velocity measurements. Here we will focus on the first method. We will still assume that the material flow is uniformly accelerated and

$$\psi^* k_m = 1{,}5\, \psi^* / d_e = const. \tag{3.58}$$

Then

$$P_{E_x} \equiv \int_0^x \psi^* \beta_1 k_m \frac{|v_1 - v_2|(v_1 - v_2)}{2}\rho_2 dx = \psi^* k_m \varepsilon \frac{G}{2a_T S_{ch}} \int_{v_{1H}}^{v_{1k}} |v_1 - v_2|(v_1 - v_2) dv_1. \tag{3.59}$$

After the integration:

$$P_{E_x} = \psi^* k_m \frac{\varepsilon G_1}{2a_T S_{ch}} \times \frac{(v_1 - v_2)^3 - (v_{1H} - v_2)^3}{3} \text{ at } v_2 < v_{1H}, \tag{3.60}$$

$$P_{E_x} = \psi^* k_m \frac{\varepsilon G_1}{2a_T S_{ch}} \times \frac{(v_1 - v_2)^3 - (v_2 - v_{1H})^3}{3} \text{ at } v_{1H} < v_2 < v_1, \tag{3.61}$$

$$P_{E_x} = \psi^* k_m \frac{\varepsilon G_1}{2a_T S_{ch}} \times \frac{(v_2 - v_1)^3 - (v_2 - v_{1H})^3}{3} \text{ at } v_2 > v_1 \tag{3.62}$$

or

$$P_{E_x} = \psi^* k_m \frac{G_1 \varepsilon}{2a_T S_{ch}} \cdot \frac{|v_1 - v_2|^3 - |v_{1H} - v_2|^3}{3}. \tag{3.63}$$

Thus, the excessive forward flow pressure (the excessive pressure at the tube inlet) is increasing along the tube length, and at the distance

$$x \equiv x_m = \left\{ v_2^2 \left[1 + \sqrt{1 - \frac{4}{(2+b)^2}} \right]^2 - v_{1H}^2 \right\} \Big/ (2a_T) \tag{3.64}$$

from the tube inlet, it reaches its maximum equal to

$$P_m - P_0 = -\lambda \frac{x_m}{D} \cdot \frac{v_2^2}{2} \rho_2 + \psi^* k_m \varepsilon \frac{G_1}{2a_T S_{ch}} \cdot \frac{|v_m - v_2|^3 - |v_{1H} - v_2|^3}{3}, \tag{3.65}$$

where (for the sake of convenience) it is assumed that

$$b = \frac{\lambda d_e v_2 \rho_1 S_{ch}}{1{,}5 D \psi^* G_1}, \tag{3.66}$$

$$v_m = \sqrt{2a_T x_m + v_{1H}^2}. \tag{3.67}$$

The excessive reverse flow pressure is increasing monotonically:

$$P - P_0 = \lambda \frac{x}{D} \frac{v_2^2}{2} \rho_2 + \psi^* k_m \varepsilon \frac{G_1}{2a_T S_{ch}} \cdot \frac{(v_1 + v_2)^3 - (v_{1H} + v_2)^3}{3}. \tag{3.68}$$

The experimental data showed that the pressure distribution along the tube length agrees satisfactorily with the estimations (Figure 3.7).

At higher airborne velocities of particles, as well as at a low mass of each particle, the medium drag forces are comparable with the weight force of particles; their motion differs noticeably from the uniformly accelerated motion. Therefore, the resulting equations apply for tubes of small length and with low inside air velocities.

Let us solve the problem for the induction pressure in general based on the dynamic equation for a stream of particles. According to Equation 3.57, and in view of Equation 3.1, we obtain:

$$P_{E_x} = \frac{G_1}{S_{ch}} \left[\left(v_{1H} + a_T \int_0^x \frac{dx}{v_1} \right) - v_1 \right]. \tag{3.69}$$

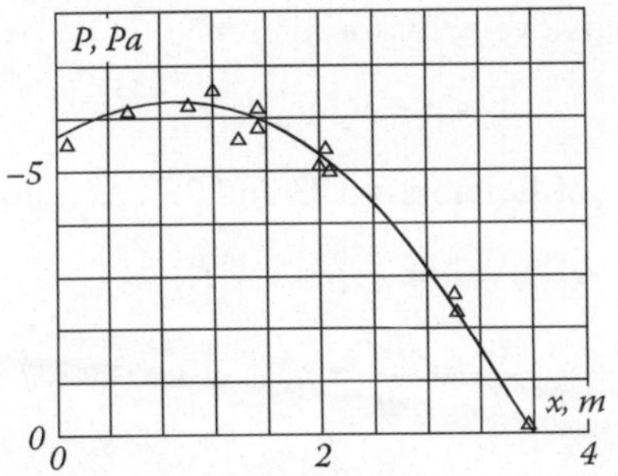

FIGURE 3.7 Chute pressure variation ($\Theta = 57$, $S_{ch} = 0.0225\ m^2$, $l = 3.6$ m) in transfer of pellets. (d = 10–20 mm, $G_1 = 1.9$ kg/s; $G_2 = 0.056$ kg/s). The solid line is a graph of Equation 3.55 (in view of Equation 3.63).

Note that the parenthetical expression

$$v_{1H} + a_T \int_0^x \frac{dx}{v_1} = v_1^0 \tag{3.70}$$

is a velocity that would be reached by a particle settling with a_T acceleration in time $\tau = \int_0^x \frac{dx}{v_1}$. In addition, τ is equal to time needed for the particle to pass x distance in the chute, accounting for the air drag force and for the wall friction force. It is apparent that when these forces are in a direction opposite to the particle motion, the velocity v_1^0 will always be higher than the material falling velocity v_1 because the latter is calculated accounting for the air drag forces but the former is not.

The differential velocity

$$\Delta v = v_1^0 - v_1 \tag{3.71}$$

will become even higher if the air motion in the chute is neglected when calculating v_1. This method may be used to estimate the induction pressure:

$$P_{E_x} \le \frac{G_1}{S_{ch}}(v_1^0 - v_{1_0}), \tag{3.72}$$

where v_{1_0} is the material velocity in x section calculated on the assumption that $v_2 = 0$.

The precise value of P_{E_x} can be determined from the equations

$$P_{E_x} = \left[v_{1H} + a_t(t - t_0) - v_1\right] G_1 / S_{ch}, \tag{3.73}$$

$$v_1 \frac{dv_1}{dx} \equiv \frac{dv_1}{dt} = a_T - \psi^* \varepsilon k_m \frac{|v_1 - v_2|(v_1 - v_2)}{2}, \tag{3.74}$$

obtained from Equations 3.69 and 3.37, respectively, by means of simple transformations.

Having taken the relative velocity for reference

$$w = v_1 - v_2,\ w_0 = v_{1H} - v_2, \tag{3.75}$$

We transform the system of Equations 3.73 and 3.74 as follows:

$$\left.\begin{aligned} P_{E_x} &= \left[w_0 - w + a_T(t - t_0)\right] G_1 / S_{ch}, \\ (w + v_2)\frac{dw}{dx} &\equiv \frac{dw}{dt} = a_T - \psi^* \varepsilon k_m |w| w / 2. \end{aligned}\right\} \tag{3.76}$$

For the sake of convenience and further comparison with the experimental data, we will transform these equations into dimensionless ones. Just as we did in Chapter 2 regarding a vertical stream of uniformly distributed particles, we introduce the so-called conventional airborne velocity

$$C_y = \sqrt{2a_T / (\psi^* \varepsilon k_m)} \tag{3.77}$$

and use it for a specific velocity. Assuming that

$$x = hl_\infty;\ v_1 = vC_y;\ v_2 = uC_y;\ w = \omega C_y; \tag{3.78}$$

$$l_\infty = C_y^2 / a_T;\ t = \tau t_\infty;\ t_\infty = C_y / a_T, \tag{3.79}$$

after simple transformations of Equation 3.76, we will obtain the following system of dimensionless equations:

$$\overline{P}_E = (\tau - \tau_0) - (\omega - \omega_0); \tag{3.80}$$

$$(\omega + u)\frac{d\omega}{dh} \equiv \frac{d\omega}{d\tau} = 1 - |\omega|\omega, \tag{3.81}$$

where

$$\overline{P}_E = P_{E_x} S_{ch} / (G_1 C_y). \tag{3.82}$$

Using the solutions for Equation 3.81 given in Table 2.2, we obtain the following functional relations

$$\tau - \tau_0 = f_\tau(\omega);\ h - h_0 = f_h(\omega). \tag{3.83}$$

Inserting this calculation into Equation 3.80, we obtain

$$\overline{P}_E = f_p(\omega). \tag{3.84}$$

Considering ω as a parameter, we will obtain the general pressure distribution along the channel length in a dimensionless form

$$\overline{P}_E = f(h). \tag{3.85}$$

For example, let us consider the reverse flow at $v_{1H} = 0$. The relative material flow velocity at the chute inlet (at the section where $v < u$) is negative; here, the particles are entrained by the air stream and the induction area occurs where a portion of settling particle energy is applied to create a positive pressure gradient and to engage air in motion. For the deceleration area, based on Equations 30 and 33 and on Table 2.2

$$\tau = \text{arctg}(v-u) + \text{arctg}u, \tag{3.86}$$

$$h = u\tau + \frac{1}{2}\ln\left\{\left[1+(v-u)^2\right]/(1+u^2)\right\}. \tag{3.87}$$

The deceleration area length is determined (on the assumption that, at the end of this area $v = u$), from the previous expression as follows:

$$h_T = u \cdot \text{arctg}u - \frac{1}{2}\ln(1+u^2), \tag{3.88}$$

and the particle residence time in this area is

$$\tau_T = \text{arctg}u. \tag{3.89}$$

For the suction area wherein $v > u$, we use Equations 16 and 20 (from the Appendix) and Table 2.2—taking into account that the initial values of τ_0 and h_0 (in our case) are the determined values of τ_T and h_T and $\omega_0 = 0$:

$$\tau - \tau_T = \frac{1}{2}\ln\left[(v-u+1)/(1-v+u)\right], \tag{3.90}$$

$$h - h_T = \frac{u}{2}\ln\left[(1+v-u)/(1-v+u)\right] - \frac{1}{2}\ln\left[1-(v-u)^2\right]. \tag{3.91}$$

Then, the pressure variation shall be described by the following equations: in the deceleration area, at $0 < h < h_T$

$$\overline{P}_E = \text{arctg}(v-u) + \text{arctg}u - v, \tag{3.92}$$

$$h = u\left[\text{arctg}(v-u) + \text{arctg}u\right] + \frac{1}{2}\ln\left\{\left[1+(v-u)^2\right]/(1+u^2)\right\}; \tag{3.93}$$

in the suction area, at $h > h_T$

$$\overline{P}_E = \frac{1}{2}\ln\left[(1+v-u)/(1-v+u)\right] - (v-u) + \overline{P}_{E_T}, \tag{3.94}$$

$$h = \frac{u}{2}\ln\left[(1+v-u)/(1-v+u)\right] + \frac{1}{2}\ln\left[1-(v-u)^2\right] + h_T, \tag{3.95}$$

where $\overline{P}_{E_T}$ is the induction pressure at the end of the deceleration area equal to

$$\overline{P}_{E_T} = \text{arctg}u - u. \quad (3.96)$$

In the absence of a directional air motion ($u = 0$), there is no deceleration area ($h_T = 0$; $\overline{P}_{E_T} = 0$) and, subject to Equations 3.94 and 3.95, the pressure distribution is described by the equation

$$\overline{P}_E = \frac{1}{2}\ln\left[(1+H)/(1-H)\right] - H; H = (1-e^{-2h})^{0,5}. \quad (3.97)$$

For comparison, let us consider a dimensionless expression relevant for this case without regard to the air drag impact on the particle stream velocity. In view of Equations (3.78) and (3.79) and subject to $v_{1H} = 0$, we obtain from Equation 3.63

$$\overline{P}_E = \frac{1}{3}\left(\left|\sqrt{2h} - u\right|^3 - u^3\right). \quad (3.98)$$

At $u = 0$, we obtain the equation

$$\overline{P}_E = (2h)^{1,5} / 3, \quad (3.99)$$

which can easily appear to be the specific case of a more general solution of Equation 3.97 at $2h \ll 1$. The graphs of these equations are shown in Figure 3.8.

As is clear from the experimental data, for $2h \leq 1$, the chute pressure forces may be calculated without regard to the medium drag impact on the material particles motion velocity.

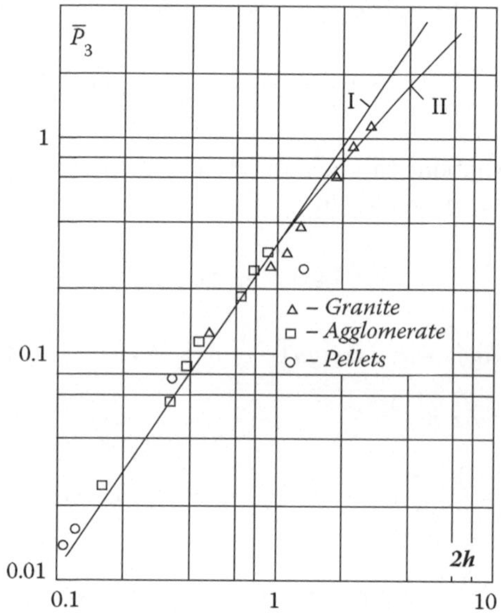

FIGURE 3.8 Induction pressure variation along the chute length ($v_{1H} = 0$; $v_2 = 0$). I is according to Equation 3.99; II is according to Equation 3.97.

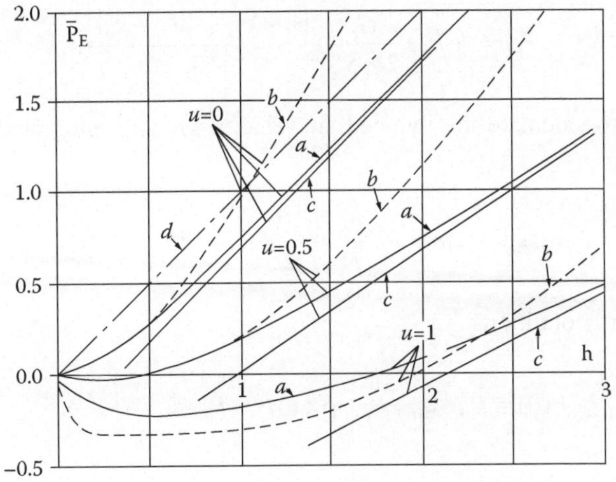

FIGURE 3.9 Induction pressure variation along the chute length at higher material drop heights; (a) according to Equations 3.92 through 3.95); (b) according to Equation 3.98; (c) according to Equation 3.100; and (d) according to Equation 3.102.

There is a different asymptotic behavior observed at higher material drop heights (Figure 3.9):

$$\bar{P}_E - \bar{P}_{E_T} = \frac{1}{u+1}(h - h_T + \ln 2) - 1. \tag{3.100}$$

The induction pressure within $u \le 0.5$; $h \ge 2$, with an accuracy of max 5%, can be calculated from Equation 3.100. An overestimated induction pressure will be obtained on the assumption that the relative material flow velocity is constant throughout the chute length and is equal to the conventional airborne velocity C_y. In this case, based on Equation 3.57

$$P_{E_x} = \int_0^x \beta_1 \rho_1 a_T dx = \frac{G_1 x}{v_2 + C_y} \cdot \frac{a_T}{S_{\text{ch}}} \tag{3.101}$$

or in a dimensionless form

$$\bar{P}_E = h/(1+u). \tag{3.102}$$

Thus, the induction pressure with airborne particles is the maximum and is equal to the weight of particle in a chute related to the chute cross-section area.

3.1.2.2 Induced Air Velocity

After determining the force behavior in the chute and the main element defining that behavior (i.e., the induction head),

$$P_{E_x} = \psi^* k_m \varepsilon \frac{G_1}{2a_T S_{ch}} \cdot \frac{|v_1 - v_2|^3 - |v_{1\text{H}} - v_2|^3}{3}, \tag{3.103}$$

it is possible to calculate the induced air velocity. Having integrated the dynamic equation

$$dp = -\lambda \frac{dx}{D} \cdot \frac{|v_2| v_2}{2} \rho_2 + dP_{E_x} \tag{3.104}$$

in the boundary conditions

$$P(0) = P_a - \zeta_\text{H} \frac{|v_2| v_2}{2} \rho_2; P(l) = P_a + \zeta_k \frac{|v_2| v_2}{2} \rho_2, \tag{3.105}$$

we obtain

$$\sum \zeta \frac{v_2^2}{2} \rho_2 = P_E \equiv \psi^* k_m \varepsilon \frac{G_1}{2a_T S_{ch}} \cdot \frac{|v_{1k} - v_2|^3 - |v_{1\text{H}} - v_2|^3}{3}, \tag{3.106}$$

where v_{1H}, v_{1k} are the material velocities at the chute inlet and outlet, m/s; P_a is the chute outside pressure, Pa; ζ_H, ζ_K are the local drag factors at the chute inlet and outlet, respectively; and $\Sigma\zeta$ is a sum of local drag factors equal to

$$\sum \zeta = \zeta_\text{H} + \zeta_k + \lambda l / D. \tag{3.107}$$

As is clear from Equation 3.106, the finite value of $\Sigma\zeta$ always results in a directional air flow in a chute. The flow direction coincides with the bulk material stream direction. For further analysis, transform Equation 3.106 into a dimensionless form

$$\frac{\varphi_k^2}{|1 - \varphi_k|^3 - |n - \varphi_k|^3} = \frac{\psi^* k_m G_1 v_{1k}}{\sum \zeta \cdot 3a_T S_{ch} \rho_1} \equiv \frac{Bu}{3}, \quad \varphi_k \equiv \frac{v_2}{v_{1k}}, \quad n = \frac{v_{1H}}{v_{1k}}. \tag{3.108}$$

The tripled right-hand side of the equation is the Butakov–Neikov criterion

$$Bu = \frac{\psi^* k_m G_1 v_{1k}}{a_T S_{ch} \rho_1 \sum \zeta}. \tag{3.109}$$

In view of the relation for the conventional airborne velocity, the following expression is obtained from the Butakov–Neikov number

$$Bu = \frac{G_1 v_{1k}}{S_{ch} \sum \zeta \frac{C_y^2}{2} \rho_2}, \tag{3.110}$$

which is the relation of the material motion quantity to the dynamic head conventional force.

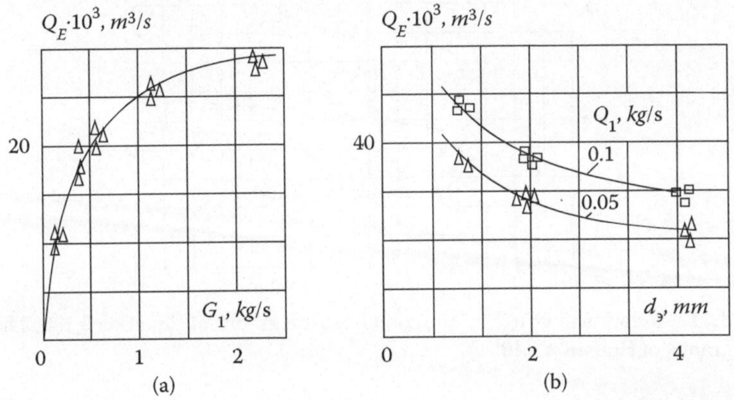

FIGURE 3.10 Relation of the induced air flow to the transferred material flow rate and fineness; (a) granite transfer, d = 1.25-2.5 mm at Θ = 45°, H = 2 m; (b) transfer of chalkstone of the same fineness at Θ = 60°, H = 3 m. The solid lines are graphs of Equation 3.111 (in view of Equation 3.108).

Analyzing the result, it may be noted that the induced air quantity

$$Q_E = \varphi_k v_{1k} S_{ch} \tag{3.111}$$

increases with the increase in the material flow rate and decreases its particle size, which agrees satisfactorily with the experimental data (Figure 3.10); φ_k is also significantly influenced by the hydraulic resistance of the chute and by the material stream velocity.

Figure 3.11 shows the graphs of Equation 3.108, which are indicative of an asymptotic nature of variation in φ_k. The area *Bu > 3* may be called the self-similarity area. Here, φ_k virtually remains unchanged and is close to the asymptotic value

$$\varphi_k = \frac{1+n}{2}, \tag{3.112}$$

and the induced air volume

$$Q_E \le \frac{1}{2}\left(v_{1H} + v_{1k}\right) S_{ch}. \tag{3.113}$$

This is explained by the deceleration area at the chute inlet where $v_2 > v_1$ and particles have a decelerating—instead of inducing—effect on the moving air.

This condition was not considered, for instance, by S. E. Butakov—who was the first to analytically study an inducing effect for a bulk material stream in chutes [15]. That is why there are no asymptotics in Butakov's formula and why φ_k in 8.7 < Bu < 13.9 is not uniquely defined, which contradicts the physical meaning (φ_k cannot have multiple values at the same parameters of a transfer group or exceed one affected only by the induction head).

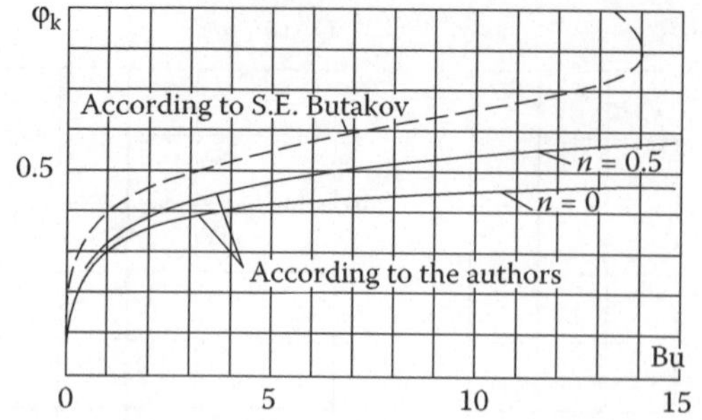

FIGURE 3.11 Relation of φ_k to Bu. The solid lines are graphs of Equation 3.108. The dashed lines are graphs of Equation 1.10.

Compared to the values that we obtained, an overestimated value of φ_k (that is determined from Butakov's formula in Bu < 13.9) can be explained by the previously noted incorrectness of the momentum conservation equation application. Using Equation 44 in the Appendix with the reductions assumed when setting up a one-dimensional problem, we obtain

$$\frac{d}{dx}\beta_1\rho_1 a_T \frac{v_1^2}{2} v_1 = \beta_1\rho_1 a_T v_1 - \frac{\beta_1}{V_p} R v_1, \tag{3.114}$$

$$\frac{d}{dx}\beta_2\rho_2 a_T \frac{v_2^2}{2} v_2 = \beta_2(\rho_0 - \rho_2) g_x v_2 - \frac{d}{dx}\beta_2 P v_2 - \frac{\lambda}{D}\cdot\frac{v_2^2}{2}\rho_2 v_2 + \frac{\beta_1}{V_p} R v_2. \tag{3.115}$$

In view of a small bulk concentration of the material

$$\beta_2 \approx 1;\ v_2 \approx \text{const at } S_{ch} = \text{const} \tag{3.116}$$

given Equations 3.1 and 3.2, we have

$$\beta_1\rho_1 v_1 v_1 \frac{dv_1}{dx} = \beta_1\rho_1 a_T v_1 - \frac{\beta_1}{V_p} R v_1, \tag{3.117}$$

$$\rho_2 v_2 v_2 \frac{dv_2}{dx} = (\rho_0 - \rho_2) g_x v_2 - v_2 \frac{dP}{dx} - \frac{\lambda}{D}\frac{v_2^2}{2}\rho_2 v_2 + \frac{\beta_1}{V_p} R v_2. \tag{3.118}$$

Dividing both sides of the equations by the corresponding velocity, we obtain the studied dynamic equations for a one-dimensional stream. Thus, we obtain similar results in a correct application of the momentum conservation equations.

The results obtained from analysis of dynamic equations for a one-dimensional stream accurately correlate with the experimental data in qualitative as well as in quantitative terms. We verified it by estimating the chute forces and comparing the induced air volumes. Figure 3.12 shows the results of a comparison of extensive experimental data with the estimated data obtained (from Equation 3.108) as well

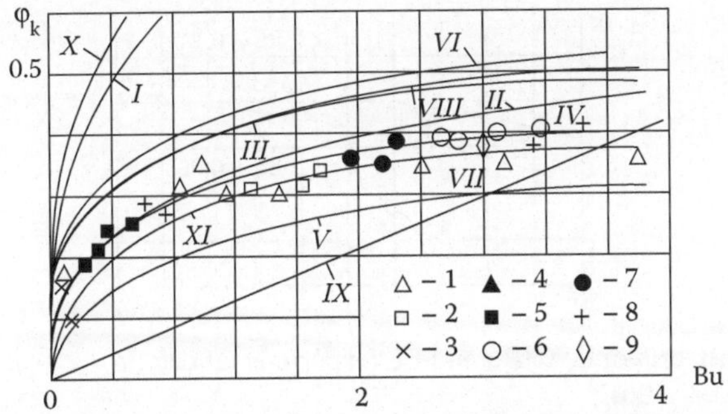

FIGURE 3.12 Relation of φ_k to Bu at Π = 0 according to (I) Hemeon, (II) Hatch, (III) Butakov, (IV) authors (by Equation 3.108), (V) Dennis and Andersen, (VI) Minko, (VII) Graschenkov, (VIII) Chulakov, (IX) Bagaevskiy and Bakirov, (X) Kilin, and (XI) Olifer. Experimental data: (1) quartzite [102]; (2,3) granite and iron ore [37]; (4) iron ore [11]; (5–9) authors' data for granite, pellets, agglomerate, charred coal, and iron ore, respectively.

as comparisons with the findings of other authors who studied the suction process. The graphs of φ_k relations to *Bu* were plotted at $\Sigma\zeta = 1{,}5$ according to Hemeon [109]; at $\sqrt[3]{E_E} = 0{,}4$ according to Hatch [108]; at $\Sigma\zeta = 1{,}5$; $\rho_1 = 3000$ kg/m³, $F_H^b = 0{,}2$ m², and $S_{ch} = 0{,}5$ m² according to Dennis and Andersen [106]; at $k^3 = 0{,}18$ according to Graschenkov and co-authors [27]; at $\Sigma\zeta = 1{,}5$; $\rho = 0{,}3$ according to Bagaevskiy and Bakirov [8]; and at $d_{av} = 10$ mm according to Olifer [71].

The experimental data of Sheleketin [102] for quartzite $d = 3$–5 mm, Kamyshenko [37] for granite $d = 22$ mm and iron ore $d = 5.6$ mm, Boshnyakov [11] for iron ore, as well as our experimental data correlate accurately with the theoretical findings concerning a one-dimensional stream. An accurate correlation with the experimental data is also obtained using formulas of Olifer in $Bu < 1$ and $d \sim 10$ mm and formulas of Graschenkov and co-authors in $Bu >$ with the introduction of a correction factor $k^3 = 0{,}18$. Results from Hemeon, Kilin [39,40], and Bagaevskiy and Bakirov yield the highest deviations.

So far, we have only considered air motion in a chute influenced by the aerodynamic interaction between the falling material and the air. Now let us evaluate the influence of local exhausts on the induced air volumes.

The induced air pressure increase is affected by the vacuum-gauge pressure that occurs due to the local exhaust operation in the lower section of the chute. Indeed, if we integrate Equation 3.104, the second boundary condition (3.105) is substituted for the following relation

$$P(l) = P_a + \zeta_k \frac{v_2^2}{2}\rho_2 - P_2, \tag{3.119}$$

where P_2 is the hood vacuum-gauge pressure under cover, Pa;
we obtain

$$\sum\zeta\frac{v_2^2}{2}\rho_2 = P_E + P_2. \tag{3.120}$$

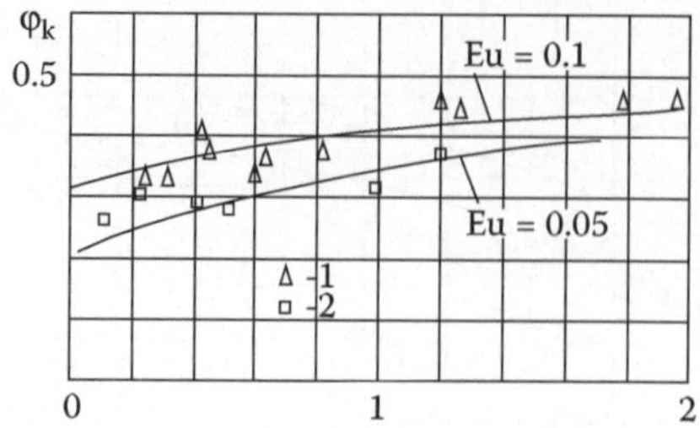

FIGURE 3.13 Relation of φ_k to Bu and Eu (experimental data: Δ – 1 – at Eu = 0.08–0.12; □ – 2 – at Eu = 0.04-0.06).

On the assumption that the material flow is uniformly accelerated, this equation can be easily transformed into the following criterion relation:

$$\varphi_k^2 = \mathrm{Bu}\left[\left|1-\varphi_k\right|^3 - \left|n-\varphi_k\right|^3\right]/3 + \mathrm{Eu}, \tag{3.121}$$

where Eu is the Euler criterion equal to

$$\mathrm{Eu} = P_2 \Bigg/ \left(\frac{v_{1k}^2}{2}\rho_2 \sum \zeta\right). \tag{3.122}$$

In this case, the comparison of the calculated induced air volumes also agrees satisfactorily with multiple experimental data (Figure 3.13).

Another approach for evaluation of the aerodynamic process in a chute is the process of air induction with a falling material as the work of a peculiar kind of a charger in the mains (chute). Let us plot the characteristic curve of this charger and estimate its efficiency factor. This charger head is nothing but the induction head. Knowing this head value, it is possible to determine the quantity of air induced. We take into account the hood vacuum-gauge pressure

$$Q_E = \sqrt{(P_E + P_2)/R_{ch}}, \tag{3.123}$$

where R_{ch} is a hydraulic characteristic of the chute

$$R_{ch} = \sum \zeta \rho_2 / (2S_{ch}^2). \tag{3.124}$$

In general, a dimensionless characteristic may be estimated using functional relations (Equations 3.80 and 3.83). Figure 3.14 shows such a characteristic curve at $v_{1H} = 0$. It was plotted by Equations 3.94 and 3.95 at $h = 1$ and 1/3. The figure also shows the experimental data for water drops of $d_E = 3$ mm settling in 6.25-m and 1.9-m long vertical tubes with a diameter of 300 mm.

The charger efficiency is expressed by the relation

$$\eta_E = P_E Q_E / W, \tag{3.125}$$

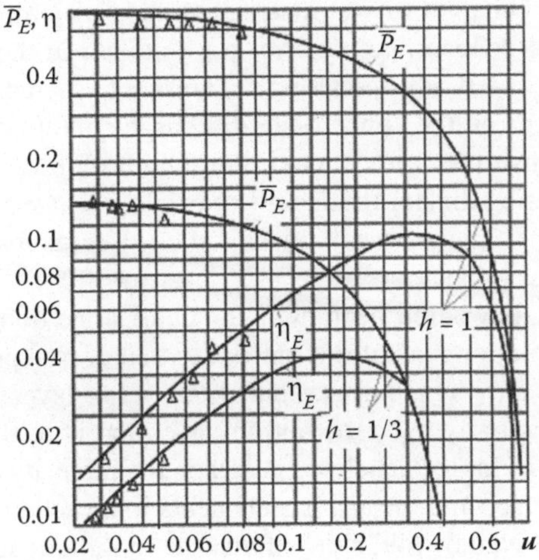

FIGURE 3.14 Aerodynamic characteristic of a "suction charger."

where P_E is a head generated by the charger at Q_E, Pa; W is the power consumed by lifting a bulk material (a portion of which is applied to create head P_E and motion Q_E of air).

In either case, the input power is

$$W = gG_1H. \tag{3.126}$$

Then Equation 3.125 will appear as

$$\eta_E = \overline{P}_E u / h. \tag{3.127}$$

The graph $\eta_E = f(u)$ shown in Figure 3.14 was plotted using this formula in view of Equations 3.92 through 3.95. As is clear from the presented data, the charger efficiency increases with an increase in the material drop height and with the airborne velocity decrease.

Having a characteristic curve for a charger and the mains (chute), we use Equation 3.123 to determine the quantity of air induced. Here we have a perfect analogy with calculating the efficiency of a fan operating with a certain hydraulic characteristic (resistance). This approach allows us to solve problems related to determination of the induced air volumes and, in more complicated cases (when spilt runners are used), those related to a cascade layout of equipment, and so on.

3.1.3 Peculiarities of the Dynamic Interaction of Air and a Bulk Material Stream in Laminar Flow in a Chute

So far, we have considered aerodynamically active or combined streams of a bulk material. The air drag of particles in a stream is comparable with the air drag force of a single particle. Now we will analyze the passive interaction of components. A good example of such interaction is streams of powder materials in chutes.

The aerodynamic behavior of wet material streams is similar to that of a lump material motion. Fine concentrate particles are retained in the form of conglomerates or "packets" due to self-adhesion forces. The size of such packets, as was proven by Popov [78], depends on the concentrate discharge conditions.* For instance, in a transfer from belt converters with a maximum width of 1000 mm, the packet size will be 10–60 mm. Using this for the specific particle size, we have a case of conventionally active streams. The air mechanics of such streams are estimated from the earlier design relations [48].

A dry concentrate, powdered bentonite, and chalkstone do not form any packets when moving. Their motion in chutes is similar to that of iron powders [32,94]. Most of the powder mass is moving along the chute bottom in a layer. Only a small portion of particles has broken away from the bottom layer and is settling above the layer's surface. The quantity of these particles is based on the intensity of the dynamic interaction of the layer surface and the air. Experiments in a vacuum chamber conducted under the guidance of Neikov [69] showed that the motion of subsieve iron powders is similar to the flow of a heavy liquid. There are no particles above the jet surface.

Thus, when considering the motion of powder materials in chutes only, the stream surface is aerodynamically active. Let us determine the interaction forces and the induced air volume.

Intercomponent interaction forces in the laminar motion of a powdered material can generally be presented as a sum of two forces, due to motion specificity:

$$P_E = P_n + P_E^*, \tag{3.128}$$

where P_n is a surface force resulting from the interaction of an irregular surface of a particle layer and air, Pa; and P_E^* is the induction pressure of particles moving above the layer.

Let us express the surface force through tangential stress τ_n:

$$P_n = \int_0^l \tau_n b dx, \tag{3.129}$$

where b is the chute width, m; and l is the chute length, m. The tangential stress can, in turn, be expressed through the dynamic head of the relative motion of particles and the hydraulic resistance factor λ^*:

$$\tau_n dx \cdot b = \lambda^* \frac{dx}{D^*} \frac{(v_1 - v_2)|v_1 - v_2|}{2} \rho_2 S_{ch} \tag{3.130}$$

or

$$\tau_n = c_\tau \frac{(v_1 - v_2)|v_1 - v_2|}{2} \rho_2, \tag{3.131}$$

* The influence of self-adhesion forces and erosion on the injection capacity of a stream of coalescent powder is explained in the study [215, 43].

where

$$c_\tau = \frac{\lambda^*}{2}(0{,}5 + \delta / b), \tag{3.132}$$

δ is a height of the chute cross-section free from any material, and D^* is the chute hydraulic diameter that is equal to

$$D^* = 4b\delta / (2\delta + b). \tag{3.133}$$

Then the surface force is

$$P_n = \int_0^l c_\tau b \frac{|v_1 - v_2|(v_1 - v_2)}{2} \rho_2 dx. \tag{3.134}$$

The induction pressure force may be written as follows:

$$P_E^* = \int_0^l \psi \frac{G_1^*}{\rho_1 v_1} k_m \frac{(v_1 - v_2)|v_1 - v_2|}{2} \rho_2 dx, \tag{3.135}$$

where G_1^* is the flow rate of particles moving above the layer, kg/s; so the total dynamic interaction force is

$$P_E = \int_0^l b\left(c_\tau + \psi \frac{G_1^* k_m}{\rho_1 v_1 b}\right) \frac{(v_1 - v_2)|v_1 - v_2|}{2} \rho_2 dx \tag{3.136}$$

or, collecting the expression in parentheses preceding the dynamic head

$$c_\tau^* = c_\tau + \psi \frac{G_1^* k_m}{\rho_1 v_1 b}, \tag{3.137}$$

we express this force as follows:

$$P_E = \int_0^l c_\tau^* b \frac{|v_1 - v_2|(v_1 - v_2)}{2} \rho_2 dx. \tag{3.138}$$

Studies with powder flows conducted by Sielberberg under the guidance of Neikov [32] have experimentally proven the prerequisites for an independent form from Equation 3.138. They established that chute length-averaged coefficient c_τ^* can be calculated from the following empirical relation

$$c_\tau^* = 0{,}09 d_{cp}(3{,}26 m_v - 3{,}6 m_v^2 + m_v^3), \tag{3.139}$$

$$m_v = 10^3 G_1 / (\rho_1 b). \tag{3.140}$$

Removing the integral in the right-hand side of the equation, on the assumption that the stream is uniformly accelerated, we will obtain the following expression for the total intercomponent interaction force

$$P_E = \frac{c_\tau^* b \rho_2}{2a_T} \cdot \frac{|v_{1k} - v_2|^3 (3v_{1k} + v_2) - |v_{1H} - v_2|^3 (3v_{1H} + v_2)}{12}. \tag{3.141}$$

The *induced air volumes* are determined if we know the intercomponent interaction force and the hydraulic characteristic of the chute. The air velocity is determined from the obvious relation

$$\sum \zeta \frac{v_2^2}{2} \rho_2 S_{ch} = P_E, \tag{3.142}$$

from which we obtain the following formula for the slip ratio of components at the chute outlet:

$$\varphi_k^2 / \left[|1 - \varphi_k|^3 (3 + \varphi_k) - |n - \varphi_k|^3 (3n + \varphi_k) \right] = A, \tag{3.143}$$

where A is a parameter that describes a transfer group and is equal to

$$A = c_\tau^* b v_{1k}^2 / (12 a_T S_{ch} \sum \zeta). \tag{3.144}$$

As in the case of pseudo-uniform particle distribution due to the existence of a deceleration area in the laminar motion of a bulk material, φ_k coefficient has a limit. In this case, the limit of φ_{lim} is determined by the equation

$$(1 - \varphi_{\text{lim}})^3 (3 + \varphi_{\text{lim}}) = (\varphi_{\text{lim}} - n)^3 (3n + \varphi_{\text{lim}}).$$

For instance, at $n = 0$, the limit value of φ_{lim} is 0.64; at $n = 0.5$, $\varphi_{\text{lim}} = 0.77$. Determining φ_k under the local exhaust influence presents no principal difficulties. The right-hand side of Equation 3.142 is determined by the addend, that is, the product of the hood vacuum-gauge pressure P_2 and the surface S_{ch}. Then, the dimensionless equation that determines φ_k will appear as follows:

$$\varphi_k^2 = A\left[|1 - \varphi_k|^3 (3 + \varphi_k) - |n - \varphi_k|^3 (3n + \varphi_k) \right] + Eu, \tag{3.145}$$

where Eu is the Euler criterion accounting for the vacuum-gauge pressure P_2 (according to Equation 3.122.

3.1.4 Air Injection in a Bin Chute with a Uniform Distribution of Particles

Another instance when it is possible to use the one-dimensional stream dynamic equations in the evaluation of aerodynamic effects is that of the uniform distribution of settling particles in a bin chute. This situation may be encountered within minus material transfer groups. When passing through a sieve screen, fine particles enter the chute across its section. The distinctive feature here is the air velocity variance.

The latter is changed along the chute length due to the cross-section variation. The variation in cross-sections of pyramid-shaped bin chutes is governed by the quadratic law

$$S = S_k \left[a + (1-a)\frac{x}{l} \right]^2, \tag{3.146}$$

where

$$a = \sqrt{S_0 / S_k},$$

and S_0, S_k are cross-section areas at the chute inlet and outlet respectively, m^2.

Based on the induced air flow equation

$$v_2 = v_{2k} \Big/ \left[a + (1-a)\frac{x}{l} \right]^2,$$

the induction head value

$$P_E = \int_{v_{1H}}^{v_{1k}} \frac{\beta_1}{V_p} \psi^* \cdot f_M \frac{|v_1 - v_2|(v_1 - v_2)}{2} \rho_2 \frac{v_1 dv_1}{g}$$

depends on the quantity of particles involved in the dynamic interaction with air. Two cases are possible here: when all particles impact the air throughout the motion path and when particles interact with the air only at the initial vertical section of sedimentation before they make contact with an angled wall. When an inelastic impact occurs, particles start sliding along the inclined surface without any noticeable interaction with air.

In the first case, we have

$$\beta_1 = G_1/(\rho_1 S v_1) = \left[a + (1-a)\frac{x}{l} \right]^{-2} \cdot G_1/(\rho_1 S_k v_1)$$

and

$$P_E \Big/ \sum \left(\zeta \frac{v_{1k}^2}{2} \rho_2 \right) = Bu_K \cdot Z^*(a, n, \varphi_k), \tag{3.147}$$

where Bu$_k$ is the Butakov–Neikov number for the final section of a bin chute

$$Bu_k = \psi^* k_m G_1 v_{1k} / \left(\sum \zeta S_k \rho_1 g\right), \tag{3.148}$$

Z^* is a coefficient subject to a, $n = v_{1H} / v_{1k}$, $\varphi_k = v_{2k} / v_{1k}$ and equal to

$$Z^* = \int_n^1 \left| z - \frac{\varphi_k}{f(z)} \right| \cdot \left(z - \frac{\varphi_k}{f(z)} \right) \frac{dz}{f(z)};\ f(z) = \left[a + (1-a)\frac{z^2 - n^2}{1 - n^2} \right]^2. \tag{3.149}$$

In the second case,

$$\beta_1 v_1 = \beta_{1\mathrm{H}} v_{1\mathrm{H}} - \text{const}$$

and

$$P_E \Big/ \sum\left(\zeta \frac{v_{1k}^2}{2}\rho_2\right) = \mathrm{Bu_H} \cdot Z, \tag{3.150}$$

where

$$\mathrm{Bu_H} = \psi^* k_m G_1 v_{1k} \Big/ \left(\sum \zeta S_{\mathrm{H}} \rho_1 g\right); Z = \int_n^1 \left| z - \frac{\varphi_k}{f(z)} \right| \left(z - \frac{\varphi_k}{f(z)} \right) dz.$$

The induced air quantity is determined by integrating the dynamic equation

$$\frac{1}{S} d(S\rho_2 v_2^2) = -\lambda \frac{dx}{D} \frac{v_2^2}{2} \rho_2 - dp + dp_E \tag{3.151}$$

if

$$P(0) = P_0 - \zeta_{\mathrm{H}} \frac{v_{2\mathrm{H}}^2}{2} \rho_2; \; P(l) = P_0 + \zeta_k \frac{v_{2k}^2}{2} \rho_2. \tag{3.152}$$

We obtain

$$\sum \zeta^* \frac{v_{2k}^2}{2} \rho_2 = P_E, \tag{3.153}$$

where

$$\sum \zeta^* = (1 - a^{-4}) + \zeta_k + \zeta_{\mathrm{H}} / a^4 + \lambda \frac{l}{D_k} \xi(a, b), \tag{3.154}$$

$$\xi(a,b) = \left[2(a-b)(a^3-1) + 3a(b-1)(a^2-1)\right] \Big/ \left[6a^3(a-1)^2\right], \tag{3.155}$$

$$b = \Pi_0 / \Pi_k, D_k = 4S_k / \Pi_k, \tag{3.156}$$

and Π_0 and Π_k are the chute inlet and outlet section perimeters, m.

Equation 3.153 will appear in a dimensionless form as:

$$\varphi_k^2 / Z^*(a, n, \varphi_k) = G_1 v_{1k} \Bigg/ \left(\sum \zeta^* S_k \frac{c_y^2}{2} \rho_2 \right) \equiv \mathrm{Bu}_k. \tag{3.157}$$

As is clear from the obtained results, φ_k variation is also asymptotic for bin chutes.

The roots of equations

$$Z^*(a, 0, \varphi_k) = 0, \tag{3.158}$$

which are nothing but the limit value of φ_{lim}, are displaced toward higher φ_k than φ_{lim} for prismatic chutes. In addition, there may be two deceleration areas in a bin chute, at the chute inlet and outlet, respectively. This is possible if $\varphi_{lim} > 1$ ($a > 1.6$). The suction area is in the middle section of the chute.

In conclusion, it should be noted that the uniform distribution of particles across a bin chute section (as described earlier) is not as common as the jet-like motion of a stream of particles. Recirculation areas may occur in this case (see Chapter 4).

3.1.5 Air Mechanics of a Stream of Particles with High Bulk Concentrations

Let us consider the air induction with a bulk material stream in the highly constrained conditions of solid particle flow-around when the material bulk concentration is so high that Equation 3.39 is inapplicable.

Let us determine the induced air quantity for the specific case of a vertical transfer, that is, the case of a gradientless particle stream. Prior to assessing the aerodynamic effects in a vertical prismatic chute having a uniform distribution of particles in the cross-section, we will first determine the formula for ψ^* in a wide range of bulk concentrations from a highly dispersed stream (the same situation for which Lyaschenko's formula is correct for a densely packed particle stream). In order to determine the drag factor of particles, we use the empirical formula from Bernstein, Pomerantsev, and Shagalova [7,34]

$$P = \frac{1,53}{(1-\beta)^{4,2}} \cdot \left(\frac{75}{\text{Re}} + \frac{15}{\sqrt{\text{Re}}} + 1 \right) \cdot \frac{H}{d_e} \cdot \frac{\left[u(1-\beta)\right]^2}{2} \rho_2, \tag{3.159}$$

which determines the drag of a dense layer (of H height) of particles of d_E size when blowing air at the filtration rate $u(1-\beta)$. Hereinafter, lower index I of β is omitted for convenience. The Reynolds number is determined by the formula

$$\text{Re} = \frac{0,45}{\beta\sqrt{1-\beta}} \cdot \frac{u(1-\beta)d_e}{v} = 0,45\frac{\sqrt{1-\beta}}{\beta}\frac{ud_e}{v}.$$

On the other hand,

$$P = \frac{H}{V_p} \psi^* f_m \frac{u^2}{2} \rho_2 = \frac{1,5}{d_E} H \psi^* \frac{u^2}{2} \rho_2. \tag{3.160}$$

Thus, for a particle in a dense layer

$$\psi^* = \frac{1}{\beta(1-\beta)^{2,2}} \left(\frac{75}{\text{Re}} + \frac{15}{\sqrt{\text{Re}}} + 1 \right) \tag{3.161}$$

or

$$\psi^* / \psi_0 = \frac{1}{\psi_0 \beta(1-\beta)^{2,2}} \left(\frac{75}{\text{Re}} + \frac{15}{\sqrt{\text{Re}}} + 1 \right). \tag{3.162}$$

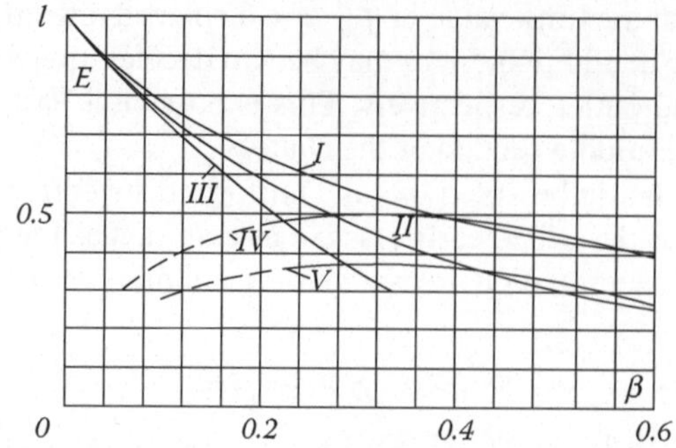

FIGURE 3.15 Dependence of coefficient E on the volume concentration of particles where they are uniformly distributed in the cross-section of prismatic chute. For sharp-grained particles: (I) according to Equation 3.164; (IV) according to Equation 3.163; for rounded particles: (II) according to Equation 3.164; (V) according to Equation 3.163; (III) Equation 2.44, according to P. V. Lyashchenko.

In the self-similarity area

$$E \equiv \sqrt{\psi_0/\psi^*} = \sqrt{\psi_0 \beta}(1-\beta)^{1,1}. \tag{3.163}$$

The linear fractional function of the form

$$E = (1 + a\beta)/(1 + b\beta) \tag{3.164}$$

smoothes the extreme areas (Figure 3.15). For sharp-grained particles, in particular, ($\psi_0 = 1.8$) $a = \frac{1}{3}$, $b = (3 + a) = \frac{10}{3}$; for rounded particles ($\psi_0 = 1$) $a = -\frac{2}{3}$, $b = \frac{7}{3}$.

Taking Equation 3.164 into account and ignoring the aerodynamic drag of the chute walls, Equation 3.38 can be written as

$$(1-\beta)v_2\rho_2 dv_2 = -(1-\beta)dP + \psi_0\left(\frac{1+b\beta}{1+a\beta}\right)^2 \beta k_m \frac{|v_1 - v_2|(v_1 - v_2)}{2}\rho_2 dx$$

or

$$\frac{\rho_2}{2} dv_2^2 = -dP + \psi_0\left(\frac{1+b\beta}{1+a\beta}\right)^2 \frac{\beta k_m}{1-\beta}\cdot\frac{|v_1 - v_2|(v_1 - v_2)}{2}\rho_2 dx \tag{3.165}$$

for the isothermal flow.

It is easy to solve Equation 3.165 for uniform bulk material motion in the chute ($v_1 = const$), for example, in case of the bound mode of motion. In this case, the volume concentration of material is not changed along the chute height, and the air flow has a constant rate ($v_2 = const$). To determine this, we will assume that $v_1 - v_2 = u > 0$.

Integrating Equation 3.165, we obtain the following

$$P_k - P_H = P_E \equiv \psi_0 \frac{\beta}{1-\beta} \cdot \left(\frac{1+b\beta}{1+a\beta}\right)^2 k_m \frac{u^2}{2} \rho_2 H, \tag{3.166}$$

which, assuming the pressure at the beginning and end of the chute is equal,

$$P_{\text{H}} = P_a - P_1 - \zeta_{\text{H}} \frac{v_2^2}{2} \rho_2; P_{\text{k}} = P_a - P_2 + \zeta_k \frac{v_2^2}{2} \rho_2, \tag{3.167}$$

can be rewritten in a dimensionless form

$$\left(\frac{\varphi}{1-\beta}\right)^2 = N + \beta \frac{B}{\sum \zeta} \left(1 - \frac{\varphi}{1-\beta}\right)^2, \tag{3.168}$$

where P_1, P_2 are rarefactions in the hoods adjacent to the upper and lower portions of the chute, respectively, Pa; H is the chute height, m;

$$\varphi = w / v_1;\ w = v_2(1-\beta) = Q_E / S_{ch}; \sum \zeta = \zeta_{\text{H}} + \zeta_{\text{k}} \tag{3.169}$$

$$N = \frac{P_2 - P_1}{\sum \zeta \frac{v_1^2}{2} \rho_2}; B = \psi_0 \frac{k_m H}{1-\beta} \cdot \left(\frac{1+b\beta}{1+a\beta}\right)^2. \tag{3.170}$$

In case of a uniformly accelerated particle flow, the value of the induction pressure, taking into account the volume concentration of the components,

$$\beta = \frac{G_1}{v_1 \rho_1 S_{ch}}; 1 - \beta = \frac{Q_2}{v_2 S_{ch}};$$

$$v_2 = w \frac{v_1}{v_1 - G_1 / (\rho_1 S_{ch})}, \tag{3.171}$$

can be represented as the following correlation

$$\overline{P}_E \equiv \frac{2 a_T \rho_1 S_{ch} P_E}{\psi_0 k_m G_1 v_{1k}^3 \rho_2} = \left[|1-\varphi|^3 - |n-\varphi|^3\right] K(n, \varphi, \beta_k) / 3, \tag{3.172}$$

$$\varphi = \frac{w}{v_1 - G_1 / (\rho_1 S_{ch})}. \tag{3.173}$$

Here, $K(n, \varphi, \beta_k)$ is a correction coefficient

$$K = \frac{3}{|1-\varphi|^3 - |n-\varphi|^3} \int_n^1 \left(\frac{z + b\beta_k}{z + a\beta_k}\right)^2 \cdot \left(\frac{z}{z - \beta_k}\right)^2 |z - \beta_k - \varphi| (z - \beta_k - \varphi) dz, \tag{3.174}$$

taking into account the constraint. According to the calculations in the area $\beta_k < 0.005$, the value K is almost equal to 1 (with the exception of a small area $\frac{1+n}{2} - 0,1 < \varphi < \frac{1+n}{2} + 0,1$, where the value $\overline{P}_E$ is small in itself), and the effect of concentration on the value of P_E can be neglected.

Equation 3.165, after integration under the conditions

$$P_{\text{k}} = P_a - P_2 + \zeta_k v_{2k}^2 \rho_2 /2, \tag{3.175}$$

$$P_{\text{H}} = P_a - P_1 + \zeta_{\text{H}} v_{2k}^2 \rho_2 /2 \tag{3.176}$$

takes the following form

$$\sum \zeta w^2 \rho_2 / 2 = P_E + P_2 - P_1, \tag{3.177}$$

where $\Sigma\zeta^*$ is the sum of the coefficients of the chute local resistances, which is equal to

$$\sum \zeta = \frac{\zeta_k + 1}{(1-\beta_k)^2} - \frac{1-\zeta_{\text{H}}}{(1-\beta_k / n)^2}. \tag{3.178}$$

Dividing both sides of Equation 3.177 by the value $\sum \zeta v_{1k}^2 \rho_2 / 2$ gives the following dimensionless ratio

$$\varphi^2 = \frac{A}{\sum \zeta} \beta_k \frac{|1-\varphi|^3 - |n-\varphi|^3}{3} K(\varphi, n, \beta) + N, \tag{3.179}$$

which defines the value of the coefficient of φ. Here

$$A = \psi_0 \frac{k_m v_{1k}^2}{g}, N = \frac{P_2 - P_1}{\sum \zeta v_{1k}^2 \rho_2 / 2}. \tag{3.180}$$

The analysis of changes in the coefficient φ with the increase of the volume concentration allows us to distinguish two characteristic areas. (In Figure 3.16, graphs of Equation 3.179 are plotted with $N = 0$, $\zeta_{\text{H}} = 1.0$; $A = 11500$.) In the area β_k <0.005–0.05, a sharp increase in the coefficient φ is seen with an increase in the volume concentration, the constraint having almost no effect. With β_k >0.005–0.05, the influence of the constraint is obvious, and the coefficient φ decreases.

3.2 EFFECT OF HEAT AND MASS EXCHANGE

The inter-components heat and mass exchange play a double role. On the one hand, an additional force—thermal pressure caused by buoyant forces—occurs in the chute. On the other hand, the mass exchange results in an additional source or outlet of the gaseous component.

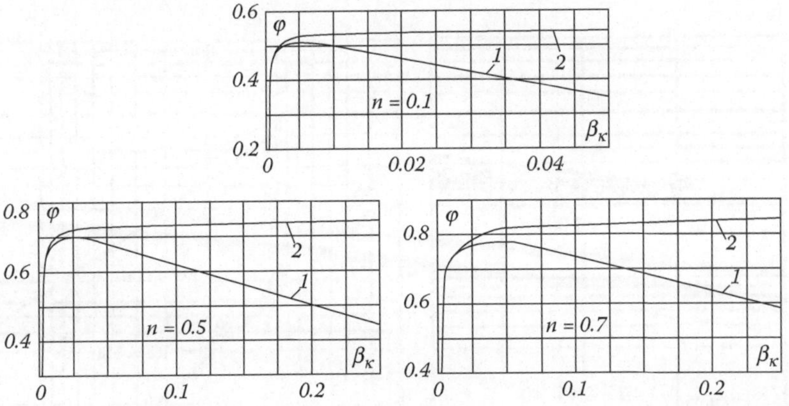

FIGURE 3.16 Change in the coefficient φ with an increase in the volume concentration in the chute: (1) taking into account the constraint of the particles (K was defined according to Equation 3.174); (2) without considering the constraint, K = 1.

3.2.1 Inter-Component Heat Exchange in an Inclined Chute

Heat exchange, as well as the force interaction between the components, is defined by the flow pattern of the particles and by the nature of their movement in the chute. An experimental study of heat exchange was carried out by using a unit to examine the inducing properties of an unheated particle flow (Figure 3.2). The value of the heat flow from the particles to the air was determined using the enthalpy method:

$$Q = c_2 G_2 (t_k - t_{\text{H}}), W, \tag{3.181}$$

where c_2 is the air heat capacity, J/kg·°; G_2 is the air mass flow rate, kg/s; and t_{H}, t_k are air inlet and outlet temperature, at the chute inlet and at its outlet, respectively (°C). The chute walls were heat sealed to prevent heat exchange with surrounding air.

Research was conducted with crushed granite (mono fraction of 1.25–2.5 mm) and iron ore (poly fraction with d_{av} – 2.5 mm, the grain composition of which is shown in Table 3.2). After it was heated up to 200–300°C, the material was transferred through a heat-sealed chute with a 0.15 × 0.15-m section at Θ = 45°, 60°, 75°. As is shown by experimental studies, the rate of heat exchange varies with the relative velocity of the particles (Figure 3.17a) and with their volume concentration (Figure 3.17b), which is consistent with a generalization about heat exchange in dispersed through flows proposed by Gorbis [24]. The established behavior of the inter-component heat exchange for an accelerated fall of particles was also confirmed by later experiments performed by Semenov [83], who studied the heat exchange between falling 10.5-mm steel balls and the air in a vertical chute with a 0.14 × 0.14 m section.

Quantitatively, however, the heat exchange in inclined chutes is significantly different from free gas suspension flows and heat exchange in a vertical chute. Here, almost every particle participates in heat exchange, and its rate is much higher than in case of particles moving in an inclined chute, where most move near the bottom

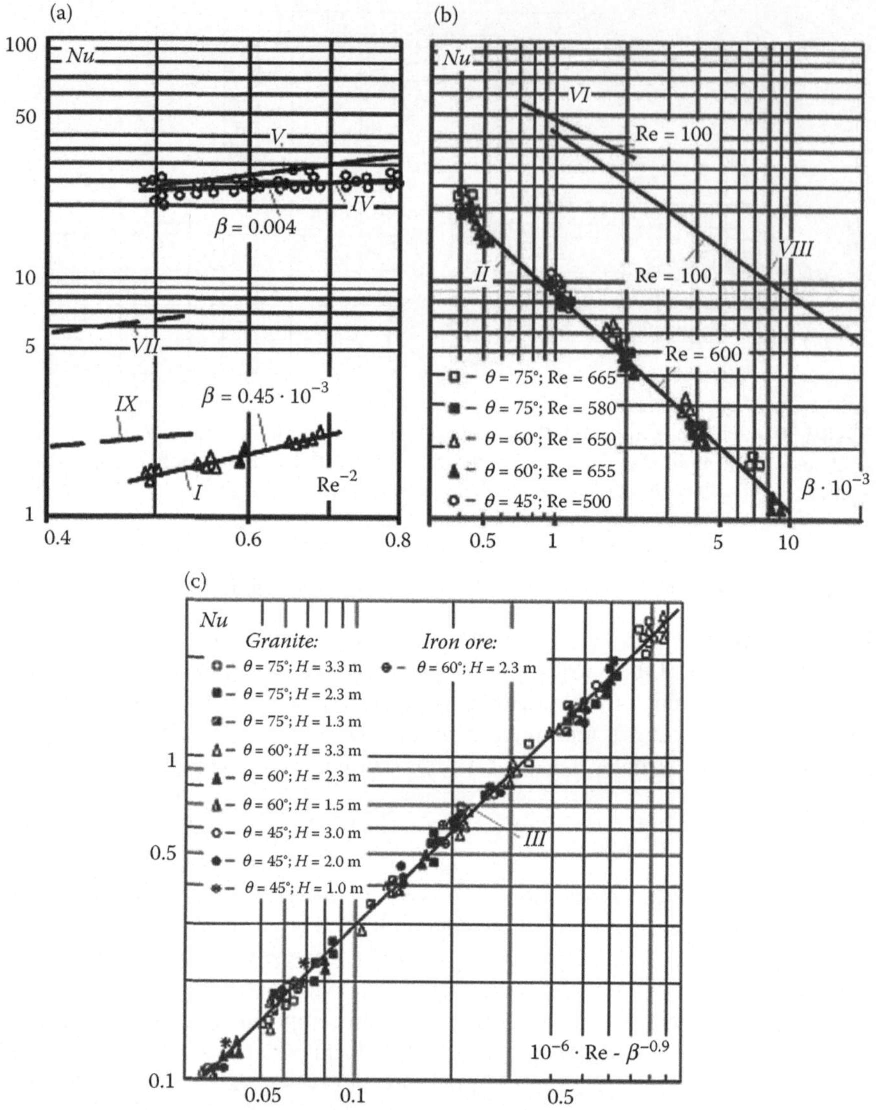

FIGURE 3.17 Influence of the number Re and volume concentration on the inter-component heat exchange in the fall of crushed granite particles in an inclined chute (I, II, III) and of steel balls in a vertical chute (IV—according to A. S. Semenov), in the flow of free (V, VI—according to Z. R. Gorbis) and stagnant gas suspension (VII, VIII—according to Z. R. Gorbis; IX—according to Morozov).

(in constraint conditions). Thus, in our case, we can speak of a conditional (apparent) heat exchange coefficient.

Here, the heat exchange process is analogous to that of a mechanically stagnated gas suspension, where we can see "dead" zones—areas of weak interaction with the air—in the flow of particles on the braking elements of mines. This can explain the near values of the Nusselt number (curves I, IX [64,24]), as well as the coincident behavior of the heat exchange with the increase of the volume concentration (slope angle of lines II and VIII).

As a result of the statistical processing of the experimental data in the range $0.0002 < \beta < 0.01$; $400 < \mathrm{Re} < 700$, the following correlation was obtained [47]:

$$Nu = 2{,}95 \cdot 10^{-6}\,\mathrm{Re} \cdot \beta^{-0{,}9}, \tag{3.182}$$

which allows us to define the inter-component heat exchange in an inclined chute. Here, the Nusselt and Reynolds numbers are expressed in terms of the average particle diameter and the average relative velocity along the chute length:

$$Nu = \frac{\alpha d}{\lambda}; \mathrm{Re} = \frac{(v_{1av} - v_2)d}{\nu}; \; v_{1av} = \frac{v_{1H} + v_{1k}}{2},$$

where α is the heat exchange coefficient, W/m^2·°; and λ is the air thermal conductivity, W/m·°.

3.2.2 Thermal Head

As a result of the heat exchange, the air density in the chute is different from the density of the surrounding air, and its unit volume is influenced by Archimedes' buoyant force. Equation 3.8, for a prismatic chute, is as follows (we suppose $\bar{v}_2 \approx const$, $\bar{\beta}_2 \approx 1$):

$$dp = -(\rho_0 - \rho_2)g_x dx - \lambda \frac{dx}{D} \cdot \frac{|v_2| v_2}{2} \rho_2 + dP_E. \tag{3.183}$$

Calculate the value

$$P_T = \int_0^l (\rho_0 - \rho_2) g_x dx; \tag{3.184}$$

this is usually called the thermal head, and the value is expressed in terms of the chute height and the averaged density of the air

$$P_T = (\rho_0 - \bar{\rho}_2) gH, \tag{3.185}$$

where

$$\bar{\rho}_2 = \frac{1}{l} \int_0^l \rho_2 dx. \tag{3.186}$$

Next, we will open the symbol of averaging and express the air density in terms of temperature. We use the thermal expansion coefficient β_T defined by the equation

$$\beta_T = \frac{1}{\rho_2}\left(\frac{\partial \rho_2}{\partial T}\right)_P, \tag{3.187}$$

to obtain

$$\rho_2 = \rho_0 \exp\left[\beta_T (T_0 - T_2)\right], \tag{3.188}$$

where T_0, ρ_0 are the temperature (°K) and the density of the surrounding air (kg/m^3); and T_2, ρ_2 are the temperature (°K) and density (kg/m^3) of the air in the chute.

To determine the temperature T_2, we use the heat-transfer equation (Equation 92) and the expression for the inter-component heat exchange (Equation 95) from the Appendix. Assuming that the process is stable, and ignoring pulse moments, this equation for a one-dimensional problem is as follows:

$$d(c_2\rho_2 T_2 v_2 S_{ch}) = \frac{\beta_1}{V_p} S_p \alpha (T_1 - T_2) S_{ch} dx. \tag{3.189}$$

In addition to these assumptions, we also assume that the temperature of the material is constant along the chute. The material is not cooled in cases of relatively low transfer heights due to a short stay in the chute (about 1 sec). Field measurements (Table 3.3) showed that the relative cooling does not exceed the accuracy of measurements and varies in the range of 1–3%.

In addition, we average the volume concentration of the material, assuming that

$$\beta_1 \approx \beta = \frac{G_1}{\rho_1 S_{ch} v_{1av}}. \tag{3.190}$$

We then integrate Equation 3.189, taking into account accepted simplifications under the condition that $T_2 = T_0$ at the beginning of the chute (at $x = 0$) to obtain

$$T_2 = T_1 - (T_1 - T_0)\exp\left(-\frac{x}{l} W_\alpha\right), \tag{3.191}$$

where

$$W_\alpha = \beta \frac{S_p}{V_p} \alpha S_{ch} l \Big/ (c_2 \rho_2 v_2 S_{ch}). \tag{3.192}$$

The expression for the air density in the chute is

$$\rho_2 = \rho_0 \exp\left[-\beta_T (T_1 - T_2)\left(1 - e^{-\frac{x}{l}W_\alpha}\right)\right], \tag{3.193}$$

and the averaged density value is

$$\bar{\rho}_2 = \rho_0 \exp\left[\beta_T (T_1 - T_0)\right] \cdot \left\{ Ei\left[\beta_T (T_1 - T_0)\right] - Ei\left[\beta_T (T_1 - T_0) e^{-W_\alpha}\right]\right\} \Big/ W_\alpha, \tag{3.194}$$

where $Ei(f)$ is an exponential integral function with the argument f.

Inserting this result for $\bar{\rho}_2$ into Equation 3.185, we obtain

$$P_T = \left(\rho_0 - \frac{\rho_0 + \rho_{2k}}{2}\Pi\right) gH, \tag{3.195}$$

TABLE 3.3
Change in Material Temperature and Chute Vapor–Air Mixture in Transfers of Heated Wet Materials

Transfer Group Name	Flow of Material G_1, kg/s	Drop Height H, m	Chute Cross-Sectional Area S_{ch}, m²	Temperature of Material, °C		Temperature of Vapor–Air Mixture, °C		
				t_{1H}	t_{1k}	t_{2k}	t_{2k}	t_0
Transfer of burnt ore material from the drum cooler to the conveyor belt	10	5.5	0.2	78	77	25	50	7
Transfer of burnt ore material from a conveyor to a conveyor	200	6.0	0.4	65	63	20	45	12
Transfer of burnt ore material from a conveyor through an intermediate bin to another conveyor	200	12.0	0.4	62	60	20	40	10
Transfer of iron-ore pellets from the drum cooler to a conveyor	8	3.0	0.2	73	70	30	35	6
Transfer of iron-ore pellets from a conveyor to another conveyor	30	3.5	0.8	70	68	30	50	10

where Π is a correction coefficient equal to

$$\Pi = 2e^{-A}\left[Ei(A) - Ei(Ae^{-W_\alpha})\right]/\left[W_\alpha \cdot (1+\rho_{2k}/\rho_0)\right], \tag{3.196}$$

$$A = (\ln \rho_{2k}/\rho_0)/(1-e^{-W_\alpha}). \tag{3.197}$$

In the area $w_\alpha < 1$; $0.6 < \rho_{2k}/\rho_0 < 1$, the coefficient Π is almost equal to 1, and the value of the averaged air density in the chute is equal to the arithmetic mean value [46]. In the general case, the thermal head equals

$$P_T = \left(\rho_0 - \frac{\rho_{2\mathrm{H}} + \rho_{2k}}{2}\right) gH. \tag{3.198}$$

Here, ρ_{2k} is the air density at the end of the chute at T_{2k} calculated taking into account the correlation obtained for the inter-component heat exchange (3.182) according to

$$T_{2k} = T_1 - (T_1 - T_{2\mathrm{H}})e^{-W_\alpha}. \tag{3.199}$$

3.2.3 Air Velocity in the Chute

We integrate Equation 3.183 to obtain

$$P_k - P_{\mathrm{H}} = -P_T - \lambda \frac{l}{D} \frac{|v_2| v_2}{2} \rho_2 + P_E \tag{3.200}$$

or, expressing the pressure at the beginning and end of the chute in terms of the coefficients of local resistances

$$P_k = P_0 + \zeta_k \frac{|v_2| v_2}{2} \rho_2, P_{\mathrm{H}} = P_0 - \zeta_{\mathrm{H}} \frac{|v_2| v_2}{2} \rho_2, \tag{3.201}$$

Equation 3.200 is as follows:

$$\sum \zeta \frac{|v_2| v_2}{2} \rho_2 = P_E - P_T. \tag{3.202}$$

This shows that the difference between the induction head and the thermal head determines the air flow and the direction of air flow in the chute, when transferring heated material. Three cases are possible in this regard.

Case 1: $P_E > P_T$. The air moves downward (forward flow). The value of the thermal head acts as an additional resistance. The volume of induced air is defined by an obvious equality:

$$Q_E = \sqrt{(P_E - P_T)/R_{ch}}. \tag{3.203}$$

Case 2: $P_E < P_T$. Air moves toward the falling material (counter flow) under the prevailing thermal head. The induction head only slows the movement:

$$Q_{ch} = \sqrt{(P_T - P_E) / R_{ch}}. \tag{3.204}$$

However, it should be noted that the sum of the coefficients of local resistances generally will not be equal to a similar amount in case of forward flow.

Case 3: $P_E = P_T$. There is no direction of air movement in the chute. Only local aerodynamically unstable air circulation can occur in this case. Consider in detail the condition of the aerodynamic instability. Designating the temperature in the chute as T_{2av} (note that in the limiting case $T_{2av} \rightarrow T_1$), the air density (according to Equation 3.188) is

$$\rho_2 = \rho_0 \exp\left(\beta_T (T_0 - T_{2av})\right) \tag{3.205}$$

or, considering that in most practical cases $(T_2 - T_0)\,\beta_T \ll 1$,

$$\rho_2 = \rho_0 \left[1 - (T_{2av} - T_0)\beta_T\right]. \tag{3.206}$$

In this case, the value of the thermal head is

$$P_T = gH\rho_0 (T_{2av} - T_0)\beta_T. \tag{3.207}$$

According to Equation 3.63, the value of the induction head is

$$P_E = k_m \psi^* \frac{\rho_2}{\rho_1} G_1 \left(v_{1k}^3 - v_{1H}^3\right) \big/ (6a_T S_{ch}). \tag{3.208}$$

Therefore, the equality of these heads takes the form of the following criterial equation

$$(1 - n^3) \cdot \mathrm{Re}_k^2 / (6Eu_0^*) = Gr, \tag{3.209}$$

where Gr is Grashof number, which characterizes the ascensional forces and equals

$$Gr = \beta_T \frac{gH^3}{\nu^2} (T_{2av} - T_0), \tag{3.210}$$

Re_k is Reynolds number, which characterizes the kinetic capacity of the particle flow at the end of the chute and equals

$$\mathrm{Re}_k = v_{1k} H / \nu, \tag{3.211}$$

and Eu_0^*– is the modified Euler criterion, which characterizes the aerodynamic drag strength of the particles and equals

$$Eu_0^* = S_{ch} \frac{C_y^{*2}}{2} \rho_0 \Big/ (G_1 v_{1k}). \tag{3.212}$$

The balance of the forces described by critical Equation 3.209 has been confirmed during an experiment involving the transfer of heated crushed granite (Figure 3.18).

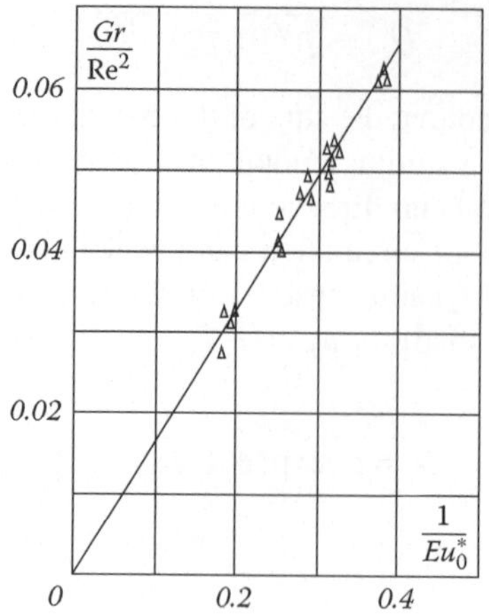

FIGURE 3.18 The balance of the forces of induction and thermal pressures in the chute when transferring heated granite (d_{av} = 1.88 mm, n ≈ 0).

Thus, in the general case, the amount of air being moved along the chute (with the local suction units operating) is equal to

$$Q_{ch} = \frac{P_E - P_T + P_2 - P_1}{\sqrt{\left|P_E - P_T + P_2 - P_1\right| \cdot R_{ch}}}, \tag{3.213}$$

where P_1, P_2 are rarefactions occurring with the local suction units in the upper hood and in the lower one (adjacent to the upper and lower ends of the chute, respectively), Pa.

Or, in a dimensionless form,

$$\varphi_k \left|\varphi_k\right| = Bu \cdot \left[\left|1 - \varphi_k\right|^3 - \left|n - \varphi_k\right|^3\right] / 3 - Eu_T, \tag{3.214}$$

where

$$Eu_T = (P_T - P_2 + P_1) \Big/ \left(\sum \zeta \frac{v_{1k}^2}{2} \rho_2 \right). \tag{3.215}$$

The minus symbol before the value φ_k (or Q_{ch}) denotes the instance of a counter flow, an instance of balance occurs with

$$(1 - n^3) Bu = 3 Eu_T. \tag{3.216}$$

3.2.4 Influence of Mass Exchange on the Volumes of Induced Air

Consider the movement of the heated wet material accompanied by moisture evaporation from the surface of falling particles. Equations of mass exchange for a one-dimensional problem will be as follows:

$$\frac{d}{dx}\beta_1\rho_1 v_1 S_{ch} = -J\cdot S_{ch};\ \frac{d}{dx}\beta_2\rho_2 v_2 S_{ch} = J\cdot S_{ch}, \tag{3.217}$$

where J is the volumetric evaporation rate, kg/(s·m³). Respectively, for an impulse

$$\frac{d}{dx}\beta_1\rho_1 v_1 v_1 S_{ch} = S_{ch}\beta_1 a_T\rho_1 - S_{ch}\beta_1 k_m\psi^* \frac{|v_1 - v_2|(v_1 - v_2)}{2}\rho_2 - Jv_1 S_{ch}, \tag{3.218}$$

$$\frac{d}{dx}\beta_2\rho_2 v_2 v_2 S_{ch} = S_{ch}\beta_2 g_x(\rho_2 - \rho_0) + S_{ch}\beta_1 k_m\psi^* \frac{|v_1 - v_2|(v_1 - v_2)}{2}\rho_2 -$$

$$-\frac{d}{dx}\beta_2 P_2 S_{ch} - \lambda\frac{S_{ch}\beta_2}{D}\frac{v_2^2}{2}\rho_2 + Jv_1 S_{ch}. \tag{3.219}$$

Assuming that the volume concentration of the material is small ($\beta_1 \ll 1$; $\beta_2 \approx 1$), the last correlation with Equation 3.217 can be written as (S_{ch} - const):

$$\frac{dp}{dx} = g_x(\rho_2 - \rho_0) + \beta_1 k_m\psi^* \frac{|v_1 - v_2|(v_1 - v_2)}{2}\rho_2 - \frac{\lambda}{D}\frac{v_2^2}{2}\rho_2 + J(v_1 - v_2). \tag{3.220}$$

The mass transport equation of the gaseous component is expressed in terms of the moisture content (m) and the flow rate of dry air (G_b)

$$J = \frac{G_b}{S_{ch}}\frac{dm}{dx}, \tag{3.221}$$

obvious equations of flow

$$\rho_2 v_2 S_{ch} = G_b(1+m), \tag{3.222}$$

and the averaged values of the velocities of the components

$$\bar{v}_1 = \frac{1}{l}\int_0^l v_1 dx = v_{1k}\cdot\frac{2}{3}\cdot\frac{1-n^3}{1-n^2}, \tag{3.223}$$

$$\bar{v}_2 = \frac{1}{l}\int_0^l v_2 dx \approx G_b(1+m_{av})/(\bar{\rho}_2 S_{ch}). \tag{3.224}$$

Assuming that the densities and velocities of the components on the right side of Equation 3.220 are averaged, after integration, on the condition that

$$P(0) = -P_1 - \zeta_{\text{H}}\frac{v_{2\text{H}}^2}{2}\rho_{2\text{H}};\ P(l) = -P_2 + \zeta_k\frac{v_{2k}^2}{2}\rho_{2k}, \tag{3.225}$$

the following equation is obtained

$$\sum\zeta^*\frac{\bar{v}_2^2}{2}\rho_2 = -P_T + P_E + P_2 - P_1 + P_J, \tag{3.226}$$

where P_J is the pressure force that occurs due to moisture evaporation from the falling particles (for brevity, we call this value the interphase pressure), which is equal to

$$P_J = \int_0^l J(v_1 - v_2)dx \approx G_1(m_k - m_{\text{H}})(\overline{v}_1 - \overline{v}_2) / S_{ch}, \tag{3.227}$$

and $\Sigma\zeta^*$ is the sum of the coefficients of local resistance, which is equal to

$$\sum \zeta^* = \zeta_{\text{H}} \left(\frac{1+m_{\text{H}}}{1+m_{av}} \right)^2 \frac{\overline{\rho}_2}{\rho_{2\text{H}}} + \zeta_k \left(\frac{1+m_k}{1+m_{av}} \right)^2 \frac{\overline{\rho}_2}{\rho_{2k}} + \lambda \frac{l}{D}. \tag{3.228}$$

In the dimensionless form, the equation can be written as

$$\overline{\varphi}^2 = Bu\left[|1-\overline{\varphi}|^3 - |n-\overline{\varphi}|^3 \right] / 3 - Eu_T + Eu_J \left(\frac{2}{3}\frac{1-n^3}{1-n^2} - \overline{\varphi} \right), \tag{3.229}$$

where Bu, Eu_T are numbers defined by correlations (3.109) and (3.215), and at $\Sigma\zeta \to \Sigma\zeta'$ and $\rho_2 \to \overline{\rho}_2$

$$Eu_J = G_b(m_k - m_{\text{H}}) \cdot v_{1k} \Big/ \left(S_{ch} \sum \zeta^* \frac{v_{1k}^2}{2} \overline{\rho}_2 \right). \tag{3.230}$$

From this, the mass flow rate of the non-condensing (dry) part of the air can be obtained

$$G_b = \overline{\varphi} \cdot v_{1k} S_{ch} \frac{\overline{\rho}_2}{1+m_{av}}. \tag{3.231}$$

Then, the amount of vapor–air mixture transferring from the chute to the lower cavity (hood) can be obtained based on the equality

$$G_{2k} = G_b(1+m_k) = \overline{\varphi} v_{1k} S_{ch} \overline{\rho}_2 \frac{1+m_k}{1+m_{av}}. \tag{3.232}$$

Thus, the amount of induced air during transfers of wet materials is increased not only due to the water vapors resulting from evaporation but also due to additional forces of the interphase pressure.

3.3 AERODYNAMICS OF AN UNSTEADY PARTICLE FLOW IN THE CHUTE

Unsteady processes occur in the chute at the equipment start-up or at the short-time loading of bulk material. We can assess the force action exerted by the flow on air for two cases: at a sudden change in the material flow and at a gradual one (smooth).

3.3.1 Sudden Change in the Material Flow

Examine the change in the forces of the induction and in thermal pressures and in the starting and stopping times of the material feeding.

Observe the change in the induction pressure using the example of the pressure distribution along the vertical pipe length of an irregular load of moderate temperature bulk material, thereby preventing the heat and mass exchange. Imagine that the lower end of the pipe is closed to the air passage (i.e., $v_2 = 0$) but open to the passage of material.

With these simplifications, on the basis of Equation 41 from the Appendix,

$$\frac{\partial v_2}{\partial t} = -\frac{\partial P}{\partial x} \cdot \frac{1}{\rho_2} + k_m \beta_1 \psi \frac{v_1^2}{2}. \tag{3.233}$$

Examine a "hard" loading of material (instant inlet of material). Then, at an arbitrary point of time $t > 0$, the change in the material flow along the length will be of a step-type shape.

$$G_1 = G_{\max} \quad \text{at} \quad x < x_0,$$

$$G_1 = 0 \quad \text{at} \quad x > x_0.$$

The kink will be moved down. Let us assume that its movement speed is

$$\overline{v}_1 = \frac{1}{l}\int_0^l v_1 dx \approx 0{,}5(v_{1H} + v_{1k}).$$

The analytical expression of the change in the material flow is as follows:

$$G_1 = G_{\max} \cdot f(x - v_1 t), \tag{3.234}$$

where

$$f(x - v_1 t) = 0 \quad \text{at} \quad x - v_1 t > 0; \tag{3.235}$$

$$f(x - v_1 t) = 1 \quad \text{at} \quad x - v_1 t < 0. \tag{3.236}$$

For simplicity, the symbol of averaging (a line above v_1) is omitted here and hereinafter.

Express the function $f(x - v_1 t)$ by means of the infinite Fourier series [95]

$$f(x - v_1 t) = \frac{1}{2} + \sum_{n=1}^{\infty} \frac{\cos \pi n - 1}{n\pi} \sin \frac{n\pi}{l_\infty}(x - v_1 t), \tag{3.237}$$

where

$$l_\infty = \lim_{t\to\infty}(x - v_1 t). \tag{3.238}$$

Then, the volume concentration of material is

$$\beta_1 = \beta_{1\infty}\left[\frac{1}{2} + \sum_{n=1}^{\infty} \frac{\cos n\pi - 1}{n\pi} \sin n \frac{\pi}{l_\infty}(x - v_1 t)\right], \tag{3.239}$$

where

$$\beta_{1\infty} = \frac{G_{1\max}}{\rho_1 S_{ch} v_1} \tag{3.240}$$

or, solving the sine function of difference of two angles,

$$\beta_1 = \beta_{1\infty}\left[\frac{1}{2} + \sum_{n=1}^{\infty}\frac{\cos n\pi - 1}{n\pi}\sin n\frac{\pi}{l_\infty}x\cos n\frac{\pi}{l_\infty}tv_1 - \sum_{n=1}^{\infty}\frac{\cos n\pi - 1}{n\pi}\cos n\frac{\pi}{l_\infty}x\sin n\frac{\pi}{l_\infty}v_1 t)\right]. \tag{3.241}$$

Using Equation 40 (mass transfer) of the Appendix for the case under consideration,

$$\frac{\partial\rho_2}{\partial t} + \frac{\partial\rho_2 v_2}{\partial x} = 0. \tag{3.242}$$

Assuming that that process is adiabatic (with factor χ)

$$\frac{P}{P_0} = \left(\frac{\rho_2}{\rho_0}\right)^{\chi}, \tag{3.243}$$

if simple transformations are made, taking into account that the density of the medium does not change significantly, Equation 3.242 can be reduced to

$$\frac{\partial v_2}{\partial x} = -\frac{\rho_0}{\bar{\rho}_2 P_0 \chi}\frac{\partial P}{\partial t}, \tag{3.244}$$

where P_0, ρ_0 are the pressure and density of air in the pipe before inputting the material.

Taking the correlation expressed in Equation 3.244 into account, after differentiation of Equation 3.233 with respect to x, we obtain an inhomogeneous equation of acoustics:

$$\frac{\partial^2 P}{\partial t^2} = v_a^2\frac{\partial^2 P}{\partial x^2} - v_a^2\frac{\partial}{\partial x}k_m\beta_1\psi\frac{v_1^2}{2}\bar{\rho}_2, \tag{3.245}$$

where $v_a = \sqrt{\chi P_0/\rho_0}$ is the propagation speed of elastic disturbances (speed of sound), m/sec.

Assume that the force of the dynamic interaction is constant

$$k_m\psi\frac{v_1^2}{2}\bar{\rho}_2 = const. \tag{3.246}$$

Under this assumption, a linear pressure distribution along the pipe would exist for stationary conditions

$$P_{st} = P_0 + k_m\beta_{1\infty}\psi\frac{v_1^2}{2}\bar{\rho}_2 x \tag{3.247}$$

or

$$P_\infty = P_l \frac{x}{l}, \tag{3.248}$$

where P_l is the excessive pressure at the end of the pipe

$$P_l = k_m \beta_{1\infty} \psi \frac{v_1^2}{2} \bar{\rho}_2 l, \tag{3.249}$$

P_{st}, P_∞ are the absolute and excessive pressures, respectively, in a stable process, Pa.

Taking into account the assumptions made and these designations, the last term of the right-hand side of Equation 3.245 can be written as

$$v_a^2 \frac{\partial}{\partial x} k_m \beta_1 \psi \frac{v_1^2}{2} \bar{\rho}_2 = v_a^2 \frac{P_l}{l l_\infty} \left[\sum_{n=1}^{\infty} (\cos n\pi - 1) \cdot \cos n\pi \frac{x}{l_\infty} \cos n\pi \frac{v_1 t}{l_\infty} + \sum_{n=1}^{\infty} (\cos n\pi - 1) \sin n\pi \frac{x}{l_\infty} \cdot \sin n\pi \frac{v_1 t}{l_\infty} \right]. \tag{3.250}$$

To formulate the initial and boundary conditions, pressure P in Equation 3.245 means excessive pressure. Because dead air was in the pipe before the input of material, and the absolute pressure in it was P_0:

$$P|_{t=0} = 0; \tag{3.251}$$

$$v_2|_{t=0} = 0. \tag{3.252}$$

Integrate Equation 3.244 to obtain

$$v_2 = v_2|_{x=0} - \frac{\rho_0}{P_0 \chi} \int_0^x \frac{\partial P}{\partial t} dx, \tag{3.253}$$

from which, considering Equation 3.252, we obtain the second initial condition

$$\left. \frac{\partial P}{\partial t} \right|_{t=0} = 0. \tag{3.254}$$

The boundary condition for the open end of the pipe (input) will be

$$P(0,t) = 0. \tag{3.255}$$

For the lower end, taking in account the air tightness of the pipe bottom,

$$v_2(l,t) = 0. \tag{3.256}$$

We integrate Equation 3.233 to obtain

$$v_2 = v_2(x,0) + \int_0^t \left(-\frac{\partial P}{\partial x} \cdot \frac{1}{\rho_2} + k_m \beta_1 \psi \frac{v_1^2}{2} \right) dt. \tag{3.257}$$

Taking into account Equation 3.256, we obtain the second boundary condition:

$$\left.\frac{\partial P}{\partial x}\right|_{x=l} = k_m \beta_1 \psi \bar{\rho}_2 \left.\frac{v_1^2}{2}\right|_{x=l}, \tag{3.258}$$

which can be written as follows (keeping Equation 3.239 and 3.249 in mind) after some transformations:

$$\left.\frac{\partial p}{\partial x}\right|_{x=l} = 2\frac{P_l}{l}\sum_{n=1}^{\infty}\frac{1-\cos n\pi}{n\pi}\cdot\cos^2 n\pi\frac{l}{2l_\infty}\sin n\pi\frac{v_1 t}{l_\infty} + \frac{P_l}{l}\sum_{n=1}^{\infty}\frac{\cos n\pi - 1}{n\pi}\sin n\pi\frac{l}{l_\infty}\cos n\pi\frac{v_1 t}{l_\infty}. \tag{3.259}$$

Thus, we have to solve the following inhomogeneous differential partial equation with second-order partial derivatives [68] with the initial conditions (Equations 3.251 and 3.254) and the boundary conditions (Equations 3.255 and 3.259)

$$\begin{aligned}\frac{\partial^2 P}{\partial t^2} = v_a^2\frac{\partial^2 P}{\partial x^2} - v_a^2\frac{P_l}{l l_\infty}\sum_{n=1}^{\infty}(\cos n\pi - 1)\cdot\cos n\pi\frac{x}{l_\infty}\cos n\pi\frac{v_1 t}{l_\infty} - \\ v_a^2\frac{P_l}{l l_\infty}\sum_{n=1}^{\infty}(\cos n\pi - 1)\sin n\pi\frac{x}{l_\infty}\sin n\pi\frac{v_1 t}{l_\infty}.\end{aligned} \tag{3.260}$$

The solution will be calculated as a sum of functions

$$P = u + \xi, \tag{3.261}$$

where u is the solution of Equation 3.260 only with the boundary conditions; ξ is the solution of this equation without a constant term with the following initial and boundary conditions:

$$\xi\big|_{t=0} = -u\big|_{t=0}; \tag{3.262}$$

$$\left.\frac{\partial \xi}{\partial t}\right|_{t=0} = -\left.\frac{\partial u}{\partial t}\right|_{t=0}; \tag{3.263}$$

$$\xi\big|_{x=0} = 0; \tag{3.264}$$

$$\left.\frac{\partial \xi}{\partial x}\right|_{x=l} = 0. \tag{3.265}$$

The task to determine u can, in turn, be divided into two sub-tasks:
(a) the solution of the equation

$$\frac{\partial^2 u}{\partial t^2} = v_a^2\frac{\partial^2 u}{\partial x^2} - v_a^2\frac{P_l}{l\cdot l_\infty}\sum_{n=1}^{\infty}(\cos n\pi - 1)\cos n\pi\frac{x}{l_\infty}\cdot\cos n\pi\frac{v_1 t}{l_\infty} \tag{3.266}$$

with

$$u\big|_{x=0} = 0; \left.\frac{\partial u}{\partial x}\right|_{x=l} = \frac{P_l}{l}\sum_{n=1}^{\infty}\frac{\cos n\pi - 1}{n\pi}\sin n\pi\frac{l}{l_\infty}\cos n\pi\frac{v_1 t}{l_\infty}; \tag{3.267}$$

and (b) the solution of the equation

$$\frac{\partial^2 u}{\partial t^2} = v_a^2\frac{\partial^2 u}{\partial x^2} - v_a^2\frac{P_l}{l\cdot l_\infty}\sum_{n=1}^{\infty}(\cos n\pi - 1)\sin n\pi\frac{x}{l_\infty}\sin n\pi\frac{v_1 t}{l_\infty} \tag{3.268}$$

with

$$u\big|_{x=0} = 0; \left.\frac{\partial u}{\partial x}\right|_{x=l} = \frac{P_l}{l}\sum_{n=1}^{\infty}\frac{\cos n\pi - 1}{n\pi}\left(1 + \cos n\pi\frac{l}{l_\infty}\right)\sin n\pi\frac{v_1 t}{l_\infty}. \tag{3.269}$$

Subtask (a): Try the solution of Equation 3.266 as follows:

$$u_a = \sum_{n=1}^{\infty} X_n(x)\cdot\cos n\pi\frac{v_1 t}{l_\infty}. \tag{3.270}$$

Inserting this solution into the initial equation after obvious reductions, we obtain

$$\frac{d^2 X_n(x)}{dt^2} + X_n(x)\cdot\left(\frac{n\pi M_v}{l_\infty}\right)^2 = \frac{P_l}{l\cdot l_\infty}(\cos n\pi - 1)\cdot\cos n\pi\frac{x}{l_\infty}, \tag{3.271}$$

where

$$M_v = v_1/v_a. \tag{3.272}$$

Integrate this equation, with

$$X_n(0) = 0,$$

$$\left.\frac{dX_n(x)}{dx}\right|_{x=l} = \frac{P_l}{l}\frac{\cos n\pi - 1}{n\pi}\cdot\sin n\pi\frac{l}{l_\infty} \tag{3.273}$$

based on the conditions (Equation 3.267), we obtain

$$X_n(x) = \Lambda\left[\Phi\sin\pi n\frac{x}{l_\infty}M_v - \cos n\pi\frac{x}{l_\infty}M_v + \cos n\pi\frac{x}{l_\infty}\right] \tag{3.274}$$

and the solution of subtask (a) is

$$u_a = \sum_{n=1}^{\infty}\Lambda\left[\Phi\sin\pi n\frac{x}{l_\infty}M_v - \cos n\pi\frac{x}{l_\infty}M_v + \cos n\pi\frac{x}{l_\infty}\right]\cdot\cos\pi n\frac{v_1 t}{l_\infty}, \tag{3.275}$$

where for simplicity of notation we assume

$$\Lambda = \frac{P_l \cdot l_\infty}{(n\pi)^2 l} \cdot \frac{\cos n\pi - 1}{M_v^2 - 1}, \tag{3.276}$$

$$\Phi = \left(M_v \sin \pi n \frac{l}{l_\infty} - \sin \pi n \frac{l}{l_\infty} M_v \right) \Big/ \cos n\pi \frac{l}{l_\infty} M_v \cdot \tag{3.277}$$

Subtask (b): Equation 3.268 is solved in the same way. Obtain

$$u_b = \sum_{n=1}^{\infty} \Lambda \left(\sin \pi n \frac{x}{l_\infty} - \Omega \sin \pi n \frac{x}{l_\infty} M_v \right) \cdot \sin \pi n \frac{v_1 t}{l_\infty}, \tag{3.278}$$

where

$$\Omega = \frac{M_v^2 - 1}{M_v} \cdot \frac{1 + \cos n\pi \dfrac{l}{l_\infty}}{\cos n\pi \dfrac{l}{l_\infty} M_v} + \frac{1}{M_v}. \tag{3.279}$$

We now calculate the function ξ, which is the solution of equation

$$\frac{\partial^2 \xi}{\partial t^2} = v_a^2 \frac{\partial^2 \xi}{\partial x^2}, \tag{3.280}$$

with the boundary conditions (Equations 3.264 and 3.265) and with the following initial conditions based on equalities (Equations 3.275 and 3.278):

$$\xi\big|_{t=0} = -(u_a + u_b)\big|_{t=o} = -\sum_{n=1}^{\infty} \Lambda \left[\Phi_n \sin \pi n \frac{x}{l_\infty} M_v - \cos n\pi \frac{x}{l_\infty} M_v + \cos n\pi \frac{x}{l_\infty} \right] = f(x); \tag{3.281}$$

$$\left.\frac{\partial \xi}{\partial t}\right|_{t=0} = -\left(\frac{\partial u_a}{\partial t} + \frac{\partial u_b}{\partial t} \right)\Bigg|_{t=0} = -\sum_{n=1}^{\infty} \Lambda n\pi \frac{v_1}{l_\infty} \left(\sin \pi n \frac{x}{l_\infty} - \Omega \sin \pi n \frac{x}{l_\infty} M_v \right) = F(x). \tag{3.282}$$

We solve it using the Fourier method—by expansion of the function in a series in the orthogonal function system $\sin \dfrac{2n+1}{2l} \pi x$ [1]. The solution is

$$\xi = \sum_{n=0}^{\infty} \left[a_k \cos \frac{2n+1}{2l} \pi v_a t + \delta_k \sin \frac{2n+1}{2l} \pi v_a t \right] \cdot \sin \frac{2n+1}{2l} \pi x, \tag{3.283}$$

where

$$a_k = \frac{2}{l} \int_0^l f(x) \sin \frac{2n+1}{2l} \pi x dx; \tag{3.284}$$

$$\delta_k = \frac{4}{(2n+1)\pi v_a} \int_0^l F(x) \sin \frac{2n+1}{2l} \pi x dx. \tag{3.285}$$

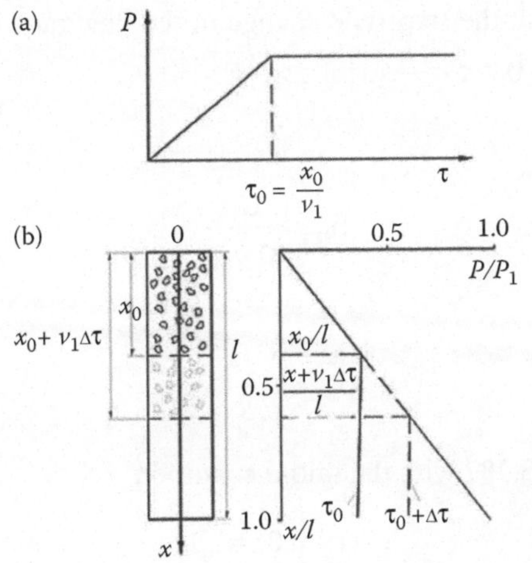

FIGURE 3.19 Change in the induction pressure (a) over time and (b) along the tube in the case of an instant loading of bulk material.

Inserting the obtained functions ξ, u_a, u_b, into Equation 3.261, we obtain the desired solution. For a small falling velocity of the material ($v_1/v_2 \ll 1$), the solution can be reduced after a number of simplifications

$$P = P_l \frac{l_\infty}{l} \sum_{n=1}^{\infty} \frac{1-\cos n\pi}{(n\pi)^2} \cos n\pi \frac{x - v_1 t}{l_\infty} + \frac{1}{2} P_l \frac{l_\infty}{l} \left(\frac{v_1 t}{l_\infty} - \frac{1}{2} \right) + \frac{1}{2} P_l \frac{x}{l}. \quad (3.286)$$

As can be seen from the graph (Figure 3.19) plotted based on this equation, the pressure in an arbitrary section x_0 increases in this section up to the maximum according to $t_0 \equiv \tau_0 = x_0/v_1$ s, (i.e., as soon as the first particles of the material reach the section under consideration).

Along the entire pipe length, the pressure reaches its maximum value as soon as the pipe is filled with the falling material. Thus, a change in the induction pressure is "rigidly" connected with a change in the material flow. The steady mode of dynamic interaction between the material and the air occurs almost simultaneously with a constant flow of material in all sections of the pipe.

In contrast to the dynamic interaction, temperature changes significantly "fall behind" the fluctuations in the material transfer mode. To see this, consider the same task after having slightly simplified it. Assume that the air's thermal conductivity is high, and suppose that the same temperature is instantly set in all pipe sections. Thus, the temperature will depend on time only. With the previous assumptions, the heat exchange equation is as follows:

$$\frac{d}{d\tau} \beta_2 \rho_2 c_2 t_2 = \alpha k_s \beta_{ll} (t_1 - t_2) - 4 \frac{k}{D} (t_2 - t_0), \quad (3.287)$$

where k is a coefficient of heat exchange with surrounding air, $B_T/(m^2 \cdot {}^\circ K)$; and β_{ll} is the volume concentration of material in the chute.

Taking into account the step-type change in the flow rate, a solution is obtained for two intervals with $0 < \tau < \frac{l}{v_1}$, with $\tau > \frac{l}{v_1}$:

in the first interval,

$$\beta_{1l} = \frac{G_1 \tau}{\rho_1 S_{ch} l}; \tag{3.288}$$

in the second interval,

$$\beta_{1l} = \frac{G_1}{\rho_1 S_{ch} v_1}. \tag{3.289}$$

Integrating Equation 3.287 with the initial condition

$$t_2(x,0) = t_0, \tag{3.290}$$

we obtain:

(a) with $0 < \tau < \frac{l}{v_1}$

$$t_2 = t_1 - (t_1 - t_0)\left\{\exp\left[-\left(R\tau + \frac{B\tau^2}{2}\right)\right] + \frac{R}{\sqrt{B}}\exp\left(-\frac{y}{2}\right)\int_{R/\sqrt{B}}^{y} erf\left(\frac{z^2}{2}\right)dz\right\}; \tag{3.291}$$

(b) with $\tau > l / v_1$

$$t_2 = t_{2\infty} - (t_{2\infty} - t_{2H})\exp\left[-\left(B\frac{l}{v_1} - R\right)\left(\tau - \frac{l}{v_1}\right)\right], \tag{3.292}$$

where t_{2H} is the air temperature in the pipe with $\tau = l/v_1$ defined by Equation 3.291; $t_{2\infty}$ is the temperature in the pipe with $\tau \to \infty$

$$t_{2\infty} = \frac{B\frac{l}{v_1}t_1 + Rt_0}{B\frac{l}{v_1} + R}; \tag{3.293}$$

and B, R are parameters introduced for simplicity of notation and equal to

$$B = \alpha k_S G_1 / (\rho_2 c_2 \rho_1 S_{ch} l); \tag{3.294}$$

$$R = 4k/(D\rho_2 c_2); \tag{3.295}$$

$$y = \sqrt{B}\tau + R/\sqrt{B}.$$

For a heat sealed pipe ($R = 0$), we have:

(a) with $0 < \tau < l / v_1$

$$t_2 = t_1 - (t_1 - t_0)\exp(-B\tau^2 / 2); \tag{3.296}$$

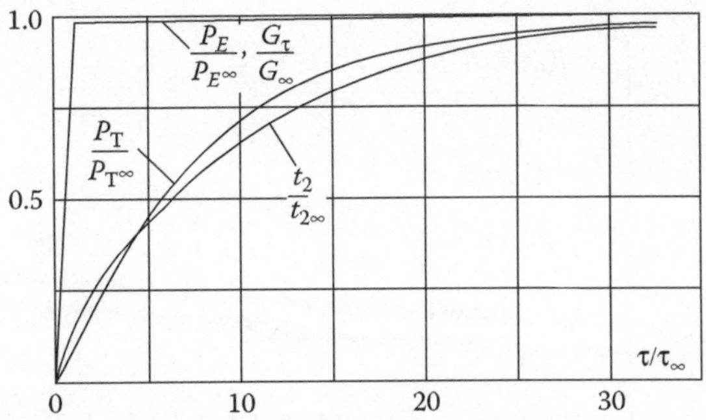

FIGURE 3.20 Temporal variation in temperature, thermal and induction pressures, and mass of particles in the chute ($P_{E\infty}$, $P_{T\infty}$ are the induction and thermal pressure with $\tau \to \infty$; $G\tau$, $G\infty$ are the masses of particles in the chute at the moments in time τ and $\tau_\infty = l / \overline{v}_1$).

(b) with $\tau > l / v_1$

$$t_2 = t_1 - (t_1 - t_0)\exp\left[-B\frac{l}{v_1}\left(\tau - \frac{l}{2v_1}\right)\right]. \tag{3.297}$$

Figure 3.20 shows the temperature curves plotted according to these formulas. It also shows the change in the thermal and induction pressures. As the curves indicate, the thermal pressure has a considerable "inertia" when compared with the induction pressure.

3.3.2 Smooth Change in the Material Flow

Given that the induction pressure is "rigidly" connected with the material flow, changes in the dynamic interaction can be assessed under conditions of a changing particle flow by means of the correlations obtained in the study of stationary flows. Thus, on the basis of Equations 3.47 and 3.29, the following equation can be used for a pressure at the end of the pipe, the lower end of which is closed to the air passage

$$P_E = \psi_0 k_m \beta e^{\frac{-1{,}8\sqrt{\beta\cdot 10^3}}{d_e\cdot 10^3}} \cdot l\frac{v_{1k}^2}{2}\rho_2\frac{1-n^3}{3(1-n)}, \tag{3.298}$$

where β is an averaged volume concentration of particles in the chute (see Equation 3.30), which varies over time due to changes in flow.

As seen from Equation 3.298, the induction pressure has a maximum at the volume concentration (β_{max}) defined by the equality

$$2 - \frac{1{,}8}{d_e \cdot 10^3}\sqrt{\beta_{max}\cdot 10^3} = 0. \tag{3.299}$$

Thus, it appears that the concentration of particles varies widely from 0 to $\beta_\infty > \beta_{max}$, and pressure surges occur during an unsteady-state process. This is clearly seen

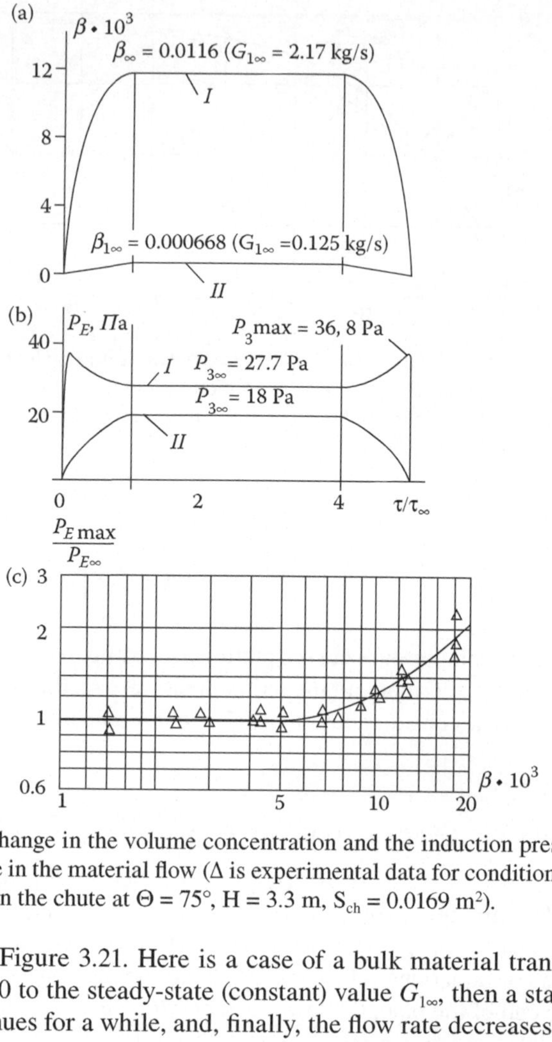

FIGURE 3.21 Change in the volume concentration and the induction pressure in the chute with a slow change in the material flow (Δ is experimental data for conditions of granite bulking $d_e = 1.88$ mm in the chute at $\Theta = 75°$, $H = 3.3$ m, $S_{ch} = 0.0169$ m^2).

in the curves in Figure 3.21. Here is a case of a bulk material transfer where flow rate varies from 0 to the steady-state (constant) value $G_{1\infty}$, then a stationary process ($G_1 = G_{1\infty}$) continues for a while, and, finally, the flow rate decreases from $G_{1\infty}$ down to 0. However, the pressure surge may be absent where the steady-state flow is so small that the volume concentration of particles in the chute is $\beta_\infty < \beta_{max}$.

The maximum value of the induction pressure, according to Equation 3.299, is

$$P_{E_{max}} = \psi_0 k_m 10^{-3}\left(\frac{2\cdot d_e\cdot 10^3}{1,8}\right)^2 e^{-2}\cdot l\frac{v_{1k}^2}{2}\rho_2\frac{1-n^3}{3(1-n)} \tag{3.300}$$

or the relative value of the pressure surge is

$$P_{E\max}/P_{E\infty} = 10^{-3}\left(\frac{2\cdot 10^3 d_e}{1,8}\right)^2 e^{-2}\Bigg/\left[\beta_\infty \exp\left(-1,8\frac{\sqrt{\beta_\infty\cdot 10^3}}{d_e\cdot 10^3}\right)\right] \text{with } \beta_\infty \geq \beta_{max};$$
$$P_{E\max}/P_{E\infty} = 1, \text{ with } \beta_\infty < \beta_{max}. \tag{3.301}$$

In studies of the induction properties of a bulk material flow in units with inclined chutes (Figure 3.2), a pressure surge was often observed when the material feeding from the upper bin began and when it stopped. The value of this surge was significant at large material flows. No pressure rise was observed with small flows. A pressure rise is absolutely in line with Equation 3.301, not only in qualitative terms but also in quantitative terms (see Figure 3.21c).

The observed fall of the thermal pressure behind the induction pressure, as well as the surge of the induction pressure during a start or stop of the process equipment must be taken into account when calculating the required volumes of aspirated air.

4 The Aerodynamics of Solid-Particle Jets

Free flows of loose matter make up the second significant common class of flows occurring in bulk material handling technology. Primarily, these are flows of dumped materials in various stockpiling systems. Free flows also occur when open-body railway cars are loaded with concentrate or with pellets from feed hoppers. In terms of dynamic interactions between solid matter and air, close approximations can be made by studying flows of material unloaded from rail cars into charging bins of crushers and of other equipment.

An equation for flow dynamics can be deduced from those formulas describing the mechanics of multi-component flows (see Equations 80 and 90 in the Appendix, ignoring pulsation moments—the latter can be successfully "smoothed out" using experimental coefficients). By concerning ourselves with overall value magnitude and ignoring minor terms (similar to deducing boundary-layer equations from the general Navier–Stokes equation), we end up with the following simultaneous equations for a planar problem:

$$\left.\begin{aligned} &v_{2(1)}\frac{\partial v_{2(1)}}{\partial x_1}+v_{2(2)}\frac{\partial v_{2(2)}}{\partial x_2}=\frac{\beta_1}{\rho_2 V_p}R(v_{2(1)}-v_{1(1)})-\frac{1}{\rho_2}\frac{\partial P}{\partial x_2}+v\frac{\partial^2 v_{2(1)}}{\partial x_2};\\ &\frac{\partial v_{2(1)}}{\partial x_1}+\frac{\partial v_{2(2)}}{\partial x_2}=0. \end{aligned}\right\} \quad (4.1)$$

These are set apart from the known Prandtl equations for isothermal jet streams by the presence of a volumetric force variable owing to the presence of falling particles in the stream.

The apparent mathematical insignificance of this difference becomes crucial in the physical sense: it is these volumetric forces, rather than initial impulse (as would be the case in many problems involving free air jets, for example), that determine the jet flow of air in the class of flows being considered here.

In our case, the following properties are relevant (from a physical sense) for the two-component free jet. First, the solid component—bulk particulate material—significantly impacts boundary layer aerodynamics and is responsible for the formation of this layer as such. Second, owing to the larger mass of particles, the solid component dynamics responsible for the jet flow mode of the gas component remain largely unaffected by airflow, setting this flow mode apart from airflows containing minute solid impurities. In other words, we are dealing with a flow where the solid component has a field of particle concentrations and velocities independent of the airflow structure. Thus, for a flow of falling particles:

$$\beta_1 = f_\beta(x_1, x_2), \tag{4.2}$$

$$v_{1(1)} = f(x_1);\; v_{1(2)} = v_n. \tag{4.3}$$

The transverse component of particle velocities v_n will be assumed to be zero in the majority of cases.

Let us consider two characteristic cases: jet motion of freely falling particles of loose matter and particulate flow in a flat duct.

4.1 AIR INJECTION IN A JET OF FREELY FALLING PARTICLES

4.1.1 Initial Equations

4.1.1.1 Changes in Volumetric Particle Concentration in a Jet of Material

If we look at symmetric flat flows with a symmetry axis OX_1, directed downward, along the path of falling particles, our primary focus will be studying the half of the jet within the first quadrant X_1OX_2 of our coordinate system (Figure 4.1).

Much like loose material traveling down a slope, a jet of freely falling particles displays an exponential distribution of particles (see, for example, studies by V. P. Pavlov [74]).

Let us assume, generally:

$$\beta = \beta_0 e^{-Ax_*^{\nu} y_*^{t}}. \tag{4.4}$$

Here, $x_* = x_1$, $y_* = x_2$ determine dimensional coordinates of a point on a plane X_1OX_2; and t, ν, A, and β_0 are certain constants.

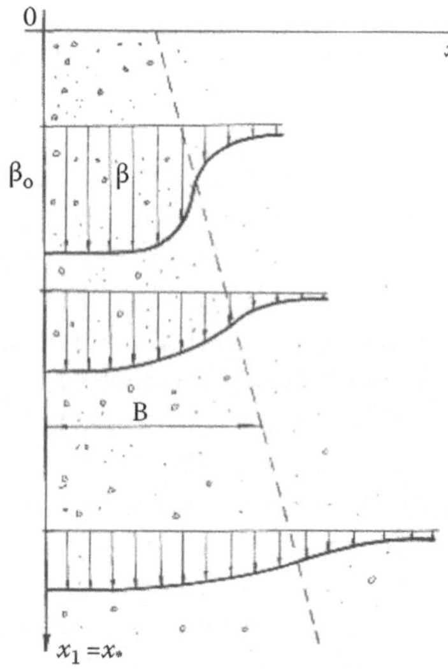

FIGURE 4.1 Changes in volumetric concentration of particles in a jet of loose matter.

To determine the material flow, we consider the live section area as a plane orthogonal to axis OX_1:

$$G_1 = 2\int_0^{\infty} \beta_0 e^{-Ax^{\nu}_{*} y_*^{t}} \nu_1 \rho_1 dy_*$$

or, considering Equation 4.3,

$$G_1 = 2v_1\rho_1\beta_0 \int_0^{\infty} e^{-Ax^{\nu}_{*} y_*^{t}} dy_* = 2v_1\rho_1\beta_0 B\cdot\Gamma\left(1+\frac{1}{t}\right), \tag{4.5}$$

where $\Gamma\left(1+\frac{1}{t}\right)$ is the gamma function of the argument $1+\frac{1}{t}$;

$$B = A^{-\frac{1}{t}} x_*^{-\frac{\nu}{t}}. \tag{4.6}$$

In order to reveal the physical sense of the value B, consider a plane-parallel flow of falling particles with $\nu = 0$. In this case:

$$B = A^{-\frac{1}{t}}, \tag{4.7}$$

$$\beta = \beta_0 e^{-\left(\frac{y_*}{B}\right)^t}. \tag{4.8}$$

Figure 4.2 visualizes changing volumetric concentrations in a plane-parallel flow with varying values of parameter t. The plots clearly show that, with $t \geq 100$, virtually all particles become confined within $0 \leq y_* \leq B$ and are evenly distributed on this interval, with a volumetric concentration of β_0. Virtually no particles occur outside this interval; hence, the volume concentration is equal to zero. Therefore, B is none

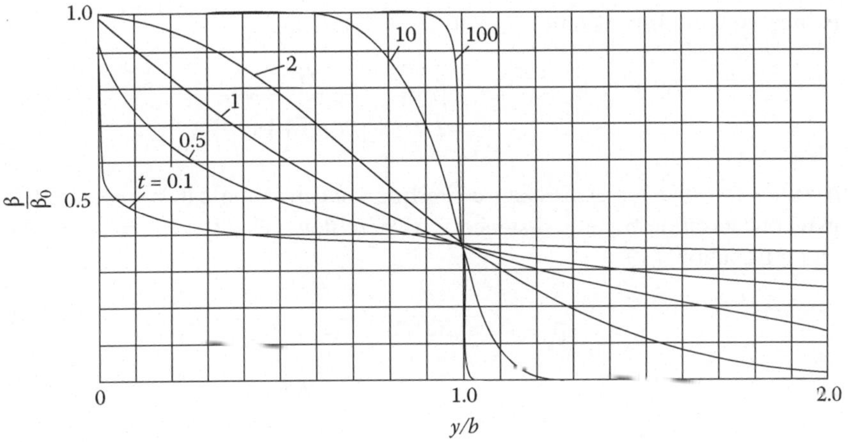

FIGURE 4.2 Changes in volumetric concentration of particles in a cross-section of a flat jet with plane-parallel motion ($\nu/t = 0$).

other than the width (or half-width, if one recalls that only one-half of the jet is being considered) of an evenly distributed particle jet at a flow rate of G_1.

With that in mind, we will categorize flows either as narrowing (those with a declining B over the height of fall with $\frac{\nu}{t} > 0$) or as widening (with the value B increasing as particles fall further owing to $\frac{\nu}{t} < 0$).

Using the notation

$$y_* / B = y_* A^{\frac{1}{t}} x_*^{\frac{\nu}{t}} = z, \tag{4.9}$$

or, in dimensionless form,

$$y a^{\frac{1}{t}} x^{\frac{\nu}{t}} = z, \tag{4.10}$$

where

$$y = y_* / l_\infty ; x = x_* / l_\infty ;$$

$$a = A l_\infty^{t+\nu}; \quad b = B / l_\infty = a^{-\frac{1}{t}} x^{-\frac{\nu}{t}}. \tag{4.11}$$

Then,

$$\beta = \beta_0 e^{-z^t} = \beta_0 e^{-a x^\nu y^t}. \tag{4.12}$$

Now, let us determine the volumetric concentration of particles on the centerline of the jet, β_0. Because of Equation 4.5,

$$\beta_0 = \frac{G_1}{2 v_1 \rho_1 B \cdot \Gamma\left(1 + \frac{1}{t}\right)} \tag{4.13}$$

or, expressed with dimensionless quantities,

$$\beta_0 = \overline{\beta}_0 a^{\frac{1}{t}} x^{\frac{\nu}{t}} / v; \quad \overline{\beta}_0 = \frac{G_1}{2 c \rho_1 l_\infty \cdot \Gamma\left(1 + \frac{1}{t}\right)}. \tag{4.14}$$

Another formula for expressing centerline concentration can be obtained for a known concentration β_{0H} at a distance x_H (where flow velocity is $v_H = v_{1H}/c$). Then, based on Equation 4.13,

$$\beta_{0H} = \frac{G_1}{2 v_{1H} \rho_1 B_H \cdot \Gamma\left(1 + \frac{1}{t}\right)}$$

and

$$\beta_0 = \beta_{0H} \frac{v_H}{v} \cdot \left(\frac{x_H}{x}\right)^{-\frac{\nu}{t}}. \tag{4.15}$$

Equations 4.14 and 4.15 can be matched to determine

$$\bar{\beta}_0 = \beta_{0H} v_H x^{-\frac{v}{t}} a^{-\frac{1}{t}}.$$

In case of a linearly accelerated flow,

$$v = \sqrt{2x}$$

and because of Equation 4.14, β_0 can be expressed as

$$\beta_0 = Kx^{\frac{v}{t}-\frac{1}{2}},$$

where

$$K = \bar{\beta}_0 a^{\frac{1}{t}} / \sqrt{2}.$$

Our further study will focus on aerodynamic properties of jets with particles distributed in a generalized exponential form (see Equation 4.12, such as reflected by even distribution (with $t \to \infty$):

$$\beta = \beta_0 = G / (2v_1 \rho_1 B) = \beta_c \text{ with } 0 \le y \le b,\ \beta = 0 \text{ with } y > b. \tag{4.16}$$

For an axially symmetric jet, the generalized exponential distribution of particles is described using the same equation (4.12), the differences being

$$\left.\begin{aligned} &\bar{\beta}_0 = \frac{G_1}{\pi l^2_{\infty} c\rho_1 \cdot \Gamma\left(1+\frac{1}{t}\right)}; \quad \beta_0 = \bar{\beta}_0 / (b^2 v); \\ &z = R / B = r / b = a^{\frac{1}{t}} x^{\frac{v}{t}} r; \quad r = R / l_{\infty}. \end{aligned}\right\} \tag{4.17}$$

With $t \to \infty$, it holds that $\beta_0 = G_1 / (\pi B^2 c\rho_1 v)$.

4.1.1.2 Volumetric Forces of Interaction between Components

The volumetric force vector of aerodynamic interaction between particles and air, caused by the different velocities of the individual components, can be expressed (in light of Equation 23 in the Appendix) as

$$\vec{F}_v = n_p \vec{R}, \tag{4.18}$$

where n_p is the count of falling particles per unit volume. Here, $\vec{R}$ is the vector of aerodynamic force exerted by a falling particle against airflow

$$\vec{R} = \psi f_M \frac{|\vec{v}_1 - \vec{v}_2|(\vec{v}_1 - \vec{v}_2)}{2} \rho_2 \tag{4.19}$$

or

$$\vec{R} = E_n |\vec{v}_1 - \vec{v}_2|^n \cdot (\vec{v}_1 - \vec{v}_2),$$

where $n = 0$, E_0 are the parameters corresponding to a viscous area of the flow surrounding the particle

$$E_0 = \nu \frac{24}{d_e} f_M \frac{\rho_2}{2}, \tag{4.20}$$

ν is the kinematic viscosity factor, m²/c; and $n = 1$; E_1 are the parameters corresponding to the area described by the square law of resistance

$$E_1 = \psi f_M \frac{\rho_2}{2}. \tag{4.21}$$

Thus assumes that the factor ψ in Equation 4.19 considers not only the mode of flow around particles but also the degree of their coupling (tightness).

Referred to as a unit mass of the medium (vector of mass forces), the force of aerodynamic interaction is equal to:

$$\vec{F}_M = n_p \vec{R} / \rho_2 = n_p E_n \left| \vec{v}_1 - \vec{v}_2 \right|^n (\vec{v}_1 - \vec{v}_2) / \rho_2. \tag{4.22}$$

Let us proceed to dimensionless quantities. We will use terminal velocity c as the characteristic velocity in the same way as we did earlier while using the inertial path length l_∞, determined by the ratio (2.67), as the characteristic size.

Dividing both sides of Equation 4.22 by $c^2 / l_\infty = g(1-\varepsilon)$ and considering $E_n c^n c = V_p \rho_1 g(1-\varepsilon)$, we will end up with

$$\vec{F} = \vec{F}_M / [g(1-\varepsilon)] = \frac{\beta}{\varepsilon} \left| \vec{v} - \vec{u} \right|^n (\vec{v} - \vec{u}) \tag{4.23}$$

or, projected to a coordinate axis:

$$F_x = \beta \left| \vec{v} - \vec{u} \right|^n (v_x - u_x) / \varepsilon, \tag{4.24}$$

$$F_y = \beta \left| \vec{v} - \vec{u} \right|^n (v_y - u_y) / \varepsilon. \tag{4.25}$$

Hereafter, we shall consider vertical flows of solid particles ($v_y = 0$) that form air currents with a pronounced longitudinal directivity ($u_x >> u_y$). The corresponding longitudinal and transverse components of the mass force vector in such flows will be equal to:

$$F_x = \beta \left| v - u_x \right|^n (v - u_x) / \varepsilon, \tag{4.26}$$

$$F_y = -\beta \left| v - u_x \right|^n u_y / \varepsilon. \tag{4.27}$$

Let us also consider a case of maximum mass forces occurring at $v >> u_x$

$$F_x = \beta v^n (v - u_x) / \varepsilon \approx \beta v^{n+1} / \varepsilon, \tag{4.28}$$

$$F_y = -\beta v^n u_y. \tag{4.29}$$

In view of Equation 4.12, mass force components (by virtue of Equations 4.26 and 4.27) will become

$$F_x = D x^\gamma \left| 1 - u_x / v \right|^n (1 - u_x / v) e^{-z'}, \tag{4.30}$$

$$F_y = -Dx^{\gamma}\left|1 - u_x / v\right|^n e^{-z'} \cdot u_y / v, \tag{4.31}$$

where

$$D = \bar{\beta}_0 a^{\frac{1}{t}} 2^{\frac{n}{2}} / \varepsilon; \gamma = \nu / t + n / 2 \tag{4.32}$$

with

$$v = \sqrt{2x}.$$

In case of an axially symmetric flow

$$F_x = Dx^{\gamma} e^{-z'}\left|1 - u_x / v\right|^n (1 - u_x / v), \tag{4.33}$$

$$F_y = -Dx^{\gamma} e^{-z'}\left|1 - u_x / v\right|^n \cdot u_r / v, \quad z = r / b, \tag{4.34}$$

where

$$D = \bar{\beta}_0 a^{\frac{2}{t}} 2^{\frac{n}{2}} / \varepsilon; \quad \gamma = 2\nu / t + n / 2. \tag{4.35}$$

4.1.1.3 Fluid Dynamics Equations

We will now put together the dynamics equations for air injected by a flow of falling particles, and we will consider a quasi-stationary turbulent flow, basing the dynamics equation on the impulse transfer equation (85) from the Appendix. Our scope will be narrowed down to flows with volumetric concentrations of solid particles small enough ($\beta_1 \ll 1$) to justify ignoring the effects of tightness and to assume $\beta_2 = 1$.

The only kinds of pulsations demanding our attention are temporary pulsations of velocities—a characteristic feature of turbulent flows. The assumptions stated earlier enable the balance equation for air to be expressed as

$$\frac{\partial}{\partial x_k}(\rho_2 \vec{v}_2 v_{2k} + \overline{\rho_2 \vec{v}''_{2\tau} v''_{k\tau}}) = \rho_2 \vec{M}_2 + \vec{r}_{21} f_{\phi} + \frac{\partial}{\partial x_k}\vec{\Pi}_{2k}. \tag{4.36}$$

Considering that the volumetric force caused by dynamic interaction of components is equal to

$$\vec{r}_{21} f_{\phi} = \vec{F}_v, \tag{4.37}$$

and the divergence of viscous stress tensor, assuming incompressible air, is ($\mathrm{div}\vec{v}_2 = 0$), owing to Equation 43 from the Appendix

$$\frac{\partial \vec{\Pi}_{2k}}{\partial x_k} = -gradP + \mu \nabla^2 \vec{v}_2, \tag{4.38}$$

the relation (4.36) for an isothermal flow can be expressed with the following Reynolds-type equation:

$$v_{2k}\frac{\partial}{\partial x_k}\vec{v}_2 = \vec{F}_M - \frac{1}{\rho_2} gradP + \nu \nabla^2 \vec{v}_2 + \frac{1}{\rho_2}\frac{\partial}{\partial x_k}\left(-\overline{\rho_2 \vec{v}''_{2\tau} v''_{2k\tau}}\right). \tag{4.39}$$

Or, with introduction of the turbulent friction stress tensor:

$$\Pi_{2\tau}=\begin{pmatrix}-\sigma_{11} & \tau_{12} & \tau_{13}\\ \tau_{21} & -\sigma_{22} & \tau_{23}\\ \tau_{31} & \tau_{32} & -\sigma_{33}\end{pmatrix}=\begin{pmatrix}\overline{\rho_2 v''_{2(1)\tau}v''_{2(1)\tau}} & \overline{\rho_2 v''_{2(1)\tau}v''_{2(2)\tau}} & \overline{\rho_2 v''_{2(1)\tau}v''_{2(3)\tau}}\\ \overline{\rho_2 v''_{2(2)\tau}v''_{2(1)\tau}} & \overline{\rho_2 v''_{2(2)\tau}v''_{2(2)\tau}} & \overline{\rho_2 v''_{2(2)\tau}v''_{2(3)\tau}}\\ \overline{\rho_2 v''_{2(3)\tau}v''_{2(1)\tau}} & \overline{\rho_2 v''_{2(3)\tau}v''_{2(2)\tau}} & \overline{\rho_2 v''_{2(3)\tau}v''_{2(3)\tau}}\end{pmatrix}, \tag{4.40}$$

the equation will take the form:

$$v_{2k}\frac{\partial}{\partial x_k}\vec{v}_2=\vec{F}_M-\frac{1}{\rho_2}gradP+\nu\nabla^2\vec{v}_2+\frac{1}{\rho_2}\frac{\partial}{\partial x_k}\vec{\Pi}_{2k\tau}. \tag{4.41}$$

The following holds for flat flows in an X_1OX_2 coordinate system:

$$v_{2(1)}\frac{\partial v_{2(1)}}{\partial x_1}+v_{2(2)}\frac{\partial v_{2(1)}}{\partial x_2}=\vec{F}_{M(1)}-\frac{1}{\rho_2}\frac{\partial}{\partial x_1}(P+\sigma_{11})+\nu\left(\frac{\partial^2 v_{2(1)}}{\partial x_1{}^2}+\frac{\partial^2 v_{2(1)}}{\partial x_2{}^2}\right)+\frac{1}{\rho_2}\frac{\partial \tau_{12}}{\partial x_2}, \tag{4.42}$$

$$v_{2(1)}\frac{\partial v_{2(2)}}{\partial x_1}+v_{2(2)}\frac{\partial v_{2(2)}}{\partial x_2}=\vec{F}_{M(2)}-\frac{1}{\rho_2}\frac{\partial}{\partial x_2}(P+\sigma_{22})+\nu\left(\frac{\partial^2 v_{2(2)}}{\partial x_1{}^2}+\frac{\partial^2 v_{2(2)}}{\partial x_2{}^2}\right)+\frac{1}{\rho_2}\frac{\partial \tau_{21}}{\partial x_1}. \tag{4.43}$$

In the vast majority of practical applications, turbulent normal stresses σ_{11} and σ_{22} are small relative to the pressure P and are therefore ignored. According to the semi-empirical turbulent momentum transfer theory put forward by L. Prandtl, tangential stresses

$$\tau_{12}=\tau_{21}=\rho_2\nu_\tau\frac{\partial v_{2(1)}}{\partial x_2}, \tag{4.44}$$

where ν_τ is the apparent kinematic viscosity of a turbulent flow, equal to

$$\nu_\tau=l^2\left|\frac{\partial v_{2(1)}}{\partial x_2}\right|, \tag{4.45}$$

and l is the length of mixing path in meters m.

In case of free turbulence, relevant to the flow of injected air in question, apparent viscosity is equal to:

$$\nu_\tau=kb_nu_0, \tag{4.46}$$

where b_n is the breadth of the turbulent mixing area in meters; u_0 is the velocity of air in the jet centerline, in meters per second; and k is the proportionality factor.

In case of free air jets, the following [104] are known

$$b_n\sim x_1^{\frac{2}{3}},\quad u_0\sim x_1^{-\frac{1}{3}},\quad \nu_\tau=0{,}037b_{\frac{1}{2}}u_0 \tag{4.47}$$

for flat jets and

$$b_n \sim x_1, \quad u_0 \sim x_1^{-1}, \quad \nu_\tau = 0{,}0175 r_{\frac{1}{2}} u_0 - const \tag{4.48}$$

for axially symmetric jets. Here, $b_{\frac{1}{2}}$ ($r_{\frac{1}{2}}$) is the distance from the centerline of the jet to a point where the longitudinal velocity of air decreases to one-half the velocity of air in the jet centerline. In our case, the velocity of injected air in the jet centerline is

$$u_0 \approx \frac{1}{2} v_1, \tag{4.49}$$

while the breadth of the mixing area is

$$b_{\frac{1}{2}} \sim B, \tag{4.50}$$

meaning the following form would be valid for the case of a free jet of solid particles

$$\nu_\tau = \frac{1}{2} k B v_1. \tag{4.51}$$

Considering that the half-width of the jet and velocity v_1 are only dependent on x_1, it holds that

$$\frac{\partial \nu_\tau}{\partial x_2} = 0,$$

and Equations 4.42, 4.43, and 4.44 can therefore be simplified as follows:

$$v_{2(1)} \frac{\partial v_{2(1)}}{\partial x_1} + v_{2(2)} \frac{\partial v_{2(1)}}{\partial x_2} = F_{M(1)} - \frac{1}{\rho_2} \frac{\partial P}{\partial x_1} + \nu \left(\frac{\partial^2 v_{2(1)}}{\partial x_1^2} + \frac{\partial^2 v_{2(1)}}{\partial x_2^2} \right) + \nu_\tau \frac{\partial^2 v_{2(1)}}{\partial x_2^2}, \tag{4.52}$$

$$v_{2(1)} \frac{\partial v_{2(2)}}{\partial x_1} + v_{2(2)} \frac{\partial v_{2(2)}}{\partial x_2} = F_{M(2)} - \frac{1}{\rho_2} \frac{\partial P}{\partial x_2} + \nu \left(\frac{\partial^2 v_{2(2)}}{\partial x_1^2} + \frac{\partial^2 v_{2(2)}}{\partial x_2^2} \right) + \frac{\partial}{\partial x_1} \nu_\tau \frac{\partial v_{2(1)}}{\partial x_2^2} \tag{4.53}$$

or, in a dimensionless notation (by dividing both sides of the equation by c_2/l_∞),

$$u_x \frac{\partial u_x}{\partial x} + u_y \frac{\partial u_x}{\partial y} = F_x - \frac{\partial \mathrm{P}}{\partial x} + N \left(\frac{\partial^2 u_x}{\partial x^2} + \frac{\partial^2 u_x}{\partial y^2} \right) + N_\tau \frac{\partial^2 u_x}{\partial y^2}, \tag{4.54}$$

$$u_x \frac{\partial u_y}{\partial x} + u_y \frac{\partial u_y}{\partial y} = F_y - \frac{\partial \mathrm{P}}{\partial y} + N \left(\frac{\partial^2 u_y}{\partial x^2} + \frac{\partial^2 u_y}{\partial y^2} \right) + \frac{\partial}{\partial x} \left(N_\tau \frac{\partial u_x}{\partial y} \right), \tag{4.55}$$

$$\frac{\partial u_x}{\partial x} + \frac{\partial u_y}{\partial y} = 0, \tag{4.56}$$

where N, N_τ are quantities inversely related to Reynolds numbers

$$N = \frac{\nu}{cl_\infty}; N_\tau = \frac{\nu_\tau}{cl_\infty}; \mathrm{P} = P/(\rho_2 c^2). \tag{4.57}$$

For an axially symmetric flow (in a cylindrical coordinate system, with radial and axial components of the velocity vector $\vec{v}_2$ expressed through v_{2r} and v_{2x} and their pulsations expressed through v''_{2r} andv''_{2x}), dynamics equations would assume the following form:

$$v_{2x}\frac{\partial v_{2x}}{\partial x_1} + v_{2r}\frac{\partial v_{2x}}{\partial x_r} = F_{MX_1} - \frac{1}{\rho_2}\frac{\partial P}{\partial x_1} + \frac{\nu}{x_r}\left[x_r\frac{\partial^2 v_{2x}}{\partial x_1^2} + \frac{\partial}{\partial x_r}\left(x_r\frac{\partial v_{2x}}{\partial x_r}\right)\right] - \frac{1}{\rho_2}\frac{\partial \sigma_{x1}}{\partial x_1} + $$
$$+ \frac{1}{\rho_2}\frac{1}{x_r}\frac{\partial}{\partial x_r}\left(-\rho_2 x_r \overline{v''_{2x}v''_{2r}}\right), \tag{4.58}$$

$$v_{2x}\frac{\partial v_{2r}}{\partial x_1} + v_{2r}\frac{\partial v_{2r}}{\partial x_r} = F_{MX_r} - \frac{1}{\rho_2}\frac{\partial P}{\partial x_r} + \nu\left(\frac{\partial^2 v_{2r}}{\partial x_1{}^2} + \frac{\partial^2 v_{2r}}{\partial x_r{}^2} + \frac{1}{x_r}\frac{\partial v_{2r}}{\partial x_r} - \frac{v_{2r}}{x_r{}^2}\right) - $$
$$- \frac{1}{\rho_2}\frac{1}{x_r}\frac{\partial}{\partial x_r}(x_r\sigma_r) + \frac{1}{\rho_2}\frac{\partial}{\partial x_1}\left(-\rho_2\overline{v''_{2x}v''_{2r}}\right), \tag{4.59}$$

where σ_{x_1} and σ_r are normal stresses of turbulent friction, correspondingly equal to

$$\sigma_{x_1} = \rho_2\overline{v''_{2x}v''_{2x}}, \quad \sigma_r = \rho_2\overline{v''_{2r}v''_{2r}}. \tag{4.60}$$

Considering that

$$-\rho_2\overline{v''_{2x}v''_{2r}} = \rho_2\nu_\tau\frac{\partial v_{2x}}{\partial x_r}, \quad \nu_\tau = f(x_1), \tag{4.61}$$

and assuming

$$P >> \sigma_{x_1}, \quad \frac{\partial P}{\partial x_r} >> \frac{1}{x_r}\frac{\partial}{\partial x_r}(x_r\sigma_r), \tag{4.62}$$

Equations 4.58 and 4.59 can be rewritten in the following dimensionless form:

$$u_x\frac{\partial u_x}{\partial x} + u_r\frac{\partial u_x}{\partial r} = F_x - \frac{\partial \mathrm{P}}{\partial x} + N\left[\frac{\partial^2 u_x}{\partial x^2} + \frac{1}{r}\frac{\partial}{\partial r}\left(r\frac{\partial u_x}{\partial r}\right)\right] + N_\tau\frac{1}{r}\frac{\partial}{\partial r}\left(r\frac{\partial u_x}{\partial r}\right), \tag{4.63}$$

$$u_x\frac{\partial u_r}{\partial x} + u_r\frac{\partial u_r}{\partial r} = F_r - \frac{\partial \mathrm{P}}{\partial r} + N\left[\frac{\partial^2 u_r}{\partial x^2} + \frac{\partial^2 u_r}{\partial r^2} + \frac{1}{r}\frac{\partial u_r}{\partial r} - \frac{u_r}{r^2}\right] + \frac{\partial}{\partial x}\left(N_\tau\frac{\partial u_x}{\partial r}\right), \tag{4.64}$$

and the continuity equation can be put together as

$$\frac{\partial r u_r}{\partial r} + \frac{\partial r u_x}{\partial x} = 0. \tag{4.65}$$

The flow of injected air in question belongs to a class of currents characterized with "free" turbulence. In this case, molecular viscosity forces are negligibly small in comparison with impulse transfer due to turbulent intermixing ($\nu_\tau >> \nu$).

For jet streams, it is true that

$$\frac{\partial^2 u_x}{\partial x^2} << \frac{\partial^2 u_x}{\partial y^2} \tag{4.66}$$

and the transverse component of the velocity of injected component is much less than the longitudinal component; therefore, dynamics equations are greatly simplified. Thus, for a flat jet, the following turbulent layer dynamics equations result

$$u_x \frac{\partial u_x}{\partial x} + u_y \frac{\partial u_x}{\partial y} = \frac{\beta}{\varepsilon}|v - u_x|^n (v - u_x) + N_\tau \frac{\partial u_x}{\partial y^2} - \frac{\partial P}{\partial x}, \tag{4.67}$$

$$\frac{\partial P}{\partial y} = -\frac{\beta}{\varepsilon}|v - u_x|^n u_y + \frac{\partial}{\partial x}\left(N_\tau \frac{\partial u_x}{\partial y}\right), \tag{4.68}$$

where $N_\tau = k_n b v$, $k_n = \dfrac{1}{2}k$ or, for a linearly accelerated flow of particles in view of (4.11):

$$N_\tau = k_n \sqrt{2} a^{-\frac{1}{t}} x^{\frac{1}{2}-\frac{v}{t}}. \tag{4.70}$$

Equation 4.67 can be written as follows in view of the continuity equation:

$$\frac{\partial}{\partial x} u_x^2 + \frac{\partial}{\partial y} u_y u_x = \frac{\beta}{\varepsilon}|v - u_x|^n (v - u_x) + N_r \frac{\partial^2 u_x}{\partial y^2} - \frac{\partial P}{\partial x}. \tag{4.71}$$

Based on Equations 4.63 and 4.64, boundary-layer equations for axially symmetric jets can be expressed as follows:

$$u_x \frac{\partial u_x}{\partial x} + u_r \frac{\partial u_x}{\partial r} = \frac{\beta}{\varepsilon}|v - u_x|^n (v - u_x) + N_\tau \frac{1}{r}\frac{\partial}{\partial r}\left(r \frac{\partial u_x}{\partial r}\right) - \frac{\partial P}{\partial x}, \tag{4.72}$$

$$\frac{\partial P}{\partial r} = -\frac{\beta}{\varepsilon}|v - u_x|^n u_r + \frac{\partial}{\partial x}\left(N_\tau \frac{\partial u_x}{\partial r}\right). \tag{4.73}$$

In addition to differential equations, we shall henceforth use an integral relation for changes in the impulse of injected air. For flat flows, this relation,

$$\int_0^\infty \frac{\partial u_x^2}{\partial x} dy = \int_0^\infty \frac{\beta}{\varepsilon}|v - u_x|^n (v - u_x)\, dy - \int_0^\infty \frac{\partial P}{\partial x} dy, \tag{4.74}$$

would result from Equation 4.71 by integrating all summands over the cross-section of the jet with the following boundary conditions:

$$u_y = 0,\ \frac{\partial u_x}{\partial y} = 0 \text{ at } y = 0; \tag{4.75}$$

$$u_x = 0,\ \frac{\partial u_x}{\partial y} = 0 \text{ at } y \to \infty. \tag{4.76}$$

The integral relation would be similar for an axially symmetric jet:

$$\int_0^\infty \frac{\partial}{\partial x} u_x^2 r dr = \int_0^\infty \frac{\beta}{\varepsilon} |v - u_x|^n (v - u_x) r dr - \int_0^\infty \frac{\partial P}{\partial x} r dr. \tag{4.77}$$

4.1.2 Structure of Air Streams in a Flat Jet of Loose Matter

4.1.2.1 Self-Similar Motion Equations

With the purpose of determining the velocity field and streamlines in a jet of injected air, we shall resort to the affine transformation method accepted in the laminar boundary layer theory. This would reduce the system of differential equations in partial derivatives to a single ordinary differential equation that would be much easier to solve. The possibility of reducing the problem in question to a self-similarity problem is facilitated by the empirical nature of the dependent variable β_1, enabling a certain degree of freedom in choosing the specific functional relation. For example, assume that the distribution of solid particles in a jet is determined by an exponential relation of the form in Equation 4.4. The hydrodynamic boundary-layer equation thus will become

$$\left.\begin{aligned} & u_x \frac{\partial u_x}{\partial x} + u_y \frac{\partial u_x}{\partial y} = Dx^\gamma e^{-z'} \left|1 - \frac{u_x}{v}\right|^n \left(1 - \frac{u_x}{v}\right) + N_\tau \frac{\partial^2 u_x}{\partial y^2}, \\ & \frac{\partial u_x}{\partial x} + \frac{\partial u_y}{\partial y} = 0. \end{aligned}\right\} \tag{4.78}$$

Considering the nature of changes in volumetric concentration (Equation 4.12), the function of current can be posited as:

$$\psi = mx^S \zeta(z), \quad z = a^{\frac{1}{t}} x^{\frac{v}{t}} y. \tag{4.79}$$

Let us express air velocities and derivatives through the functions introduced thereto:

$$u_x \equiv \frac{\partial \psi}{\partial y} = mx^{S+\frac{v}{t}} a^{\frac{1}{t}} \zeta'; \tag{4.80}$$

$$u_y \equiv -\frac{\partial \psi}{\partial x} = -mx^{S-1}(s\zeta + \frac{v}{t}\zeta' z); \tag{4.81}$$

$$\frac{\partial u_x}{\partial x} = ma^{\frac{1}{t}} x^{s+\frac{\nu}{t}-1} \left[\left(s + \frac{\nu}{t} \right) \zeta' + \frac{\nu}{t} \zeta'' z \right]; \tag{4.82}$$

$$\frac{\partial u_x}{\partial y} = ma^{\frac{2}{t}} x^{s+2\frac{\nu}{t}} \zeta''; \tag{4.83}$$

$$\frac{\partial^2 u_x}{\partial y^2} = ma^{\frac{3}{t}} x^{s+3\frac{\nu}{t}} \zeta'''. \tag{4.84}$$

Substitution of these relations into the system equation (4.78) would yield:

$$m^2 a^{\frac{2}{t}} x^{2\left(s+\frac{\nu}{t}\right)-1} \cdot \left\{ \zeta' \left[\left(s + \frac{\nu}{t} \right) \zeta' + \frac{\nu}{t} \zeta'' z \right] - \zeta'' \left[s\zeta + \frac{\nu}{t} \zeta' z \right] \right\} =$$
$$= mN_\tau a^{\frac{3}{t}} x^{s+\frac{3}{t}} \zeta''' + Dx^\gamma e^{-z^t} \left| 1 - m\frac{x^{s+\frac{\nu}{t}}}{\nu} a^{\frac{1}{t}} \zeta' \right|^n \cdot \left(1 - m\frac{x^{s+\frac{\nu}{t}}}{\nu} a^{\frac{1}{t}} \zeta' \right). \tag{4.85}$$

Then, introducing the following substitution for clarity

$$mx^{s+\frac{\nu}{t}} a^{\frac{1}{t}} / \nu = K(x), \tag{4.86}$$

$$mN_\tau a^{\frac{3}{t}} x^{s+3\frac{\nu}{t}-\gamma} / D = N(x), \tag{4.87}$$

with

$$m = \sqrt{\frac{D}{s+\frac{\nu}{t}}} a^{-\frac{1}{t}}, \tag{4.88}$$

$$s + \frac{\nu}{t} = \frac{1+\gamma}{2} = \frac{1}{2} + \frac{1}{2}\left(\frac{\nu}{t} + \frac{n}{2} \right), \tag{4.89}$$

we would end up with the following differential equation:

$$\zeta'^2 - \frac{s}{s+\frac{\nu}{t}} \zeta\zeta'' = e^{-z^t} \left| 1 - K(x)\zeta' \right|^n (1 - K(x)\zeta') + N(x) \cdot \zeta''', \tag{4.90}$$

which, with

$$K(x) \approx \overline{K(x)} = \frac{1}{x} \int_0^x K(x) dx = K, \tag{4.91}$$

$$N(x) \approx \overline{N(x)} = \frac{1}{x} \int_0^x N(x) dx = N \tag{4.92}$$

would turn into an ordinary equation of self-similar motion (or, more precisely, an equation of quasi-self-similar motion, considering the approximate character of Equations 4.91 and 4.92):

$$\zeta'^2 - \varepsilon_v \zeta\zeta'' = e^{-z^t} \left|1 - K\zeta'\right|^n (1 - K\zeta') + N\zeta''', \tag{4.93}$$

where

$$\varepsilon_v = \frac{s}{s + \frac{v}{t}}. \tag{4.94}$$

The parameter N describing the relation of turbulent viscosity forces to aerodynamic forces can be represented in view of Equations 4.70, 4.88, and 4.89 with the following ratio:

$$N = \frac{2k_n}{\sqrt{(1+\gamma)D}} \cdot \frac{x^{1-\frac{\gamma}{2}+\frac{v}{t}}}{a^{-\frac{1}{t}}}, \tag{4.95}$$

and the value K, accounting for the ratio of air velocity to solid particle velocity, is equal to:

$$K = \sqrt{\frac{D}{1+\gamma}} x^{\frac{\gamma}{2}}. \tag{4.96}$$

Notice that appropriately chosen constants v, t, s, and γ would cancel out the explicit dependence of N and K on x, and Equation 4.93 would indeed describe strictly self-similar motion in the class of power functions (see Equation 4.79) in question. So, for expanding flows with,

$$\frac{v}{t} = -1 \text{ and } n = 2;\ \gamma = 0;\ s = \frac{3}{2};\ \varepsilon_v = 3$$

the self-similar motion can be expressed with the equation:

$$\zeta'^2 - 3\zeta\zeta'' = e^{-z^t} (1 - K\zeta')^2 + N\zeta''', \tag{4.97}$$

where parameters N and K are constant and equal to the following:

$$N = 2k_n a^{\frac{1}{t}} / \sqrt{D};\ K = \sqrt{D}. \tag{4.98}$$

Within the class of plane-parallel flows ($v/t = 0$), the following holds in the case of viscous flow around particles ($n = 0$):

$$K = \sqrt{D};\quad N \approx \overline{N} = \frac{2k_n}{\sqrt{D}bx_k} \int_0^{x_k} x\,dx = \frac{k_n x_k}{\sqrt{D}b}; \tag{4.99}$$

$$\zeta'^2 - \zeta\zeta'' = e^{-z^t}(1 - K\zeta') + \overline{N}\zeta''', \tag{4.100}$$

while the following will hold in the self-similar area of flow ($n = 1, \gamma = 1/2$)

$$K \approx \overline{K} = \sqrt{\frac{D}{1,5}}\frac{1}{x_k}\int_0^{x_k} x^{\frac{1}{4}}dx = \frac{4}{5}\sqrt{\frac{D}{1,5}}x_k^{\frac{1}{4}}, \tag{4.101}$$

$$N \approx \overline{N} = \frac{2k_n}{\sqrt{1,5Db}}\cdot\frac{1}{x_k}\int_0^{x_k} x^{\frac{3}{4}}dx = \frac{8k_n}{7b\sqrt{1,5D}}x_k^{\frac{3}{4}}, \tag{4.102}$$

$$\zeta'^2 - \zeta\zeta'' = e^{-z^t}(1 - \overline{K}\zeta')^2 + \overline{N}\zeta'''. \tag{4.103}$$

Unlike Equation 4.97, Equations 4.100 and 4.103 describe quasi-self-similar motion because parameters $\overline{N}$ and $\overline{K}$ depend on the height of solid particle fall.

Let us now modify the expressions for injected air component velocities using the notation introduced earlier in Equations 4.80, 4.81, 4.82, 4.88, and 4.89

$$u_x = \sqrt{\frac{2D}{1+\gamma}}x^{\frac{1+\gamma}{2}}\zeta', \tag{4.104}$$

$$u_y = -\sqrt{\frac{2D}{1+\gamma}}bx^{\frac{\gamma-1}{2}-\frac{\nu}{t}}\cdot\left[\left(\frac{1+\gamma}{2} - \frac{\nu}{t}\right)\zeta + \frac{\nu}{t}\zeta' z\right], \tag{4.105}$$

$$\frac{\partial u_x}{\partial y} = \sqrt{\frac{2D}{1+\gamma}}\frac{1}{b}x^{\frac{1+\gamma}{2}+\frac{\nu}{t}}\zeta''. \tag{4.106}$$

Boundary conditions:

$$u_y = 0, \ \frac{\partial u_x}{\partial y} = 0 \text{ at } y = 0; \tag{4.107}$$

$$u_x = 0 \text{ at } y \to \infty \tag{4.108}$$

would take the following form with the new notation

$$\zeta = 0, \zeta'' = 0 \text{ at } z = 0; \tag{4.109}$$

$$\zeta' = 0 \text{ at } z \to \infty. \tag{4.110}$$

The integral relation (Equation 4.74) in view of Equations 4.80 and 4.82 would become

$$2m^2x^{2\left(s+\frac{\nu}{t}\right)-1}\cdot a^{\frac{2}{t}}\int_0^\infty\left[\left(s + \frac{\nu}{t}\right)\zeta'^2 + \frac{\nu}{t}\zeta'\zeta'' z\right]dz = Dx^\gamma\int_0^\infty e^{-z^t}\left|1 - K(x)\zeta'\right|^n(1 - K(x)\zeta')dz \tag{4.111}$$

or, by means of Equations 4.88 and 4.89,

$$2\int_0^\infty (\zeta'^2 + \delta_\nu \zeta'\zeta'' z)dz = \int_0^\infty e^{-z^t} \left|1 - K\zeta'\right|^n (1 - K\zeta')dz, \tag{4.112}$$

where

$$\delta_\nu = \frac{\nu / t}{s + \nu / t}. \tag{4.113}$$

In case of a plane-parallel flow of solid particles ($\nu / t = 0$), the expressions for air velocity and the integral relation for changing impulse are greatly simplified. So, in the case of viscous flow around particles ($n = 0$; $\gamma = 0$), the following form would be operative:

$$u_x = \sqrt{2D} x^{\frac{1}{2}} \zeta', \quad u_y = -\sqrt{2D} \frac{b}{2\sqrt{x}} \zeta, \tag{4.114}$$

$$2\int_0^\infty \zeta'^2 dz = \int_0^\infty e^{-z^t} (1 - K\zeta')dz, \tag{4.115}$$

with aerodynamic resistance of particles distributed according to a square law ($n = 1$; $\gamma = 1/2$)

$$u_x = \sqrt{\frac{4}{3} D} x^{\frac{3}{4}} \zeta', \quad u_y = -\sqrt{\frac{4}{3} D} b x^{-\frac{1}{4}} \frac{3}{4} \zeta; \tag{4.116}$$

$$2\int_0^\infty \zeta'^2 dz = \int_0^\infty e^{-z^t} (1 - K\zeta')^2 dz. \tag{4.117}$$

Expressions obtained thus enable the physical sense of ζ' and ζ to be discovered easily. Let

$$\zeta' \to \alpha \quad \text{at} \quad z \to 0, \tag{4.118}$$

then the velocity of air in the jet centerline would be

$$u_m = \sqrt{\frac{4}{3} D} x^{\frac{3}{4}} \alpha, \tag{4.119}$$

and the change in the relative longitudinal velocity component in a cross-section of the jet would be

$$u_x / u_m = \zeta' / \alpha. \tag{4.120}$$

Thus, numerically ζ' is characteristic of changes in relative velocity.

Given a known longitudinal component of velocity vector, the flow of injected air can be determined by integrating along the *OY axis:*

$$q_E = 2\int_0^y u_x dy, \quad q_E \equiv Q_E / (cl_\infty). \tag{4.121}$$

Owing to Equations 4.10 and 4.11,

$$dy = a^{-\frac{1}{t}} x^{-\frac{\nu}{t}} dz = bdz, \tag{4.122}$$

and considering Equation 4.104, we would end up with

$$q_E = 2bu_0\zeta, \tag{4.123}$$

where

$$u_0 = \sqrt{\frac{2D}{1+\gamma}} x^{\frac{1+\gamma}{2}} \tag{4.124}$$

or

$$\frac{q_E}{2bu_0} = \frac{Q_E}{2Bcu_0} = \zeta. \tag{4.125}$$

Therefore, ζ is specifically the relative airflow in a $2z$ wide jet. The total flow rate of the injected air would be equal to

$$Q_E = 2Bcu_0\zeta_\infty, \tag{4.126}$$

where

$$\zeta_\infty = \lim_{z\to\infty} \zeta. \tag{4.127}$$

Let us illustrate another possible way of deriving Equation 4.93. Changing the choice of m and N parameters in a particular fashion would result in a general equation that, in turn, would yield that of a free air jet as its special case. Let us convert the initial equation (4.85). Positing

$$N_\tau = 1/\mathrm{Re} - const, \quad \mathrm{Re} = cl_\infty / \nu$$

and introducing a new independent variable

$$\eta = a^{-\frac{1}{t}} z = x^{\frac{\nu}{t}} y,$$

the equation will take the form:

$$m^2 x^{2s-1+2\frac{\nu}{t}} \left\{ \zeta'^2 \left(s + \frac{\nu}{t} \right) - s\zeta\zeta'' \right\} = \frac{m}{\mathrm{Re}} x^{s+3\frac{\nu}{t}} \zeta''' + Dx^\gamma e^{-a\eta^t} \left| 1 - m \frac{x^{s+\frac{\nu}{t}}}{v} \zeta' \right| \left(1 - m \frac{x^{s+\frac{\nu}{t}}}{v} \zeta' \right).$$

Next, let us choose the m that would make the following relation hold true for forces:

$$m^2 x^{2s-1+2\frac{\nu}{t}} \bigg/ \left(\frac{m}{\mathrm{Re}} x^{S+3\frac{\nu}{t}}\right) = 1; \quad Dx^{\gamma} \bigg/ \left(\frac{m}{\mathrm{Re}} x^{s+3\frac{\nu}{t}}\right) = G.$$

Then, positing

$$s = 1 + \frac{\nu}{t}; \quad m = 1/\mathrm{Re},$$

the previously introduced parameters N and K can be converted into new variables

$$G = D \cdot \mathrm{Re}^2 x^{\gamma-1-4\frac{\nu}{t}} \approx D\,\mathrm{Re}^2 \overline{x^{\gamma-1-4\frac{\nu}{t}}}, \tag{4.128}$$

$$K = \frac{1}{\sqrt{2}\,\mathrm{Re}} x^{\frac{1}{2}+2\frac{\nu}{t}} \approx \frac{1}{\sqrt{2}\,\mathrm{Re}} \overline{x^{\frac{1}{2}+2\frac{\nu}{t}}}, \tag{4.129}$$

and the differential equation

$$\left(1 + 2\frac{\nu}{t}\right)\zeta'^2 - \left(1 + \frac{\nu}{t}\right)\zeta\zeta'' = Ge^{-a\eta'} \left|1 - K\zeta'\right|^n (1 - K\zeta') + \zeta''' \tag{4.130}$$

will describe a quasi-self-similar flow.

With

$$\frac{\nu}{t} = -\frac{1}{4}; \quad \gamma = 0; \quad v = \sqrt{2x}, \tag{4.131}$$

this flow will be strictly self-similar, as

$$\gamma - 1 - 4\frac{\nu}{t} = 0 \quad \text{and} \quad G = D \cdot \mathrm{Re}^2; \quad K = (\sqrt{2}\,\mathrm{Re})^{-1}.$$

With

$$\gamma = 1 + 4\frac{\nu}{t}; \tag{4.132}$$

$$G = D \cdot \mathrm{Re}^2; \quad K = \frac{1}{\mathrm{Re}} \left(\overline{\frac{x^{1+2\frac{\nu}{t}}}{v}}\right) \tag{4.133}$$

Equation 4.130 will describe quasi-self-similar motion because the parameter K is generally dependent on x.

Velocity vector components assume the following form in this notation:

$$u_x = \frac{1}{\mathrm{Re}} x^{1+2\frac{\nu}{t}} \cdot \zeta', \tag{4.134}$$

$$u_y = -\frac{x^{\frac{\nu}{t}}}{\mathrm{Re}} \left[\left(1 + \frac{\nu}{t}\right)\zeta + \frac{\nu}{t}\zeta'\eta\right], \tag{4.135}$$

$$\frac{\partial u_x}{\partial y} = \frac{1}{\text{Re}} x^{1+3\frac{\nu}{t}} \zeta'', \quad \frac{\partial u_x}{\partial x} = \frac{x^{2\frac{\nu}{t}}}{\text{Re}} \left[\left(1 + 2\frac{\nu}{t}\right)\zeta' + \frac{\nu}{t}\zeta''\eta \right]. \tag{4.136}$$

The boundary conditions become

$$\zeta = 0; \quad \zeta' = \alpha; \quad \zeta'' = 0 \quad \text{at} \quad \eta = 0; \tag{4.137}$$

$$\zeta' = 0 \quad \text{at} \quad \eta \to \infty. \tag{4.138}$$

The integral condition

$$2\int_0^\infty u_x \frac{\partial u_x}{\partial x} \partial y = D \int_0^\infty x^\gamma e^{-a\eta^t} \left|1 - K\zeta'\right|^n (1 - K\zeta')dy \tag{4.139}$$

in light of Equations 4.134 and 4.136 becomes

$$\int_0^\infty \left[\left(1 + 2\frac{\nu}{t}\right)\zeta'^2 + \frac{\nu}{t}\zeta'\zeta''\eta \right] d\eta = \frac{1}{2} G \int_0^\infty e^{-a\eta^t} \left|1 - K\zeta'\right|^n (1 - K\zeta')d\eta \tag{4.140}$$

or*

$$\left(1 + \frac{3}{2}\nu\right) \int_0^\infty \zeta'^2 d\eta = \frac{1}{2} G \int_0^\infty e^{-a\eta^t} \left|1 - K\zeta'\right|^n (1 - K\zeta')d\eta. \tag{4.141}$$

A somewhat different form of integral relation would result if Equation 4.139 is written as

$$\left(\int_0^\infty u_x^2 dy \right)_x - \frac{1}{2} I_0 = \int_0^x \left(D \int_0^\infty x^\gamma e^{-a\eta^t} \left|1 - K\zeta'\right|^n (1 - K\zeta')dy \right) dx \tag{4.142}$$

or, having expressed velocity through newly introduced functions and considering Equation 4.132, we end up with

$$\int_0^\infty \zeta'^2 d\eta = \frac{1}{2} I x^{-\left(2+3\frac{\nu}{t}\right)} + \frac{G}{2 + 3\frac{\nu}{t}} \int_0^\infty e^{-a\eta^t} \left|1 - K\zeta'\right|^n (1 - K\zeta')d\eta, \tag{4.143}$$

$$I = I_0 \text{Re}^2. \tag{4.144}$$

This makes it evident that with $G \neq 0$ and $I_0 \neq 0$ there is no way of arriving at an equation describing a strictly self-similar flow in the power function class being considered. Such an equation can only be obtained with $G = 0$ and $\nu/t = -2/3$. That is, for an immersed jet with an initial impulse I_0 (as well as with $I_0 = 0$ and $\nu/t = -1/4$, $\gamma = 0$)). The first case would turn Equation 4.143) into

* The left sides of Equations 4.140 and 4.141 are equal by identity as $\left(1 + 2\frac{\nu}{t}\right)\zeta'^2 + \frac{\nu}{t}\zeta'\zeta''\eta =$ 2
$\left(1 + \frac{3}{2}\frac{\nu}{t}\right)\zeta'^2 + \frac{1}{2}\frac{\nu}{t}\left(\zeta'^2\eta\right)'$ and $\int_0^\infty \left(\zeta'^2\eta\right)' d\eta = 0$ in view of boundary conditions (Equations 4.137 and 4.138).

$$\int_0^{\infty} \zeta'^2 d\eta = \frac{1}{2} I, \tag{4.145}$$

and the differential equation (4.130) will take the form

$$\zeta''' + \frac{1}{3}(\zeta'^2 + \zeta\zeta'') = 0, \tag{4.146}$$

resolving [163] into

$$\zeta = c\,th \frac{c}{6} \eta, \tag{4.147}$$

$$\zeta' = \frac{c^2}{6}\left(1 - th^2 \frac{c}{6} \eta\right), \tag{4.148}$$

$$\zeta'' = -\frac{c^3}{18}\left(1 - th^2 \frac{c}{6} \eta\right) \cdot th \frac{c}{6} \eta, \tag{4.149}$$

where $c = \sqrt[3]{4{,}5 I_0}$.

Owing to Equation 4.134, the air flow in a jet is generally equal to

$$Q_c = 2\int_0^{\infty} u_x dy = \frac{2}{\text{Re}} x^{1+\frac{\nu}{t}} \int_0^{\infty} \zeta' d\eta = \frac{2}{\text{Re}} x^{1+\frac{\nu}{t}} \cdot \zeta(\infty). \tag{4.150}$$

For a flat immersed jet, considering Equation 4.147,

$$Q_c = \frac{2}{\text{Re}} x^{-\frac{1}{3}} \cdot c. \tag{4.151}$$

4.1.2.2 Approximate Solution of Self-Similar Motion Equation

Equation 4.93 is used to determine the structure of jet streams and the contribution of viscous forces. Considering that a precise solution is only available for a few special cases and that approaches using numerical methods face difficulties,* we will resort to an approximate solution using the Blasius method. We will find solutions for small and large values of the independent variable (in zero and infinity areas) and then "splice" both solutions together in a certain specially selected point z_0. For an appraisal of this method, consider the equation of an immersed air jet (Equation 4.146). To that end, Equation 4.137 will be expressed for smaller η based on a Maclaurin series as follows:

$$\zeta_0 = a_1\eta - a_2 \frac{\eta^3}{3} + \ldots, \tag{4.152}$$

$$\zeta_0' = a_1 - a_2\eta^2 + \ldots, \tag{4.153}$$

$$\zeta_0'' = -2a_2\eta + \ldots, \tag{4.154}$$

* Difficulties arise due to the nonlinearity of the boundary problem, although solving the Cauchy problem requires fitting the value of α.

where

$a_1 = \alpha; \quad a_2 = \alpha^2 / 6.$

Assume that the solution at infinity takes the form

$$\zeta_\infty = B + u(\eta), \tag{4.155}$$

where B is a certain variable much larger than the function $u(\eta)$ with $\eta \to \infty$. Substitution of values for ζ_∞ into Equation 4.146 will result in the following differential equation

$$u''' = -\frac{1}{3}(u'^2 + Bu'') \tag{4.156}$$

or, assuming the following with $\eta \to \infty$

$$u'^2 << Bu'', \tag{4.157}$$

$$\frac{u'''}{u''} = -B/3, \tag{4.158}$$

whence

$$u'' = -Ae^{-\frac{B}{3}\eta}, \quad A = \text{const} \tag{4.159}$$

and then

$$\zeta_\infty = B - \frac{9}{B^2} Ae^{-\frac{B}{3}\eta}, \tag{4.160}$$

$$\zeta'_\infty = \frac{3}{B} Ae^{-\frac{B}{3}\eta}, \tag{4.161}$$

$$\zeta''_\infty = -Ae^{-\frac{B}{3}\eta}. \tag{4.162}$$

The splicing condition calls for the values of ζ_∞ and ζ_0, as well as ζ'_∞ and ζ'_0, and ζ''_∞ and ζ''_0 to be equal in point $\eta = z_0$, that is,

$$\left.\begin{aligned} a_1 z_0 - a_2 \frac{z_0^3}{3} &= B - \frac{9}{B^2} Ae^{-\frac{B}{3}z_0}, \\ a_1 z_0 - a_2 z_0^2 &= \frac{3}{B} Ae^{-\frac{B}{3}z_0}, \\ -2a_2 z_0 &= -Ae^{-\frac{B}{3}z_0}. \end{aligned}\right\} \tag{4.163}$$

The splicing point can be chosen, for example, from an integral relation (such as Equation 4.145):

$$\int_0^{z_0} \zeta_0'^2 d\eta + \int_{z_0}^{\infty} \zeta_\infty'^2 d\eta = \frac{1}{2} I, \tag{4.164}$$

whence, considering

$$\zeta'_\infty = b_1 e^{-b_2(\eta - z_0)}, \quad b_1 = \frac{6}{b} a_2 z_0, \quad b_2 = \frac{B}{3}, \tag{4.165}$$

the following equation is obtained:

$$a_1^2 - 2a_1 a_2 \frac{z_0^3}{3} + a_2^2 \frac{z_0^5}{5} + \frac{54}{c^3} a_2^2 z_0^2 = \frac{1}{2} I, \tag{4.166}$$

closing the set (Equation 4.163) and enabling us, given a known impulse I, to find the constants α, c, A, and z_0 and, therefore, to determine the coefficients a_1, a_2, b_1, and b_2 (Table 4.1).

In accordance with Equation 4.150, the air flow rate in a jet would equal

$$Q_0 = \frac{2}{\text{Re}} x^{-\frac{1}{3}} B. \tag{4.167}$$

As the data in the final columns of Table 4.1 indicate, air flow rates calculated using Equation 4.167, in the area $10 < I < 10^6$, are virtually identical to values determined using a more precise formula (Equation 4.151). Relative error never exceeds 10%. There is a similar satisfactory alignment between results for the velocity diagram in a cross-section of the jet (Figure 4.3). Due to identical behavior of integral curves, similar satisfactory application of the Blasius method may be expected for solving the equation for a jet of loose matter as well.

4.1.2.3 Uniformly Distributed Particles

Using the method described earlier, let us first consider the simplest but the most characteristic case of a uniform distribution of particles in a jet, that is, the limit case (with $t \to \infty$) of Equation 4.93). In this case,

$$e^{-z^t} = \begin{cases} 1 & \text{at} \quad z < 1, \\ 0 & \text{at} \quad z > 1, \end{cases} \tag{4.168}$$

TABLE 4.1
Coefficients for Equations 4.153, 4.165, and 4.167

I	a_1	a_2	z_0	b_1	B	C
1	0.196	0.006	3.24	0.129	0.963	1.65
5	1.20	0.242	1.31	0.792	2.39	2.82
10	2.18	0.790	0.97	1.433	3.21	3.56
20	3.74	2.33	0.74	2.46	4.20	4.48
50	7.32	8.92	0.53	4.82	5.89	6.08
100	11.9	23.7	0.42	7.84	7.51	7.66
10^3	57.2	54.6	0.19	37.7	16.5	16.5
10^4	$2.68 \cdot 10^2$	$1.19 \cdot 10^4$	$8.76 \cdot 10^{-2}$	176	35.6	35.6
10^5	$1.24 \cdot 10^3$	$2.58 \cdot 10^5$	$4.06 \cdot 10^{-2}$	818.5	76.7	76.6
10^6	$5.78 \cdot 10^3$	$5.56 \cdot 10^5$	$1.88 \cdot 10^{-2}$	3800	165	165

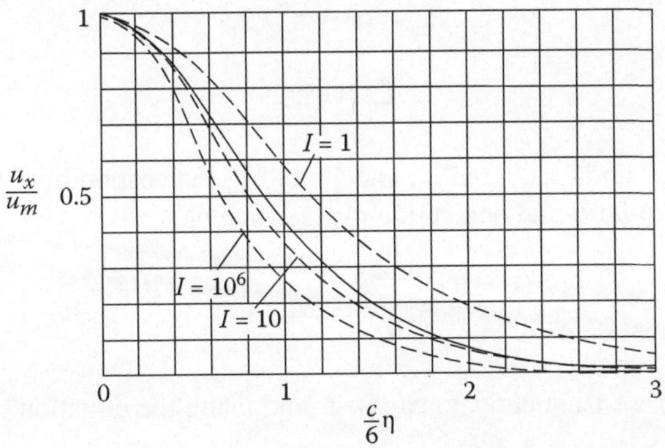

FIGURE 4.3 Changes in relative air velocities through the cross-section of the jet (u_m is the velocity of air along the centerline of the jet; the solid line corresponds to the plot of Equation 4.148; and the dashed lines correspond to plots using Equations 4.153 and 4.165).

and, therefore, plane-parallel flows ($\varepsilon_v = 1, \quad \delta_v = 0$) will have the following equations describing the structure of air currents in the "zero" area (at $z < 1$):

$$\zeta'^2 - \zeta\zeta'' = |1 - K\zeta'|^n (1 - K\zeta') + N\zeta''', \tag{4.169}$$

$$\zeta(0) = 0; \zeta'(0) = \alpha; \zeta''(0) = 0 \tag{4.170}$$

and in the "infinity" area

$$\zeta'^2 - \zeta\zeta'' = N\zeta''', \tag{4.171}$$

$$\zeta(\infty) = B; \zeta'(\infty) = 0; \zeta''(\infty) = 0. \tag{4.172}$$

The solution of Equation 4.171 takes the form

$$\zeta_\infty = B - A\frac{N^2}{B^2}e^{-\frac{B}{N}z}, \tag{4.173}$$

$$\zeta'_\infty = A\frac{N}{B}e^{-\frac{B}{N}z}, \tag{4.174}$$

$$\zeta''_\infty = -Ae^{-\frac{B}{N}z}. \tag{4.175}$$

Accordingly, at "zero," owing to Equation 4.170 and

$$\zeta'''(0) = \frac{1}{N}\left[\alpha^2 - (1 - K\alpha)^2\right] = -\varepsilon, \tag{4.176}$$

it holds that

$$\zeta_0 = \alpha z - \varepsilon z^3 / 6, \tag{4.177}$$

$$\zeta_0' = \alpha - \varepsilon z^2 / 2, \tag{4.178}$$

$$\zeta_0'' = -\varepsilon z. \tag{4.179}$$

By equating ζ_∞ and ζ_0, ζ_∞' and ζ_0', and ζ_∞'' and ζ_0'', correspondingly, at point $z_0 = 1$, we can find α, A, and B and determine the coefficients

$$\frac{B}{N} = \frac{2\varepsilon}{2\alpha - \varepsilon}; A = \varepsilon e^{\frac{B}{N}}, \tag{4.180}$$

where the α (and, consequently, ε) can be found using the equation

$$N\varepsilon^2 - \left(\alpha - \frac{\varepsilon}{2}\right)^3 - \varepsilon\left(\alpha - \frac{\varepsilon}{2}\right)\left(\alpha - \frac{\varepsilon}{6}\right) = 0. \tag{4.181}$$

The value of ζ_∞', considering Equation 4.180, can be found using the equation

$$\zeta_\infty' = \varepsilon \frac{N}{B} e^{-\frac{B}{N}(z-1)}. \tag{4.182}$$

Values of α, ε, and B are listed in Table 4.2.

A comparison of the obtained results with the numerical solutions of Equations 4.169 and 4.171 illustrates (Figures 4.4 through 4.7) that the Blasius method works satisfactorily in the area $N \geq 1$ when turbulent viscosity forces are comparable to or larger than aerodynamic forces.

Relative error in calculating longitudinal velocity components in a transverse cross-section of jet at $N \geq 1$ and $K \leq 1$ stays below 5%, decreasing even further below 3% for injected air flow rate. In this area, the velocity diagram becomes flattened further as viscosity forces increase. The longitudinal component of air velocity outside the jet is practically equal to the velocity inside the jet.

In the area of small viscosity forces (at $N < 1$), the velocity diagram is characterized by a notable velocity gradient at the boundary of the material jet. The boundary of the flow of injected air moves closer to the jet centerline as N decreases. Let us assume the boundary to be the distance z_c to a point where the longitudinal velocity component is equal to 10% of centerline velocity. While at $N = 1$, the half-width of the air stream is $z_c = 3.5$; at $N = 0.1$, it becomes three times narrower, $z_c = 1.2$. Finally, at $N \leq 0.01$, the air jet boundary becomes almost ideally aligned with the boundary of the plane-parallel flow of material particles. There is virtually no accompanying airflow outside of the jet. Expressing ζ_∞ at $N \to 0$ via ζ_∞^0, let us follow changes in the relation $\zeta_\infty / \zeta_\infty^0$ as viscosity forces decrease (Figure 4.8a, dashed curves). There is a clear asymptotic character of changes in this variable. In the area of greater viscosity forces, the ratio $\zeta_\infty / \zeta_\infty^0$, sharply increasing at $N > 1$, becomes much larger than one. Air injection outside of the jet rises to a significant proportion. Within the $N < 0.5$ area, the ratio $\zeta_\infty / \zeta_\infty^0$ is virtually equal to one.

TABLE 4.2
Parameters Used in Equations 4.173 through 4.182

No.	α	ε	*B*	α	ε	*B*
$K=0$				$K=0.5$		
10^{-4}	1.000	1.999	0.667	0.667	1.333	0.445
10^{-3}	0.999	1.992	0.667	0.666	1.326	0.445
10^{-2}	0.990	1.923	0.670	0.660	1.864	0.950
10^{-1}	0.925	1.445	0.713	0.621	0.891	0.507
10^{0}	0.711	0.494	1.065	0.505	0.304	0.863
10^{1}	0.421	0.082	2.164	0.334	0.058	1.914
10^{2}	0.211	0.010	4.643	0.185	0.008	4.351
10^{3}	0.100	0.001	10.00	0.093	0.001	9.686
10^{4}	0.046	0.0001	21.54	0.045	0.0001	21.22
$K=1$				$K=1.5$		
10^{-4}	0.500	0.999	0.333	0.400	0.799	0.267
10^{-3}	0.500	0.993	0.334	0.400	0.793	0.267
10^{-2}	0.495	0.935	0.340	0.396	0.739	0.274
10^{-1}	0.469	0.626	0.403	0.377	0.474	0.339
10^{0}	0.393	0.213	0.744	0.323	0.161	0.663
10^{1}	0.278	0.044	1.739	0.239	0.036	1.607
10^{2}	0.166	0.007	4.114	0.150	0.006	3.915
10^{3}	0.088	0.001	9.404	0.083	0.001	9.149
10^{4}	0.044	0.0001	20.91	0.042	0.0001	20.62

4.1.2.4 One-Dimensional Problem

Thus, at smaller viscosity forces, air injection is confined within the boundaries of material flow. In this case, the longitudinal component of air velocity only sharply changes at the boundary of the air current, where it becomes virtually equal to zero. Therefore, the flow of the injected air may be considered devoid of gradient (i.e., cross-sectional velocity changes are similar to changes in concentration). For example, consider an axially symmetric jet of radius b with uniformly distributed particles

$$\beta = \frac{\overline{\beta}_0}{b^2 v} \gamma(r), \tag{4.183}$$

where

$$\gamma(r) = 1 \quad \text{at} \quad 0 \le r \le b, \tag{4.184}$$

$$\gamma(r) = 0 \quad \text{at} \quad r > b. \tag{4.185}$$

Assuming that air velocity varies similarly to

$$u_x = \omega(x) \cdot \gamma(r), \tag{4.186}$$

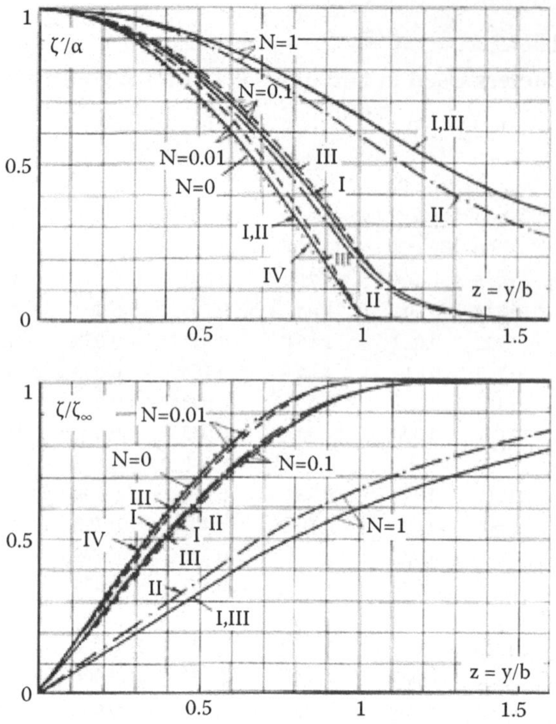

FIGURE 4.4 Changes in air velocities and flow rates at $N \leq 1$ and $K = 0$: (I) numerical method; (II) solution without considering air viscosity forces in a stream of material; (III) approximate solution allowing for viscosity forces; (IV) limit case ($N \to 0$).

let us formulate the one-dimensional problem using the momentum conservation law.

Equation 4.77 would then become

$$\frac{\partial}{\partial x}\int_0^{\infty} u_x^2 r dr = \int_0^{\infty} \frac{\bar{\beta}_0 \gamma(r)}{b^2 \varepsilon v} \left|v - u_x\right|^n \left(v - u_x\right) r dr. \tag{4.187}$$

In view of

$$\int_0^{\infty} \omega^2 \gamma^2(r) r dr = \omega^2 b^2 / 2 \tag{4.188}$$

$$\int_0^{\infty} \frac{\bar{\beta}_0 \gamma(r)}{b^2 \varepsilon \upsilon} \left|v - \omega \cdot \gamma(r)\right|^n \left(v - \omega \cdot \gamma(r)\right) r dr = \frac{\bar{\beta}_0}{2\varepsilon v}\left|v - \omega\right|^n (v - \omega),$$

the integral relation (Equation 4.187) could be reduced to the following differential equation:

$$\frac{d\omega^2}{dx} = \frac{\bar{\beta}_0}{b^2 \varepsilon v}\left|v - \omega\right|^n (v - \omega). \tag{4.189}$$

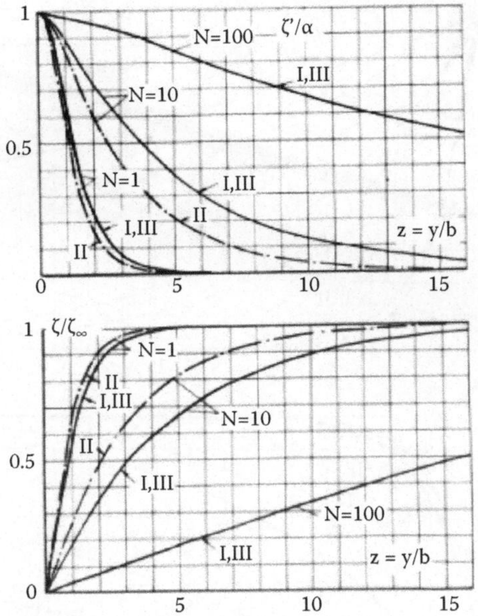

FIGURE 4.5 Changes in air velocities and flow rates at $N \geq 1$ and $K = 0$: (I) numerical method; (II) solution without considering air viscosity forces in a stream of material; (III) approximate solution allowing for viscosity forces.

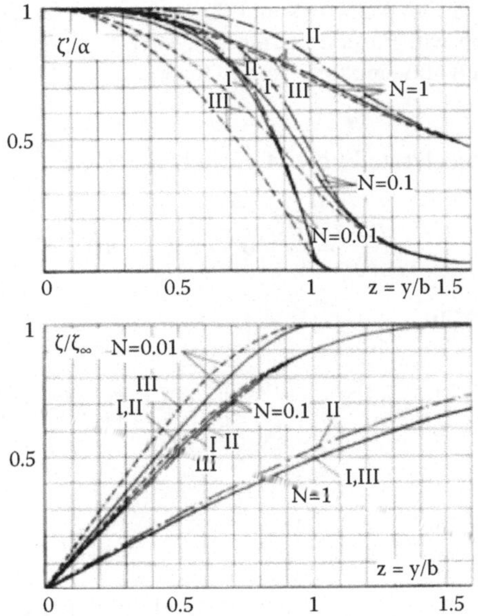

FIGURE 4.6 Changes in air velocities and flow rates at $N \leq 1$ and $K = 1$: (I) numerical method; (II) solution without considering air viscosity forces in a stream of material; (III) approximate solution allowing for viscosity forces.

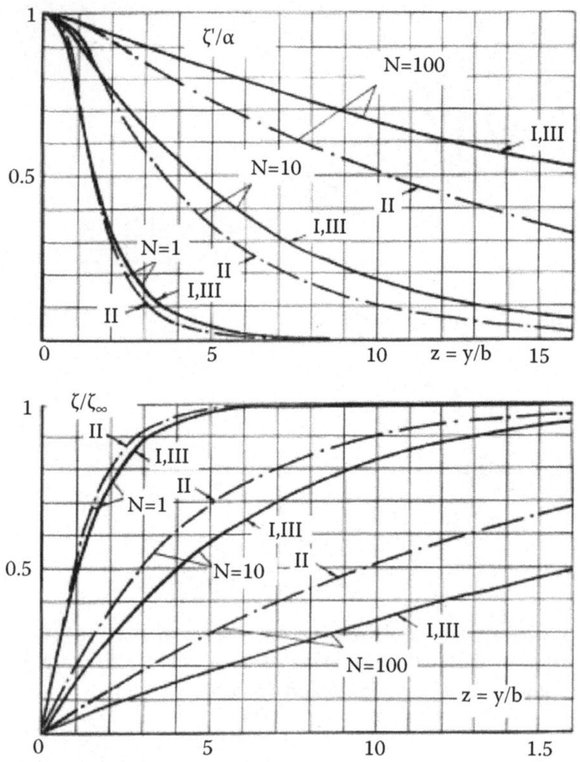

FIGURE 4.7 Changes in air velocities and flow rates at $N \geq 1$ and $K = 1$: (I) numerical method; (II) solution without considering air viscosity forces in a stream of material; (III) approximate solution allowing for viscosity forces.

Assuming a linearly accelerated flow of particles $\left(\upsilon = \sqrt{2x}\right)$, we end up with

$$\frac{d\omega^2}{d\upsilon} = \frac{\bar{\beta}_0}{b^2\varepsilon}|\upsilon - \omega|^n (\upsilon - \omega) \tag{4.190}$$

or, in dimensionless form,

$$u^* \frac{du^*}{d\upsilon^*} = \frac{\bar{\beta}_0}{2b^2\varepsilon} \cdot \left(\frac{2b^2\varepsilon}{\bar{\beta}_0}\right)^n |\upsilon^* - u^*|^n (\upsilon^* - u^*), \tag{4.191}$$

where

$$u^* = \omega \frac{\bar{\beta}_0}{2b^2\varepsilon}; \qquad \upsilon^* = \upsilon \frac{\bar{\beta}_0}{2b^2\varepsilon}. \tag{4.192}$$

The flow rate of air injected by a linearly accelerated material flow is equal to

$$q_E = 2\pi \int_0^\infty u_x r dr \tag{4.193}$$

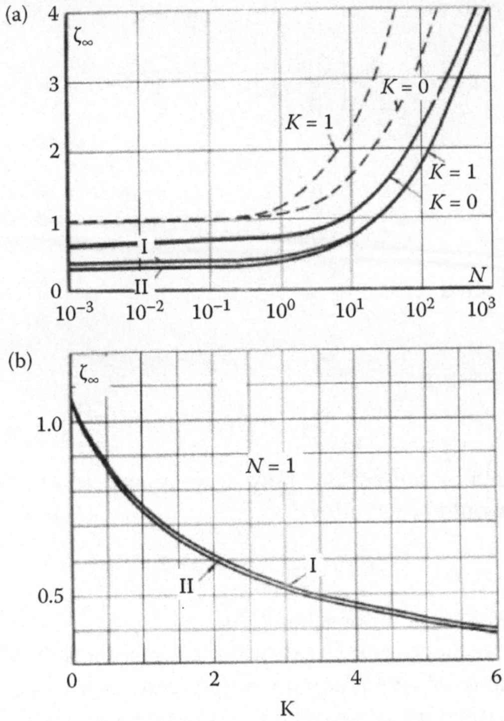

FIGURE 4.8 Changes in airflow within a flat jet of freely falling particles depending on (a) the viscosity force and (b) the force of interaction between components.

or, in view of Equation 4.186

$$q_E = \pi b^2 \omega = \frac{2\pi b^4 \varepsilon}{\bar{\beta}_0} u^*. \tag{4.194}$$

Considering that

$$u^* / \upsilon^* = \varphi, \tag{4.195}$$

the flow rate of injected air can be expressed by means of components' slip ratio

$$q_E = \pi b^2 \varphi \upsilon. \tag{4.196}$$

For a flat jet with a half-breadth b, the corresponding ratios would become

$$\beta = \frac{\bar{\beta}_0}{b\upsilon} \gamma(r), \qquad \bar{\beta}_0 = \frac{G_1}{2l_\infty c\rho_1}, \tag{4.197}$$

and, based on the integral relation (Equation 4.74), with a linearly accelerated flow of material $(dx = \upsilon d\upsilon)$,

$$\omega \frac{d\omega}{d\upsilon} = \frac{\bar{\beta}_0}{2b\varepsilon} |\upsilon - \omega|^n (\upsilon - \omega) \tag{4.198}$$

or, in dimensionless form,

$$u^* du^* = \frac{\bar{\beta}_0}{2b\varepsilon}\left(\frac{2b\varepsilon}{\bar{\beta}_0}\right)^n \left|\upsilon^* - u^*\right|^n \left(\upsilon^* - u^*\right) dv^*, \tag{4.199}$$

where

$$u^* = \omega\frac{\bar{\beta}_0}{2b\varepsilon}, \qquad \upsilon^* = \upsilon\frac{\bar{\beta}_0}{2b\varepsilon}. \tag{4.200}$$

The flow rate of injected air is

$$q_E = 4b^2\varepsilon u^* / \bar{\beta}_0 \tag{4.201}$$

or

$$q_E = 2b\varphi\upsilon. \tag{4.202}$$

At $n = 1$ (in the area of self-similar flow around particles) in Equation 4.191, the resulting equation would be as follows:

$$u^* du^* = \left(\upsilon^* - u^*\right)^2 d\upsilon^* \tag{4.203}$$

as first studied by Neikov and Logachev [70].

Despite its simple appearance, Equation 4.203 is unsolvable in quadratures. Approximate solutions involving certain simplifications are listed in Table 4.3. The estimation of these solutions is provided by a comparison with a numerical solution for Equation 4.203 in homogeneous initial conditions (Table 4.4).

Integral curves of the equation

$$\frac{du^*}{d\upsilon^*} = \left|\upsilon^* - u^*\right|\left(\upsilon^* - u^*\right) / u^* \tag{4.204}$$

become "forgetful" of their initial conditions rather soon, tending toward a "zero-level" integral curve (Figure 4.9). The equation of the latter may be described with enough accuracy using the equation for inflection points of integral curves:

$$\upsilon^* = 2\varphi^2 / \left[(1-\varphi)(1-\varphi^2)\right], \quad \varphi = \mathrm{u}^* / \upsilon^*. \tag{4.205}$$

The values of φ calculated using this formula are somewhat less than the values obtained with the "zero-level" integral curve. Relative error at $\upsilon^* = 0.1$ amounts to 6.6% and falls to zero due to decreasing absolute value with increasing υ^* (reaching 0.3% already at $\upsilon^* = 4$).

A similar situation may be noted in the numerical solution of the equation

$$\frac{d\varphi}{d\upsilon^*} = \frac{|1-\varphi|(1-\varphi)}{\varphi} - \frac{\varphi}{\upsilon^*}, \tag{4.206}$$

which results from Equation 4.204 by replacing $u^* = \varphi\upsilon^*$. Integral curves of the second equation rapidly follow the curve passing through coordinate origin (Figure 4.11a). Conspicuously, rapid changes of φ along the "zero-level" curve are confined to

TABLE 4.3
Solutions to Equation for Air Injection Induced by Solid Material Jet with Uniformly Distributed Particles (One-Dimensional Problem)

No	Differential Equations	Solutions
1	2	3
I. Exact solutions		
1.	$\frac{d\omega^2}{dx} = \frac{\beta_c}{\varepsilon}/(1+\omega);\ \beta_c = \bar{\beta}_0/b^2$ Obtained from the input equation (4.189) for a case when the relative velocity of flow of material is fixed and equal to its terminal velocity (i.e., $\upsilon - \omega = 1$)	$\omega^2 - \omega_0^2 + 2\frac{\omega^3 - \omega_0^3}{3} = \frac{\beta_c}{\varepsilon}(x - x_0)$
2.	Equation 4.203 The flow of material is linearly accelerated.	$u^* = \sum_{n=0}^{\infty} a_n(\upsilon^* - \upsilon_0^*)^n$, $a_n = \frac{1}{na_0}\left[\sum_{i=0}^{n-1} b_{n-1-i}\cdot b_i - \sum_{i=0}^{n-2} a_{n-1-i}\cdot a_{i+1}\cdot(i+1)\right]$, where $a_0 = u_0^*$; $b_0 = a_0 - \upsilon_0^*$; $b_1 = a_1 - 1$; $b_2 = a_2$; ... $b_k = a_k$ $(k = 2, 3, 4 \ldots)$
II. Approximate solutions		
3.	$\frac{dz}{dt} = (t - z)^2$ Obtained from Equation 4.190, assuming $n = 1$, $d\omega^2 \approx 2\bar{\omega}d\omega$, where $\bar{\omega}$ is the average air velocity within [0, x] and introducing new variables $z = \omega A^{-1}; t = \upsilon A^{-1};\ A = \sqrt{\frac{2\varepsilon b^2\bar{\omega}}{\bar{\beta}_0}}$.	$z = t\text{-th}(t - t_0 + \delta);\quad \delta = \frac{1}{2}\ln\frac{1+(t_0 - z_0)}{1-(t_0 - z_0)}$. With $\omega_0 = \upsilon_0 = 0$ and $\bar{\omega} = \frac{1}{2}\omega$ $\varphi = 1 - \sqrt{\frac{\varphi}{2\upsilon^*}}\,th\sqrt{\frac{2\upsilon^*}{\varphi}}$, $\varphi = u^*/\upsilon^*$. The relative error of this relation stays below + 10%.
4.	Averaging air velocities on the right side of Equation 4.203 leads to the following equation with separable variables: $u^* du^* = (\upsilon^* - \bar{u}^*)^2\, d\upsilon^*$	$(u^*)^2 - (u_0^*)^2 = \frac{2}{3}\left[(\upsilon^* - \bar{u}^*)^3 - (\upsilon_0^* - \bar{u}_0^*)^3\right]$. with $u_0 = \upsilon_0 = 0, \bar{u}^* = 0,5u^*$ $\upsilon^* = \frac{6\varphi^2}{4 - 6\varphi + 3\varphi^2}$, $\varphi = u^*/\upsilon^*$. Relative error of this ratio increases along with υ^*: at $\upsilon^* < 2$ it remains below + 20% while at $\upsilon^* < 4$ it is 30%.
5.	Averaging material flow velocities on the right side of Equation 4.203 leads to the following equation with separable variables: $\frac{u^* du^*}{(\bar{\upsilon}^* - u^*)^2} = d\upsilon^*$	$\upsilon^* - \upsilon_0^* = \bar{\upsilon}^*\frac{u^* - u_0^*}{(\bar{\upsilon}^* - u^*)(\bar{\upsilon}^* - u_0^*)} + \ln\frac{\bar{\upsilon}^* - u^*}{\bar{\upsilon}^* - u_0^*}$. With $u_0 = \upsilon_0 = 0$, $\bar{\upsilon}^* = \frac{1}{2}\upsilon^*$ $\upsilon^* = \frac{2\varphi}{1 - 2\varphi} + \ln(1 - 2\varphi)$, $\varphi = u^*/\upsilon^*$. The relative error of this formula in the $\upsilon^* < 2$ area may be as high as 30%.

(continued)

TABLE 4.3 (Continued)
Solutions to Equation for Air Injection Induced by Solid Material Jet with Uniformly Distributed Particles (One-Dimensional Problem)

No	Differential Equations	Solutions
6.	Replacing the right side of Equation 4.190 $\frac{\bar{\beta}_0}{b^2 \varepsilon \upsilon}(\upsilon-u)^2 \approx \frac{\bar{\beta}_0}{b^2 \varepsilon}\bar{\upsilon}\left(1-\frac{u}{\upsilon}\right)^2,$ where $\bar{\upsilon}$ is the average value of flow velocity, we end up with $\varphi+\upsilon\frac{d\varphi}{d\upsilon}=P\frac{(1-\varphi)^2}{\varphi},$ Where $P=\frac{\bar{\beta}_0\bar{\upsilon}}{2b^2\varepsilon}, \quad \varphi=u/\upsilon, \quad \varphi_0=u_0/\upsilon_0$	$\ln\frac{\upsilon}{\upsilon_0}=-\frac{1}{2\sqrt{P}}\left[a\ln\frac{\varphi+a}{\varphi_0+a}-c\ln\frac{\varphi+c}{\varphi_0+c}\right],$ where $a=-\frac{\sqrt{P}}{\sqrt{P}-1}; \quad c=-\frac{\sqrt{P}}{\sqrt{P}+1}.$ With $\upsilon_0^*=10^{-4}$ and $P=0{,}5\upsilon^*$ $\varphi=\frac{\sqrt{0{,}5\upsilon^*}}{\sqrt{0{,}5\upsilon^*}+1}\left(1-\exp\left\{\left(\sqrt{0{,}5\upsilon^*}+1\right)\left[\frac{\ln\left(1-\varphi+\varphi/\sqrt{0{,}5\upsilon^*}\right)}{\sqrt{0{,}5\upsilon^*}-1}-2\ln 10^4\upsilon^*\right]\right\}\right)$ Relative error of this formula falls with increasing values of υ^*. In the $\upsilon^* > 0.1$ area the error does not exceed 13%.
7.	Linearization of the right side of Equation 4.190 $\frac{\bar{\beta}_0}{b^2 \varepsilon \upsilon}(\upsilon-u)^2 \approx \frac{\bar{\beta}_0}{b^2 \varepsilon \upsilon}(\upsilon-u)\left(\overline{\upsilon-u}\right),$ yields $\varphi+\upsilon\frac{d\varphi}{d\upsilon}=S\frac{(1-\varphi)^3}{\varphi},$ Where $S=\frac{\bar{\beta}_0}{b^2\varepsilon}\left(\overline{\upsilon-u}\right); \ \varphi=\frac{u}{\upsilon}; \ \varphi_0=\frac{u_0}{\upsilon_0}$	$\ln\frac{\upsilon}{\upsilon_0}=-\frac{1}{c-a}\left[a\ln\frac{\varphi+a}{\varphi_0+a}-c\ln\frac{\varphi+c}{\varphi_0+c}\right],$ Where $a=\frac{S}{2}\left(1+\sqrt{1+\frac{4}{S}}\right), \quad c=\frac{S}{2}\left(1-\sqrt{1+\frac{4}{S}}\right).$ With $\upsilon_0^*=10^{-4}$ and $S=0{,}5\left(\upsilon^*-u^*\right)$ $\varphi=-c\left\{1-\exp\left[\frac{a}{c}\ln\left(1+\frac{\varphi}{a}\right)+\frac{a-c}{c}\ln 10^4\upsilon^*\right]\right\}.$ At $\upsilon^* = 0.1$, error amounts to 13% and declines in absolute value with increasing υ^*
8.	With the assumption that $\frac{\bar{\beta}_0}{b^2 \varepsilon \upsilon}(\upsilon-u)^2 \approx \frac{\bar{\beta}_0}{b^2 \varepsilon}\upsilon\overline{\left(1-\frac{u}{\upsilon}\right)^2}$ the Equation 4.190 would become $\frac{u}{\upsilon}\frac{du}{d\upsilon}=S\upsilon, \quad S=\frac{\bar{\beta}_0}{2b^2\varepsilon}\overline{\left(1-\frac{u}{\upsilon}\right)^2}$	$\varphi=\sqrt{\frac{2}{3}S\upsilon\left(1-\frac{\upsilon_0^3}{\upsilon^3}\right)+\varphi_0^2\frac{\upsilon_0^2}{\upsilon^2}},$ Where $\varphi=u/\upsilon, \ \varphi_0=u_0/\upsilon_0.$ With $\varphi_0=0$ and $\overline{(1-\varphi)^2}=(1-\varphi)^2: \ \varphi=\frac{\sqrt{2\upsilon^*/3}}{1+\sqrt{2\upsilon^*/3}}.$ This formula produces an error below –4.5%.
9.	Linearization of the right side $\frac{\bar{\beta}_0}{b^2 \varepsilon \upsilon}(\upsilon-u)^2 \approx \frac{\bar{\beta}_0}{b^2 \varepsilon}\upsilon\left[1-u\overline{\left(\frac{2}{\upsilon}-\frac{u}{\upsilon^2}\right)}\right]$ enables conversion of	$\upsilon^{*3}-\upsilon_0^{*3}=\frac{3}{T^2}\left[T\left(u_0^*-u^*\right)-\ln\frac{1-Tu^*}{1-Tu_0^*}\right].$ With $u_0^*=\upsilon_0^*=0$ and $T\approx\frac{2-\varphi}{\upsilon^*}$

TABLE 4.3 (Continued)
Solutions to Equation for Air Injection Induced by Solid Material Jet with Uniformly Distributed Particles (One-Dimensional Problem)

No	Differential Equations	Solutions
	Equation 4.190 into the following easily integrable equation: $u^* \frac{du^*}{d\upsilon^*} = \upsilon^{*2}\left(1-Tu^*\right)$, where $T=\frac{2b^2\varepsilon}{\beta_0}\left(\frac{2}{\upsilon}-\frac{u}{\upsilon^2}\right)$	it holds that $\frac{\upsilon^*}{3} = -\frac{\varphi}{2-\varphi} - \frac{2\ln(1-\varphi)}{(2-\varphi)^2}$, making it evident that $\varphi = 1$ at $\upsilon^* \to \infty$. The relative error of this formula in the $\upsilon^* < 4$ area remains below +20%.
10.	Linearization of the left and right sides of Equation 4.203 reduces the latter to a simple equation $\frac{du^*}{d\upsilon^*} = \left(\upsilon^* - u^*\right)\left(\frac{\bar{\upsilon}^*}{\bar{u}^*_x} - 1\right)$	$u^* = \upsilon^* - \frac{1}{n}\left(1 - Ce^{-n\upsilon^*}\right)$, where $n = \bar{\upsilon}^*/\bar{u}^* - 1,\ C = \left[1 + n\left(u_0^* - \upsilon_0^*\right)\right]e^{n\upsilon_0^*}$, among other, at $\upsilon_0^* = \upsilon_0^* = 0,\ (C = 1)$ and, considering $n \approx \varphi^{-1} - 1$, we obtain the following equation $\varphi = 1 - \frac{1-\exp\left[-\frac{\upsilon^*}{\varphi}(1-\varphi)\right]}{\upsilon^*(1-\varphi)}\varphi$, Error in the $\upsilon^* > 0.01$ area is limited to 10%.
III.	Solutions at small air velocities ($\upsilon >> \omega$)	
12.	Ignoring the value of ω in the right side of Equation 4.190 (i.e., assuming $\upsilon - \omega \approx \upsilon$), we end up with $\frac{d\omega^2}{d\upsilon} = \frac{\beta_0}{b^2\varepsilon}\upsilon^2$. Solutions that meet this equation indicate the maximum speed of injected air.	$\varphi^2 = \varphi_0^2\left(\frac{\upsilon_0^*}{\upsilon^*}\right)^2 + \frac{2}{3}\left[\upsilon^* - \upsilon_0^*\left(\frac{\upsilon_0^*}{\upsilon^*}\right)^2\right]$. at low initial velocities $\left(\upsilon_0^* = 0\right)$ $\varphi = \sqrt{2\upsilon^*/3}$. This formula produces an error of +22.3% at $\upsilon^* = 0.1$ ($\varphi = 0.211$) and 73.7% at $\upsilon^* = 1$ ($\varphi = 0.47$). Injected air volume $Q_E = \frac{2}{3}\frac{G_M}{\rho_2}\left(\frac{\upsilon_1}{c}\right)^2\frac{1}{2\varphi} = \frac{\upsilon_1}{c}\sqrt{\frac{1}{3}\frac{G_M}{\rho_2}}S\upsilon_1$, where S is the cross-sectional area of the jet
13.	Replacing the right side of Equation 4.190 $\frac{\bar{\beta}_0}{b^2\varepsilon\upsilon}(\upsilon-\omega)^2 \approx \bar{\upsilon}\cdot\upsilon\frac{\bar{\beta}_0}{b^2\varepsilon}$, yields $\frac{d\omega^2}{d\upsilon} = \bar{\upsilon}\frac{\bar{\beta}_0}{b^2\varepsilon}\upsilon$, where $\bar{\upsilon} = \frac{\upsilon+\upsilon_0}{2}$ – const	$\varphi^2 = \varphi_0^2\left(\frac{\upsilon_0^*}{\upsilon^*}\right)^2 + \frac{1}{2}\left[\left(\upsilon^* - \upsilon_0^*\right)\left(\frac{\upsilon^* + \upsilon_0^*}{\upsilon^*}\right)^2\right]$. at low initial velocities $\varphi = \sqrt{0{,}5\upsilon^*}$. This formula gives an error of + 6% at $\upsilon^* = 0.1$ ($\varphi = 0.211$) and 26.3% at $\upsilon^* = 0.4$ ($\varphi = 0.354$). Injected air volume $Q_E = \frac{1}{2}\frac{G_M}{\rho_2}\left(\frac{\upsilon_1}{c}\right)^2\frac{1}{2\varphi} = \frac{1}{2}\frac{\upsilon_1}{c}\sqrt{\frac{G_M}{\rho_2}}S\upsilon_1$

TABLE 4.4
Values of φ

υ^*	u^*	φ	υ^*	u^*	Φ
10^{-8}	10^{-8}	1	0.4	0.1414	0.3535
10^{-7}	10^{-8}	0.1	0.5	0.1903	0.3806
10^{-6}	$1.003 \cdot 10^{-8}$	0.010	0.6	0.2420	0.4034
10^{-5}	$2.762 \cdot 10^{-8}$	$2.762 \cdot 10^{-3}$	0.7	0.2961	0.4230
10^{-4}	$8.109 \cdot 10^{-7}$	$8.109 \cdot 10^{-3}$	0.8	0.3521	0.4402
0.001	$2.526 \cdot 10^{-5}$	$2.526 \cdot 10^{-2}$	0.9	0.4099	0.4555
0.002	$7.081 \cdot 10^{-5}$	$3.540 \cdot 10^{-2}$	1	0.4692	0.4692
0.003	$1.292 \cdot 10^{-4}$	$4.306 \cdot 10^{-2}$	2	1.1205	0.5603
0.004	$1.978 \cdot 10^{-4}$	$4.944 \cdot 10^{-2}$	3	1.8363	0.6121
0.005	$2.750 \cdot 10^{-4}$	$5.500 \cdot 10^{-2}$	4	2.5899	0.6475
0.006	$3.598 \cdot 10^{-4}$	$5.997 \cdot 10^{-2}$	5	3.3691	0.6738
0.007	$4.516 \cdot 10^{-4}$	$6.451 \cdot 10^{-2}$	6	4.1672	0.6945
0.008	$5.496 \cdot 10^{-4}$	$6.870 \cdot 10^{-2}$	7	4.9801	0.7114
0.009	$6.534 \cdot 10^{-4}$	$7.260 \cdot 10^{-2}$	8	5.8049	0.7256
0.01	$7.627 \cdot 10^{-4}$	$7.627 \cdot 10^{-2}$	9	6.6396	0.7377
0.02	$2.099 \cdot 10^{-3}$	0.1050	10	7.4827	0.7483
0.03	$3.779 \cdot 10^{-3}$	0.1260	15	11.790	0.7860
0.04	$5.719 \cdot 10^{-3}$	0.1430	20	16.203	0.8101
0.05	$7.876 \cdot 10^{-3}$	0.1575	30	25.211	0.8404
0.06	$1.022 \cdot 10^{-2}$	0.1703	40	34.372	0.8593
0.07	$1.272 \cdot 10^{-2}$	0.1817	50	43.631	0.8726
0.08	$1.537 \cdot 10^{-2}$	0.1921	60	52.96	0.8827
0.09	$1.815 \cdot 10^{-2}$	0.2017	70	62.343	0.8906
0.1	$2.105 \cdot 10^{-2}$	0.2105	80	71.768	0.8971
0.2	$5.522 \cdot 10^{-2}$	0.2761	90	81.227	0.9025
0.3	$9.602 \cdot 10^{-2}$	0.3201	100	90.716	0.9076

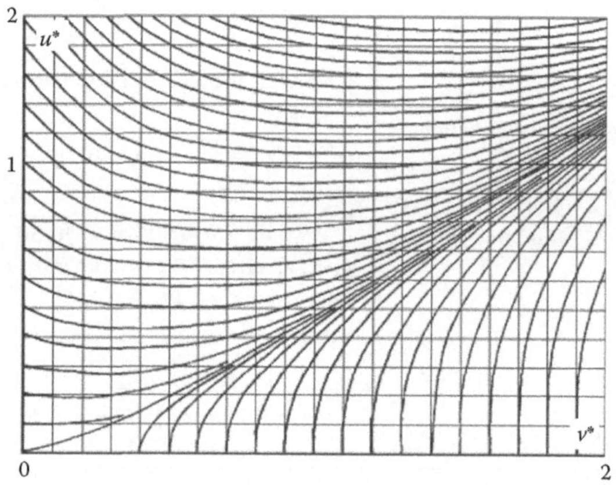

FIGURE 4.9 Changes in airflow velocity along the jet of material (solutions to Equation 4.204).

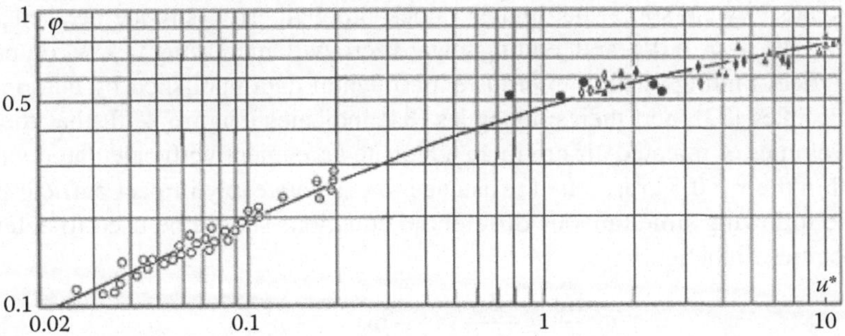

FIGURE 4.10 Figure 4.10 Behavior of the factor φ along the jet (solid line corresponds to data from Table 4.4):
O – water droplets (d = 3.4 mm; D_O, = 0.8 m; G_1 = 0.07–0.16 kg/s; H = 0.5–3.0 m)
Θ – water droplets (d = 3 mm; D_O = 0.22 m; G_1 = 0.08–0.14 kg/s; H = 1–2.7 m)
◊ – iron ore (d = 10–20 mm; D_O = 0.1–0.15 m; H = 1.35–2.5 m; G_1 = 5.2–18.6 kg/s)
♦– iron ore (d = 0.3–20 mm; D_O = 0.08–0.15 m; H = 1.35–2.5 m; G_1 = 4.4–23 kg/s)
● – pellets (d = 13.8 mm; D_O = 0.8–0.15 m; H = 1.5–2,5 m; G_1 = 3–19 kg/s)
Δ – granite (d = 5–10 mm; D_O = 0.8–0.1 m; H = 1.35–2.5 m; G_1 = 3.3–5.4 kg/s)
▲ – granite (d = 10–20 mm; D_O = 0.8–0.15 m; H = 1.35–2.5 m; G_1 = 4.4–23 kg/s)

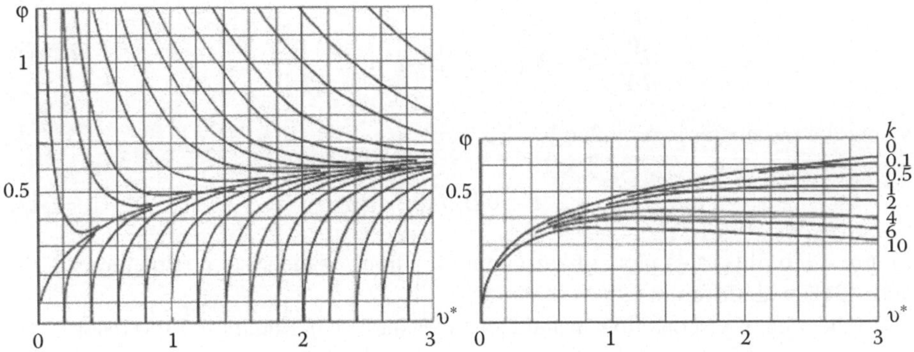

FIGURE 4.11 Changes in slip ratios of components along the jet (a) without accounting for resistance forces—the solution to Equation 4.206, (b) accounting for resistance forces—solutions to Equations 4.210 and 4.211).

short distances from the origin of the jet ($\upsilon^* < 0.1$). Growth of the phase slip ratio consequently slows down, remaining for $\upsilon^* < 3$ within 0,4–0,6 (0.5 on average). The latter condition accounts for adequate accuracy of the approximate equations of the form

$$u^* du^* \approx \upsilon^{*2} \left(1-\bar{\varphi}\right)^2 d\upsilon^*, \tag{4.207}$$

the solution of which is illustrated in Table 4.3 (see Item 8).

The zero-level integral curve is described accurately enough by the equation of inflection points

$$\upsilon^* = \frac{\varphi^3}{(1-\varphi)(1-\varphi^2)}\left(1+\sqrt{\frac{2}{\varphi^2}-1}\right), \tag{4.208}$$

which tends to yield somewhat higher values although error is moderate, remaining below + 2.4% at $\upsilon^* \geq 0{,}1$ (and tending toward zero with increasing υ^*). A comparison of analytical findings with numerous experimental data published by authors [69], their disciples [23], and their colleagues [84] indicates (Figure 4.10) that the measured volumes of injected air are in an adequate agreement with calculated values.

Within the $x > 0.5$ area, when resistance forces noticeably impact particle velocities, the following simultaneous differential equations should be used to determine air velocities in the jet:

$$\left.\begin{aligned} \omega\frac{d\omega}{dx} &= \frac{\beta_0}{2\upsilon_c\varepsilon}(\upsilon_c-\omega)|\upsilon_c-\omega|, \\ \upsilon_c\frac{d\upsilon_c}{dx} &= 1-(\upsilon_c-\omega)|\upsilon_c-\omega|, \end{aligned}\right\}, \tag{4.209}$$

where υ_c is the speed of particles accounting for resistance forces (introducing this variable would enable one to differentiate from the speed of linearly accelerated motion υ).

By employing the velocity υ as a measure of distance and introducing variables u^*, υ^* in accordance with Equation 4.192, these equations can be rewritten as follows:

$$u^*\frac{du^*}{d\upsilon^*} = \frac{\upsilon^*}{t}(t-u^*)|t-u^*|, \tag{4.210}$$

$$\frac{t}{\upsilon^*}\frac{dt}{d\upsilon^*} = 1-K(t-u^*)|t-u^*|,\ K = \left(\frac{2\varepsilon}{\beta_c}\right)^2, \tag{4.211}$$

where the parameter is provided by

$$t = \upsilon_c\beta_c/(2\varepsilon). \tag{4.212}$$

Figure 4.11b illustrates plots of zero integral curves of the system (Equations 4.210 and 4.211) with different values of K.

In the area of viscous flow around particles (at $n = 0$), Equation 4.190 resolves as

$$\left(\frac{\upsilon}{\upsilon_0}\right)^2 = \frac{0{,}5a(1-\varphi_0)-\varphi_0^2}{0{,}5a(1-\varphi)-\varphi^2}\cdot\left[\frac{(\varphi-p)(\varphi_0-q)}{(\varphi-q)(\varphi_0-p)}\right]^{\frac{a}{2(p-q)}}, \tag{4.213}$$

where

$$p = -\frac{a}{4}\left(1+\sqrt{1+\frac{8}{a}}\right);\quad q = \frac{a}{4}\left(\sqrt{1+\frac{8}{a}}-1\right); \tag{4.214}$$

$$a = \frac{\bar{\beta}_0}{b^2\varepsilon};\quad \varphi = \frac{\omega}{\upsilon};\quad \varphi_0 = \frac{\omega_0}{\upsilon_0}. \tag{4.215}$$

Similar findings could be obtained for a flat jet with a half-breadth b. One should just replace $\frac{\bar{\beta}_0}{b^2}$ with $\bar{\beta}_c$ in formulas as defined by Equation 4.16.

4.1.2.5 Exponentially Distributed Particles

Maintaining our assumption that pressure gradients are absent, let us consider a generalized exponential distribution of particles across the jet. From now on, without detriment to generality, we will consider the example of a flat jet with volumetric concentration of particles determined by Equation 4.12.

The splicing method will be used to bring together zero and asymptotic solutions of Equation 4.93. Consider the most characteristic case of a self-similar mode of plane-parallel flow around particles ($n = \varepsilon_v = 1$) with its velocity exceeding that of injected air. Let us find the solution in "zero" and "infinity" areas for the equation

$$\zeta'^2 - \zeta\zeta'' = e^{-z^t}\left(1 - K\zeta'\right)^2 + N\zeta''' \tag{4.216}$$

with the following boundary conditions

$$\zeta(0) = 0; \quad \zeta'(0) = \alpha; \quad \zeta''(0) = 0; \quad \zeta'(\infty) = 0. \tag{4.217}$$

Similar to the case of uniformly distributed particles, searches for $\zeta(z)$ function for smaller z will be based on a Maclaurin's series, limited to four initial terms. Considering that, for the case in question,

$$\zeta'''(0) = \lim_{z\to 0}\zeta''' = -\varepsilon, \tag{4.218}$$

the solution at zero is in no way different from the "zero" solution of Equation 4.169. The desired function ζ as well as its derivatives ζ' and ζ'' are determined by Equations 4.179, 4.180, and 4.181, accordingly. The solution at infinity will be determined in the form

$$\zeta = B + u(z), \tag{4.219}$$

where B is a certain constant, with greater z having $B \gg u(z)$.

The function $u(z)$ can be deduced from the equation

$$u''' = \frac{1}{N}\left[u'^2 - Bu'' - e^{-z^t}\left(1 - Ku'\right)^2\right], \tag{4.220}$$

that can be simplified assuming

$$Ku' \ll 1,\ u'^2 \ll e^{-z^t} \tag{4.221}$$

at large z. In this case, Equation 4.220 is linearized as follows:

$$u''' = \frac{1}{N}\left[-Bu'' - e^{-z^t}\right], \tag{4.222}$$

and its solution becomes

$$u'' = -Ae^{-\frac{B}{N}z} + \frac{1}{N}\int_z^{\infty} e^{\frac{B}{N}(x-z)} e^{-x^t}\,dx, \tag{4.223}$$

$$u' = A\frac{N}{B}e^{-\frac{B}{N}z} - \frac{1}{N}\int_z^{\infty}\left(\int_z^{\infty} e^{\frac{B}{N}(x-z)} e^{-x^t}\,dx\right)dz, \tag{4.224}$$

$$u = -A\left(\frac{N}{B}\right)^2 e^{-\frac{B}{N}z} + \frac{1}{N}\int_z^{\infty}\int_z^{\infty}\left(\int_z^{\infty} e^{\frac{B}{N}(x-z)} e^{-x^t}\,dx\right)dz dz. \tag{4.225}$$

The results differ from asymptotic solution of Equation 4.171 by having integral components accounting for forces of interaction between components (given that a fraction of falling particles occurs "outside" of the jet of material). These components take their simple forms at purely exponential distribution of particles ($t = 1$). The asymptotic solution becomes

$$\zeta_\infty = B - A\left(\frac{N}{B}\right)^2 e^{-\frac{B}{N}z} + e^{-z}/(N-B), \tag{4.226}$$

$$\zeta'_\infty = A\frac{N}{B}e^{-\frac{B}{N}z} - e^{-z}/(N-B), \tag{4.227}$$

$$\zeta''_\infty = -Ae^{-\frac{B}{N}z} + e^{-z}/(N-B). \tag{4.228}$$

The splicing condition at z_0 produces

$$\varphi_1(z_0) \equiv \alpha z_0 - \varepsilon z_0^3/6 = B - A\left(\frac{N}{B}\right)^2 e^{-\frac{B}{N}z_0} + e^{-z_0}/(N-B), \tag{4.229}$$

$$\varphi_2(z_0) \equiv \alpha - \varepsilon z_0^2/2 = A\frac{N}{B}e^{-\frac{B}{N}z_0} - e^{-z_0}/(N-B), \tag{4.230}$$

$$\varphi_3(z_0) \equiv -\varepsilon z_0 = -Ae^{-\frac{B}{N}z_0} + e^{-z_0}/(N-B), \tag{4.231}$$

and the integral relation (Equation 4.117) enables the following equation to be deduced (the case of $K = 0$ being considered, for example):

$$f(z_0) \equiv \alpha^2 z_0 - \alpha\varepsilon z_0^3/3 + \varepsilon^2 z_0^5/20 + \left(A\frac{N}{B}e^{-\frac{B}{N}z_0}\right)^2 N/(2B) - \left(A\frac{N}{B}e^{-\frac{B}{N}z_0}\right)\cdot\frac{e^{-z_0}}{N-B}\cdot 2/\left(1+\frac{B}{N}\right) + \left(\frac{e^{-z_0}}{N-B}\right)^2\cdot\frac{1}{2} = \frac{1}{2}, \tag{4.232}$$

closing the set of combined Equations 4.229 through 4.231 and enabling the constants B, A, α, and z_0 to be obtained for a given N. To that end, the equation

$$\left(e^{-z_0} - \varphi_3 N\right)^2 + N\cdot\varphi_1\cdot\varphi_2\cdot\varphi_3 - \left[(\varphi_1+\varphi_2)\varphi_2 e^{-z_0} + N\varphi_2^3\right] = 0 \tag{4.233}$$

must be solved together with Equation 4.232. It should be noted that, owing to the presence of two exponential functions in an asymptotic solution, joint solution is not always possible. Therefore, the integral relation (Equation 4.232) can be changed with the requirement $f(z_0) - \frac{1}{2} = \min$.

The values of ζ and ζ' will be determined by relations:

$$\zeta = \begin{cases} \alpha z - \frac{\varepsilon}{6} z^3, & \text{if} \quad z \le z_0, \\ B - A_3 e^{-\frac{B}{N}(z-z_0)} + A_2 e^{-(z-z_0)}, & \text{if} \quad z \ge z_0, \end{cases} \tag{4.234}$$

$$\zeta' = \begin{cases} \alpha - \varepsilon \frac{z^2}{2}, & \text{if} \quad z \le z_0, \\ A_1 e^{-\frac{B}{N}(z-z_0)} - A_2 e^{-(z-z_0)}, & \text{if} \quad z \ge z_0, \end{cases} \tag{4.235}$$

where

$$A_1 = A\frac{N}{B} e^{-\frac{B}{N} z_0}; \quad A_2 = e^{-z_0} / (N - B); \quad A_3 = \frac{N}{B} A_1. \tag{4.236}$$

Table 4.5 lists constant values for some N for the case of maximum forces of interaction between components ($K = 0$).

As the data here indicates, the longitudinal velocity component in the $N \ll 1$ area almost perfectly reproduces changes in solid particle concentration across flow. The solution in this case would become* ($z_0 \to 0$)

$$\zeta = B - e^{-z} / (B - N), \tag{4.237}$$

$$\zeta' = e^{-z} / (B - N), \tag{4.238}$$

and, in view of the integral relation

$$\int_0^\infty \zeta'^2 dz = 0.5, \tag{4.239}$$

it holds that

$$B = 1 + N. \tag{4.240}$$

In the extreme case $N \to 0$, the following relation would result:

$$\zeta = 1 - e^{-z}, \quad \zeta' = e^{-z}, \tag{4.241}$$

as a solution of a degenerate differential equation (Equation 4.93) at $N = K = 0$.

---

* Generally, the longitudinal velocity component can be posited as $u_x = u_0 e^{-\zeta y}$ with u_0 and ζ being certain functions of x [45].

TABLE 4.5
Values of Parameters Used in Calculating the Structure of Airflow Injected into a Flat Jet with Exponentially Distributed Falling Particles

N	z_0	α	ε	B	B/N	A1	A2	A3
0.001	0.010	0.991	18.391	1.000	1000	−0.0008	−0.9907	-10^{-6}
0.01	0.093	0.954	8.9270	1.004	100.4	−0.0009	−0.9166	-10^{-5}
0.05	0.183	0.907	3.5383	1.020	20.40	−0.0103	−0.8583	$-5\cdot10^{-4}$
0.10	0.234	0.877	2.3146	1.040	10.396	−0.0289	−0.8422	$-2.8\cdot10^{-4}$
0.22	0.292	0.832	1.3992	1.083	4.924	−0.0927	−0.8650	$-1.9\cdot10^{-3}$
0.44	0.328	0.782	0.8837	1.155	2.629	−0.2736	−1.0078	−0.104
0.6	0.334	0.755	0.7157	1.201	2.002	−0.4754	−1.1909	−0.237
0.8	0.333	0.729	0.5866	1.254	1.568	−0.8816	−1.5776	−0.562
1.0	0.333	0.706	0.5014	1.303	1.303	−1.6873	−2.3661	−1.295
2.0	0.300	0.628	0.3026	1.500	0.750	2.0975	1.4827	2.796
4.0	0.240	0.542	0.1756	1.779	0.445	0.8917	0.3544	2.004
8.0	0.180	0.455	0.0991	2.155	0.269	0.5968	0.1429	2.216
10	0.167	0.429	0.0816	2.299	0.230	0.5373	0.1099	2.337
20	0.110	0.351	0.0438	2.831	0.142	0.4027	0.0522	2.845
40	0.075	0.284	0.0230	3.514	0.088	0.3091	0.0254	3.518
80	0.047	0.228	0.0119	4.384	0.055	0.2404	0.0126	4.386
100	0.046	0.212	0.0096	4.712	0.047	0.2220	0.0100	4.712

With greater speed, the diagram of longitudinal velocities of injected air noticeably differs from exponential distribution of particle concentration (Figure 4.12). Quantitatively, the field of velocities agrees rather well with data calculated using splicing.

Thus, at small viscosity forces, the equation of a gradient-free boundary layer (Equation 4.67) can be greatly simplified

$$u_x \frac{\partial u_x}{\partial x} + u_y \frac{\partial u_x}{\partial y} = F_x. \tag{4.242}$$

Considering that the transverse velocity component is small ($u_x << u_y$), it is possible to posit, for the first approximation, that

$$u_x \frac{\partial u_x}{\partial x} \approx F_x \tag{4.243}$$

or, at low air velocities ($u_x << \upsilon$), in accordance with Equation 4.28, at $n = 0$

$$u_x \frac{\partial u_x}{\partial x} \approx \beta \upsilon^2 / \varepsilon. \tag{4.244}$$

In that case, we could end up with somewhat higher values for the longitudinal component of speed and flow rate. Let us make an estimation of this approximation for the generalized exponential distribution (Equation 4.12). Using Equations 4.80,

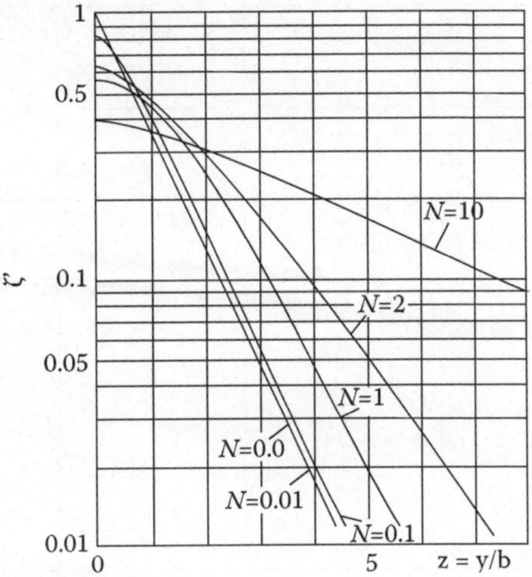

FIGURE 4.12 Changes in function ζ' for a flat jet with exponentially distributed particles.

4.82, 4.88, and 4.89 at $\frac{\nu}{t} = 0$, Equation 4.244 would be transformed into the following simple equation of self-similar motion:

$$\zeta'^2 = e^{-z'}, \tag{4.245}$$

that should not be very difficult to solve.

Figure 4.13 provides plots of functions ζ and ζ' based on these solutions (lower indexes "T" apply to ζ_T and ζ'_T values) and Equation 4.245 (ζ_n and ζ'_n values). As evident from these plots, within the limits of $z < 0.5$, $t \geq 1$, the distribution curve of the longitudinal airspeed component can be described by Equation 4.245. In a limiting case ($t \geq 10$), the curve described by this equation will agree adequately with a precise solution over the entire breadth of the jet.

Therefore, with uniform distribution of particles inside the jet ($\beta = \beta_c - const$), the structure of injected air current can be described with an approximate equation (Equation 4.244) that could be written as follows:

$$u_x \frac{\partial u_x}{\partial \upsilon} = \frac{\beta_c}{\varepsilon}\upsilon^2, \qquad \frac{1}{\upsilon}\frac{\partial u_x}{\partial \upsilon} + \frac{\partial u_y}{\partial y} = 0, \tag{4.246}$$

where it can be deduced that

$$\frac{u_x^2}{2} = \frac{\beta_c}{\varepsilon}\frac{\upsilon^3}{3} + S(y). \tag{4.247}$$

The function $S(y)$ could be determined using the distribution of velocities defined at initial cross-section (at $x = 0$). For example, assuming uniform initial conditions

$$u_x = 0 \text{ at } x = 0 \ (\upsilon = \upsilon_0). \tag{4.248}$$

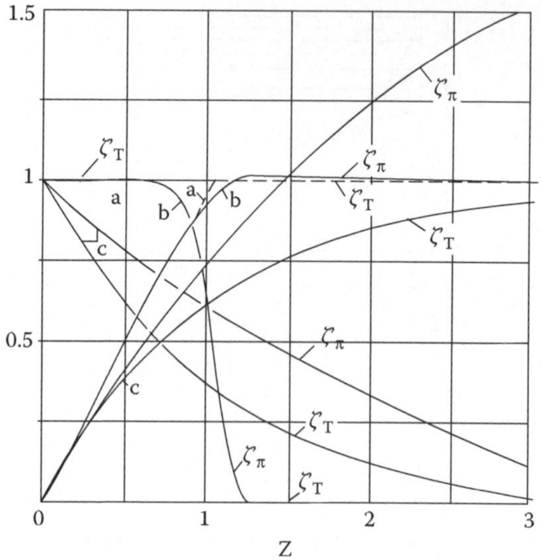

FIGURE 4.13 Changes in functions ζ and ζ′ for a flat jet with exponentially distributed particles and small viscosity forces; $N \to 0$, $K = 0$; (a) $t \to \infty$; (b) $t = 10$; (c) $t = 1$.

In this case,

$$S(y) = -\frac{\beta_c}{\varepsilon} \cdot \frac{\upsilon_0^3}{3} \tag{4.249}$$

and

$$u_x = \sqrt{2\frac{\beta_c}{\varepsilon}\frac{\upsilon^3 - \upsilon_0^3}{3}}. \tag{4.250}$$

The transverse velocity component

$$u_y = -\int_0^y \frac{1}{\upsilon}\frac{\partial u_x}{\partial \upsilon}\,dy + p(x), \tag{4.251}$$

where $p(x)$ is the function determining changes in velocity u_y along the jet (or along a straight line parallel to the centerline). In symmetric jets $u_y = 0$ with $y = 0$, that is

$$p(x) = 0. \tag{4.252}$$

Considering that the longitudinal component u_x is only a function of x, Equation 4.251, accounting for Equation 4.252, would become

$$u_y = -\frac{1}{\upsilon}\frac{du_x}{d\upsilon} \cdot y \tag{4.253}$$

or, considering Equation 4.250, we would end up with the following explicit expression for the transverse component

$$u_y = \sqrt{\frac{3}{2}\frac{\beta_c}{\varepsilon}} \cdot \frac{\upsilon \cdot y}{\sqrt{\upsilon^3 - \upsilon_0^3}}. \tag{4.254}$$

The differential equation for a flow line

$$\frac{dy}{u_y} = \frac{\upsilon d\upsilon}{u_x}, \tag{4.255}$$

in view of Equation 4.253, takes the form

$$\frac{dy}{y} = -\frac{du_x}{u_x}, \tag{4.256}$$

and could be integrated into a flow function

$$\psi = yu_x \tag{4.257}$$

or, explicitly in view of Equation 4.250,

$$\psi = y\sqrt{2\frac{\beta_c}{\varepsilon} \cdot \frac{\upsilon^3 - \upsilon_0^3}{3}}.$$

4.1.2.6 Effect of Pressure Gradient

Now we will consider the effect of the pressure gradient on the injecting capability of a flat jet of falling particles. Consider a linearly accelerated flow of material of the breadth 2b between two horizontal airtight planes set apart by the distance l. Let us further assume that particles are uniformly distributed across it. Axis OX is oriented along the axis of this flow (Figure 4.14). In order to analyze aerodynamic processes inside this flow, we shall use boundary-layer equations (Equations 4.67 and 4.68). That being said, we will ignore convectional acceleration of air flow and will express mass forces of interaction between components with linearized relations:*

$$F_x = \bar{D}(\upsilon - u_x),\ F_y = -\bar{D}u_y, \tag{4.258}$$

where

$$\bar{D} = \frac{\bar{\beta}_0}{b\varepsilon}\left|1 - \overline{\left(\frac{u}{\upsilon}\right)}\right| = \frac{D}{\sqrt{2}}\left|1 - \overline{\left(\frac{u}{\upsilon}\right)}\right|. \tag{4.259}$$

Here, the parameters $\bar{\beta}_0$ and D are calculated using Equations 4.14 and 4.32 at $n = 1$; $\frac{\nu}{t} = 0$. Boundary layer equations are thereby greatly simplified:

* These simplifications are unlikely to affect estimation of the pressure gradient effect in a significant sense. On the other hand, accounting for convectional acceleration and aerodynamic resistance forces determined by quadratic rather than linear law would preclude any analytical solution of the problem in question.

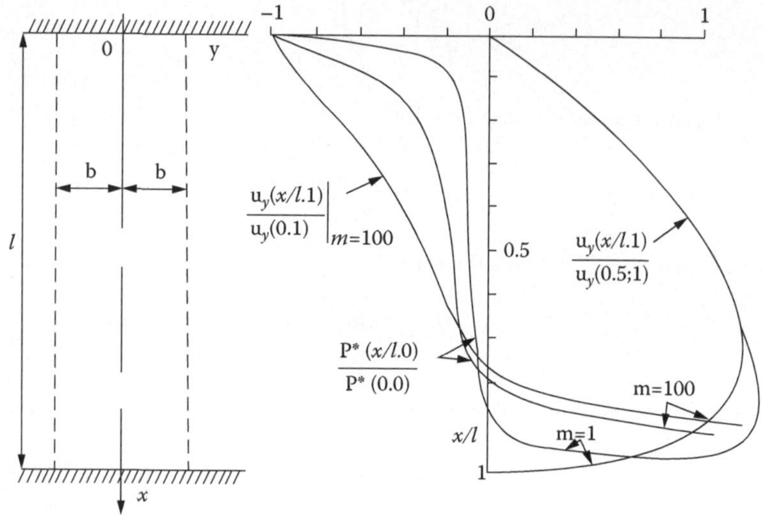

FIGURE 4.14 Changes in air pressure and velocity in a flat jet of freely falling particles (b = 0.08; l = 0.4).

$$\frac{\partial \mathbf{P}}{\partial x} = \bar{D}(\upsilon - u_x) + N_\tau \frac{\partial^2 u_x}{\partial y^2}; \qquad \frac{\partial \mathbf{P}}{\partial y} = -\bar{D}u_y. \tag{4.260}$$

We will define our boundary conditions by requiring impermeable planes

$$u_x(0, y) = u_x(l, y) = 0, \tag{4.261}$$

defining a flow symmetry condition

$$u_y(x,0) = 0; \qquad \left.\frac{\partial u_x}{\partial y}\right|_{y=0} = 0, \tag{4.262}$$

and by assuming that excess pressure and the longitudinal component of the air velocity vector on the jet boundary are equal to zero:

$$\mathbf{P}(x,b) = 0, \tag{4.263}$$

$$u_x(x,b) = 0. \tag{4.264}$$

By integrating Equation 4.260,

$$\mathbf{P}(x,y) - \mathbf{P}(x,b) = \int_b^y \bar{D}u_y dy, \tag{4.265}$$

contingent on Equation 4.263, we would end up with the following relation

$$\mathbf{P}(x,y) = -\bar{D}\int_b^y u_y dy, \tag{4.266}$$

that defines changes in pressure in a material flow.

Substitution of these relations into the first equation (Equation 4.260) would yield the relation

$$-\bar{D}\int_b^y \frac{\partial u_y}{\partial x} dy = \bar{D}(\upsilon - u_x) + \bar{N}_\tau \frac{\partial^2 u_x}{\partial y^2}, \tag{4.267}$$

binding together component velocities of injected air. By bringing into consideration a current function ψ, and by differentiating over *y*, this relation could be transformed into a simple uniform fourth-order equation:

$$\frac{\partial^4 \psi}{\partial y^4} = m\left(\frac{\partial^2 \psi}{\partial x^2} + \frac{\partial^2 \psi}{\partial y^2}\right), \tag{4.268}$$

where

$$m = \bar{D} / N_\tau. \tag{4.269}$$

We shall solve this equation using the Fourier method. Let us assume that

$$\psi = S(x) \cdot R(y), \tag{4.270}$$

where $S(x)$, $R(y)$ are some functions of x and y.

Then,

$$u_x = \frac{\partial \psi}{\partial y} = SR', \qquad u_y = -\frac{\partial \psi}{\partial x} = -S'R, \tag{4.271}$$

and Equation 4.268 would be converted into the following ordinary differential equation:

$$(R'''' - mR'')/R = mS''/S = -4k^4, \tag{4.272}$$

where k is a certain constant. The solution of Equation 4.272 could be found by solving two boundary value problems. The first being:

$$S'' + \frac{4k^4}{m} S = 0, \tag{4.273}$$

where, owing to Equation 4.261,

$$S(0) = S(l) = 0. \tag{4.274}$$

The second boundary value problem is

$$R'''' - mR'' + 4k^4 R = 0, \tag{4.275}$$

$$R(0) = R'(b) = R''(0) = 0. \tag{4.276}$$

Here, boundary conditions are specified by Equations 4.262 and 4.264.

The first problem is relatively easy to solve. Special solutions assume the form

$$S = C_2 \sin \frac{2k^2}{\sqrt{m}} x, \tag{4.277}$$

$$\frac{2k^2}{\sqrt{m}} = \frac{n\pi}{l}, \qquad n = 1,2,3\ldots \tag{4.278}$$

A general solution of the simple equation (Equation 4.275) could be expressed with the following equation:

$$R = A_1 \text{ch}k\alpha y \sin k\beta y + A_2 \text{sh}k\alpha y \sin k\beta y + A_3 \text{sh}k\alpha y \cos k\beta y + A_4 \text{ch}k\alpha y \cos k\beta y, \tag{4.279}$$

that, conditional on Equation 4.276, would assume the form

$$R = A_1 \left(\text{ch}k\alpha y \sin k\beta y + a\,\text{sh}k\alpha y \cos k\beta y \right), \tag{4.280}$$

where

$$a = (\alpha + \varepsilon\beta)/(\beta - \varepsilon\alpha), \tag{4.281}$$

$$\varepsilon = \text{cth}k\alpha b \cdot \text{ctg}k\beta b, \tag{4.282}$$

$$\alpha = \sqrt{1+\delta}, \qquad \beta = \sqrt{1-\delta}, \tag{4.283}$$

$$\delta = m/\left(4k^2\right). \tag{4.284}$$

Thus, Equation 4.268 would have a special solution

$$\psi_n = B(n) \cdot \sin \frac{2k^2}{\sqrt{m}} x \left(\text{ch}k\alpha y \sin k\beta y + a\,\text{sh}k\alpha y \cos k\beta y \right), \tag{4.285}$$

where $B(n)$ is a constant dependent on n.

Owing to the linearity and uniformity of the original equation and its boundary conditions, its solutions may be provided by any sums of the following special solutions:

$$\psi = \sum_{n=1}^{\infty} B(n) \sin \frac{2k^2}{\sqrt{m}} x.\left(\text{ch}k\alpha y \cdot \sin k\beta y + a\,\text{sh}k\alpha y \cdot \cos k\beta y \right). \tag{4.286}$$

The value $B(n)$ could be derived by requiring this solution to meet Equation 4.268. After some superficial (if tedious) transformations, we would end up with

$$B(n) = b_n k / \left\{ \left(\frac{n\pi}{l} \right)^2 \cdot \left(\frac{\alpha\beta - \delta\varepsilon}{\beta - \alpha\varepsilon} \text{sh}k\alpha b \cdot \sin k\beta b + \frac{\delta + \alpha\beta\varepsilon}{\beta - \alpha\varepsilon} \text{ch}k\alpha b \cos k\beta b \right) \right\}, \tag{4.287}$$

$$b_n = \frac{2}{l}\int_0^l \upsilon \sin n\pi \frac{x}{l}\, dx. \tag{4.288}$$

The latter equation determines the factors for decomposing the flowing material velocity into a sine Fourier series. In particular, in the case of a linearly accelerated flow with an initial velocity of υ_0, it holds that

$$b_n = \upsilon_k \cdot \xi\left(n, \frac{\upsilon_0}{\upsilon_k}\right), \qquad \upsilon_k = \sqrt{2l + \upsilon_0^2} \tag{4.289}$$

$$\xi\left(n, \frac{\upsilon_0}{\upsilon_k}\right) = \frac{4}{1-(\upsilon_0/\upsilon_k)^2} \int_{\upsilon_0/\upsilon_k}^{1} \left[\sin n\pi \frac{z^2-(\upsilon_0/\upsilon_k)^2}{1-(\upsilon_0/\upsilon_k)^2}\right] z^2 dz. \tag{4.290}$$

Tables 4.6 and 4.7 list values of the function $\xi\left(n, \frac{\upsilon_0}{\upsilon_k}\right)$, determined numerically.

One should keep in mind that Fourier series used by us (Equation 4.286) have a peculiar feature where the manifestation is conditional upon

$$\delta = \frac{\sqrt{m}}{2}\,\frac{l}{n\pi} > 1. \tag{4.291}$$

Let us designate the closest natural number (other than zero) meeting this condition as n_0, that is,

$$n_0 = \left[\frac{\sqrt{ml}}{2\pi}\right]. \tag{4.292}$$

Then, with $n < n_0$ (in light of Equation 4.291), it follows that

$$\beta = \sqrt{1-\delta} = i\sqrt{|1-\delta|} = i\hat{\beta} \tag{4.293}$$

and series for the flow function would become

$$\psi = \sum_{n=1}^{n_0} \hat{B}(n) \sin\frac{n\pi}{l}x\left(\hat{a}\mathrm{sh}k\alpha y\mathrm{ch}k\hat{\beta}y - \mathrm{ch}k\alpha y \cdot \mathrm{sh}k\hat{\beta}y\right) + \sum_{n=n_0+1}^{\infty} B(n)\sin\frac{n\pi}{l}x\left(\mathrm{ch}k\alpha y \sin k\beta y + a\mathrm{sh}k\alpha y \cdot \cos k\beta y\right), \tag{4.294}$$

where

$$\hat{B}(n) = b_n k \Big/ \left[\left(\frac{n\pi}{l}\right)^2 \alpha\hat{\beta}\frac{\hat{\varepsilon}^2 - 1}{\hat{\beta} - \alpha\hat{\varepsilon}}\mathrm{sh}k\alpha b\,\mathrm{sh}k\hat{\beta}b\right], \tag{4.295}$$

$$\hat{\beta} = \sqrt{|1-\delta|}; \quad \hat{\varepsilon} = \mathrm{cth}k\alpha b \cdot \mathrm{cth}k\hat{\beta}b, \tag{4.296}$$

$$\hat{a} = \left(\alpha + \hat{\beta}\hat{\varepsilon}\right) / \left(\hat{\beta} + \alpha\hat{\varepsilon}\right). \tag{4.297}$$

TABLE 4.6
Values of ξ (*n*, 0)

n	ξ (*n*, 0)	*n*	ξ (*n*, 0)	*n*	ξ (*n*, 0)
1	0.874705	31	0.021843	61	0.010931
2	−0.240602	32	−0.018654	62	−0.009830
3	0.256098	33	0.020483	63	0.010579
4	−0.131267	34	−0.017593	64	−0.009533
5	0.147583	35	0.019280	65	0.010250
6	−0.090863	36	−0.016646	66	−0.009254
7	0.103145	37	0.018211	67	0.009941
8	−0.069662	38	−0.015797	68	−0.008991
9	0.079094	39	0.017253	69	0.009651
10	−0.056561	40	−0.015032	70	−0.008743
11	0.064056	41	0.016391	71	0.009378
12	−0.047647	42	−0.014338	72	−0.008510
13	0.053780	43	0.0I5611	73	0.009120
14	−0.041182	44	−0.013706	74	−0.008290
15	0.046321	45	0.014901	75	0.008876
16	−0.036276	46	−0.013127	76	−0.008080
17	0.040663	47	0.014253	77	0.008646
18	−0.032424	48	−0.012596	78	−0.007882
19	0.036227	49	0.013660	79	0.008428
20	−0.029317	50	−0.012108	80	−0.007694
21	0.032657	51	0.013113	81	0.008222
22	−0.026758	52	−0.011656	82	−0.007516
23	0.029722	53	0.012609	83	0.008025
24	−0.024614	54	−0.011237	84	−0.007347
25	0.027268	55	0.012143	85	0.007839
26	−0.022790	56	−0.010848	86	−0.007185
27	0.025185	57	0.011709	87	0.007663
28	−0.021220	58	−0.010485	88	−0.007032
29	0.023396	59	0.011306	89	0.007494
30	−0.019834	60	−0.010147	90	−0.006885

Narrowing down the scope to a case of $n_0 = 0$ and using Equations 4.286 through 4.290, we can express calculated ratios for projections of the injected air velocity vector as follows:

$$u_x = \sum_{n=1}^{\infty} 2kB(n)\sin n\pi\frac{x}{l}(\mathrm{ch}k\alpha y\cdot\cos k\beta y - \varepsilon\mathrm{sh}k\alpha y\cdot\sin k\beta y)/(\beta-\alpha\varepsilon)\ ; \quad (4.298)$$

$$u_y = -\sum_{n=1}^{\infty} B(n)\frac{n\pi}{l}\cos n\pi\frac{x}{l}(\mathrm{ch}k\alpha y\cdot\sin k\beta y + a\mathrm{sh}k\alpha y\cdot\cos k\beta y), \quad (4.299)$$

TABLE 4.7
Values $\xi(n, \upsilon_0/\upsilon_k)$

	$\xi(n, \upsilon_0/\upsilon_k)$ at υ_0/υ_k equal to					
n	**0.2**	**0.4**	**0.6**	**0.8**	**0.95**	**1.0**
1	0.89676	0.95665	1.04427	1.15164	1.24174	1.27324
2	−0.22339	−0.18073	−0.12494	−0.06342	−0.01591	0
3	0.26992	0.30179	0.34080	0.38220	0.41382	0.42441
4	−0.11944	−0.09362	−0.06332	−0.03180	−0.00796	0
5	0.15793	0.17947	0.20401	0.22923	0.24828	0.25465
6	−0.08159	−0.06303	−0.04233	−0.02121	−0.00531	0
7	0.11154	0.12781	0.14562	0.16372	0.17735	0.18189
8	0.06195	−0.04746	−0.03179	−0.01591	−0.00398	0
9	0.08622	0.09927	0.11323	0.12739	0.13793	0.14147
10	−0.04992	−0.03804	−0.02544	−0.01273	−0.00318	0
11	0.07027	0.08116	0.09263	0.10418	0.11286	0.11575
12	−0.04179	−0.03174	−0.02121	−0.01061	−0.00265	0
13	0.05931	0.06864	0.07837	0.08815	0.09549	0.09784
14	−0.03594	−0.02723	−0.01818	−0.00909	−0.00227	0
15	0.05131	0.05947	0.06792	0.07640	0.08276	0.08488
16	−0.03152	−0.02383	−0.01591	−0.00796	−0.00199	0
17	0.04521	0.05246	0.05992	0.06741	0.07302	0.07490
18	−0.02806	−0.02119	−0.01414	−0.00707	−0.00177	0
19	0.04041	0.04694	0.05362	0.06031	0.06534	0.06701
20	−0.02529	−0.01908	−0.01273	−0.00637	−0.00159	0
21	0.03654	0.04246	0.04851	0.05457	0.05911	0.06063
22	−0.02301	−0.01735	−0.01157	−0.00579	−0.00145	0
23	0.03334	0.03877	0.04429	0.04982	0.05397	0.05531
24	−0.02111	−0.01590	−0.01061	−0.00531	−0.00133	0
25	0.03066	0.03566	0.04075	0.04584	0.04966	0.05093
26	−0.01950	−0.01468	−0.00979	−0.00490	−0.00123	0
27	0.02837	0.03302	0.03773	0.04244	0.04598	0.04716
28	−0.01812	−0.01363	−0.00909	−0.00455	−0.00114	0
29	0.02641	0.03074	0.03513	0.03951	0.04281	0.04391
30	−0.01692	−0.01273	−0.00849	−0.00424	−0.00106	0
31	0.02470	0.02876	0.03286	0.03697	0.04010	0.04107

and for changes of pressure within it:

$$\mathbf{P}(x, y) = -\bar{D}\sum_{n=1}^{\infty} l\frac{b_n}{n\pi}\cos n\pi\frac{x}{l}\cdot\Pi(n, y), \tag{4.300}$$

$$\Pi(n, y) = 1 - \frac{(\alpha\beta - \delta\varepsilon)\,\mathrm{sh}k\alpha y\sin k\beta y + (\delta + \alpha\beta\varepsilon)\,\mathrm{ch}k\alpha y\cos k\beta y}{(\alpha\beta - \delta\varepsilon)\,\mathrm{sh}k\alpha b\sin k\beta b + (\delta + \alpha\beta\varepsilon)\,\mathrm{ch}k\alpha b\cos k\beta b}. \tag{4.301}$$

Let us proceed with analyzing pressure changes along the centerline of the jet. This enables Equations 4.300 and 4.301 to be simplified somewhat

$$\mathbf{P}(x,0) = -\bar{D}\sum_{n=1}^{\infty} l\frac{b_n}{n\pi}\cos n\pi\frac{x}{l}\cdot\Pi(n,0), \tag{4.302}$$

$$\Pi(n,0) = 1 - \frac{\left(\delta/\sqrt{1-\delta^2}\right)\mathrm{sh}k\alpha b\sin k\beta b + \mathrm{ch}k\alpha b\cos k\beta b}{\mathrm{sh}^2 k\alpha b + \cos^2 k\beta b}. \tag{4.303}$$

Figure 4.14 illustrates changes in relative pressure

$$\mathbf{P}^*\left(\frac{x}{l},0\right) = \mathbf{P}(x,0)/\frac{\bar{D}l\upsilon_k}{\pi} = -\sum_{n=1}^{\infty}\frac{\xi(n,0)}{n}\Pi(n,0)\cos n\pi\frac{x}{l} \tag{4.304}$$

and relative velocity of air along the jet and on its boundary.

This data indicates that the jet could be separated in two parts lengthwise. The upper part (about 80–90% of the entire length) displays negative pressure, which causes the injection of air. The lower part experiences negative pressure, with airflow escaping the stream of particles. The negative pressure in the origin of the jet and the excess pressure at its end numerically depend, ceteris paribus, on the parameter m (Figure 4.15), and there is a clearly observable area of self-similarity (at $m < 0.1$) where the proportionality factor

$$\mathrm{k}_0 = \mathbf{P}^*/m - const \tag{4.305}$$

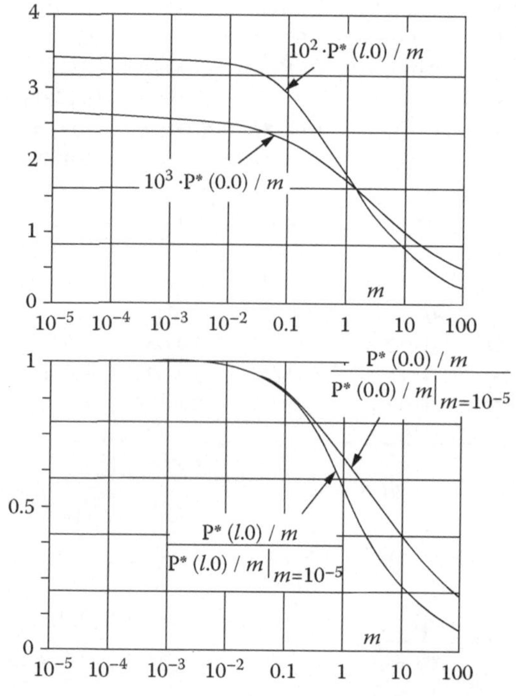

FIGURE 4.15 Changes in pressure in the origin and in the end of a flat jet with increasing values of m.

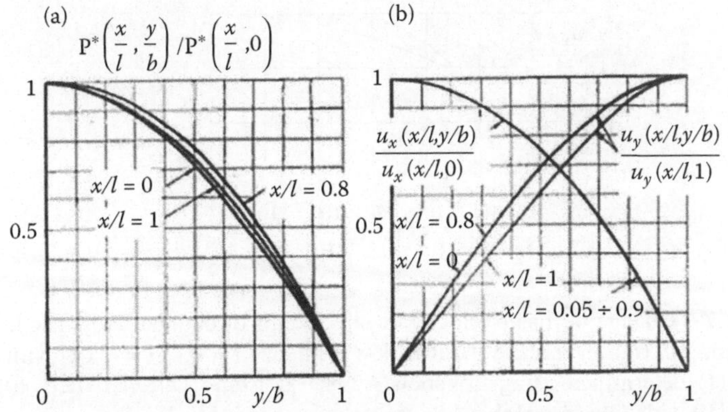

FIGURE 4.16 Changes in (a) air pressure and (b) velocity over the cross-section of a flat jet of loose solid matter (at m = 100).

is independent of m. For the upper point of the jet, this ratio is equal to

$$k_0^B = \left.\frac{\mathbf{P}^*(0,0)}{m}\right|_{m=10^{-5}} = 2,6\cdot10^{-3} \tag{4.306}$$

and the magnitude of rarefaction is determined by the simple equation

$$\mathbf{P}(0,0) = -2,6\cdot10^{-3} m\bar{D}l\upsilon_k/\pi. \tag{4.307}$$

Accordingly, for the lower point,

$$k_0^H = \left.\frac{\mathbf{P}^*(l,0)}{m}\right|_{m=10^{-5}} = 3,4\cdot10^{-2}, \tag{4.308}$$

$$P(l,0) = 3,4\cdot10^{-2} m\bar{D}l\upsilon_k/\pi. \tag{4.309}$$

The jet manifests a parabolic cross-sectional pressure profile (Figure 4.16), which is notably identical in its character:

$$\mathbf{P}\left(\frac{x}{l},\frac{y}{b}\right) \approx \mathbf{P}^*\left(\frac{x}{l},0\right)\cdot\left[1-\left(\frac{y}{b}\right)^2\right]. \tag{4.310}$$

With $0.05 < \frac{x}{l} < 0.9$, the longitudinal component of the air velocity vector develops similarly (Figure 4.16b),

$$u_x\left(\frac{x}{l},\frac{y}{b}\right) \approx u_x\left(\frac{x}{l},0\right)\cdot\left[1-\left(\frac{y}{b}\right)^2\right]. \tag{4.311}$$

At the origin and end points of the jet, this velocity is equal to zero. The transverse component of the velocity vector exhibits a somewhat different behavior. Although varying in direction and absolute value and reaching its maxima at the end of the jet and on its boundaries, the value u_y is self-similar for all sections:

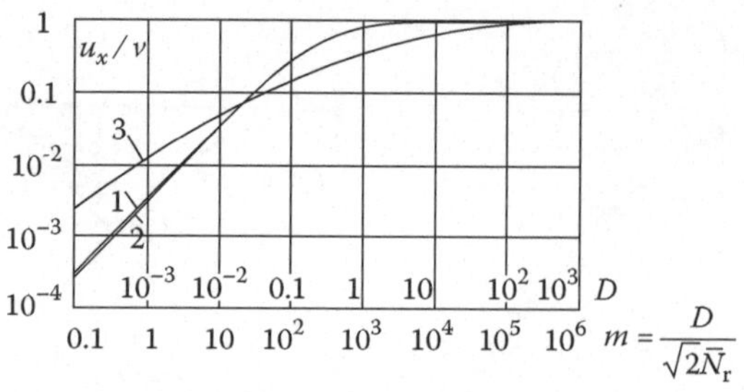

FIGURE 4.17 Change in linear velocity of injected air throughout cross-section $x/l = 0.5$ with increasing m (D) for a plane-parallel flow with size $l = 0.4$; $b = 0.08$ (with $K_n = 0.04$; $\bar{N}_\tau \approx 10^{-3}$): (1) determined using Equation 4.298; (2) from Table 4.3 (item 6); (3) using Equation 4.119 with data from Table 4.2.

$$u_y\left(\frac{x}{l},\frac{y}{b}\right) \approx u_y\left(\frac{x}{l},1\right)\cdot\left[1-\left(1-\frac{y}{b}\right)^2\right], \tag{4.312}$$

including the critical section where the injection area transitions into an injection area.

It would be of interest to analyze changes in the speed of injected air as a function of the parameter $\bar{D}$ (Figure 4.17). We can notice asymptotic increases in u_x as the volumetric concentration of particles in the flow rises. A comparison of these findings with solutions found earlier reveals that the pressure gradient has virtually no bearing on absolute velocities of injected air, and it can be ignored when calculating the amount of air being injected. Of significant impact is the magnitude of convectional acceleration (curve 3 as contrasted with curves 1 and 2 in view of the value of $\frac{du_x}{dt}$).

Figure 4.18 illustrates the structure of air flow surrounding a flat jet. Flow lines inside a stream of particles have been determined using Equation 4.294. Airflow patterns outside of the jet have been modeled electrically with an EGDA 9-60 integrator. As illustrated by the data, flow rates of circulating air decrease sharply with increasing distance from the jet centerline. A considerable proportion of air (in excess of 80%) circulates in an area limited by the jet centerline and a parallel straight line running at a distance of $7b$. Thus, isolating a flat jet of material with vertical walls set apart by six to seven gauge distances fails to affect jet structure in a significant way. This fact is in qualitative and quantitative agreement with experimental findings [44, 45].

4.1.3 Injection of Air in an Axially Symmetric Jet of Freely Falling Particles

Hydrodynamic equations for an axially symmetric jet of injected air without accounting for pressure gradients, based on Equations 4.63, 4.65, and 4.33 with $N_\tau \gg N$, could be expressed as follows:

$$u_x\frac{\partial u_x}{\partial x} + u_r\frac{\partial u_x}{\partial r} = Dx^{\gamma}\left|1-\frac{u_x}{\upsilon}\right|^n\left(1-\frac{u_x}{\upsilon}\right)e^{-z'} + N_\tau\frac{1}{r}\frac{\partial}{\partial r}\left(r\frac{\partial u_x}{\partial r}\right), \tag{4.313}$$

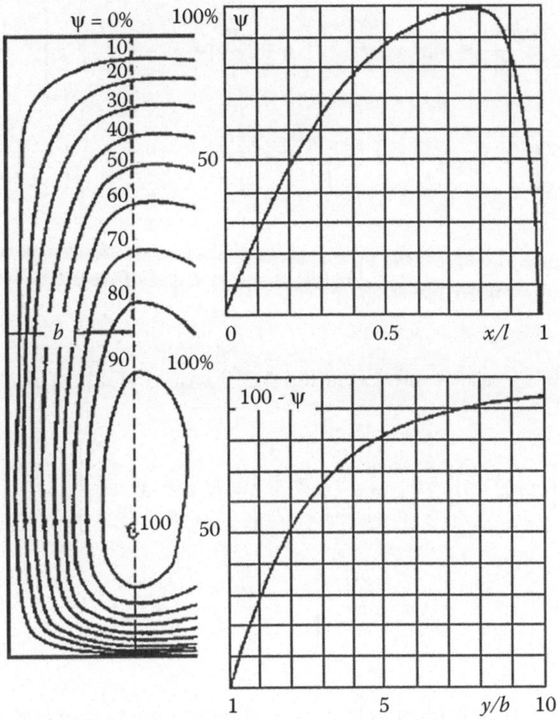

FIGURE 4.18 Structure of airflows adjacent to a flat jet of loose matter (m = 100; b = 0.08; l = 0.4).

$$\frac{\partial ru_x}{\partial x} + \frac{\partial ru_r}{\partial r} = 0. \tag{4.314}$$

Consider again a flow with a generalized exponential distribution of particles described by Equation 4.12.

4.1.3.1 Self-Similar Motion Equations

Similar to the case with a flat jet, let us introduce a flow function

$$\left.\begin{aligned} \psi &= mx^s \zeta(z), \\ z = r/b, \quad b &= a^{-\frac{1}{t}} x^{-\frac{\nu}{t}} \end{aligned}\right\}. \tag{4.315}$$

Velocity vector projections $\vec{u}$ and their derivatives would then assume the following form:

$$u_x = \frac{1}{r}\frac{\partial \psi}{\partial r} = ma^{\frac{2}{t}} x^{s+2\frac{\nu}{t}} \cdot \zeta' / z; \tag{4.316}$$

$$u_r = -\frac{1}{r}\frac{\partial \psi}{\partial x} = -ma^{\frac{1}{t}} x^{s+\frac{\nu}{t}-1} \cdot \left(s\frac{\zeta}{z} + \frac{\nu}{t}\zeta' \right); \tag{4.317}$$

$$\frac{\partial u_x}{\partial x} = ma^{\frac{2}{t}} x^{s+2\frac{\nu}{t}-1} \cdot \left[\left(s + 2\frac{\nu}{t} \right) \frac{\zeta'}{z} + \frac{\nu}{t} \left(\frac{\zeta'}{z} \right)' z \right]; \tag{4.318}$$

$$\frac{\partial u_x}{\partial r} = ma^{\frac{3}{t}} x^{s+3\frac{\nu}{t}} \cdot \left(\frac{\zeta'}{z} \right)'; \quad r\frac{\partial u_x}{\partial r} = ma^{\frac{2}{t}} x^{s+2\frac{\nu}{t}} \cdot z \left(\frac{\zeta'}{z} \right)'; \tag{4.319}$$

$$\frac{1}{r}\frac{\partial}{\partial r}\left(r\frac{\partial u_x}{\partial r} \right) = ma^{\frac{4}{t}} x^{s+4\frac{\nu}{t}} \cdot \left[z \cdot \left(\frac{\zeta'}{z} \right)' \right]' \frac{1}{z}. \tag{4.320}$$

Substitution of these values into Equation 4.313 would yield

$$m^2 a^{\frac{4}{t}} x^{2\left(s+2\frac{\nu}{t}\right)-1} \cdot \left[\left(s + 2\frac{\nu}{t} \right) \left(\frac{\zeta'}{z} \right)^2 - s\frac{\zeta}{z} \left(\frac{\zeta'}{z} \right)' \right] = N_\tau m a^{\frac{4}{t}} x^{s+4\frac{\nu}{t}} \cdot \left[z \left(\frac{\zeta'}{z} \right)' \right]' \frac{1}{z} + \\ + Dx^\gamma \left| 1 - \frac{ma^{\frac{2}{t}} x^{S+2\frac{\nu}{t}}}{\upsilon} \cdot \frac{\zeta'}{z} \right|^n \cdot \left(1 - \frac{ma^{\frac{2}{t}} x^{S+2\frac{\nu}{t}}}{\upsilon} \cdot \frac{\zeta'}{z} \right) e^{-z'}. \tag{4.321}$$

It should be noted that in the absence of flowing particles ($D = 0$), as well as in case of $s = 1$; $\frac{\nu}{t} = -1$; and $m = N_\tau$ – const, this relation could be transformed into the equation

$$\left(\frac{\zeta'}{z} \right)'' + \frac{1+\zeta}{z} \left(\frac{\zeta'}{z} \right)' + \left(\frac{\zeta'}{z} \right)^2 = 0, \tag{4.322}$$

describing the structure of untwisted round air jet. The following requirements must be satisfied so that Equation 4.321 could be reduced to a self-similar flow equation:

$$Nm^2 a^{\frac{4}{t}} x^{2\left(s+2\frac{\nu}{t}\right)-1} \cdot s = N_\tau m a^{\frac{4}{t}} x^{s+4\frac{\nu}{t}}; \tag{4.323}$$

$$m^2 a^{\frac{4}{t}} x^{2\left(s+2\frac{\nu}{t}\right)-1} \cdot s = D\mathrm{x}^\gamma; \tag{4.324}$$

$$ma^{\frac{2}{t}} x^{s+2\frac{\nu}{t}} = K\upsilon, \tag{4.325}$$

where N and K are parameters in the equation

$$\left(1 + 2\frac{\nu}{ts} \right) \left(\frac{\zeta'}{z} \right)^2 - \frac{\zeta}{z} \left(\frac{\zeta'}{z} \right)' = N \left[z \cdot \left(\frac{\zeta'}{z} \right)' \right]' \frac{1}{z} + e^{-z'} \left| 1 - K\frac{\zeta'}{z} \right|^n \left(1 - K\frac{\zeta'}{z} \right). \tag{4.326}$$

Equation 4.324 resolves to

$$s+2\frac{\nu}{t}=\frac{1+\gamma}{2}; \tag{4.327}$$

$$m=\sqrt{D/s}\,a^{-\frac{2}{t}} \tag{4.328}$$

and parameters

$$K=\sqrt{D/s}\,x^{\frac{1+\gamma}{2}}/\upsilon=\sqrt{D/(2s)}\,x^{\frac{\gamma}{2}},\qquad \left(\upsilon=\sqrt{2x}\right); \tag{4.329}$$

$$N=\frac{N_\tau}{sm}x^{\frac{1}{2}+2\frac{\nu}{t}-\frac{\gamma}{2}} \tag{4.330}$$

or, in view of Equations 4.70 and 4.328,

$$N=\frac{2k_n}{\sqrt{2sD}}a^{\frac{1}{t}}x^{1+\frac{\nu}{t}-\frac{\gamma}{2}}. \tag{4.331}$$

Therefore, an equation of strictly self-similar motion could only be obtained with

$$\gamma=0;\quad \frac{\nu}{t}=-1;\quad s=\frac{5}{2}, \tag{4.332}$$

when parameters N and K are not explicitly dependent on x

$$N=2k_n a^{\frac{1}{t}}/\sqrt{5D},\qquad K=\sqrt{D/5}. \tag{4.333}$$

Unfortunately, for the important case of a plane-parallel particle flow (at $\nu/t=0,\quad b=a^{-\frac{1}{t}}$), it is impossible to formulate a strictly self-similar problem in the class of power functions being considered (see Equation 4.315). Parameters N and K would have to be averaged over jet length as illustrated in the case of a flat jet. The following inequalities should be met as consistently as possible in order to minimize error with this method:

$$0<\gamma/2<1. \tag{4.334}$$

It should be easy to verify whether these conditions are valid within the area of quadratic resistance law ($n=1$) and the area of viscous flow around particles ($n=0$).

In the first case ($\gamma=1/2$; $s=3/4$), it becomes

$$m=b^2\sqrt{\frac{4}{3}D},\quad N-\frac{2k_n x^{\frac{3}{4}}}{b\sqrt{1,5D}}\sim\frac{8}{7\sqrt{1,5D}}\frac{k_n}{b}x_h^{\frac{3}{4}}, \tag{4.335}$$

$$K=\sqrt{\frac{2}{3}D}\,x^{\frac{1}{4}}\approx\frac{4}{5}\sqrt{\frac{2}{3}D}\,x_k^{\frac{1}{4}}; \tag{4.336}$$

and, for viscous mode ($n = 0$; $\gamma = 0$; $s = \frac{1}{2}$),

$$m = b^2\sqrt{2D}, \quad N = \frac{2k_n x}{b\sqrt{D}} \approx \frac{k_n}{b\sqrt{D}} x_k, \tag{4.337}$$

$$K = \sqrt{D}. \tag{4.338}$$

When it is necessary to ensure that parameter N remains constant, it should be enough to modify Equations 4.323 through 4.325. In particular, replacing Equation 4.323 with the following:

$$m^2 a^{\frac{4}{t}} x^{2\left(S+2\frac{\nu}{t}\right)-1} \cdot s = N_\tau m a^{\frac{4}{t}} x^{S+4\frac{\nu}{t}} \tag{4.339}$$

or, in view of Equation 4.70,

$$ms = \sqrt{2}k_n a^{-\frac{1}{t}} x^{\frac{3}{2}-S-\frac{\nu}{t}}, \tag{4.340}$$

leads to the following requirements in place of Equations 4.327 and 4.328:

$$m = 2k_n a^{-\frac{1}{t}} / \left(s\sqrt{2}\right), \tag{4.341}$$

$$s = \frac{3}{2} - \frac{\nu}{t}. \tag{4.342}$$

Let us change Equation 4.324 as follows:

$$G m^2 a^{\frac{4}{t}} x^{2\left(S+2\frac{\nu}{t}\right)-1} \cdot s = D x^\gamma. \tag{4.343}$$

In that case, the parameter N would be supplanted with the parameter G characterizing the number of falling particles. In light of Equations 4.341, 4.342, and 4.343, it could be expressed as follows:

$$G = 0{,}5 s D a^{-\frac{2}{t}} x^{\gamma-2\left(1+\frac{\nu}{t}\right)} / k_n^2. \tag{4.344}$$

The previous requirement (Equation 4.325) for the purpose of determining K may be left intact. In view of Equations 4.341 and 4.342, it becomes

$$K = \frac{k_n}{s} a^{\frac{1}{t}} x^{1+\frac{\nu}{t}}, \tag{4.345}$$

and thus Equation 4.321 would become

$$\left(1+2\frac{\nu}{ts}\right)\left(\frac{\zeta'}{z}\right)^2 - \frac{\zeta}{z}\left(\frac{\zeta'}{z}\right)' = \left[z \cdot \left(\frac{\zeta'}{z}\right)'\right]' \frac{1}{z} + G e^{-z^t} \left|1 - K\frac{\zeta'}{z}\right|^n \left(1 - K\frac{\zeta'}{z}\right). \tag{4.346}$$

Apparently Equation 4.346, when bounded with these conditions, may only describe strictly self-similar motion at $\gamma = 0$; $\nu/t = -1$; and s = −5/2. In this case, the parameters G and K are constants and are equal to

$$G = \frac{5D}{4k_n^2} a^{-\frac{2}{t}}, \quad K = 2k_n a^{\frac{1}{t}} / 5. \tag{4.347}$$

In case of a plane-parallel flow ($\nu/t = 0$), the values m, K, and G are correspondingly equal to the following ($s = 1.5$)

$$m = \frac{2k_n}{1{,}5\sqrt{2}} a^{-\frac{1}{t}}, \quad K = \frac{2k_n}{3} a^{\frac{1}{t}} x \approx \frac{k_n}{3} a^{\frac{1}{t}} x_k, \tag{4.348}$$

$$G = 0.75 D a^{-\frac{2}{t}} x^{\gamma-2} / k_n^2 \approx 0.75 D b^2 x_k^{\gamma-2} / \left[k_n^2 (\gamma - 1)\right]. \tag{4.349}$$

Thus, with the goal of improving calculation accuracy, it is preferable to analyze Equation 4.346 rather than resolve the equality condition (Equation 4.326). This is because turbulent viscosity forces are well accounted for in the first case. However, one should keep in mind that these forces are much weaker than the volumetric forces of interaction between components that are less accurately reflected in Equation 4.346. Equations 4.348 and 4.349 also fare worse in terms of approximation accuracy because the absolute power value for x in expressions for parameters G and K never falls below one. Therefore, we shall subsequently analyze Equation 4.326, positing

$$\zeta = z(u - \xi), \qquad \xi = \frac{1}{z}\int_0^z u(z)\,dz, \tag{4.350}$$

$$\zeta' = u'z, \tag{4.351}$$

would appear as

$$\varepsilon_\nu u'^2 - u''\left(u + \frac{N}{z} - \xi\right) = Nu''' + e^{-z^t} |1 - Ku'|^n (1 - Ku'), \tag{4.352}$$

where

$$\varepsilon_\nu = 1 + \frac{2}{s} \cdot \frac{\nu}{t}. \tag{4.353}$$

We will be using a new function $u(z)$ to express air velocity projections and their derivatives. Based on Equations 4.316 through 4.320, for a plane-parallel flow ($\nu/t = 0$) with a quadratic law of resistance ($\gamma = 1/2$; $s = 3/4$; $m = \sqrt{\frac{4}{3} D b^2}$), we can write:

$$u_x = \sqrt{\frac{4}{3} D} x^{\frac{3}{4}} u'; \tag{4.354}$$

$$u_r = -\frac{3}{4}\sqrt{\frac{4}{3}}Dbx^{-\frac{1}{4}}(u-\xi); \tag{4.355}$$

$$\frac{\partial u_x}{\partial x} = \frac{3}{4}\sqrt{\frac{4}{3}}Dx^{-\frac{1}{4}}u'; \tag{4.356}$$

$$\frac{\partial u_x}{\partial r} = \sqrt{\frac{4}{3}}Db^{-1}x^{\frac{3}{4}}u''; \quad r\frac{\partial u_x}{\partial r} = \sqrt{\frac{4}{3}}Dx^{\frac{3}{4}}zu''. \tag{4.357}$$

Then the boundary conditions

$$u_x = u_0,\ u_r = 0,\ \frac{\partial u_x}{\partial r} = 0 \quad \text{at} \quad r = 0; \tag{4.358}$$

$$u_x = 0,\ u_r = 0,\ \frac{\partial u_x}{\partial r} = 0 \quad \text{at} \quad r \to \infty \tag{4.359}$$

would appear as

$$u(0) = c, \quad u'(0) = \alpha, \quad u''(0) = 0, \tag{4.360}$$

$$u(\infty) = B, \quad u'(\infty) = 0, \quad u''(\infty) = 0, \tag{4.361}$$

and where B, c, and α are constants to be determined. In particular, similar to the case of a flat flow, the integral relation (Equation 4.77) could be used for that purpose. For a gradient-free flow in view of Equations 4.354 and 4.33, it could be represented as follows:

$$\int_0^\infty u'^2 z dz = \frac{1}{2}\int_0^\infty e^{-z^t}(1-Ku')^2 z dz. \tag{4.362}$$

4.1.3.2 Solving the Self-Similar Equation

Let us solve the problem for a characteristic case of uniform particle concentration across the jet (with $t \to \infty$). In this case,

$$u'^2 - u''\left(u + \frac{N}{z} - \xi\right) = Nu''' + (1-Ku')^2 \quad \text{at} \quad z < 1, \tag{4.363}$$

$$u'^2 - u''\left(u + \frac{N}{z} - \xi\right) = Nu''' \quad \text{at} \quad z > 1. \tag{4.364}$$

Considering boundary conditions (Equation 4.360) in the area of smaller z, the solution would appear as follows:

$$u_0 = c + \alpha z - \frac{\varepsilon}{2}\frac{z^3}{6}, \tag{4.365}$$

$$u_0' = \alpha - \frac{\varepsilon}{2}\cdot\frac{z^2}{2}, \tag{4.366}$$

$$u_0'' = -\frac{\varepsilon}{2}z, \tag{4.367}$$

where

$$\varepsilon = \frac{1}{N}\left[(1 - K\alpha)^2 - \alpha^2\right]. \tag{4.368}$$

In order to determine an asymptotic solution, we shall simplify Equation 4.364. Suppose that in the area of large values of z

$$u = B + \zeta(z), \tag{4.369}$$

where $\zeta(z)$ is the desired function

$$\zeta(z) << B, \tag{4.370}$$

then Equation 4.364 would become

$$\zeta'^2 - \zeta''\left(B + \frac{N}{z} + \zeta - \xi\right) = N\zeta'''. \tag{4.371}$$

Let us further suppose that

$$\left|B + \frac{N}{z}\right| >> \zeta - \xi; \tag{4.372}$$

$$\zeta'^2 << \left|\zeta''\left(B + \frac{N}{z}\right)\right|, \tag{4.373}$$

then

$$\zeta''' = -\zeta''\left(\frac{B}{N} + \frac{1}{z}\right) \tag{4.374}$$

and

$$\zeta'' = -Ae^{-\frac{B}{N}z} / z; \tag{4.375}$$

$$\zeta' = -A\int_{\infty}^{z} \frac{1}{z} e^{-\frac{B}{N}z} dz \le A\frac{N}{B}e^{-\frac{B}{N}z} \cdot \frac{1}{z}; \tag{4.376}$$

$$\zeta = -A\int_{\infty}^{z}\left(\int_{\infty}^{z} \frac{1}{z} e^{-\frac{B}{N}z} dz\right)dz \le -A\frac{N^2}{B^2}e^{-\frac{B}{N}z} \cdot \frac{1}{z}. \tag{4.377}$$

Therefore,

$$u_\infty \approx B - A\frac{N^2}{B^2}e^{-\frac{Bz}{N}} / z; \tag{4.378}$$

$$u'_{\infty} \approx A\frac{N}{B}e^{-\frac{Bz}{N}}/z; \tag{4.379}$$

$$u''_{\infty} \approx -Ae^{-\frac{Bz}{N}}/z. \tag{4.380}$$

Noticeably, our findings do not contradict the accepted provisions in Equations 4.370, 4.372, and 4.373. Combined Equations 4.363 and 4.364 could be solved approximately by splicing an asymptotic solution and a solution "at zero." Assuming $z_0 = 1$ as the splicing point, we would end up with the following combined equations for determining B, A, and α (the constant c could be assumed equal to zero):

$$\alpha - \frac{\varepsilon}{12} = B - A\frac{N^2}{B^2}e^{-\frac{B}{N}}, \tag{4.381}$$

$$\alpha - \frac{\varepsilon}{4} = A\frac{N}{B}e^{-\frac{B}{N}}, \tag{4.382}$$

$$-\frac{\varepsilon}{2} = -Ae^{-\frac{B}{N}}. \tag{4.383}$$

The last relation enables us to deduce

$$A = \frac{\varepsilon}{2}e^{\frac{B}{N}}, \tag{4.384}$$

and Equation 4.382 results in

$$\frac{B}{N} = \frac{2\varepsilon}{4\alpha - \varepsilon} = \frac{\varepsilon/2}{\alpha - \varepsilon/4}. \tag{4.385}$$

Given N and K, the value of α (and ε accordingly, in light of Equation 4.368) could be found using Equation 4.381, which would change as follows by applying Equations 4.384 and 4.385:

$$N\left(\frac{\varepsilon}{2}\right)^2 - \left(\alpha - \frac{\varepsilon}{4}\right)^3 - \frac{\varepsilon}{2}\left(\alpha - \frac{\varepsilon}{12}\right)\left(\alpha - \frac{\varepsilon}{4}\right) = 0. \tag{4.386}$$

Constants determined this way would enable us to determine the function $u(z)$, its derivatives, and—considering Equation 4.354—ultimately to find the value of the longitudinal component of injected air velocity:

$$u_x = \sqrt{\frac{4}{3}Dx^{\frac{3}{4}}} \cdot \left(\alpha - \frac{\varepsilon}{4}z^2\right) \quad \text{at} \quad z \le 1; \tag{4.387}$$

$$u_x = \sqrt{\frac{4}{3}Dx^{\frac{3}{4}}} \cdot \left(\alpha - \frac{\varepsilon}{4}\right) \cdot \frac{1}{z}e^{-\frac{B}{N}(z-1)} \quad \text{at} \quad z \ge 1 \tag{4.388}$$

or

$$\frac{u_x}{u_m} = 1 - \delta z^2 \quad \text{at} \quad z \le 1; \tag{4.389}$$

$$\frac{u_x}{u_m} = (1-\delta)\frac{1}{z}e^{-\frac{B}{N}(z-1)} \quad \text{at} \quad z \geq 1, \tag{4.390}$$

where u_m is the axial velocity of air, equal to

$$u_m = \sqrt{\frac{4}{3}Dx^{\frac{3}{4}}\alpha}, \tag{4.391}$$

and δ is a factor dependent on parameters N and K and equal to

$$\delta = \varepsilon/(4\alpha). \tag{4.392}$$

Table 4.8 lists values of α, $\varepsilon/2$, B, and δ, determined in accordance with Equation 4.386.

TABLE 4.8
Parameters of Axially Symmetric Air Jet Injected by Freely Falling Particle Flow

N	α	$\varepsilon/2$	B	$\alpha - \varepsilon/4$	δ	$\alpha\xi$
			$K = 0$			
10^{-4}	0.9998	1.999	0.666	$3 \cdot 10^{-4}$	0.999	0.5004
10^{-2}	0.9808	1.904	0.667	$2.87 \cdot 10^{-2}$	0.971	0.5055
0.1	0.8579	1.320	0.668	0.198	0.769	0.5876
1	0.5535	0.347	0.912	0.380	0.313	1.3015
10	0.2841	$4.6 \cdot 10^{-2}$	1.760	0.261	$8.09 \cdot 10^{-2}$	3.2391
10^2	0.1349	$4.9 \cdot 10^{-3}$	3.707	0.132	$1.82 \cdot 10^{-2}$	7.2784
10^3	0.0629	$5 \cdot 10^{-4}$	7.948	$6.27 \cdot 10^{-4}$	$3.96 \cdot 10^{-3}$	15.821
			$K = 0.5$			
10^{-4}	0.6665	1.333	0.444	$3 \cdot 10^{-4}$	0.999	0.3336
10^{-2}	0.6541	1.252	0.446	$2.8 \cdot 10^{-2}$	0.957	0.3424
0.1	0.5818	0.822	0.480	0.171	0.706	0.4477
1	0.4136	0.229	0.766	0.299	0.277	1.1368
10	0.2401	$3.58 \cdot 10^{-2}$	1.612	0.222	$7.46 \cdot 10^{-2}$	2.9874
10^2	0.1239	$4.32 \cdot 10^{-3}$	3.551	0.121	$1.74 \cdot 10^{-2}$	6.9979
10^3	0.0604	$4.68 \cdot 10^{-4}$	7.786	$6.01 \cdot 10^{-2}$	$3.88 \cdot 10^{-3}$	15.507
			$K = 1$			
10^{-4}	0.4999	0.999	0.333	$3 \cdot 10^{-4}$	0.999	0.2502
10^{-2}	0.4907	0.927	0.337	$2.75 \cdot 10^{-2}$	0.943	0.2610
0.1	0.4419	0.581	0.384	0.151	0.658	0.3751
1	0.3325	0.168	0.674	0.249	0.252	1.0288
10	0.2088	$2.91 \cdot 10^{-2}$	0.150	0.194	$6.97 \cdot 10^{-2}$	2.7942
10^2	0.1147	$3.85 \cdot 10^{-3}$	3.416	0.113	$1.68 \cdot 10^{-2}$	6.7136
10^3	0.0581	$4.42 \cdot 10^{-4}$	7.636	$5.79 \cdot 10^{-2}$	$3.8 \cdot 10^{-3}$	15.231

The findings are in a sound qualitative and quantitative agreement with data provided over time by various researchers experimenting with injection properties of water droplets (Figures 4.19 and 4.20) including V. P. Gaiduk and A.M. Golyshev [23] and these authors [70]. A significant discrepancy between experimental and theoretical data for axial velocity (dashed line on Figure 4.19) is explained by aerodynamic resistance influencing the velocity of falling particles in the area corresponding to $x > 0.5$. In reality, these particles fall slower than $\sqrt{2x}$.

The flow rate of injected air is determined by an obvious equation

$$q_E = 2\int_0^\infty u_x r dr = 2b^2\int_0^\infty u_x z dz \tag{4.393}$$

or, in view of Equations 4.389 and 4.390,

$$q_E = \sqrt{\frac{4}{3}D x^{\frac{3}{4}} b^2 \alpha\xi}, \tag{4.394}$$

where

$$q_E = Q_E / \left(\pi l_\infty^2 c\right), \tag{4.395}$$

$$\xi = 1 - \frac{\delta}{2} + \frac{(1-\delta)^2}{\delta}. \tag{4.396}$$

As evident from Table 4.8, at small viscosity forces ($N \to 0$) within the area of maximum volumetric forces ($K = 0$), it holds that $\alpha \approx 1$, $\delta \approx 1$. Let us designate injected air flow rate corresponding to this case using q_E (0, 0), while designating air flow rate determined by the relation (4.394) using q_E (N, K).

Based on Equation 4.394,

$$q_E(0,0) = \sqrt{D/3x^{\frac{3}{4}}} b^2. \tag{4.397}$$

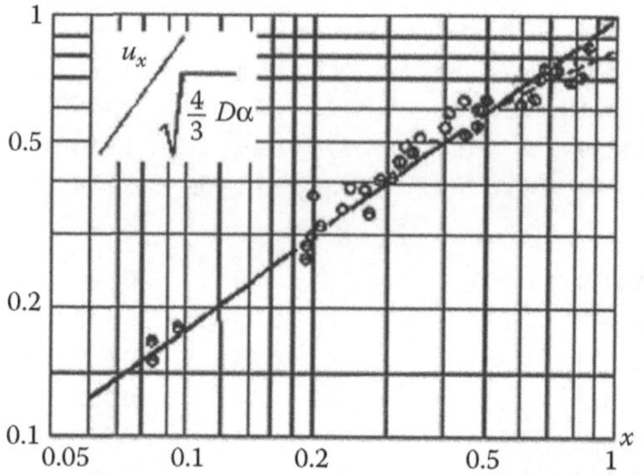

FIGURE 4.19. Changes in linear velocity along jet (experimental data; ⦶ published by V.P. Gaiduk; O published by A.M. Golyshev; ⊖ measured by authors).

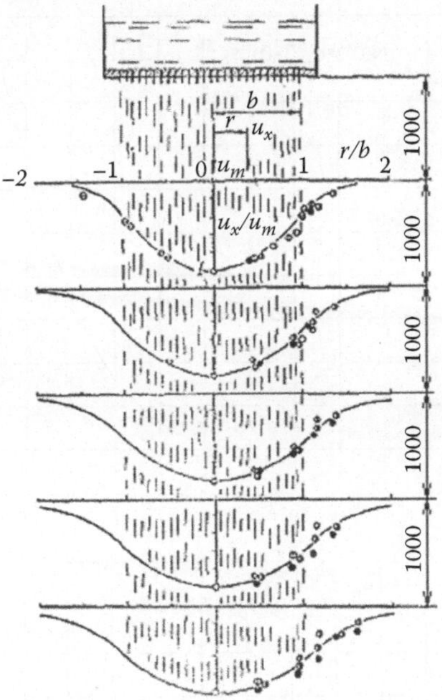

FIGURE 4.20 Velocity diagram of air injected by an axially symmetric jet of freely falling droplets (experimental data: ●: G = 0.25 kg/s, ◑: G = 0.277 kg/s, ⦶: G = 0.166 kg/s, published by V. P. Gaiduk; O: G = 0.14 kg/s, published by A. M. Golyshev; ⊖: G = 0.129 kg/s as measured by authors).

Figure 4.21 illustrates plots of the function q_E $(N, 0)/q_E$ (0, 0) and q_E $(1,K)/q_E$ (1.0). As can be seen on these plots, viscosity forces significantly influence injected air volumes beyond $N > 0.1$.

It would be instructive to compare these equations for q_E with earlier calculations when solving the one-dimensional problem (Equation 4.189). According to the definition (Equation 4.395), the dimensionless flow rate of injected air (Equation 4.196) would then become:

$$q_E = b^2 \varphi \upsilon. \tag{4.398}$$

Figure 4.22 provides plots of axial and average speeds of injected air based on Equations 4.391 and 4.394, correspondingly (curves marked with Roman numeral I), and Equation 4.398 in view of Equation 4.195 and data from Table 4.4 (curves marked with Roman numeral II) at $D = 3/4$, $2k_n/b = 1$.

As the plots show, for smaller vertical distances traveled by material ($x < 0.5$), the average injected air flow rate model is adequate. Viscosity forces are muted in this case ($N \leq 0.1$). With increasing travel path height ($x > 1$), the influence of viscosity forces becomes noticeable.

Therefore, it should be possible to use Equations 4.196 and 4.208 when calculating injected air flow rates for axially symmetric jets of freely falling material particles dumped from a shorter free-fall height ($x < 0.1$; $N < 0.1$). In the case of a greater

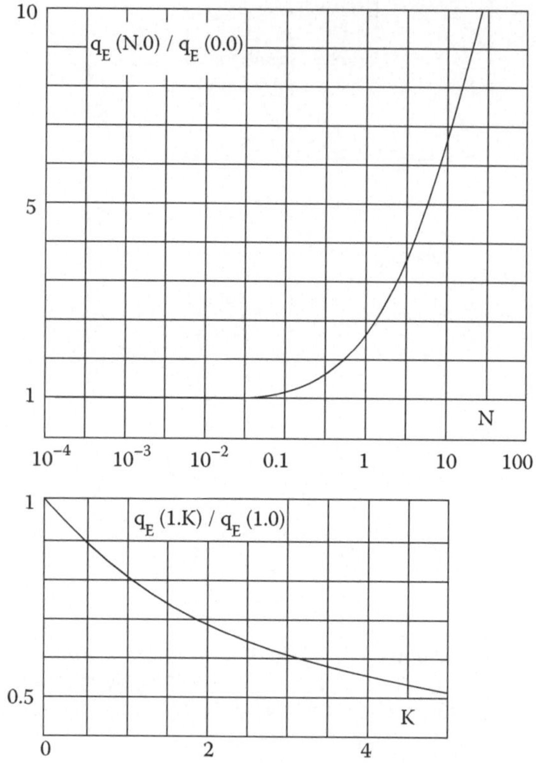

FIGURE 4.21 Changes in injected air flow rate with increasing N and K.

free-fall height, viscosity forces should be accounted for, and injected air volumetric flow rates should be calculated, using Equation 4.394, in view of data provided in Table 4.8.

4.2 THE AERODYNAMICS OF A JET OF PARTICLES IN A CHANNEL

So far we have studied solid particles flowing in a chute and in a jet of loose matter. Both situations represent extreme cases of the more general problem of material flowing through a duct with different distances between flow boundaries and duct walls. Now we shall consider a flat flow limited by vertical walls. The flow would be symmetrical with respect to centerline axis *OX* with positive direction of the axis corresponding to the direction of flowing particles. Because of the symmetry of the aerodynamic field, we shall only study the airflow pattern in the first quadrant, *XOY*, of the coordinate system we have selected. Basic calculations for studying aerodynamic processes will be determined by dimensionless dynamics equations (Equations 4.54–4.56) that could be expressed as follows, provided that $N_\tau >> N$:

$$u_x \frac{\partial u_x}{\partial x} + u_y \frac{\partial u_x}{\partial y} = F_x - \frac{\partial \mathbf{P}}{\partial x} + N_\tau \frac{\partial^2 u_x}{\partial y^2}, \tag{4.399}$$

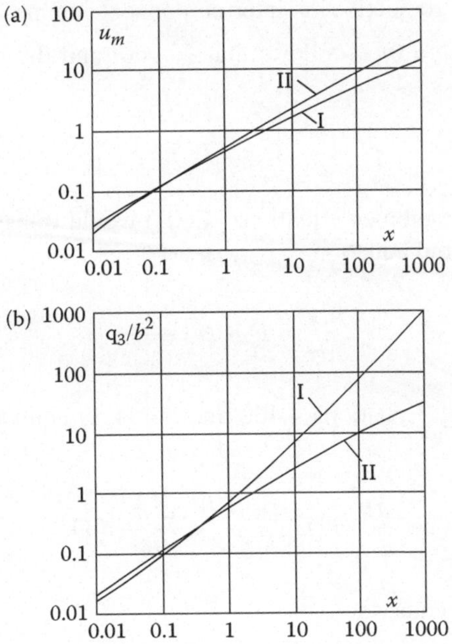

FIGURE 4.22 Changes of (a) axial and (b) cross-sectional average flow rates of injected air.

$$u_x \frac{\partial u_y}{\partial x} + u_y \frac{\partial u_y}{\partial y} = F_y - \frac{\partial \mathbf{P}}{\partial y} + \frac{\partial}{\partial x}\left(N_\tau \frac{\partial u_x}{\partial y}\right), \tag{4.400}$$

$$\frac{\partial u_x}{\partial x} + \frac{\partial u_y}{\partial y} = 0. \tag{4.401}$$

4.2.1 Plane-Parallel Flow

In case of a plane-parallel motion of solid particles, the airflow initiated by the particles inside the duct could be represented by plane-parallel motion ($u_y = 0$). Combined Equations 4.399 through 4.401 are thereby greatly simplified. Due to the continuity equation, $\frac{\partial u_x}{\partial x} = 0$; that is, the velocity u_x, although remaining constant over the flow line, depends only on the ordinate

$$u_x = f_u(y). \tag{4.402}$$

The first of the combined equations will, therefore, assume the following form:

$$\frac{\partial \mathbf{P}}{\partial x} = F_x + N_\tau \frac{d^2 u_x}{dy^2}; \qquad \frac{\partial \mathbf{P}}{\partial y} = F_y. \tag{4.403}$$

Considering that the projection of inter-component interaction force on OY axis equals zero ($\upsilon_y = 0$; $u_y = 0$), it follows that $\frac{\partial \mathbf{P}}{\partial y} = 0$; and the pressure only changes along the duct

$$\mathbf{P} = f_p(x). \tag{4.404}$$

Then, the first of combined equations (4.403) would transform into an ordinary second-order differential equation

$$\frac{d\mathbf{P}}{dx} = F_x + N_\tau \frac{d^2 u_x}{dy^2}, \tag{4.405}$$

that, as determined by Equations 4.402 and 4.404, is equivalent to the following combined equation

$$\frac{d\mathbf{P}}{dx} = \Pi; \quad F_x + N_\tau \frac{d^2 u_x}{dy^2} = \Pi, \tag{4.406}$$

where Π is a constant equal to

$$\Pi = \left(\mathbf{P}_K - \mathbf{P}_H\right) / l$$

and $\mathbf{P}_H$, $\mathbf{P}_K$ is the pressure at the beginning and at the end of a duct of length l.

It should be noted that, generally, $\upsilon = f\upsilon(x)$ and the projection of the vector of inter-component interaction force onto OX depends on x and y. This fact contradicts the initial equation (4.402). Hence, the supposition about the plane-parallel character of the injected airflow inside the duct and the accelerated movement of material particles makes no sense.

Uniform movement of particles should be assumed in order to eliminate this inconsistency. Because that would significantly restrict the application of our findings, let us consider one special case where material velocity $\upsilon = \upsilon_0$ – const greatly exceeds air velocity; Equation 4.28, for generalized exponential distribution of particles, then results in

$$F_x \approx \beta_0 e^{-\left(\frac{y}{b}\right)^t} \cdot \upsilon_0^2, \tag{4.407}$$

and Equation 4.406 becomes

$$N_\tau \frac{d^2 u_x}{dy^2} = -\beta_0 \upsilon_0^2 e^{-\left(\frac{y}{b}\right)^t} + \Pi. \tag{4.408}$$

Under boundary conditions

$$u_x(b_0) = 0; \quad \left.\frac{du_x}{dy}\right|_{y=0} = 0, \tag{4.409}$$

the solution has the form

$$u_x = B_k \int_y^{b_0} \left(\int_0^y e^{-\left(\frac{y}{b}\right)^t} dy \right) dy - \Pi_k \frac{b_0^2 - y^2}{2}, \tag{4.410}$$

where b_0 is a dimensionless duct breadth variable; and B_k, Π_k are dimensionless complexes

$$B_k = \frac{\beta_0 \upsilon_0^2}{N_\tau}; \quad \Pi_k = \frac{\Pi}{N_\tau}.$$

In particular, with particles uniformly distributed inside the duct (t → ∞),

$$u_x = \left(B_k - \Pi_k\right) \frac{b_0^2 - y^2}{2}. \tag{4.411}$$

In this case, the airflow direction, being the same across the entire duct, is determined with a summation sign $B_k - \Pi_k$; $\Pi_k > B_k$ gives rise to a counterflow, and $\Pi_k < B_k$ corresponds to a direct flow. A perfect analogy would be the one-dimensional movement of material in a chute.

In this case, the air flow rate would be determined by

$$q_E = 2\int_0^{b_0} u_x dy = 2\left(B_k - \Pi_k\right)\frac{b_0^3}{3}. \tag{4.412}$$

When the concentration of material is not constant throughout its cross-section, but it varies (e.g., according to exponential law, $t = 1$), at a certain Π_k, the airflow could delaminate so that some air would flow downward (along the centerline with its greater concentration of particles) and the remainder would be displaced upward. Indeed, the solution for Equation 4.410 at $t = 1$ would assume the form

$$u_x = B_k b\left(b_0 - y\right) - \Pi_k \frac{b_0^2 - y^2}{2} - B_k b^2 \left(e^{-\frac{y}{b}} - e^{-\frac{b_0}{b}} \right). \tag{4.413}$$

And

$$\frac{\Pi_k}{B_k} < 2\left(\frac{b}{b_0}\right)^2 \cdot \left(\frac{b_0}{b} - 1 + e^{-\frac{b_0}{b}}\right) \tag{4.414}$$

the velocity $u_x(0) > 0$ along the centerline would correspond to a direct flow zone.

Along the straight line $y = y_0$ where y_0 is the ordinate meeting the equation

$$B_k b\left(b_0 - y_0\right) = \Pi_k \frac{b_0^2 - y_0^2}{2} + B_k b^2 \left(e^{-\frac{y_0}{b}} - e^{-\frac{b_0}{b}} \right), \tag{4.415}$$

the velocity u_x becomes equal to zero. Finally, the $y_0 < y < b_0$ area manifests countercurrent airflow ($u_x < 0$).

In this case, the straight $y = y_0$ becomes a dividing line between direct flow and counterflow. The equality condition determining the first type of airflow is

$$q_E^{\downarrow} = \int_0^{y_0} u_x dy = B_k b \frac{2b_0 y_0 - y_0^2}{2} - \Pi_k \frac{3b_0^2 y_0 - y_0^3}{6} + B_k b^3 \left(e^{-\frac{y_0}{b}} - 1 + \frac{y_0}{b} e^{-\frac{b_0}{b}} \right), \qquad (4.416)$$

but, for the second type, it would be

$$q_E^{\uparrow} = \int_{y_0}^{b_0} u_x dy = B_k b \frac{(b_0 - y_0)^2}{2} - \Pi_k (b_0 - y_0) \frac{2b_0^2 - b_0 y_0 - y_0^2}{6} + B_k b^3 \left[e^{-\frac{b_0}{b}} \left(1 + \frac{b_0 - y_0}{b} \right) - e^{-\frac{y_0}{b}} \right]. \qquad (4.417)$$

4.2.2 One-Dimensional Flow

In practice, the plane-parallel flow pattern considered before is extremely unlikely to occur. Transverse overflow of air—the key necessary condition for such currents—is hardly conceivable. Solving the generalized problem analytically would pose insurmountable difficulties at $u_y \neq 0^*$. Nor is it easy to solve hydromechanical equations numerically due to nonlinearity [80]. A possible alternative approach may involve equations that bind cross-sectional averages of various flow parameters. As illustrated earlier, one-dimensional problems yield satisfactory outcomes often enough. Thus, we could formulate a one-dimensional problem for a jet of loose matter confined to a duct with its wall set apart by the distance b_0 from the centerline. Let us denote the half-breadth of such a jet as b_n. Consequently, there would be two flows: air moving together with material inside a band $0 \le y \le b_n$ corresponding to an inner dual-component flow and air flowing through a gap between the wall and jet boundary surface corresponding to an outer single-component flow.

Let us suppose that falling particles are distributed uniformly across the jet. In order to obtain average equations featuring dynamics of air in these fields, we will integrate Equation 4.71 along the OY axis. For the inner flow ($0 \le y \le b_n$), the equation would appear as

$$\frac{\partial}{\partial x} \int_0^{b_n} u_x^2 dy + u_y u_x \Big|_0^{b_n} = \frac{D}{\sqrt{2}\upsilon} \int_0^{b_n} (\upsilon - u_x)^2 dy - \frac{\partial}{\partial x} \int_0^{b_n} \mathbf{P} dy + N_\tau \frac{\partial u_x}{\partial y} \Bigg|_0^{b_n}. \qquad (4.418)$$

For the outer flow ($b_n \le y \le b_0$), the equation would appear as

$$\frac{\partial}{\partial x} \int_{b_n}^{b_0} u_x^2 dy + u_y u_x \Big|_{b_n}^{b_0} = -\frac{\partial}{\partial x} \int_{b_n}^{b_0} \mathbf{P} dy + N_\tau \frac{\partial u_x}{\partial y} \Bigg|_{b_n}^{b_0}. \qquad (4.419)$$

To perform the averaging, suppose that pressure remains constant throughout the cross-section of the duct. Thus,

$$\int_0^{b_n} \mathbf{P} dy \approx \mathbf{P} b_n; \quad \int_{b_n}^{b_0} \mathbf{P} dy = \mathbf{P}(b_0 - b_n). \qquad (4.420)$$

Air velocity in the inner flow, averaged by flow rate, will be designated using u, and that in the outer flow will be designated using ω (the positive direction matching the direction of *OX* axis):

$$\int_0^{b_n} u_x dy = b_n u; \quad \int_{b_n}^{b_0} u_x dy = (b_0 - b_n)\omega. \tag{4.421}$$

Positing that

$$\int_0^{b_n} u_x^2 dy \approx b_n u^2; \quad \int_{b_n}^{b_0} u_x^2 dy \approx (b_0 - b_n)\omega^2, \tag{4.422}$$

$$\int_0^{b_n} (\upsilon - u_x)^2 dy \approx b_n (\upsilon - u)^2 \tag{4.423}$$

and assuming normal admission of air on the boundary between the inner and outer streams, that is,

$$u_x(x, b_n) = 0; \quad \left.\frac{\partial u_x}{\partial y}\right|_{y=b_n} = 0, \tag{4.424}$$

and further considering the consequence to flow symmetry,

$$u_y(x, 0) = 0; \quad \left.\frac{\partial u_x}{\partial y}\right|_{y=0} = 0, \tag{4.425}$$

and accounting for a frictional shear stress at the duct wall,

$$-\overline{\tau}_{cm} = N_\tau \left.\frac{\partial u_x}{\partial y}\right|_{y=b_0}; \quad \overline{\tau}_{cm} = \frac{\tau_{cm}}{\rho_2 c^2} \tag{4.426}$$

integral relations (Equations 4.418 and 4.419) would lead us to the following system of ordinary differential equations:

$$\frac{du^2}{dx} = \frac{D}{\sqrt{2\upsilon}}(\upsilon - u)^2 - \frac{d\mathbf{P}}{dx} \quad \text{at } 0 \le y \le b_n, \tag{4.427}$$

$$\frac{d\omega^2}{dx} = -\frac{d\mathbf{P}}{dx} - \varepsilon_\tau \quad \text{at} \quad b_n \le y \le b_0, \tag{4.428}$$

$$u + \omega(r-1) = u_0 + \omega_0(r-1) = u_m - \text{const}, \tag{4.429}$$

where

$$r = b_0 / b_n; \quad \varepsilon_\tau = \overline{\tau}_{cm} / (b_0 - b_n). \tag{4.430}$$

The latter equation expresses a cross-sectional flow rate conservation law in a duct with impervious walls.

In view of Equations 4.428 and 4.429, Equation 4.427 could be expressed in the following form, making it easier to integrate:

$$\frac{du^2}{dx}+\frac{2(u_m-u)}{(r-1)^2}\cdot\frac{du}{dx}-\varepsilon_\tau=\frac{D}{\sqrt{2}\upsilon}(\upsilon-u)^2 \tag{4.431}$$

or

$$\left[\frac{2r(r-2)}{(r-1)^2}u+\frac{2u_m}{(r-1)^2}\right]\frac{du}{dx}=\frac{D}{\sqrt{2}\upsilon}(\upsilon-u)^2+\varepsilon_\tau. \tag{4.432}$$

This equation could be applied for analyzing the simplest case—when $\upsilon=\upsilon_0-\text{const}$ and forces of friction against duct walls are negligibly small. Equation 4.432 would thus become

$$\frac{du}{dx}=\frac{D(\upsilon_0-u)^2}{\sqrt{2}\upsilon_0\cdot R(r,u)}; \tag{4.433}$$

$$R(r,u)=2\left[r(r-2)u+u_m\right]/(r-1)^2 \tag{4.434}$$

and would resolve at initial conditions as follows:

$$u=u_H \quad \text{at} \quad x=x_H,$$

assuming the form

$$\left[\frac{2r(r-2)}{(r-1)^2}\upsilon_0+\frac{2u_m}{(r-1)^2}\right]\cdot\frac{u-u_H}{(\upsilon_0-u)(\upsilon_0-u_H)}+\frac{2r(r-2)}{(r-1)^2}\ln\frac{\upsilon_0-u}{\upsilon_0-u_H}=\frac{D}{\sqrt{2}\upsilon_0}(x-x_H). \tag{4.435}$$

Let us analyze the behavior of u and ω along the duct with different b_0/b_n ratios characterizing flow restriction by the duct walls. The following values will be assumed as known initial data:

$$u_H=u_0; \quad \omega_H=\omega_0 \quad \text{at } x_H=0. \tag{4.436}$$

Then Equation 4.435 could be transformed into

$$\frac{D}{\sqrt{2}\upsilon_0}x=\left[\frac{2r(r-2)}{(r-1)^2}\upsilon_0+\frac{2u_m}{(r-1)^2}\right]\cdot\frac{u-u_0}{(\upsilon_0-u)(\upsilon_0-u_0)}+\frac{2r(r-2)}{(r-1)^2}\ln\frac{\upsilon_0-u}{\upsilon_0-u_0} \tag{4.437}$$

or

$$\bar{x}=\left[\frac{2r(r-2)}{(r-1)^2}+\frac{2\bar{u}_m}{(r-1)^2}\right]\cdot\frac{\bar{u}-\bar{u}_0}{(1-\bar{u})(1-\bar{u}_0)}+\frac{2r(r-2)}{(r-1)^2}\ln\frac{1-\bar{u}}{1-\bar{u}_0}, \tag{4.438}$$

where

$$\bar{u}_m = u_m / \upsilon_0 = \bar{u}_0 + \bar{\omega}_0 (r-1); \tag{4.439}$$

$$\bar{u}_0 = u_0 / \upsilon_0; \quad \bar{\omega}_0 = \omega_0 / \upsilon_0; \quad \bar{u} = u / \upsilon_0; \tag{4.440}$$

$$\bar{x} = xD / \left(\sqrt{2}\upsilon_0\right). \tag{4.441}$$

The distance between the duct origin and a cross-section where the inner air flow velocity becomes equal to

$$u = u_m, \tag{4.442}$$

and will be expressed using x_m. This cross-section will henceforth be considered as critical, and x_m will be regarded as the initial run of the duct. In the critical section, the continuity equation (Equation 4.429) would make the outer flow velocity equal to zero. Due to the equality condition (Equation 4.438), the relative length of the initial run will be

$$\bar{x}_m = \left[\frac{2r(r-2)}{(r-1)^2} + \frac{2\bar{u}_m}{(r-1)^2}\right] \cdot \frac{\bar{u}_m - \bar{u}_0}{(1-\bar{u}_m)(1-\bar{u}_0)} + \frac{2r(r-2)}{(r-1)^2} \ln \frac{1-\bar{u}_m}{1-\bar{u}_0}. \tag{4.443}$$

Figure 4.23 plots the dependence of this length on r in various initial conditions. As it can be seen, the value $\bar{x}_m$ will increase when the flow centerline is moved away from the duct walls (with increasing r) and when initial velocities $\bar{u}_0$ and $\bar{\omega}_0$ are increased. Additional air volume is necessary to ensure increased air velocities.

Beyond the critical section lies a zone of upward outer flow ($\omega < 0$). As air moves further away from the critical section, the upward outer flow will experience increasing flow rates until a maximum is reached at a certain spot that we will call the extreme cross-section. As Equation 4.434 hints, the presence of an extreme cross-section is conditional on

$$R(r, u_e) = 0 \tag{4.444}$$

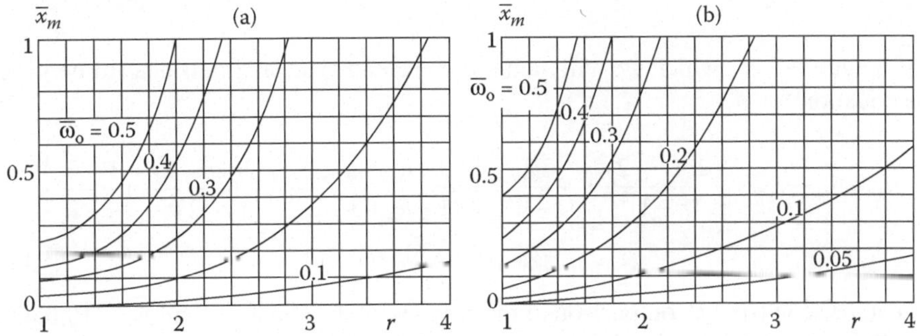

FIGURE 4.23 Relative length of the initial run as a function of flow restriction at (a) $\bar{u}_0 = 0$ and (b) $\bar{u}_0 = 0,2$.

or

$$u_E = -\frac{u_m}{r(r-2)}, \qquad (u_e \le \upsilon_0). \tag{4.445}$$

As we can see, in case of a downward initial flow in the duct, it would only be possible at restriction degrees

$$r < 2. \tag{4.446}$$

The length of the zone $x_e - x_m$ (which we will name the initial eddy run length) is determined using Equation 4.438

$$\bar{l}_H \equiv \bar{x}_e - \bar{x}_m = \left[\frac{2r(r-2)}{(r-1)^2} + \frac{2\bar{u}_m}{(r-1)^2}\right] \cdot \frac{\bar{u}_e - \bar{u}_m}{(1-\bar{u}_e)(1-\bar{u}_m)} + \frac{2r(r-2)}{(r-1)^2} \ln\frac{1-\bar{u}_e}{1-\bar{u}_m}. \tag{4.447}$$

The equality condition (4.438) determines changes in velocity on this run. Further velocity increases u become impossible because the function $R(r,u)$ turns negative; therefore,

$$\frac{du}{dx} < 0,$$

that is, air begins to escape the inner flow. Air flow rate in the outer counterflow decreases to zero in the next critical section.

In this case, the differential equation (4.433) would be rewritten as

$$\frac{du}{dx} = -\frac{D(\upsilon_0 - u)^2}{\sqrt{2}\upsilon_0 |R(r,u)|}, \tag{4.448}$$

and, in the initial condition,

$$u = u_e \quad \text{at} \quad x = x_e$$

would become

$$\bar{x} - \bar{x}_e = \left|\left[\frac{2r(r-2)}{(r-1)^2} + \frac{2\bar{u}_m}{(r-1)^2}\right] \cdot \frac{\bar{u} - \bar{u}_e}{(1-\bar{u})(1-\bar{u}_e)} + \frac{2r(r-2)}{(r-1)^2} \ln\frac{1-\bar{u}}{1-\bar{u}_e}\right|. \tag{4.449}$$

The length $\bar{x}_m - \bar{x}_e$, which we will name the final eddy run length, is determined with the equation

$$\bar{l}_k \equiv \bar{x}_m - \bar{x}_e = \left|\left[\frac{2r(r-2)}{(r-1)^2} + \frac{2\bar{u}_m}{(r-1)^2}\right] \cdot \frac{\bar{u}_m - \bar{u}_e}{(1-\bar{u}_m)(1-\bar{u}_e)} + \frac{2r(r-2)}{(r-1)^2} \ln\frac{1-\bar{u}_m}{1-\bar{u}_e}\right|. \tag{4.450}$$

As can be seen from a comparison of the result with the equality condition (Equation 4.447),

$$\bar{l}_H = \bar{l}_k, \tag{4.451}$$

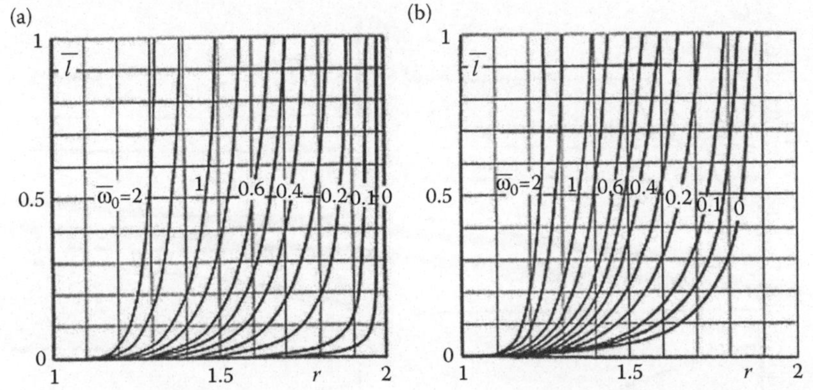

FIGURE 4.24 Variation in relative eddy length as a function of restricting the flow of loose matter at (a) $\bar{u}_0 = 0$ and (b) $\bar{u}_0 = 0,2$.

which can be explained by a constant velocity of falling particles. The total length of an eddy, resulting from the relation

$$\bar{l} = 2\bar{l}_H = 2\bar{l}_k, \tag{4.452}$$

decreases with decreasing initial airflow velocity in the inner and outer flows (Figure 4.24) with the relative duct size kept constant. Lower values of r would produce more eddies in the outer flow (Figure 4.25). Absolute velocity in a flow of particles fluctuates around average value. At the limit $r \to 1$, it becomes equal to u_0. In this case, we are considering a one-dimensional problem for a chute. The other extreme case could be observed with increasing r. Increasing distances between the flow and the duct wall reduces the occurrence of eddies until counterflow could only be observed near the end of the duct. Finally, further increases of r result in an exclusively direct flow of air along the entire duct with increasing velocities in the inner flow and decreasing velocities in the outer flow. The limit case of $r \to \infty$ corresponds to a free flow of particles for which the air injection at $\upsilon = \upsilon_0$ – const could be described in view of Equations 4.433 and 4.434 by

$$\frac{du}{dx} = \frac{D}{\sqrt{2}\upsilon_0} \cdot \frac{(\upsilon_0 - u)^2}{2u}, \tag{4.453}$$

which resolves at $u = u_0$ and at $x = 0$ as

$$\frac{\upsilon_0}{\upsilon_0 - u} - \frac{\upsilon_0}{\upsilon_0 - u_0} + \ln\frac{\upsilon_0 - u}{\upsilon_0 - u_0} = \frac{D}{2\sqrt{2}\upsilon_0}x. \tag{4.454}$$

Figure 4.26 shows how duct breadth may change the final velocity of material injected from the duct with the flow. This change is notably asymptotic in nature. Velocity almost stabilizes when duct walls become spaced by 5–7 b_n. Walls produce no braking effect on the velocity of injected air. As material comes closer to the

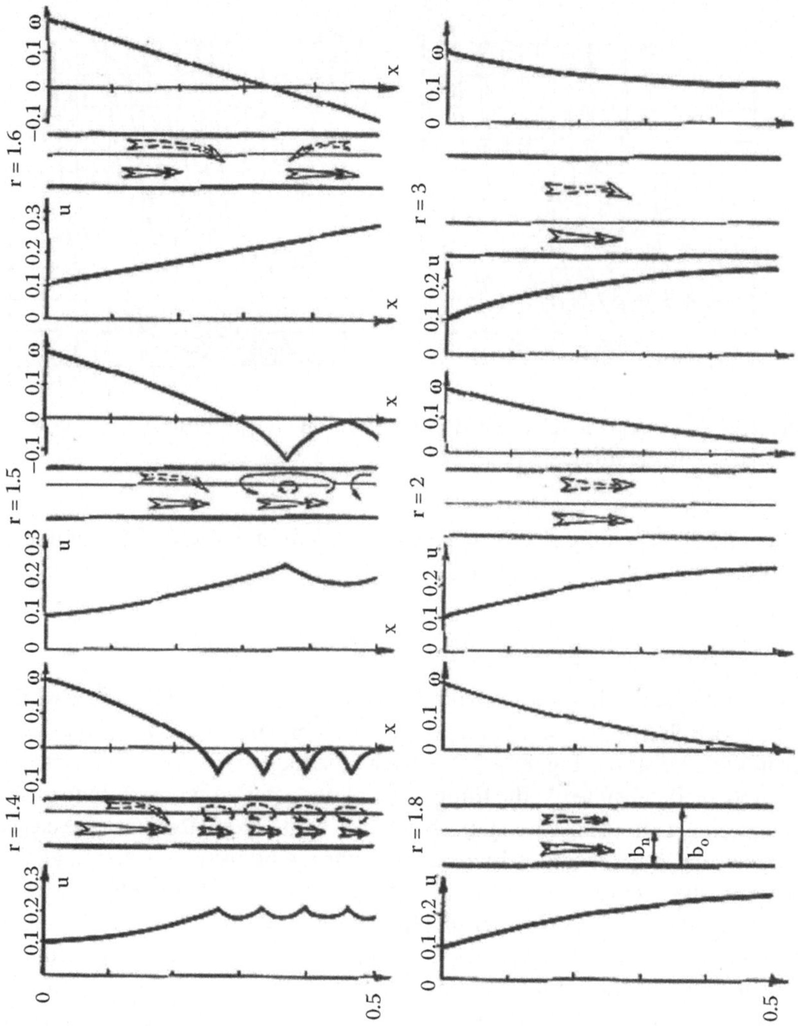

FIGURE 4.25 Variation in relative air velocity inside duct for uniformly distributed falling particles of loose matter ($D = \sqrt{2}$; $\upsilon_0 = 0.5$; $u_0 = 0{,}1$ with $\omega_0 = 0{,}2$).

flow, the quantity of injected air noticeably drops. This happens due to impaired air overflow conditions from the outer into the inner flow.

A similar flow pattern could be observed with linearly accelerated particles of loose matter. The differential equation (4.432) describing changes in air velocity in the inner flow at the duct walls (at negligibly small frictional forces) could be rewritten as

$$\left(a_1 u + b_1\right)\frac{du}{d\upsilon} = \frac{D}{\sqrt{2}}(\upsilon - u)^2, \tag{4.455}$$

where

$$a_1 = 2r(r-b)/(r-1)^2; \quad b_1 = 2u_m/(r-1)^2. \tag{4.456}$$

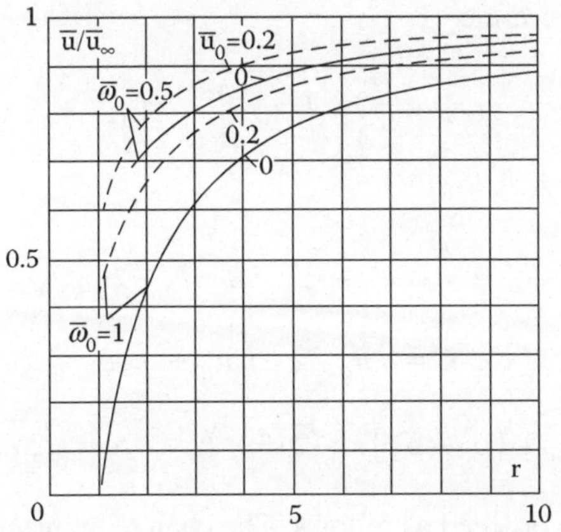

FIGURE 4.26 Variation in relative velocity of injected air at the end of the duct ($\overline{x} = 0{,}5$) as a function of flow restriction ($\overline{u}_\infty$ is the injected airflow velocity at the end of the jet at r → ∞).

Substituting for variables

$$\hat{\upsilon} = D\frac{\upsilon a_1 + b_1}{\sqrt{2a_1^2}}; \qquad \hat{u} = D\frac{ua_1 + b_1}{\sqrt{2a_1^2}}, \tag{4.457}$$

Equation 4.455 could be reduced to the form

$$\hat{u}\frac{d\hat{u}}{d\hat{\upsilon}} = (\hat{\upsilon} - \hat{u})^2, \tag{4.458}$$

considered by us when solving the problem of air injection with a free jet.

Analytical relations could be derived either from the data listed in Table 4.4 or from the approximate equalities in Table 4.3. As an example, we can plot calculated ratios using the approximation

$$(\upsilon - u)^2 \approx \upsilon^2\overline{\left(1 - \frac{u}{\upsilon}\right)}^2, \qquad \overline{\left(1 - \frac{u}{\upsilon}\right)}^2 \approx \left(1 - \frac{u}{\upsilon}\right)^2, \tag{4.459}$$

producing satisfactory results for a free jet (see 8 in Table 4.3). Equation 4.455 would be easy to integrate in view of this approximation. At initial conditions

$$u = u_H, \qquad \upsilon = \upsilon_H \quad \text{at} \quad x = x_H$$

it holds that

$$u_1\frac{u^2 - u_H^2}{2} + h_1(u - u_H) = \gamma_a\frac{D}{\sqrt{2}}\cdot\frac{\upsilon^3 - \upsilon_H^3}{3\upsilon^2}(\upsilon - u)^2, \tag{4.460}$$

where

$$\gamma_a = \begin{cases} 1 \\ -1. \end{cases}$$

Therefore, we can determine

$$u = \frac{B}{2A}\left(\sqrt{1+\frac{4AC}{B^2}} - 1\right), \tag{4.461}$$

where

$$A = a_1/2 - z; \qquad B = b_1 + 2\upsilon z; \tag{4.462}$$

$$C = a_1 u_H^2/2 + b_1 u_H + \upsilon^2 z; \tag{4.463}$$

$$z = \gamma_a D\cdot\left(\upsilon^3 - \upsilon_H^3\right)/\left(\sqrt{2}\cdot 3\upsilon^2\right); \qquad \upsilon = \sqrt{2x + \upsilon_0^2}. \tag{4.464}$$

Calculation should proceed as follows. The change in injected air velocity on the initial run is determined:

$$x_H = 0; \quad x_m \geq x \geq x_H; \qquad \gamma_a = 1; \quad u_H = u_0; \quad \upsilon_H = \upsilon_0. \tag{4.465}$$

Equation 4.461 is used to calculate velocity u. Its value grows from u_0 to u_m. By further increasing x, we transition into the initial run of the first eddy. Without changing initial values of u_H, υ_H, and x_H, we end up with

$$x_m \leq x \leq x_e, \; u_m \leq u \leq u_e = -u_m/\left[r(r-2)\right]$$

if $r < 2$ (the center of the eddy will not be reachable with $r \geq 2$). Further increases of x lead to a transition into the final run of the first eddy. Changes in velocity u are determined by the same Equation 4.461 with different initial values

$$\upsilon_H = \upsilon_{He} = \sqrt{2x_e + \upsilon_0^2}; \quad u_H = u_e; \quad \gamma_a = -1. \tag{4.466}$$

In this area, the velocity u decreases from u_e down to u_m (as x increases from x_e to x_m^I). The initial run of the second eddy occurs here. Changes in the velocity u on this spot could be determined using Equation 4.461, adjusted for different initial conditions

$$x_H = x_m^I; \quad \upsilon_H = \sqrt{2x_m^I + \upsilon_0^2}; \quad u_H = u_m; \quad \gamma_a = 1. \tag{4.467}$$

Velocity increases again from u_m to u_e. After that, the final run of the second eddy begins, so that initial conditions must be adjusted again in order to calculate velocities

$$x_H = x_e^I; \qquad \upsilon_H = \sqrt{2x_e^I + \upsilon_0^2}; \quad u_H = u_e; \quad \gamma_a = -1. \tag{4.468}$$

The calculation procedure is repeated. As we can see, $\gamma_a = +1$ should be posited at initial runs of eddies and $\gamma_a = -1$ should be posited at final runs. These runs differ in length because of equal acceleration of the particle flow. Unlike in the case of uniform motion considered earlier, the initial run is longer than the final run and the second eddy is longer overall than the first one. This becomes evident in Figure 4.27, which shows calculated flow patterns for a jet in a duct using the same initial

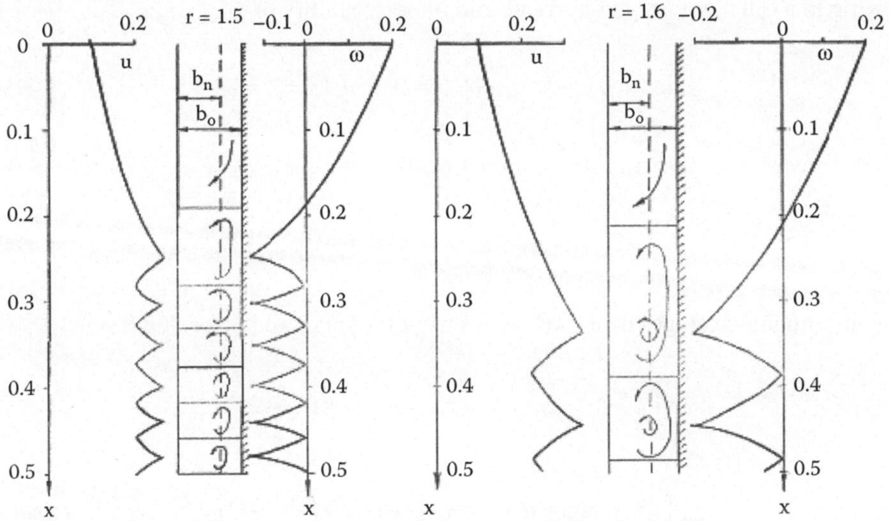

FIGURE 4.27 Variation in relative air velocity inside duct for linearly accelerated falling particles of loose matter ($D=\sqrt{2}$; $\upsilon_0 = 0.5$; $u_0 = 0{,}1$; $\omega_0 = 0{,}2$).

parameters that were used to produce the flow pattern for a uniformly moving flow of loose matter (see Figure 4.25).

For airflows inside a cylindrical duct where a stream of falling particles is located coaxially, integral dynamics equations could be written based on Equations 4.72 and 4.433 as follows:

$$\frac{\partial}{\partial x}2\pi\int_0^{r_n} u_x^2 r dr + 2\pi r u_r u_x\Big|_0^{r_n} = \frac{D}{\sqrt{2}\upsilon}2\pi\int_0^{r_n}(\upsilon - u_x)^2 r dr - \frac{\partial}{\partial x}\int_0^{r_n} 2\pi \mathbf{P} r dr + N_\tau 2\pi r\frac{\partial u_x}{\partial r}\Big|_0^{r_n} \tag{4.469}$$

at $0 \le r \le r_n$;

$$\frac{\partial}{\partial x}2\pi\int_{r_n}^{r_0} u_x^2 r dr + 2\pi r u_r u_x\Big|_{r_n}^{r_0} = -\frac{\partial}{\partial x}2\pi\int_{r_n}^{r_0} \mathbf{P} r dr + N_\tau 2\pi r\frac{\partial u_x}{\partial r}\Big|_{r_n}^{r_0} \tag{4.470}$$

at $r_n \le r \le r_0$ where r_n, r_0 are dimensionless radii of particles and duct boundaries.

Based on the same assumptions for simplification, namely that the static pressure is constant throughout the cross-section of the duct

$$\int_0^{r_n} \mathbf{P} r dr = \mathbf{P}\frac{r_n^2}{2}; \quad \int_{r_n}^{r_0} \mathbf{P} r dr = \mathbf{P}\frac{r_0^2 - r_n^2}{2}; \tag{4.471}$$

air admission at the boundary of the outer and inner channels occurs radially

$$u_x(x, r_n) = 0; \quad \frac{\partial u_x}{\partial r}\Big|_{r=r_n} = 0; \tag{4.472}$$

owing to axial symmetry of currents and impermeability of duct walls

$$u_r(x,0)=0;\quad \left.\frac{\partial u_x}{\partial r}\right|_{r=0}=0;\quad u_r(x,r_0)=0; \tag{4.473}$$

in the presence of shearing stress at duct walls

$$-\bar{\tau}_w=N_\tau \left.\frac{\partial u_x}{\partial r}\right|_{r=r_0},\quad \bar{\tau}_w=\tau_w/\left(\rho_2 c^2\right) \tag{4.474}$$

by introducing averages over cross-sections of the inner and outer flows

$$2\pi\int_0^{r_n} u_x r dr=\pi r_n^2 u;\quad 2\pi\int_{r_n}^{r_0} u_x r dr=\pi\left(r_0^2-r_n^2\right)\omega, \tag{4.475}$$

$$2\pi\int_0^{r_n} u_x^2 r dr\approx\pi r_n^2 u^2;\quad 2\pi\int_{r_n}^{r_0} u_x^2 r dr\approx\pi\left(r_0^2-r_n^2\right)\omega^2, \tag{4.476}$$

$$2\pi\int_0^{r_n}\left(\upsilon-u_x\right)^2 r dr\approx\pi r_n^2(\upsilon-u)^2 \tag{4.477}$$

integral relations are reduced to differential equations of one-dimensional streams

$$\frac{du^2}{dx}=\frac{D}{\sqrt{2\upsilon}}(\upsilon-u)^2-\frac{d\mathbf{P}}{dx}\quad \text{at}\quad 0\le r\le r_n, \tag{4.478}$$

$$\frac{d\omega^2}{dx}=-\frac{d\mathbf{P}}{dx}-\varepsilon_\tau\quad \text{at}\quad r_n\le r\le r_0, \tag{4.479}$$

$$u+\omega\left(n^2-1\right)=u_0+\omega_0\left(n^2-1\right)=u_m, \tag{4.480}$$

where n is the ratio among radii of boundaries surrounding the jet of material

$$n=r_0/r_n; \tag{4.481}$$

$$\varepsilon_\tau=\bar{\tau}_w 2r_0/\left(r_0^2-r_n^2\right). \tag{4.482}$$

Therefore, combined equations for an axially symmetric flow would differ from similar equations of a plane problem only in the equation for airflow (4.480) that depends on relative duct size, squared. The resulting numerical relationships of the planar problem are valid for the axially symmetric problem as well. In this case, it is just enough to replace r with n^2 in formulations.

These findings are in qualitative and quantitative agreement with experimental data. Indeed, the described turbulent flows were observed for the first time by A. S. Serenko, who researched currents in a sand layer moving along the bottom wall of a one-meter-long square pipe [87]. It was noted that air countercurrents did not always occur; they only occurred at certain locations on the upper duct wall (with respect to flowing material).

With a clearance height of 40 mm, a unidirectional current of injected air was observed in the duct. In this case, flowing particles filled the entire cross-section of the duct ($r \to 1$). Countercurrents arose when duct clearance height increased. Notably, air moved in line with the particle layer at the beginning but reversed into a counterflow toward the end of the duct. A similar pattern was reported by Neikov and Zilberberg who researched the aerodynamics of iron powder streams in a tilted chute [67].

A. S. Serenko's experiments have shown that the distance from the duct inlet to the point where an air countercurrent arises could be reduced almost to zero by obstructing the inlet with a gate valve. In other words, the initial run becomes shorter as the original rate of the outer flow diminishes—this agrees with our findings.

Circulation inside a duct filled with material flowing throughout the entire cross-section (which we will label "natural" circulation) is likely only in exceptional cases. Natural circulation is hindered by a number of factors. First of all, when particles are lumpy and grainy, they occupy virtually the entire duct clearance area, and the inherent transverse gradient of particle concentration slightly changes the longitudinal velocity profile of injected air. When aspiration develops in a descending pattern in a hollow duct area not filled with material, there is an outside positive gradient precluding the occurrence of a countercurrent. The opposite effect occurs when handling heated material—a thermal head produced by inter-component heat exchange will promote formation of natural circulation.

5 Engineering Solutions for Dust Release Containment and Air Dedusting

Aspiration is the most universal and common dedusting method used for handling loose materials at ore pretreatment plants. It ensures containment of released dust by using aspirated cowls with subsequent separation of dust from air evacuated by suction. This method remains the only dedusting option for sintering processes and for pelletization of iron-ore concentrates when the alternative method of wet dedusting becomes impossible due to high airborne dust content and thermal breakdown of hot sinter and fired pellets by water.

Equipment design problems can only be solved successfully when specific material handling technology issues and the peculiarities of process equipment operation are fully addressed. Optimization of these solutions calls for a close study of the aerodynamic processes that are part of dust-laden air stream formation, the patterns of dust particle origination, and precipitation of particles (from the air) in all elements of containing devices (chutes, cowls, and dust collection funnels). Benefits of lower initial dust concentrations extend beyond reducing the complexity and costs of air cleaning in centralized dust collection plants. Initial air cleaning in cowls having coarse dust removal improves the reliability of air duct systems, reduces the probability of coarse particles clogging horizontal runs of the network, and mitigates the abrasive wear of air duct walls, thus conferring greater overall aspiration system efficiency.

The pressing need to reduce aspiration unit power consumption necessitates more accurate calculation of capacity requirements for local suction units and drives adoption of special techniques for reducing aspiration volumes. Therefore, equipment design must be predicated on a number of requirements:

- Aspiration cowl types and primary structural members must be chosen based upon analysis of dust-generating equipment performance—its process-specific features and its design aspects.
- Decisions on optimum capacity of local suction units and rational dust receiver layout must account for the aerodynamic coupling between cowls and air injection processes in chutes.

- Analysis of aerodynamic processes involving dust must determine the type and layout of dust release mitigation equipment selected as well as the early dust settling devices in cowls and dust receivers.

We will study ore handling facilities at pretreatment plants as an example for analyzing practical implementation of these guidelines. Special attention will be given to surface sources of dust release—the longtime primary culprit of fugitive dust emissions—and to dumping/reclaiming facilities and their handling of pellets at outdoor storage sites and in railway car loading stations.

5.1 BASIC PREMISES FOR CALCULATING LOCAL SUCTION CAPACITY

5.1.1 Initial Equations

Calculations of local suction capacity will be based on the air balance equation. The amount of air evacuated from the cowl (Q_a), given isothermic conditions, is equal to the amount of air entering the cowl through chutes, process openings, and leaky joints [3,116–118],

$$Q_a = \sum_{i=1}^{N} Q_i . \tag{5.1}$$

Thus, a certain negative pressure must be maintained in the cowl to induce a counterflow of air in leaky joints and in openings that would prevent any dust from escaping. For brevity, we will use the term "optimum" when referring to the minimum value of this negative pressure and to the pump performance required for maintaining it inside the cowl.

Considering the turbulent character of air flow in chutes and openings, air losses can be expressed as follows:

$$\Delta P_i = R_i Q_i^2, \qquad R_i = \zeta_i \rho_2 / \left(2S_i^2\right), \tag{5.2}$$

where ΔP_i is pressure loss in i-th chute/opening (Pa); R_i is the hydraulic performance property of i-th chute/opening/vent (Pa·s²/m⁶); ζ_i is the local resistance coefficient of i-th chute/opening/vent; and S_i is the cross-section area of i-th chute/opening/vent (m²).

On the other hand, the available pressure drop in i-th chute is generally determined by injection pressure (P_{ei}), thermal head (P_{Ti}), the pressure (P_{eqi}) produced by moving parts of equipment (e.g., rotary crusher hammers), and different negative pressures in cowls (ΔP_{ai}):

$$\Delta P_{chi} = P_{ei} - P_{Ti} \pm P_{eqi} + \Delta P_{ai}, \tag{5.3}$$

and the airflow through the i-th chute is obviously equal to

$$Q_i = \frac{\Delta P_{chi}}{\left|\Delta P_{chi}\right|} \sqrt{\frac{\left|\Delta P_{chi}\right|}{R_i}} . \tag{5.4}$$

Field measurements of aerodynamic properties indicate a value $\zeta_i = 2.4$ for openings and for open apertures in the majority of aspiration cowls. The total incoming airflow through M openings and apertures in aspiration cowls is equal to

$$Q_H = 0,65\sum_{i=1}^{M} F_i\sqrt{2P_i/\rho_2} \tag{5.5}$$

where F_i is the area of the i-th opening m^2; M is the total number of openings; and P_i is the negative pressure in a cowl near the i-th opening (Pa).

Dividing leaky areas into segments with equal live section areas and measuring the negative pressure on the inner surface of their adjacent cowl walls enables a simpler formula to be used for determining suction airflow:

$$Q_H = 0,65F_H\sqrt{2P/\rho_2}, \tag{5.6}$$

where F_H is the total leaky area in the cowl ($F_H = MF_0$); F_0 is the area of a single, equally sized opening (m^2); and P is the medium pressure in the cowl, equal to

$$P = \left(\frac{1}{M}\sum_{i=1}^{M}\sqrt{P_i}\right)^2. \tag{5.7}$$

Optimum suction capacity and negative pressure depend on the process and design parameters of the dust-generating equipment and its layout within an equipment train.

5.1.2 Determining the Minimum Negative Pressure

Negative pressure in a cowl is determined by the magnitude and behavior of pressure change at the interior surface of its walls. Static pressure at walls depends, in turn, on aerodynamic and thermal processes within the cowl. Let us consider these processes in a representative cowl at a conveyor loading location.

5.1.2.1 Interaction between an Injected Air Jet and Suction Spectrum of a Local Suction Unit

Airflow dynamics inside the cowl were studied using a laboratory simulation of the cowl (Figure 5.1) with a gap imitating leakiness between the bottom of the cowl and its side walls. Side walls were made of transparent plastic for visualization of air streams. The flow of injected air was imitated in this case with an airflow from a fan routed into the cowl via its chute. Air velocities in the opening and in various cross-sections have been measured using a thermoelectric anemometer. Cowl designs allowed for variable geometrical properties: height, length, and location of the aspiration funnel.

Testing has shown that the character of airflow distribution within the conveyor loading the cowl is determined by the interaction between the incoming jet of injected air and the suction spectrum of the funnel. When the jet of injected air leaves the

FIGURE 5.1 Laboratory installation for studying the aerodynamics of cowls: 1 = cowl model; 2 = chute; 3 = local suction; 4 = fan.

chute, it spreads around on the conveyor belt, running against vertical walls of the cowl in a fanning jet. Walls convert the dynamic head of the jet into a static one, thus accounting for uneven pressure on the interior surfaces of walls. In this case, the maximum pressure can be measured on the vertical wall sections closest to the chute (where the fanning jet flows faster).

Pressure diminishes toward the funnel. This pressure distribution behavior persists with changing flow rates of supplied and evacuated air. However, the coefficient of variation between pressures measured in N points

$$r_p = \sqrt{\sum_{i=1}^{N}\left(P_i - \bar{P}\right)^2 / \left[(N-1)\bar{P}^2\right]} \tag{5.8}$$

will be changing. It will fall when the outlet section of the chute is lifted above the conveyor belt or when jet velocity at the chute outlet decreases. It can be explained by a drop of the dynamic head of the jet at the cowl walls.

To prevent the jet of injected air from escaping to the outside, pump capacity must be increased so that negative pressure is maintained throughout the entire surface area of the vertical walls. This negative pressure must exceed the dynamic head of the jet at the wall closest to the chute [69,70].

$$P_{\min} \geq \frac{\upsilon_{2e}^2}{2}\rho_2. \tag{5.9}$$

Pressure becomes increasingly negative toward the aspiration cowl: the mean negative pressure $\bar{p}$ is equal to the negative pressure p_∞ within the low air mobility area at the cowl ceiling, between the chute and the aspiration flange (Figure 5.2). At the same time, the optimum negative pressure is proportional to the dynamic head created by airflow at the final section of the chute (Figure 5.3)

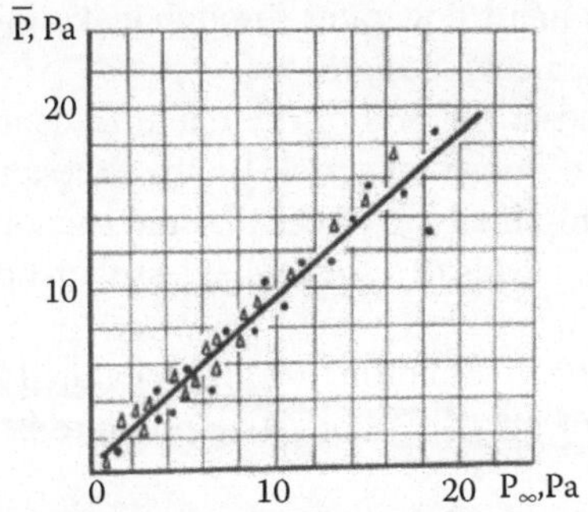

FIGURE 5.2 Plot of relationship between mean negative pressure along the cowl perimeter and negative pressure at its ceiling.

$$\overline{P}_{opt} = k_y \upsilon_{2k}^2 \rho_2 / 2 \tag{5.10}$$

or, considering that υ_{2k} is proportional to the fall speed of the material,

$$\overline{P}_{opt} \approx k \upsilon_{1k}^2 \rho_2 / 2, \tag{5.11}$$

where k_y, k are proportionality ratios. This was confirmed in the subsequent research of other authors [32,119,120].

Upon analyzing velocity fields plotted using a thermoelectric anemometer (Figure 5.4.), it was determined that the aspiration flange should be installed preferably at a distance 1.2 to 1.3 times B_k away from the chute (B_k is the conveyor belt width in meters). This is confirmed with data provided by O. D. Neikov and E. N. Boshnyakov [121,124] and by subsequent research from other authors [2,59,122,123]. Closer installation of the flange with respect to the chute causes deformation of the suction spectrum, producing a highly uneven velocity field across the inlet section of the flange (Figure 5.4 a, b), thereby impairing dust evacuation into the aspiration network. Deeper negative pressure from the chute toward the aspiration flange led

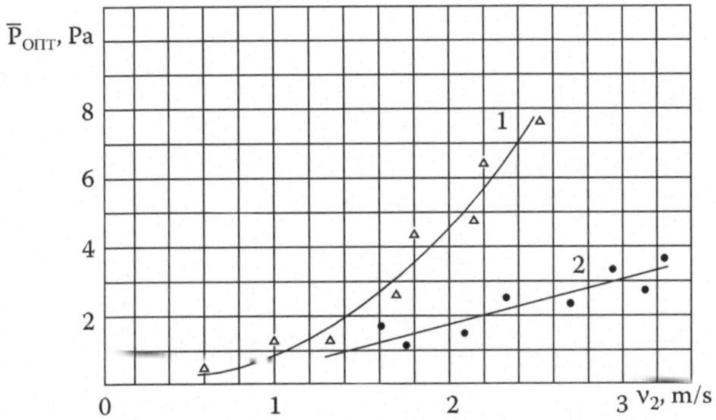

FIGURE 5.3 Change in optimum negative pressure as a function of injected air velocity: 1 = findings from authors' experiment; 2 = data published by O. D. Neikov and E. N. Boshnyakov [121].

to unwarranted increases in air flow rates through leaky joints—one of the primary drawbacks of single-walled cowl designs.

A baffle plate installed at the end of the chute to decrease conveyor belt wear caused by falling pieces of loose matter shifts the excess pressure zone toward the aspiration flange. This amplifies the effect that the suction spectrum of the funnel has on incoming air jet and leads to a somewhat less than optimum negative pressure in the cowl.

The incoming jet can be diverted further and separated from the conveyor belt in a shorter time if a zone of intensified negative pressure is maintained in the upper part of the cowl.

To that end, the upper part of the cowl should be separated from the lower part with a horizontal partition wall having a trapezoidal slot at its middle that will allow any settled dust to sift through. The slot narrows down toward the flange, enabling greater volumes of air to be evacuated from the chute area rather than at the end of the cowl, where large leaky areas occur. Such a structure decreases the area of eddy

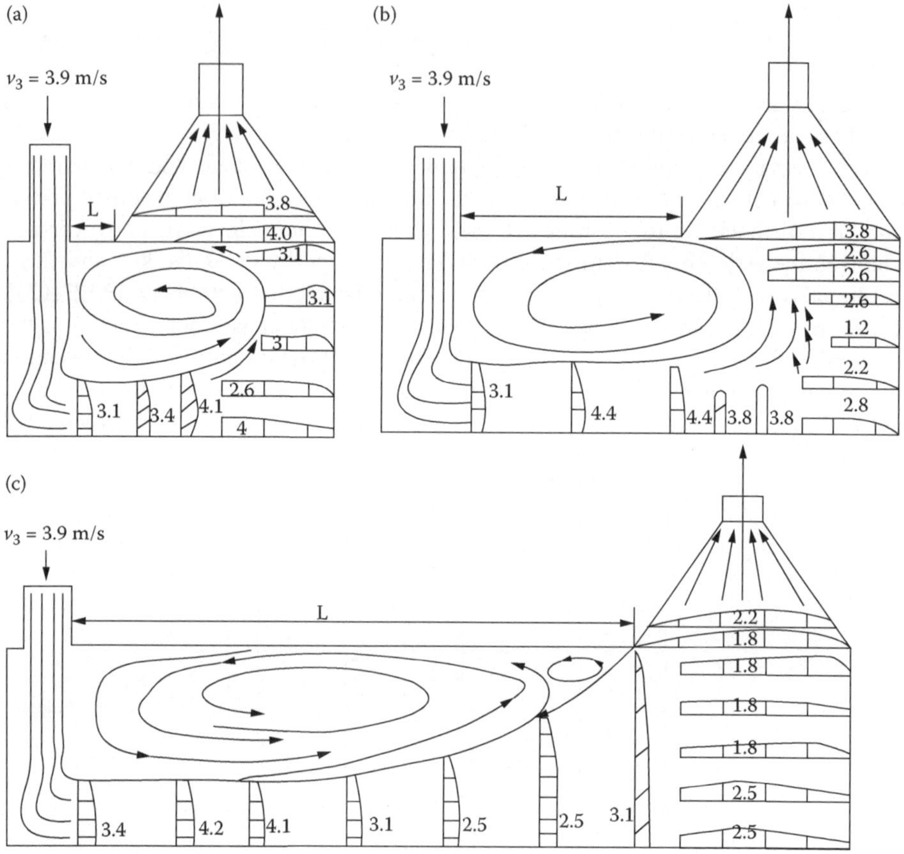

FIGURE 5.4 Airflow motion and field diagram within cowls at (a) L = 0.1 B_k; (b) L = 0.45 B_k ; (c) L = 1.3 B_k.

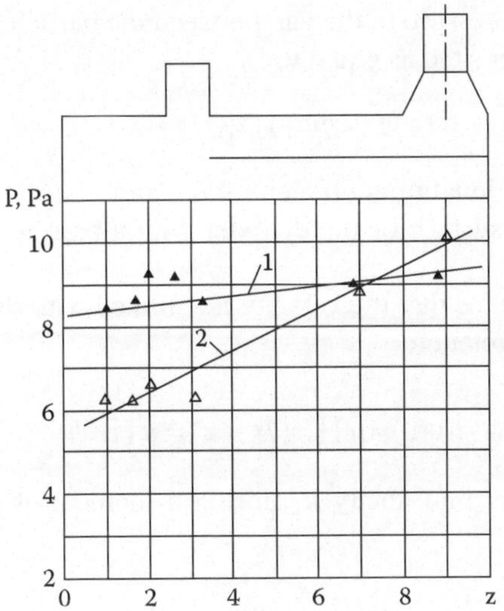

FIGURE 5.5 Distribution of negative pressure along the perimeter of the cowl bottom:1 = with a horizontal partition; 2 = without any partition.

formation and prevents flow deformation at the aspiration flange inlet. All things being equal, the negative pressure is more evenly distributed along the vertical walls and reaches greater values than with single walls (Figure 5.5).

Negative pressure is the most uniform in double-walled cowls. Here, side walls of the inner chamber disrupt the fan jet of injected air, considerably weakening the direct impact of the incoming jet on the exterior walls.

5.1.2.2 Compressive Effect

As the falling particulate material hits the conveyor belt, air is compressed and squeezed out mechanically. To estimate this effect, consider the example of falling particles shaped as ellipsoids of rotation (Figure 5.6). Ignoring the aerodynamic resistance force, a falling particle can be described with the equation

$$m\frac{d\upsilon}{dt} \equiv m\upsilon\frac{d\upsilon}{dx} = mg - f_M p, \tag{5.12}$$

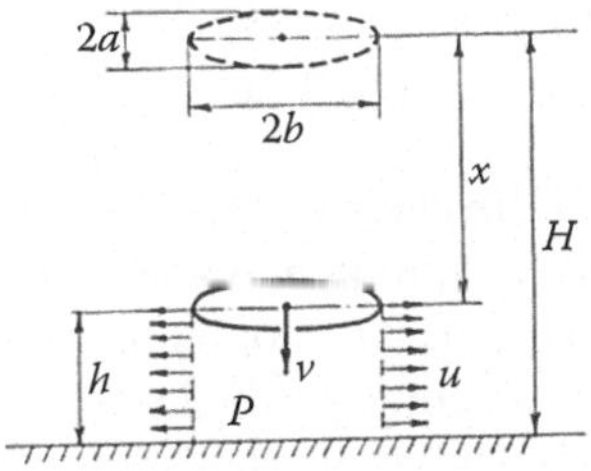

FIGURE 5.6 Diagram of air displacement by a falling particle.

where p is the excess pressure in the gap between the particle and the conveyor belt. This pressure can be posited as equal to

$$P = ku^2\rho_2/2, \tag{5.13}$$

where k is the proportionality coefficient ($k \approx 1$); and u is the mean velocity of displaced air at the side surface of the cylinder with a base area f_M (and a perimeter equal to Π_M).

Flow continuity means that the velocity u is linked with the particle fall velocity υ through an obvious relation

$$\upsilon f_M = u\Pi_M (H - x) = u\Pi_M h. \tag{5.14}$$

By choosing displaced air velocity u_∞ during a moment of sustained particle fall (with $d\upsilon/dx = 0$)

$$u_\infty = \sqrt{2mg/(f\,k\rho_2)} \tag{5.15}$$

and length l_∞ is equal to

$$l_\infty = f_M^3 k\rho_2/(m\Pi_M) = 2f_M^2 g/(u_\infty \Pi_M)^2, \tag{5.16}$$

as characteristic variables, we end up with a dimensionless form of the initial equation (Equation 5.12):

$$\frac{d\varpi^2}{dz} - \frac{1-2z}{z^2}\varpi^2 = -1/z^2, \tag{5.17}$$

where $\varpi = u/u_\infty$; $z = h/l_\infty$. This equation can be solved for initial conditions $\varpi_0^2 \to 0$ at $z \to z_0 = H/l_\infty$ as

$$\varpi^2 = \left[z_0 e^{\frac{1}{z_0} - \frac{1}{z}} - z + e^{-\frac{1}{z}} Ei\left(\frac{1}{z}\right) - e^{-\frac{1}{z_0}} Ei\left(\frac{1}{z_0}\right) \right] / z^2. \tag{5.18}$$

For globular particles, in particular, the velocity of displaced air at the time a particle collides with an obstacle equals

$$u = 0.5\upsilon_k, \tag{5.19}$$

where $\upsilon_k = \sqrt{2gH}$ is the free fall velocity of the particle (m/s).

For flattened particles with a volume equivalent to a sphere of diameter d_e, the displacement velocity is much higher ($u > \upsilon_k$) (Figure 5.7). When a particle appears near an opening when it hits the conveyor belt, the displaced air will leak outside the cowl. To avoid this leak, negative pressure must be maintained at particle fall location.

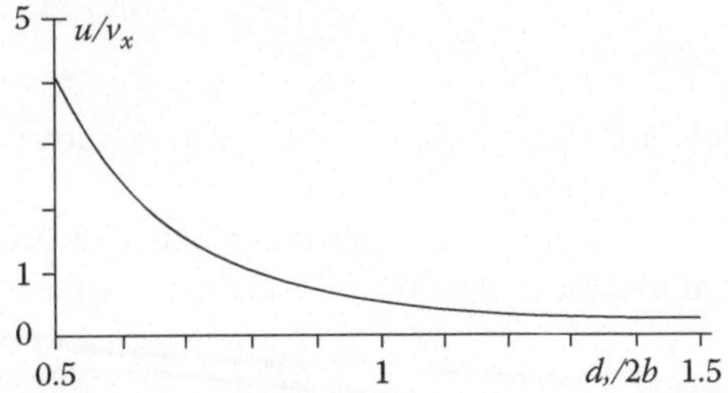

FIGURE 5.7 Changes in relative velocity of displaced air as a function of flattened particle size.

$$P \geq n \frac{\upsilon_k^2}{2} \rho_2, \tag{5.20}$$

where n is the coefficient accounting for particle shape and dampening of displaced air velocity.

An accepted practice in filling station design is to move the location of where the particles are dumped onto the conveyor belt away from any leaky joints (e.g., by installing a chute shoe with side borders, by folding vertical walls of the chute inward, etc.).

5.1.2.3 Thermal Pressure in the Cowl

When handling heated materials, heat exchange causes the air temperature inside the cowl to increase above the ambient air temperature. The result would be higher pressure toward the top of the thermal head walls (located in the upper part of the cowl) than at the bottom. If the temperature is distributed evenly inside the cowl, the pressure difference is equal to

$$\Delta P_T = H_y \left(\rho_0 - \rho_y\right) g, \tag{5.21}$$

where H_y is cowl height (m); and ρ_y, ρ_0 are air densities inside and outside of the cowl (kg/m^3).

The upper part of the cowl is particularly prone to air leakage. To prevent the escape of dust-laden air, negative pressure exceeding the Δp_T pressure must be maintained inside the cowl.

5.1.2.4 Optimizing the Choice of Negative Pressure

When one of these excess pressure formation factors occurs inside the cowl, Equation 5.11, 5.20, or 5.21 is used to determine the optimum negative pressure. For example, drive drum cowls, as a rule, do not experience overlaying air jets. Sustained negative pressures caused by air injection are observed in such cowls when handling non-heated materials. Considering the high volume of cowl inner space (one-second air replacement rate is below one), negative pressure is evenly distributed. Its magnitude depends on the velocity of air u_s entering the cowl through leaky joints and process openings

$$P_y = \zeta_{in} \frac{u_s^2}{2} \rho_2. \tag{5.22}$$

When heated materials are introduced, a downward negative pressure gradient is established.

When installing local suction units, optimum negative pressure values should account for cowl height and air temperature inside the cowl

$$P_y = P_{\min} + \frac{\Delta P_T}{2}, \tag{5.23}$$

where P_{min} is the minimum negative pressure prohibiting dust escape (usually P_{min} = 2 Pa [u_s = 1 m/s]).

Table 5.1 lists examples of optimum negative pressure values; 5 Pa has been established as the optimum negative pressure for heated material conveyor drive drums.

Usually a combination of the aforementioned excess pressure formation processes could be observed inside the cowl, and optimum negative pressure can be determined by field testing. All parameters of the dust-producing assembly must be considered in this case. So, for the conveyor loading cowl (Table 5.2) parameters to be taken into account, one must include: cowl type (and, therefore, the peculiar aerodynamic interaction between injected air flow and the suction spectrum of the aspiration flange); material coarseness (and, therefore, the character and intensity of mechanical air displacement by a particle falling onto the conveyor belt); and the material temperature (and, therefore, the magnitude of the thermal head).

5.1.3 Choosing an Aspiration Layout and Calculating the Performance of Local Suction Units at Handling Facilities

Handling facilities can be classified into three groups by methodological approaches to calculating the required aspiration volume: handling facilities where air discharge

TABLE 5.1
Optimum Negative Pressure inside Conveyor Drive Drum Cowl*

Air Temperature inside Cowl, °C	Optimum Negative Pressure (Pa) at H_y			
	1 m	1.5 m	2 m	2.5 m
30	2.2	2.3	2.4	2.5
40	2.4	2.6	2.8	3.0
50	2.5	2.8	3.1	3.4
60	2.7	3.1	3.4	3.8
80	3.0	3.5	4.0	4.5
100	3.3	3.9	4.5	5.2

*At t_0 = 20°C.

TABLE 5.2
Values of Optimum Negative Pressure in Belt Conveyor Loading Location Cowl

Cowl Type	Optimum Negative Pressure		
	$d_e < 0.2$ mm	$d_e = 0.2$–3 mm	$d_e > 3$ mm
Cowl with uniform walls	$\frac{8}{6}$	$\frac{10}{9}$	$\frac{12}{10}$
Double-walled cowl	$\frac{6}{4}$	$\frac{7}{5}$	$\frac{8}{6}$
Cowl equipped with a horizontal partition	$\frac{7}{5}$	$\frac{8}{6}$	$\frac{10}{8}$

Numerators indicate optimum negative pressures for handling unheated materials; denominators refer to handling heated materials.

is predominantly injection-driven; handling facilities in systems generating directed air currents; and handling facilities operating with heated materials.

Handling facilities with injection-driven air discharge occur at non-heated material processing trains. They primarily include conveyor-to-conveyor transfer facilities, crushing and screening assemblies, and bin loading and stockpiling facilities at storage locations. In this case, air currents in chutes and cowls arise as a result of dynamic interaction between air and flowing material particles in chutes, as well as from operation of local suction units.

The aspiration layout provides for installation of aspiration flanges on cowls of downstream equipment. The flow rate of air entering the cowl via its chute is determined either by injection pressure

$$Q_{ch} = \sqrt{(P_e + P_2 - P_1) / R_{ch}}, \tag{5.24}$$

where P_1, P_2 are optimum negative pressure values (Pa) in the upper and lower cowls, correspondingly, or by component slip ratios

$$Q_{ch} = \varphi_{ch} S_{ch} \upsilon_{1k}. \tag{5.25}$$

In this case, calculation relations for P_e and φ_{ch} should be chosen depending on structural and process features of the handling facility.

The most convenient parameter determining flow restriction rate for conveyor-to-conveyor transfers is the relation S_c^* / S_{ch}, where S_c^* is the cross-sectional area of material layer on the upper conveyor belt, determined by

$$S_c^* = G_1 / (\rho_{1H} \upsilon_b), \tag{5.26}$$

ρ_{1H} is the bulk density of the material (kg/m^3); and υ_b is the supply conveyor belt velocity (m/s).

When the cross-section of the chute is largely free of loose material $\left(S_c^* / S_{ch} < 0.2\right)$, the flow rate of air entering the lower aspirated cowl is the sum total of airflow coming together with a material

$$Q_c = \varphi_c \upsilon_{1k} S_c \tag{5.27}$$

and the amount of air flowing through the free cross-section of the chute

$$Q_0 = (P_2 - P_1) / \sqrt{R_0 |P_2 - P_1|}, \qquad R_0 = \zeta_{ch} \rho_2 / 2\left(S_{ch} - S_c\right)^2, \tag{5.28}$$

where ζ_{ch} is the local resistance factor of the chute, and S_c is the cross-section area of the material jet entering the lower chute (m^2).

When particles are thrown off a driving drum rotating with a linear velocity υ_b, the value of S_c is determined by trajectories of marginal flow particles that can be calculated using the formula*

$$S_c = S_c^* \frac{\upsilon_\pi}{h} \sqrt{\frac{2H}{g}} \left(\frac{h+R}{R} \sqrt{1 + \frac{h}{H}} - 1 \right), \tag{5.29}$$

where R is drive drum radius (m); H is the material fall height (m); and h is the thickness of material layer on the conveyor, equal to

$$h \approx \sqrt{S_c^*} / 2. \tag{5.30}$$

Component slip factor ϕ_c within the jet of material can be determined using the data provided in Chapter 4.

5.1.3.1 Conveyor-to-Conveyor Transfers

The most common assemblies are those transferring loose materials from one conveyor to another These assemblies will be designated as conveyor-to-conveyor transfers for brevity.

Air suction volume is determined at conveyor loading locations (Figure 5.8) using the air balance condition:

$$Q_a = Q_{ch} + Q_H \tag{5.31}$$

Prismatic and bin-shaped chutes with a quasi-uniform distribution of particles across flowing material are the most common variety. Fall height is quite modest ($h \le 0.5$); therefore, resistance of the medium is not accounted for in calculations of particle velocities.

The sum total of local resistance coefficients is determined using the formula

$$\sum \zeta = \zeta_{in} + \zeta_{ch}, \tag{5.32}$$

* If $S_c < 1.5S_c^*$, subsequent calculations should assume $S_c = 1.5S_c^*$.

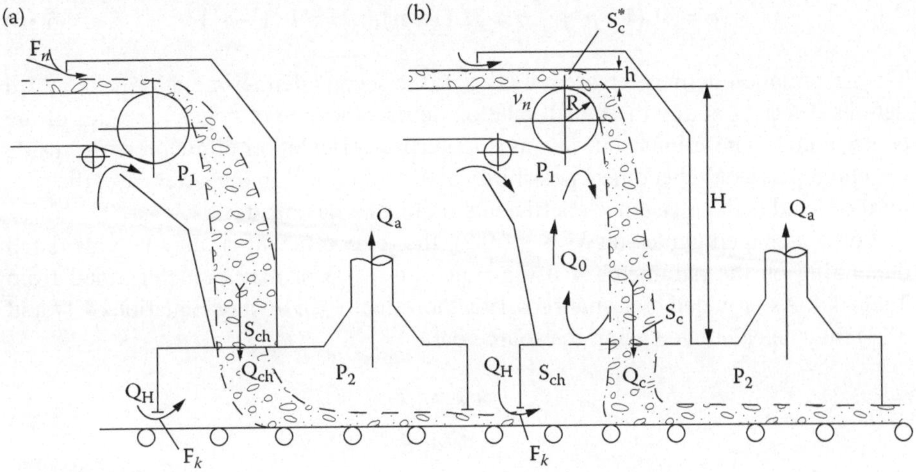

FIGURE 5.8 Aspiration layout for a conveyor-to-conveyor cold transfer facility: (a) transfer via ordinary chute; (b) transfer via bin-shaped chute with $P_c > P_2$.

where ζ_{in} is the resistance coefficient for air entering into the drive drum cowl of the upper conveyor with a total area of leaky joints and process openings F_{Hf}

$$\zeta_{in} = 2,4\left(S_{ch} / F_{Hf}\right)^2, \tag{5.33}$$

where ζ_{ch} is the chute resistance coefficient, accepted as $\zeta_{ch} = 1.5$ for vertical chutes and $\zeta_{ch} = 2.5$ for inclined chutes.

When particles of material are transferred through prismatic chutes, calculations for the coefficient φ in light of Equation 3.121 assume the form

$$\varphi = \begin{cases} \dfrac{a}{2(1+b)}\left[1-\sqrt{1-\dfrac{4(m-N)(1+b)}{a^2}}\right] & \text{at } m \geq N-1; \\ \sqrt{L(1-\varphi)^3-(\varphi-n)^3+N} & \text{at } n^2-N \leq m \leq N-1; \\ \dfrac{a}{2(b-1)}\left[1-\sqrt{1-\dfrac{4(m+N)(b-1)}{a^2}}\right] & \text{at } -N\dfrac{(1-n)^3}{1-n^3} \leq m \leq n^2-N; \\ \dfrac{a}{2(b+1)}\left[1-\sqrt{1-\dfrac{4(m+N)(1+b)}{a^2}}\right] & \text{at } L \leq -\dfrac{N}{1-n^3}, \end{cases} \tag{5.34}$$

with the following simplifications:

$$L = \frac{1}{3}Bu, \quad N = Eu, \tag{5.35}$$

$$a = 3L\left(1-n^2\right), \quad b = 3L(1-n), \quad m = L\left(1-n^3\right). \tag{5.36}$$

The first relation defines the area of $\varphi_{ch} > 1$; the second defines $n \le \varphi_{ch} \le 1$; the third defines $0 \le \varphi_{ch} \le n$; and the fourth relation defines the area of negative values of the coefficient ϕ. The minus sign means counterflow orientation of air pressure inside the chute that would become a possibility at $N \le -L(1-n^3)$. In the latter case, the sum total of local chute resistance coefficients requires adjustment.

For bin-shaped chutes (at $S_c^*/S_{ch} \le 0.2$), the jet coefficient φ_c may be calculated depending on the parameter υ^* using Equation 4.205 or may be determined from Table 4.4. As provided by Equation 4.192, the value υ^* (in light of Equations 4.17 and 3.77) may be calculated using the source data

$$\upsilon^* = \frac{\psi^* k_m \upsilon_{1k} G_1}{4 S_c \rho_1 g}. \tag{5.37}$$

Given a known coefficient φ_c we can use Equation 5.27 to determine the airflow Q_c in the jet.

In order to determine the air flow rate through the free cross-section of the chute, the negative pressure value in the unaspirated upper cowl* must be known.

$$P_1 = R_u \cdot Q_{ch}^2; \qquad Q_{ch} = Q_0 + Q_c; \tag{5.38}$$

$$R_u = 2{,}4\rho_2 / \left(2F_{Hf}^2\right). \tag{5.39}$$

In this case, the Equations 5.28 and 5.38 can be solved jointly for a flow rate expression Q_0

$$Q_0 = \begin{cases} \dfrac{P_c}{Q_c\left(R_0+R_u\right)}\left[\sqrt{1+\dfrac{P_2-P_c}{P_c^2}Q_c^2\left(R_0+R_u\right)}-1\right] & \text{at } P_2 \ge P_c, \\ -\dfrac{P_c}{Q_c\left(R_0-R_u\right)}\left[\sqrt{1+\dfrac{P_c-P_2}{P_c^2}Q_c^2\left(R_0-R_u\right)}-1\right] & \text{at } P_2 \le P_c, \end{cases} \tag{5.40}$$

where P_c is the negative pressure that would occur in the upper cowl were air evacuated from it at a flow rate Q_c,

$$P_c = R_u Q_c^2; \tag{5.41}$$

And R_0 is the hydraulic characteristic of the chute

* When this cowl is aspirated (e.g., when heated materials are being handled), the values P_1 as well as P_2 shall be conditional on ensuring complete dust release containment. In this case, Equation 5.38 would make no sense.

$$R_0 = \zeta_{ch}\rho_2 / \left[2\left(S_{ch} - S_c\right)^2\right]. \tag{5.42}$$

Thus, a number of intermediate calculations (Table 5.3) must be performed to arrive at air balance components Q_{ch} and Q_H.

The latter, given a specific negative pressure in the aspirated cowl (P_2) and the total area of leaky joints and process openings (F_{Hb}), can be numerically determined in accordance with Equation 5.6 using the formula

$$Q = 0{,}65 F_{Hb}\sqrt{2P_2 / \rho_2}. \tag{5.43}$$

The following parameters of the handling facility serve as source data for calculations: material flow rate (G_1) and its particle-size composition (m_i, d_i), density of particles proper (ρ_1) and their bulk density (ρ_{1H}), structural dimensions of the chute (H, S_{ch}, Θ_i, l_i) and of areas of leaky joints and process openings in the cowls (F_{Hf}, F_{Hb}), and upper conveyor belt speed (υ_b) and the radius of its drive drum (R).

A comparison of calculated and measured aspiration volumes provides evidence for the acceptability of the aforementioned calculation method. Quantitative deviations remain within the relative error of the field experiment.

TABLE 5.3
Procedure for Flow Rate Calculation of Air Evacuated by Suction from Conveyor Loading Cowl

Design Variables	Relations for Calculation	
	at $S_c^* / S_{ch} \le 0.2$	**at $S_c^* / S_{ch} \ge 0.2$**
Cross-section of material on conveyor S_C^*	(5.26)	
Material flow velocities in the chute υ_{1H}, υ_{1K}, $n = \upsilon_{1H}/\upsilon_{1K}$	(2.20), (2.29), and (2.32)	
Cross-section of material flow in the chute S_C	$S_c = S_{ch}$	(5.29)
Average particle diameter – d (d_e)	(3.36)	
Volumetric concentration of material	(3.30)	
Reduced particle resistance coefficient – ψ^*	(3.29)	
Sum total of mass resistance coefficients – $\sum\zeta$	(5.32) and (5.33)	
Negative pressure inside aspirated cowl – P_2	Data from Table 5.2	
Bu parameter	(3.109)	Not calc.
Parameter υ^*	Not calc.	(5.37)
Parameter *Eu*	(3.122)	
φ coefficient for the chute	(5.34)	Not calc.
φ coefficient for jet of material	Not calc.	(4.205) and Table 4.4
The injected air flow rate is Q_{ch}	(5.25)	(5.27), (5.40), and (5.38)
Flow rate of air entering through leaky joints in the lower cowl – Q_H	(5.43)	
Flow rate of air evacuated from the conveyor loading location – Q_a	(5.31)	

Measured readings of evacuated air flow rates significantly exceeded calculated values for assemblies 3, 4, and 14 because of a failure to fine-tune these assemblies, resulting in their suboptimal operation throughout our studies. The existing capacity of aspiration systems was insufficiently high enough to ensure optimum negative air pressure air in the cowls tested. Some air leaked from the inner space of the cowls to the outside.

To clarify the applicability limits of these methods, changes in calculated injected air flow rates with increased material flow rates in industrial transfer chutes were analyzed. Calculations of injected air flow rates were performed twice, at different flow rates (in other words, at different S_c^*).

The first calculation assumed that the entire cross-section area of the chute is occupied with material (calculated airflow coming through the chute in this case was denoted as Q_{ch}). The second calculation allowed for material flow to occupy a significant section of the chute (calculated airflow entering the cowl through the chute in this case was denoted as Q_{ch}^c). These calculations identify the restriction coefficient area** $S_c^* / S_{ch} \approx 0.2$ as the optimal "bridge" for the majority of industrial transfers. The deviation among airflow volumes calculated using both methods remains below ± 20% in this area (Figure 5.9).

Jet-like flows of material are a feature of spacious chutes where flowing particles of material largely escape collisions with chute walls and do not spread throughout its cross-section. Such a flow is unlikely to occur in inclined prismatic chutes. When handling lumpy multi-fraction or single-fraction materials, or when the free fall

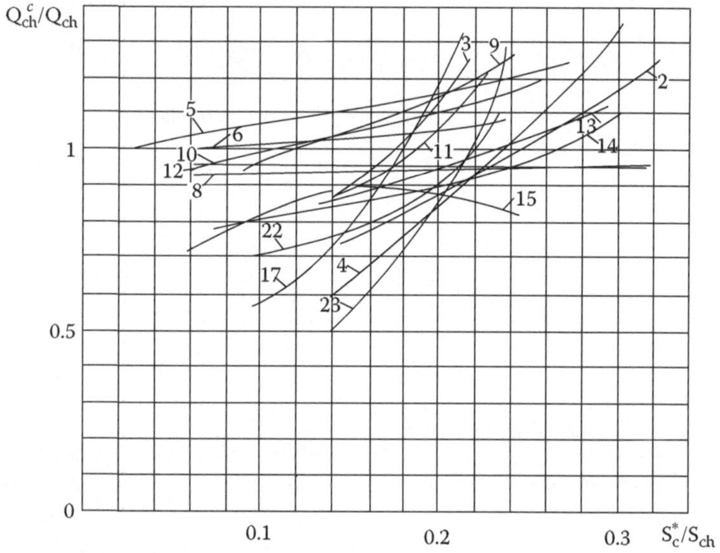

FIGURE 5.9 Changes in Q_{ch}^c / Q_{ch} with increased flow rate of transferred material for industrial handling facilities (curve numbers correspond to assembly numbers).

**S_c^*/S_c averages 0.4 for the transfer facilities we studied. Therefore, $S_c^*/S_{ch} = 0.2$ is posited for the "bridge" area.

height from the upper conveyor to the bottom of the inclined chute section is large enough ($H > 1$ m), particles ricochet and spread through the entire cross-section. If the total length of the inclined section is greater than one-half of the fall height, such a flow shall be aerodynamically considered a pseudo-uniform flow of particulate matter in a chute.

5.1.3.2 Conveyor (Feeder)–Crusher–Conveyor

A handling facility with a separate crusher unit (without preliminary screening) differs from an ordinary conveyor-to-conveyor design mainly by a significant change in a material's particle-size composition as it passes through the crusher.

Generally, the aerodynamic resistance of crushing devices and of the material adhering to them impedes the flow of injected air being discharged into the chute, possibly causing excess pressure in the loading (upper) part of the crusher. Another likely location of excess pressure occurs at the point where the crushed material falls onto the conveyor belt. Therefore, local suction units should be installed at the top of the crusher and at the cowl of the shoe under the discharge chute (Figure 5.10).

In jaw crushers, the upper air suction path begins at the feeder cowl. Keeping dust-laden air out when the bin is excessively emptied and unblocking the jaw crusher are

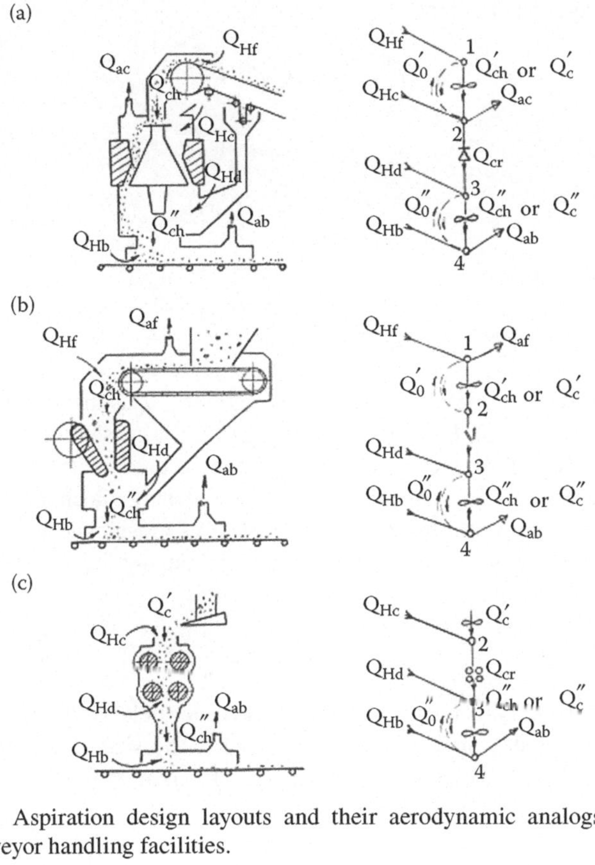

FIGURE 5.10 Aspiration design layouts and their aerodynamic analogs for feeder-to-crusher-to-conveyor handling facilities.

two requirements leading to the choice of local suction layout. Four-roll crushers are usually loaded using a belt feeder, whereby material falls from a small height, and air is only sucked away at the cowl shoe of the bottom conveyor.

Air balance of aspirated cowls determines the capacity of their suction units. To arrive at balance equations, individual sections of the design layout are studied to assess their aerodynamic performance and to prepare the corresponding air flow rate equations.

The aerodynamic performance of an area transferring air inside the crusher is calculated using the formula

$$R_{cr} = \zeta_{cr}\rho_2 / (2f_{cr}^2). \tag{5.44}$$

The local resistance coefficient is taken as $\zeta_{cr} = 1.4$ for gyratory and jaw crushers and as $\zeta_{cr} = 2.3$ for four-roll crushers. Calculated section area for air transit via the crusher is determined by the following empirical relations: for fine and medium grade cone crushers

$$f_{cr} = 0.5\pi D_K\left(b_0 + 0.0227D_K\right); \tag{5.45}$$

for coarse grade cone crushers

$$f_{cr} = 0.5\pi D_K\left(b_0 + r\right); \tag{5.46}$$

for jaw crushers

$$f_{cr} = 0.5L\left(b_0 + 0.5s\right); \tag{5.47}$$

and for four-roll crushers

$$f_{cr} = 2\left(L + 2D_r\right)\delta_g, \tag{5.48}$$

where D_K is the base diameter of the crusher cone (m); b_0 is the width of the unloading slot of the crusher (m); L is roll (jaw) length (m); r is crusher eccentricity (m); s is jaw pitch (m); D_r is roll diameter (m); and δ_g is the average width of the gap between crusher housing and rolls. Transit flow rates of air passing via chutes are determined similarly to that of conveyor-to-conveyor transfers of loose material.

The cross-sectional size of a coarse-material jet during the loading of jaw crushers was accepted as equal to

$$S_c = 1.2d \cdot B_n \tag{5.49}$$

where d is the average piece size (m); and B_n is feeder belt width (m).

For discharge chutes of cone and jaw crushers, the size of the flow of crushed material was accepted as equal to the discharge slot area, that is

$$S_c = 2f_{cr}. \tag{5.50}$$

High rotation speeds of rolls in four-roll crushers result in a significant expansion of the jet. Expansion angles can reach as much as 4° to 5°, and the cross-sectional area of the jet of material entering the cowl is determined by the formula

$$S_c = L(b_0 + 0.15H), \tag{5.51}$$

where H is the fall height of the material as the distance between the centerline of bottom rolls and the cowl entrance (m); and b_0 is the width of the discharge slot on the lower pair of rolls (m).

The initial velocity of the material in the jet is accepted as equal to the linear velocity of the lower rotating rolls.

Air balance equations are solved first to determine the negative pressure in non-aspirated cowls and then to find out the amount of air passing through. As an example, consider the case of aspiration in a cone crusher with a jet of material flowing through chutes. For certainty, the position of the chutes will be deemed vertical. We will use the following bottom indices to designate pressure in confluence locations: P_1 = negative pressure in feeder cowl, P_2 = in cowl chute, P_3 = in crushed material bin, and P_4 = in lower conveyor chute. In addition, we will use a single upper prime symbol when referring to the parameters of the loading chute and a double prime when referring to the parameters of the discharge chute. Aerodynamic behavior of cowls and chutes can be expressed using local resistance coefficients

$$R_{Hf} = 2.4\frac{\rho}{2F_{Hf}^2};\ R_{Hc} = 2{,}4\frac{\rho}{2F_{Hc}^2};\ R_{Hd} = 2.4\frac{\rho}{2F_{Hd}^2};\ R_{Hb} = 2.4\frac{\rho}{2F_{Hb}^2};$$
$$R_0' = \varsigma_{ch}'\rho/\left[2\left(S_{ch}' - S_c'\right)^2\right];\ R_0'' = \varsigma_{ch}''\rho/\left[2\left(S_{ch}'' - S_c''\right)^2\right]. \tag{5.52}$$

The flow rate of air injected by a jet of loose material will, accordingly, be designated as

$$Q_c' = \varphi_c' v_{1k}' S_c', \quad Q_c'' = \varphi_c'' v_{1k}'' S_c''. \tag{5.53}$$

Obvious air balance equations are

$$\begin{aligned} Q_{Hf} &= Q_c' + Q_0' = Q_{ch}'; \\ Q_{cr} + Q_{Hd} &= Q_{ch}'' = Q_c'' + Q_0'' \end{aligned} \tag{5.54}$$

or, expressed through aerodynamic performances and pressures,

$$\left.\begin{aligned} \sqrt{P_1}/\sqrt{R_{Hf}} &= Q_c' + (P_2 - P_1)/\left[\sqrt{R_0'|P_2 - P_1|}\right], \\ (P_3 - P_2)/\sqrt{R_\partial|P_3 - P_2|} + \sqrt{P_3/R_{Hd}} &= Q_c'' + (P_4 - P_3)/\left[\sqrt{R_0''|P_4 - P_3|}\right]. \end{aligned}\right\} \tag{5.55}$$

In this case, negative pressure in aspirated cowls is defined by P_2 and P_4. A solution of combined equations (Equation 5.55) leads to the values P_1 and P_3 that, in turn, make derivation of balance components easy enough.

The following equations describe suction efficiency:

$$Q_{ac} = Q_{Hc} + Q'_{ch} - Q_{cr}; \tag{5.56}$$

$$Q_{ab} = Q''_{ch} + Q_{Hb}. \tag{5.57}$$

If $Q_{ac} < 0$, then the necessary negative pressure will be maintained inside the cowl of the crusher even when suction is absent. Moreover, this negative pressure will not be below optimum. In this case, P_2 would remain unknown. To determine it, Equation 5.55 must be solved together with the air balance equation for the cowl of the crusher

$$Q_{Hc} + Q'_{ch} = Q_{cr}, \tag{5.58}$$

which can be rewritten in the following form:

$$\sqrt{P_2 / R_{Hc}} + Q'_c + (P_2 - P_1) / \left[\sqrt{R'_0 |P_2 - P_1|}\right] = (P_3 - P_2) / \sqrt{R_{cr} |P_3 - P_2|}. \tag{5.59}$$

By repeating the calculation of air balance for the lower aspirated cowl, the new value of Q_{ab} is lower than the previous one. Thus, if a certain margin is desirable, it is reasonable to skip recalculating the capacity of this suction unit.* Calculations for production facilities tested at a number of ore pretreatment plants across the former Soviet Union testify to the applicability of the described methodology. Calculated aspiration volumes agree with measured readings. Relative error stays within 15%.

5.1.3.3 Conveyor–Screen–Conveyor

In terms of aspiration, there is no qualitative difference between the aforementioned crushing assemblies and bulk sizing assemblies in freestanding (as opposed to integration with crusher) screening units. In a quantitative sense, the latter differ by having fewer local suction units at the bottom of the assembly. In this case, aspiration cowls with air suction units are provided for all conveyors taking up sized material (Figure 5.11). Airflows in the loading chute (denoted as #1 for certainty) and in the screen over the chute (#2) are calculated in the same way, as in the case of conventional transfer facilities. For the screen-through chute (#3), two calculations are possible: particles distributed quasi-uniformly and particles flowing as a jet. The following formula is used for calculating the cross-section size of the overjet:

$$S_c = kS_c^*, \quad S_c^* = G_1 / (\rho_{1H} u_{overs}), \tag{5.60}$$

where u_{overs} is the velocity of particles as they return from the screen deck, equal to

$$u_{overs} = 0.05\sqrt{d_2}, \tag{5.61}$$

* The previous value of Q_{ab} results in P_1, P_2, P_3 and P_4 remaining unknown. They can be determined by jointly solving Equations 5.55, 5.59, and 5.57, demonstrating that negative pressure in the lower cowl (at this value of Q_{ab}) would be somewhat above optimum.

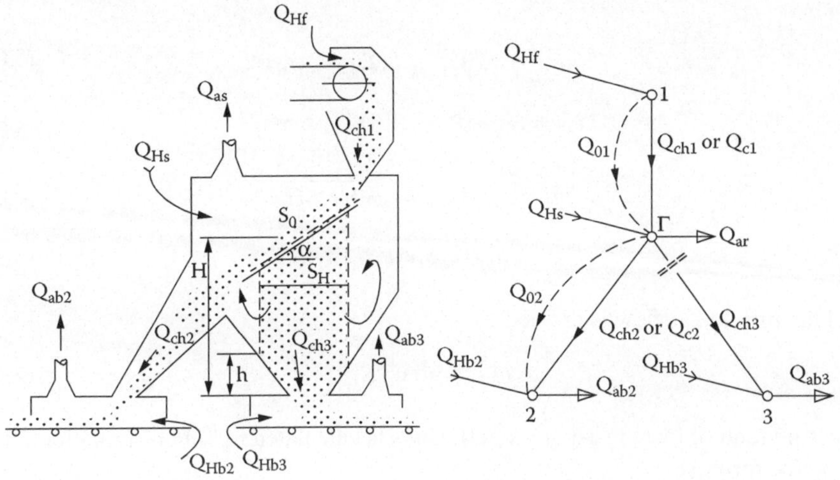

FIGURE 5.11 Aspiration design layout for screening of unheated material.

d_2 is the medium diameter of the particles of the material being handled (mm); and k is the jet expansion coefficient, accepted as $k = 2$ for a free flow and as $k = 2.5$ for impact against inclined chute walls.

A special consideration for this assembly is the evaluation of the dynamic behavior between air and particles as they pass through the screen grate. The cross-section of the flow of these particles depends on grate area and its inclination to the horizon

$$S_H = k_e S_p \cos\alpha, \quad S_p = ab, \tag{5.62}$$

where a, b are screen grate dimensions (m); k_e is the factor to adjust for effective operating area of the grate, accepted as equal to 0.9; and α is the grate inclination angle (°).

Considering that grates of vibration screen units are equipped with sufficiently spacious cowls ($S_y >> S_p$ where S_y is the plan area of the cowl [m²]), circulation airflows arise in gaps between the cowl walls and the outer surface of the flow before it meets the inclined walls of the bin-shaped chutes. Injection of air into the container cowl is only affected by the final part of the flow through a bin-shaped chute with a height

$$h = H(\sqrt{S_H / S_{ch3}} - 1) / (\sqrt{S_y / S_{ch3}} - 1). \tag{5.63}$$

In addition, only the interaction between vertically falling particles is accounted for, that is, the effect of particles reflected from chute walls is ignored. In light of Equations 3.150 and 3.157, the injection coefficient obeys the formula

$$\varphi_K = \sqrt{N_3 + Bu_3 \cdot Z(a_3, n_3, \varphi_K)}, \tag{5.64}$$

where

$$Bu_3 = \psi^* K_m G_3 v_{3k} / (\Sigma\zeta_3 S_{ch3} \rho_M g), \tag{5.65}$$

$$N_3 = (P_3 - P_r) / (0,5\Sigma\zeta_3 v_{3K}^2 \rho_b), \tag{5.66}$$

$$n_3 = v_{3H} / v_{3K} = \sqrt{1 - h / H}, \tag{5.67}$$

$$a_3 = \sqrt{S_{H3} / S_{ch3}}, \tag{5.68}$$

and the injected airflow is defined as

$$Q_{ch3} = \varphi_K v_{3K} S_{ch3}. \tag{5.69}$$

The sum total of local resistance coefficients is calculated in light of Equation 3.154 using the formula

$$\Sigma\zeta_3 = 2.5 + \zeta_{sc} / a_3^4, \tag{5.70}$$

where ζ_{sc} is the resistance factor of the screen deck:

$$\zeta_{sc} = \zeta_p \cdot \zeta_0 / (\sqrt{\zeta_p} + \sqrt{\zeta_0})^2; \tag{5.71}$$

$$\zeta_p = 3300 \frac{G_3}{\rho_{MH} b u_{overs}} \left(\frac{S_{ch3}}{S_p} \right)^2; \quad \zeta_0 = 1,5 + \left(\frac{S_{ch3}}{S_g} \right)^2, \tag{5.72}$$

where S_g is the area of gaps between the housing walls and screen deck

$$S_g = 2(a + b)\delta_g \tag{5.73}$$

and δ_g is the gap width (m). This calculation procedure yields satisfactory results, with relative error not exceeding 20%.

5.1.3.4 Dry Magnetic Separation Assembly

The aforementioned calculations for injection-driven discharge of air via chutes perform equally well for estimating air replacement in magnetic separator cowls for dry beneficiation of iron ores. The practical aspiration layout (Figure 5.12) provides for evacuation of air by suction either from the upper part of the separator cowl or from a feeder cowl. (In the absence of a significant vertical path of the material, it is possible to view the feeder and separator cowls as a common integral cowl.) Assemblies for conveyor loading with separated concentrate and nonmagnetic tails are also subject to aspiration.

Calculating the required aspiration volume takes into account the linear velocity of the lower separation drums and the hydraulic resistance displayed by flowing loose matter at its confluence in the intersected concentrate chutes (typical of three-drum 168A-SE separators) or nonmagnetic product chutes (typical of four-drum 189A-SE separators).

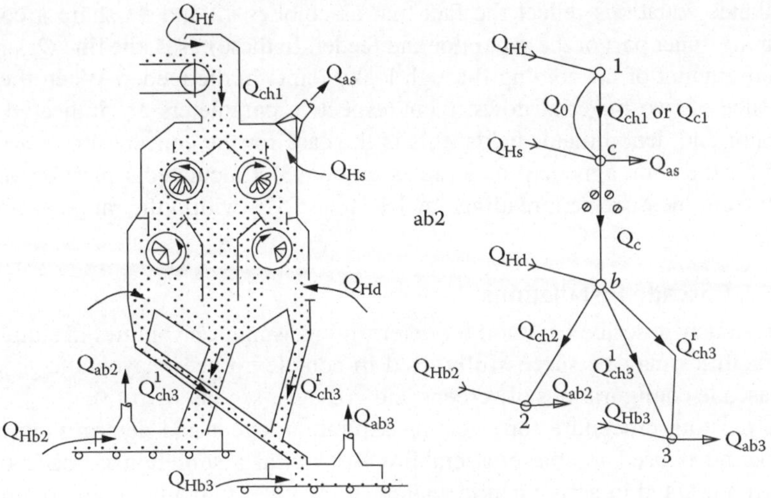

FIGURE 5.12 Aspiration design layout for a four-drum magnetic separator (Type 189A-SE).

Exit velocities in a 168A-SE separator were assumed to be equal to 6.72 m/s for concentrate returning from the upper drum and 4.93 m/s for concentrate mixture returning from the upper and lower drums. Exit velocity for nonmagnetic material leaving the separator was assumed to be equal to 4.79 m/s in view of the linear velocity of the lower drum (at n_{rot} = 25 rpm). Exit velocity for material leaving a 189A-SE separator was equal to 4.05 m/s for concentrate and 4.25 m/s for a nonmagnetic product (rotation speed of the lower drums was n_{rot} = 49 rpm).

Hydraulic resistance of a cross-current of loose matter was determined using the empirical relation

$$\zeta_n = 1.7 \cdot (\beta \cdot 10^3)^{0.3} + 1.9, \tag{5.74}$$

that was found by processing the findings of G. V. Slyusarenko's postgraduate experiment.

A special survey of air replacement behavior within the separator housing (assembly #1) in the absence of material resulted in a factor of local resistance (by separator rolls) to air flowing through from the upper cowl into the bin. Referred to as dynamic airflow in the cross-section of the central concentrate chute ($F_c = 0.732$ m^2), this coefficient is equal to

$$\zeta_c = \frac{2}{\rho}(P_b - P_c) / (Q / F_c)^2 = 1.44, \tag{5.75}$$

where Q is the total amount of air leaving through discharge chutes.

Airflow from the upper cowl into the bin of the separator obeys the formula:

$$Q_c = 0.788\sqrt{P_b - P_c}. \tag{5.76}$$

Air balance equations reflect the fact that assemblies #2 and #3 share a common cowl in the upper part of the separator and feeder. In these cases, the line Q_{ch1} reports the total amount of air coming through leaky joints in the feeder. When there is a confluence of two material flows, their respective parameters are indicated in the numerator and denominator fields (this is the case for the concentrate in assembly #2 and for the nonmagnetic product in assembly #3). Calculated aspiration volumes deviate from measurement readings by less than 15% (within the range of field test fidelity).

5.1.3.5 Cascade Installations

The previously described method for determining aspiration volumes in simple handling facilities may be successfully used in complex process layouts (e.g., widely used cascade configurations of screens and crushers, see Figure 5.13).

The existing procedure for selecting aspiration layout and determining aspiration volume is used for these assemblies. An initial assumption is made that all cowls (represented in aerodynamic analogs as confluence points of air streams) are equipped with local suction units. The specified optimum negative pressure must be maintained in the cowls. The flow rate of air in chutes is determined. After air

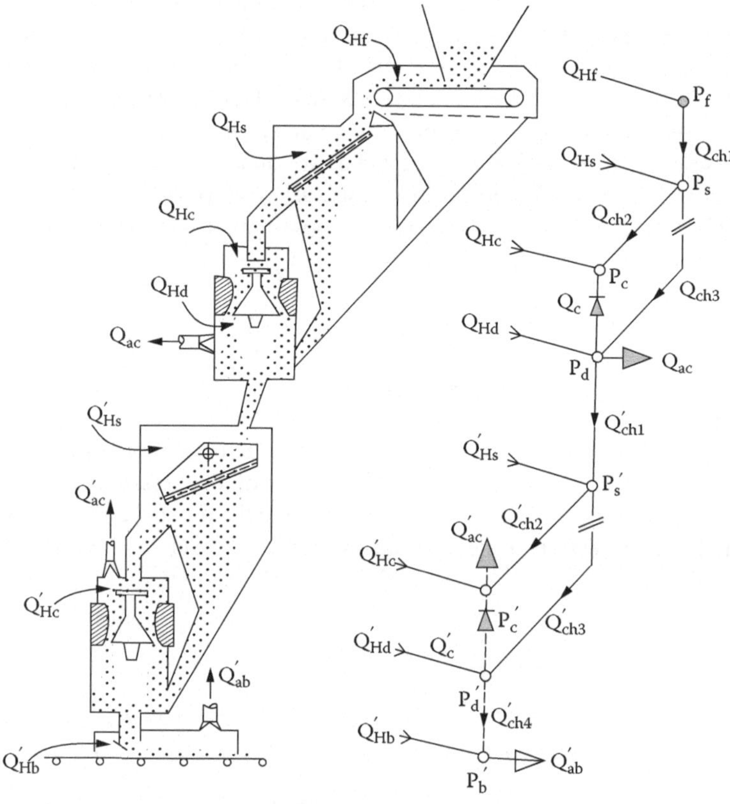

FIGURE 5.13 Aspiration design layout for cone crushers.

balance is prepared for each cowl (confluence node), local suction capacities can be estimated. A negative calculated aspiration volume means that the assembly in question does not require suction, that is, airflow in the chutes will be enough to keep it at or below its specified negative pressure. After refining the aspiration layout (leaving local suction units only in locations with positive aspiration volumes), the calculation repeats.

For example, consider the refined aspiration layout as depicted in Figure 5.13. Let us proceed with air balance equations for all non-aspirated cowls

$$Q_{\mathrm{Hf}}(P_{\mathrm{f}}) = Q_{\mathrm{ch1}}(P_{\mathrm{s}} - P_{\mathrm{f}}), \tag{5.77}$$

$$Q_{\mathrm{Hs}}(P_{\mathrm{s}}) + Q_{\mathrm{ch1}}(P_{\mathrm{s}} - P_{\mathrm{f}}) = Q_{\mathrm{ch2}}(P_{\mathrm{c}} - P_{\mathrm{s}}) + Q_{\mathrm{ch3}}(P_{\mathrm{d}} - P_{\mathrm{s}}), \tag{5.78}$$

$$Q_{\mathrm{Hc}}(P_{\mathrm{c}}) + Q_{\mathrm{ch2}}(P_{\mathrm{c}} - P_{\mathrm{s}}) = Q_{\mathrm{c}}(P_{\mathrm{d}} - P_{\mathrm{c}}), \tag{5.79}$$

$$Q'_{\mathrm{Hs}}(P'_{\mathrm{s}}) + Q'_{\mathrm{ch1}}(P'_{\mathrm{s}} - P_{\mathrm{d}}) = Q'_{\mathrm{ch2}}(P'_{\mathrm{c}} - P'_{\mathrm{s}}) + Q'_{\mathrm{ch3}}(P'_{\mathrm{d}} - P'_{\mathrm{s}}), \tag{5.80}$$

$$Q'_{Hd}(P'_{\mathrm{d}}) + Q'_{\mathrm{c}}(P'_{\mathrm{d}} - P'_{\mathrm{c}}) + Q'_{\mathrm{ch3}}(P'_{\mathrm{d}} - P'_{\mathrm{s}}) = Q_{\mathrm{ch4}}(P'_{\mathrm{b}} - P'_{\mathrm{d}}). \tag{5.81}$$

Here, the expressions in parentheses indicate negative pressures (or their difference). The pressure magnitude determines the flow rate in question.

Five combined equations contain five unknown variables: P_{f}, P_{s}, P_{c}, P'_{s}, and P'_{d} (values of P_{d} and P'_{c} are considered known because the respective cowls are provided with suction units to maintain the specified negative pressure). Moreover, the former three equations describe the interrelation among negative pressures in the upper part of the assembly while the latter two do that for the lower part. Once these negative pressures are determined, the respective air flow rates will not be hard to determine. A solution of air balance equations for aspirated cowls leads us to the required air suction capacity in the refined aspiration layout for the process facility in question. Calculations for real-world aspiration layouts for industrial screens and crushers show a satisfactory agreement between calculated and measured aspiration volumes.

5.1.3.6 Specific Issues of Injection-Driven Air Discharge Calculations for Complex Configurations of Chutes

Until now, only the simplest transfer chutes have been considered. Their respective aerodynamic processes can be represented with one of the classical models: the case of uniformly distributed particles in a straight chute or the case of a particle jet in a spacious chute.

Practical cases of material handling, however, often involve rather complicated chute configurations where no single approach exists for describing air injection behavior. This calls for a combination of models that rests heavily on the correct choice of calculation parameters for the flow of particulate matter.

Variations in cross-sections of a free flow of solid particles are determined primarily by the kinematics of the initial particle release. Collisions between particles, as well as airflows surrounding the particle stream from the outside, are less decisive processes and may be ignored. Consider, for example, the process of dropping off loose matter from the drive drum in a belt feeder. Let angle α' determine the point where the upper layer of particles departs from the drum, and let α'' be the same for the lower layer (Figure 5.14a). Conditional on the equality between pressure forces and centrifugal process in the departure point,

$$\cos\alpha' = (v_0')^2 / [g(R+\delta)], \quad \cos\alpha'' = (v_0'')^2 / (gR) = v_l^2 / (gR). \tag{5.82}$$

These combined equations can be appended by calculating velocities for a rotating solid body as determined by theoretical mechanics:

$$\frac{v}{r} = \frac{v_l}{R} = \frac{v_0'}{R+\delta}, \tag{5.83}$$

and an obvious relation for the flow continuity condition

$$\int_R^{\delta+R} v dr = v_l h_0, \tag{5.84}$$

where h_0 is the depth of the material layer on the conveyor belt (Figure 5.14b). With a triangular section of the transported layer,

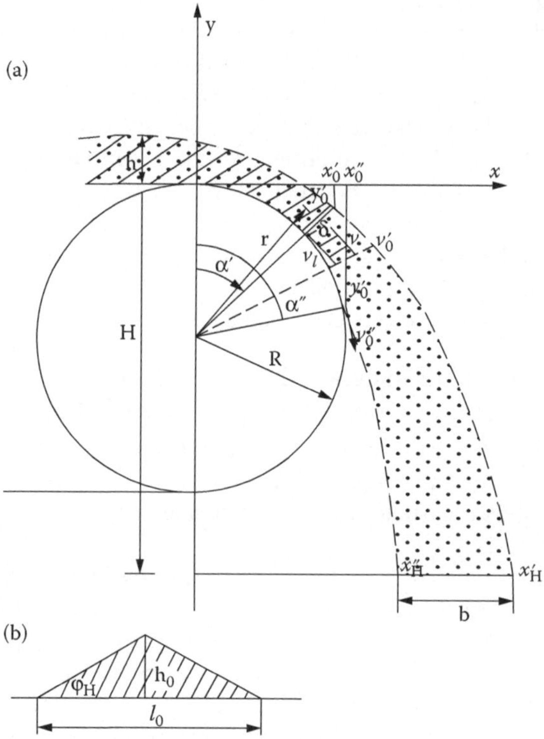

FIGURE 5.14 Illustration for calculating the geometry of a jet of particles departing from a conveyor drive drum.

$$h_0 = \sqrt{S_c^*/\mathrm{ctg}\varphi_H}, \quad l_0 = 2\sqrt{S_c^*\mathrm{ctg}\varphi_H}, \tag{5.85}$$

where φ_H is the angle at which the material hits the moving conveyor belt, accepted as 75% of the angle of rest [125]. For friable rocks $\varphi_H = 30^0$, the geometry of the initial layer may be calculated using formulas:

$$h_0 = \sqrt{S_c^* / \sqrt{3}}, \quad l_0 = 2\sqrt{3}h_0. \tag{5.86}$$

The integral relation (Equation 5.84) would yield

$$\delta = R(\sqrt{1 + 2h_0 / R} - 1) \tag{5.87}$$

or, at $h_0 / R << 0,5$

$$\delta = h_0(1 - 0,5h_0 / R). \tag{5.88}$$

In view of the calculation for velocities (Equation 5.83), cosines of departure angles relate as

$$\frac{\cos\alpha'}{\cos\alpha''} = 1 + \delta / R, \tag{5.89}$$

that is, $\alpha' < \alpha''$.

The cross-section of a stream of dropped particles will change accordingly following the trajectories of the top and bottom layers of material

$$\left.\begin{aligned} x' &= x_0' + v_{0x}'t, \\ y' &= y_0' + v_{0y}'t - 0,5gt^2; \end{aligned}\right\} \tag{5.90}$$

$$\left.\begin{aligned} x'' &= x_0'' + v_{0x}''t, \\ y'' &= y_0'' + v_{0y}''t - 0,5gt^2, \end{aligned}\right\} \tag{5.91}$$

where

$$x_0' = (R + \delta)\sin\alpha'; y_0' = (R + \delta)\cos\alpha' - R; \tag{5.92}$$

$$x_0'' = R\sin\alpha''; y_0'' = -R(1 - \cos\alpha''); \tag{5.93}$$

$$v_{0x}' = v_0'\cos\alpha'; \tag{5.94}$$

$$v_{0y}' = -v_0'\sin\alpha', \tag{5.95}$$

$$v_{0x}'' = v_L\cos\alpha''; \tag{5.96}$$

$$v_{0y}'' = -v_L\sin\alpha''. \tag{5.97}$$

The time required for a particle to fall down to level $y = -H$ is defined as

$$t'_H = \frac{v'_{0y}}{g} + \sqrt{\left(\frac{v'_{0y}}{g}\right)^2 + \frac{2(H + y'_0)}{g}} \tag{5.98}$$

for the top layer and

$$t''_H = \frac{v''_{0y}}{g} + \sqrt{\left(\frac{v''_{0y}}{g}\right)^2 + \frac{2(H + y''_0)}{g}} \tag{5.99}$$

for the bottom layer. The cross-section height at level $y = -H$ is obviously equal to

$$b_H = x'_H - x''_H = x'_0 - x''_0 + v'_{0x}t'_H - v''_{0x}t''_H. \tag{5.100}$$

Cross-sectional widening of the jet progresses as fast as particles roll down the slope of the initial triangular section. Assuming the roll-down height to be equal to one-half of h_0, we have an approximate formula

$$v_{ck} = \sqrt{gh_0}\cos\varphi_H \approx \sqrt{0,75gh_0}, \tag{5.101}$$

and, therefore, the cross-section width at a level $y = -H$ is equal to

$$a = l_0 + v_{ck}\cdot t_H = 2\sqrt{\sqrt{3}S_c^*} + \sqrt{1,5h_0H}, \tag{5.102}$$

where $t_H = \sqrt{2H/g}$ is the time required for particles to fall.

Assuming an elliptical shape for the cross-section, the area of the jet cross-section at level $y = -H$ would equal

$$S_c = \pi ab/4\cdot \tag{5.103}$$

Using the concept of reduced expansion angle γ for a jet of particles,

$$tg\frac{\gamma}{2} = \left(\sqrt{\frac{4S_c}{\pi}} - \sqrt{\frac{4S_c^*}{\pi}}\right)/H,$$

the required cross-sectional area of the jet may be determined using the formula

$$S_c = \frac{\pi}{4}\left(\sqrt{4\frac{S_c^*}{\pi}} + Htg\frac{\gamma}{2}\right)^2\cdot \tag{5.104}$$

Past research indicates that the expansion angle for a jet of particles falling out of a round-shaped hole equals $\gamma = 2 - 4^0$ and, for one falling out of a cylindrical flange, $\gamma = 1 - 2^0$.

Particularly complex aerodynamic calculations are associated with two process groups: transfer processes with descent channels gradually narrowing downward

(Figure 5.15a, b, c) and transfers through expanding chutes and passages (Figure 5.15d, e, f). In all cases, the flowing material, starting off as a free jet of particles, injects air toward the narrow duct. Here, the two airflows either separate or blend together.

The airflow at the entrance of the narrow channel can be denoted as

$$Q_c + Q_0 = Q_{ch}. \tag{5.105}$$

The air flow rate outside the free jet could be positive (forward flow) and negative (counterflow), depending on the sign of the difference between negative pressures at the channel inlet P_{ch} and the negative pressure in the upper cowl (the drive drum cowl) P_1

$$Q_0 = (P_{ch} - P_1) / \sqrt{R_0 \left| P_{ch} - P_1 \right|}. \tag{5.106}$$

Air is subsequently injected in a narrow channel (chute). The flow rate of air passing through this channel is determined by the injection head P_{ech} and by the difference between negative pressures at its ends

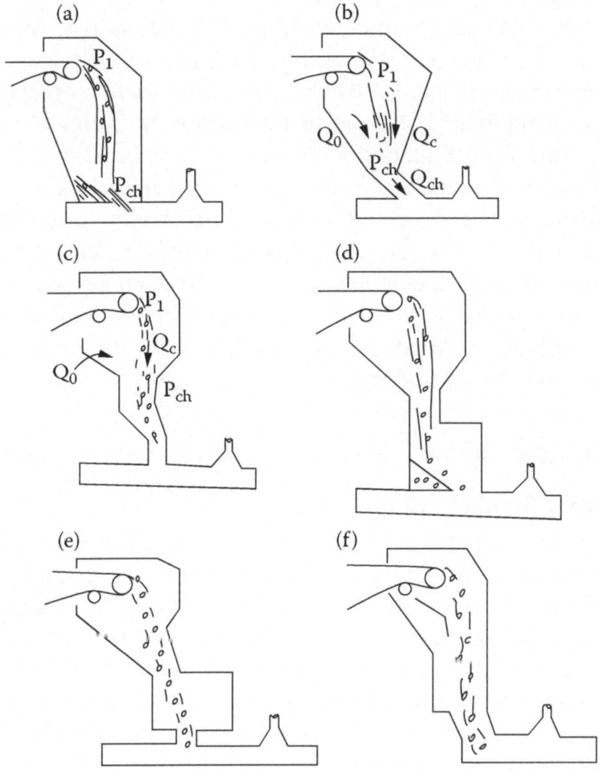

FIGURE 5.15 Transfer layouts involving variable-section chutes.

$$Q_{ch} = (P_{ech} + P_k - P_{ch}) / \sqrt{R_{ch} |P_{ech} + P_k - P_{ch}|}, \tag{5.107}$$

where P_k is the negative pressure at the channel outlet (Pa).

For material leaving the channel in an aspirated lower cowl, $P_k = P_2$. P_{ch}, P_1, Q_0, Q_{ch} are unknown variables. They can be resolved using three equations. The set of combined equations is closed using the equations for the airflow coming in through leaky joints in the upper cowl:

$$Q_{ch} = P_1 / \sqrt{R_H |P_1|}. \tag{5.108}$$

When the material passes from the first channel into another channel instead of an aspirated cowl, P_k becomes an unknown variable as well. The set of equations is closed by writing a sequence of equations for airflows in the second, third, and last channels and equating them together because the value of Q_{ch} would be the same.

When two adjacent channels have a leaky joint, their respective airflows will differ by the flow rate value

$$Q_{Hch} = P_c / \sqrt{R_{Hch} |P_c|}, \tag{5.109}$$

where P_c is the negative pressure in joint (Pa).

A feature of the second group of handling facilities is that the process of air injection in a narrow channel is followed by a process of air injection by a stream of material in an expanded channel where two parallel airflows reappear inside and outside the jet. Calculations have the same form as for the initial section, except that the negative pressure P_1 in Equation 5.106 should be replaced with P_k (the negative pressure at the outlet of the narrow channel). When there is a leak in the expanded channel (Figure 5.14d), the amount of in-leaking air should be accounted for in the negative pressure value P_k. Despite their apparent simplicity, calculations of the flow rate of air entering through a complex configuration of chutes are computationally intensive tasks, unimaginable without a computer. Due to its inherent universality, the described calculation method can also be used for the more simplistic classes of transfer facilities considered earlier.

5.1.4 Calculations of Air Replacement in High-Speed Machinery

5.1.4.1 Hammer Breakers as Fans

High speed working components produce airflows inside cowls and adjacent chutes. These airflows should also be accounted for in aspiration design decisions and in volume calculations. Hammer and rotary crushers are common classes of machines at ore pretreatment factories that operate with positive pressure inside cowls. Crusher rotors are aerodynamically similar to centrifugal fan impellers. Flow directions are determined by the structural design of the rotor and of the crusher chamber. For example, in non-reversible rotary and hammer crushers, airflows are directed downward (along with the moving material). An opposing counterflow of air occurs during operation of reversible hammer crushers.

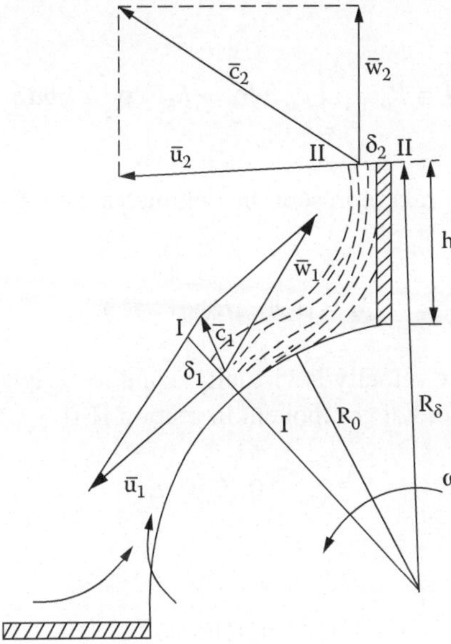

FIGURE 5.16 Diagram of air currents at the running rotor of a crusher.

The current inside the gap between two symmetric crusher hammers (Figure 5.16) can be represented as follows. Air from behind the first hammer and from the side cavities of the crusher enclosure appears as a flat jet lying atop the drum and is then thrown away by the frontal part of the next hammer. A constant rotor running speed results in a steady airflow with the resulting moment M of external surface and volume forces relative to rotation axis equal to [126]

$$M = \int_{\sigma} r c_u \rho c_n d\sigma, \tag{5.110}$$

where r is the distance away from rotation axis (m); c_u is the tangential component of air velocity (m/s); and c_n is the projection of air velocity onto an outer normal (to the surface) σ, constraining the volume of the environment in question (m/s).

Applying this definition to the flow area between cross-section I-I and II-II leads to

$$M = \int_{\sigma_i} r_1 c_{1u} \rho c_{1n} b d\delta_1 + \int_{\sigma_{ii}} r_2 c_{2u} \rho c_{2n} b d\delta_2, \tag{5.111}$$

where b is rotor length (m); and δ_1, δ_2 is airflow thickness in cross-sections I-I and II-II, correspondingly (m).

Replacing momentary component values with average values

$$c_{1u} \approx \overline{c}_{1u}, c_{2u} \approx \overline{c}_{2u}, \tag{5.112}$$

and considering

$$r_1 = R_0, r_2 = R_\delta, \tag{5.113}$$

we can find

$$M = R_0\bar{c}_{1u}\rho\int_{\sigma_I}\bar{c}_{1n}bd\delta_1 + R_\delta\bar{c}_{2u}\rho\int_{\sigma_{II}}c_{2n}bd\delta_2 \tag{5.114}$$

or, considering that integrals represent the volumetric flow Q, a more compact notation can be used:

$$M = \rho Q(\bar{c}_{1u}R_0 + \bar{c}_{2u}R_\delta). \tag{5.115}$$

Because the relative flow velocity is virtually equal to peripheral velocity in section I-I while it matches the radial component in section II-II,

$$\bar{c}_{1u} = 0, \bar{c}_{2u} = u_2, \tag{5.116}$$

therefore, in our case,

$$M = \rho QR_\delta u_2, \tag{5.117}$$

and the theoretical pressure produced by the crusher rotor, as a fan impeller of its own kind, equals

$$H_T = \omega M / Q = \rho u_2^2. \tag{5.118}$$

Just like fans, the following holds true for rotary crushers

$$Q' / [n'(D_2')^3] = Q'' / [n''(D_2'')^3] = c_Q, \tag{5.119}$$

$$P' / [\rho'(n'D_2')^2] = P'' / [\rho''(n''D_2'')^2] = c_p, \tag{5.120}$$

and they are used for recalculating centrifugal blower properties.

A generalization of empirical data (Figure 5.17) shows an agreement between dimensionless characteristics of rotary crushers and parabolic law (solid lines on plots)

$$\bar{P} = \bar{P}_{max} - a\bar{Q}^2, \tag{5.121}$$

where $\bar{P} = P / H_T$ is the ratio of full pressure to theoretical head defined by Equation 5.118; $\bar{P}_{max}$ is the maximum head ratio (at $\bar{Q} = 0$); and $\bar{Q} = Q / (\pi R_\delta^2 u_2)$ is dimensionless airflow discharged by the crusher rotor. Values of $\bar{P}_{max}$ and a are determined by structural properties of the rotor (Table 5.4).

The effective head increases as hammers become closer aerodynamically to the fan impeller blades. Particularly for hammer mills and crushers, $\bar{P}_{max}$ increases with the number of hammers on the rotor. Using experimental data published by V. P. Osokin [127] for a laboratory model of a self-ventilated hammer mill results in the following:

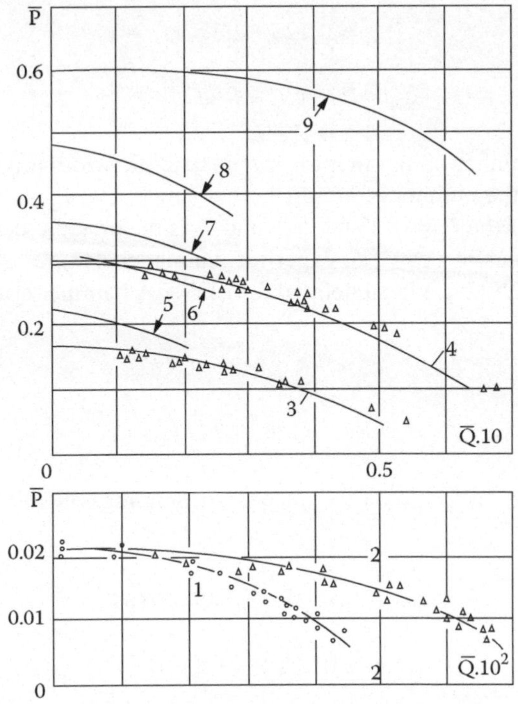

FIGURE 5.17 Dimensionless aerodynamic characteristics of crushers. Experimental data was provided by I. I. Afanasyev and I. N. Logachyov et al. for LDM-1A hammer crushers ($m_D = 2$, $m_p = 4$): with grid iron, curve 1 (°); without grid iron, curve 2(Δ); and for SM-937 disintegrators: dual-basket design, curve 3(Δ); and single-basket design, curve 4(Δ). V. P. Osokin provided experimental data for a model of a self-ventilated hammer mill $m_p = 3$ – curves 5 (at $m_D = 2$), 6 ($m_D = 4$), 7 ($m_D = 6$), and 8 ($m_D = 12$). Curve 9 is the performance characteristic of a Ts5-31 centrifugal fan [128].

TABLE 5.4
Dimensionless Aerodynamic Parameters of Crushers

Crusher Type	$\bar{P}_{max}$	a
LDM 1—A hammer crusher		
(a) With grid iron	0.022	200
(b) Without grid iron	0.022	80
SM-937 disintegrator [129]		
(a) With fixed basket	0.17	50
(b) Without fixed basket	0.3	50
Hammer mill models with varying number of hammers in a row [127]:		
$m_D = 2$	0.23	150
$m_D = 4$	0.31	150
$m_D = 6$	0.365	150
$m_D = 12$	0.485	150

$$\bar{P}_{max} = 0,125 + 2,3\xi_m m_p \frac{l_m}{l_p}, \quad (5.122)$$

$$\xi_m = \bar{P}_\infty - (\bar{P}_\infty - \bar{P}_2)e^{-0,17(m_D-2)}, \quad (5.123)$$

where m_p is the number of hammer rows; l_m is hammer width (m); l_p is crusher rotor length (m); m_D is the number of hammers in a single row; $\bar{P}_2$ is the pressure coefficient at $m_D = 2$ (for the $\bar{P}_2 = 0,23$ model); and $\bar{P}_\infty$ is the pressure coefficient for larger numbers of hammers in a row (for experimental model at $m_D \geq 12, \bar{P}_\infty = 0,5$).

Scale factors [129] have been defined for LDM-1A hammer crushers

$$c_p = \frac{\varphi_{cr}}{\rho}\rho_0 m_p m_D l_m / l_p, \quad (5.124)$$

$$c_Q = c_{cr} m_p m_D l_m / l_p, \quad (5.125)$$

where φ_{cr}, c_{cr} are proportionality factors depending on the degree of grill perforation σ

$$\varphi_{cr} = 0,0375 + 0,0192\cos\pi\sigma, \quad (5.126)$$

$$c_{cr} = 0,0133[1 - \exp(-7\sigma^{1,4})], \quad (5.127)$$

ρ_0 is air density at 20°C ($\rho_0 = 1,213$ kg/m^3).

As a result of Equations 5.119 and 5.120,

$$\bar{P} = c_p / \pi^2; \bar{Q} = 4c_Q / \pi^2,$$

then, in view of Equation 5.121,

$$\bar{P}_{max} = \frac{c_p}{\pi^2} + a\left(\frac{4}{\pi^2}c_Q\right)^2, \quad (5.128)$$

and the dimensionless characteristic of this crusher would be

$$\bar{P} = \frac{c_p}{\pi^2} - a\left[\bar{Q}^2 - \left(\frac{4}{\pi^2}c_Q\right)^2\right]. \quad (5.129)$$

5.1.4.2 Aspiration Volumes

Operation of rotary and hammer crushers is characterized by $P_{eq} >> P_e$; therefore, the velocity of air delivered by their rotors exceeds that of the material. An additional hydraulic resistance is provided by the flow of falling particles. If no material is present, crushers inject a maximum amount of air through their adjacent chutes. Therefore, aspirated airflow calculations consider the worst case (idle running equipment).

Reversible hammer crushers inject air from the lower and into the upper cowl; therefore, local suction is provided in the feeder cowl. The amount of air entering the feeder cowl through the chute is determined with the ratio

$$Q_{ch} = \sqrt{(P_{eq} + P_n) / R_{ch}}, \; R_{ch} = \Sigma\zeta\rho / (2S_{ch}^2 \qquad (5.130)$$

where P_{eq} is the pressure produced by the crusher when the specified volume of air is injected (Pa); S_{ch1} is the cross-sectional area of the upper (loading) chute (m^2); and $\Sigma\zeta$ is the sum total of local resistance coefficients of chutes related to the dynamic head in the upper chute:

$$\Sigma\zeta = \zeta_{ch1} + \zeta_{ch2}\left(\frac{S_{ch1}}{S_{ch2}}\right)^2 + 2.4\left(\frac{S_{ch1}}{F}\right)^2; \qquad (5.131)$$

P_n is the negative pressure maintained by a local suction unit in the feeder cowl (P_a); ζ_{ch1}, ζ_{ch2} are local resistance coefficients of the upper and lower (unloading) chutes; S_{ch2} is the cross-sectional area of the lower chute (m^2); F_{Hk} is the area of leaky joints in the cowl for the lower chute shoe (m^2); and ρ is air density (kg/m^3).

In view of Equation 5.121, the head created by the crusher is equal to

$$P_{eq} = P_{\max} - \xi Q_{ch}^2, \qquad (5.132)$$

$$\xi = a\rho / \left(\frac{\pi D^2}{4}\right)^2, P_{\max} = \bar{P}_{\max}\rho(n\pi D)^2, \qquad (5.133)$$

and the calculated relation (Equation 5.130) can be represented as

$$Q_{ch} = \sqrt{\frac{P_{\max} + P_n}{R_{ch} + \xi}}, \qquad (5.134)$$

where n is crusher rotor rotation speed (revolutions per second); and D is rotor diameter adjusted for hammer length (m).

In an operating hammer crusher, the asymmetric position of the grid iron causes air from the upper chute to be injected into the lower chute and thus into the aspirated cowl of the lower conveyor. Calculation formulas have the following form: Aerodynamic characteristics of chutes are determined in view of the total of local resistance coefficients related to a dynamic head of air in the lower chute and with substitution of optimum negative pressure in the aspirated cowl of the lower conveyor in place of negative pressure in feeder cowl.

The calculation method was validated and proved acceptable in trials of industrial crushing assemblies. Relative deviation of design aspiration volumes from measured ones stays within the industrial experiment error margin.

Amounts of injected air can be reduced by ducting the positive pressure area (normally, the aspiration cowl housing) together with the negative pressure area (the cavity at crusher rotor shaft). The resulting inside circulation of air would significantly reduce aspiration volumes.

Let us analyze the effect of an operating bypass air duct system. Two bypass layouts are generally possible (Figure 5.18). Assuming the crusher cowl as leak-tight, we arrive at obvious relations between differential pressure and air flow rate.

For crushers equipped with double bypass ducts (Figure 5.18a), the following equations result:

$$\left.\begin{aligned} &P_0 - P_1 = R_1 L^2; P_2 - P_0 = R_0 (L + Q_3)^2; \\ &P_0 - P_4 = R_3 Q_3^2; P_5 - P_3 = R_2 (L - Q_4)^2 = R_2 Q_2^2; \\ &P_5 - P_4 = R_4 Q_4^2; P_a - P_5 = R_y L^2, \end{aligned}\right\} \tag{5.135}$$

where P_a is atmospheric pressure (Pa); $P_0, P_1, P_2, P_3, P_4, P_5$ are absolute pressures (Pa); $R_y, R_0, R_1, R_2, R_3, R_4$ are respective aerodynamic properties of leaky joints in the lower cowl, loading/unloading chute sections, and air ducts ($P_a/(\text{m}^3/\text{s})^2$); L is the flow rate of air entering the aspirated feeder cowl through the loading chute (m^3/s); and Q_3, Q_4 is the flow rate of circulating air inside the upper and lower bypass ducts (m^3/s).

As the surveys show, the greatest negative pressure inside the crusher housing occurs at the rotor shaft, and its magnitude is proportional to the total head

$$P_a - P_4 = kP_{eq}. \tag{5.136}$$

In our case,

$$P_{eq} = P_{\max} - \xi(Q_3 + L)^2. \tag{5.137}$$

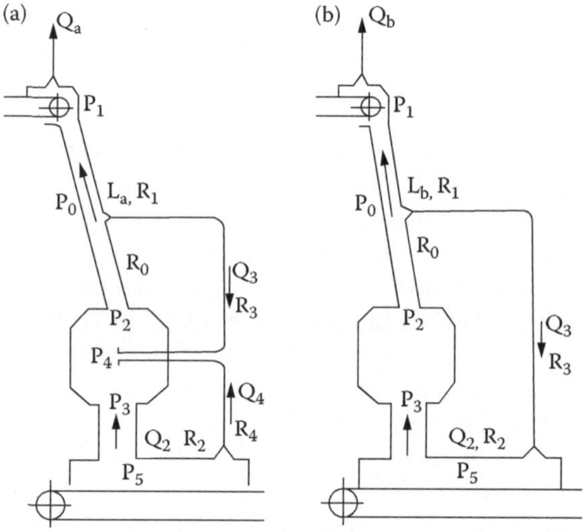

FIGURE 5.18 Aspiration design layouts for a hammer-type reversible crusher with a bypass air duct system.

In view of Equations 5.136 and 5.137, the combined equations (Equation 5.135) can be rewritten as follows:

$$\hat{P}_1 + P_{max} = (R_1 + R_y)L^2 + (R_0 + \xi)(L + Q_3)^2 + R_2(L - Q_4)^2; \tag{5.138}$$

$$\hat{P}_1 = (R_1 + R_y)L^2 + R_4 Q_4^2 - R_3 Q_3^2; \tag{5.139}$$

$$kP_{max} - \hat{P}_1 = k\xi(Q_3 + L)^2 + R_3 Q_3^2 - R_1 L^2, \tag{5.140}$$

enabling air flow rates to be determined for a given negative pressure inside the aspirated cowl $\hat{P}_1$ and for a known structural design of a bypass-equipped crusher.

From Equation 5.138 we can deduce

$$L = \sqrt{\frac{\hat{P}_1 + P_{max}}{R_{ch} + \xi} - \frac{\xi + R_0}{R_{ch} + \xi} Q_3(Q_3 + 2L) - \frac{R_2}{R_{ch} + \xi} Q_4(Q_4 - 2L)}, \tag{5.141}$$

$$R_{ch} = R_0 + R_1 + R_2 + R_y, \tag{5.142}$$

that is,

$$L < L_\infty = \sqrt{\frac{\hat{P}_1 + P_{max}}{R_{ch} + \xi}}. \tag{5.143}$$

Thus, the amount of air injected into an aspirated cowl decreases as the amount of air circulating in a bypass duct system increases. Industrial tests of a DMRIE 1450*1300 hammer crusher with a rotor running at 985 rpm indicate that, with the upper bypass duct ($\zeta_3 = 3.09$) sized for a diameter of 400 mm and the lower duct ($\zeta_4 = 1{,}52$) sized for twice less a diameter, the amount of injected air was brought down twofold from 3 m^3/s to 1.6 m^3/s. It was found that the coefficient $k \approx 5$, and $\xi = 27$. The following aerodynamic characteristics were observed: $R_0 = 0.2$ Pa $/(m^3/s)^2$, $R_1 = 1.4$ Pa $/(m^3/s)^2$, $R_2 = 5.625$ Pa $/(m^3/s)^2$, $R_y = 5.76$ Pa $/(m^3/s)^2$, $R_3 = 117.4$ Pa $/(m^3/s)^2$, $R_4 = 924$ Pa $/(m^3/s)^2$, $P_{max} = 358$ Pa, and $\hat{P}_1 = 8$ Pa. Figure 5.19a illustrates calculated changes in airflow as a function of increased bypass duct resistance. Decreasing R_3 and R_4 brings about significantly lower injected air volumes. Notably, this decrease is more pronounced with both circulation rings in operation.

Let us compare these findings with the example of a single air duct connecting the loading chute and the chute of the discharge trough shoe. The initial combined equations have the form

$$\left.\begin{array}{l} P_0 - P_1 = R_1 L^2; P_2 - P_0 = R_0(Q_3 + L)^2; P_5 - P_3 = R_2(Q_3 + L)^2; \\ P_0 - P_5 = R_3 Q_3^2; P_a - P_5 = R_y L^2, \end{array}\right\} \tag{5.144}$$

and the solution appears as

$$L = \sqrt{\frac{\hat{P}_1 + P_{max}}{R_{ch} + \xi} - \frac{(R_1 + R_y)(R_0 + R_2 + \xi)}{(R_{ch} + \xi)^2}} - \frac{R_0 + R_2 + \xi}{R_{ch} + \xi} Q_3, \tag{5.145}$$

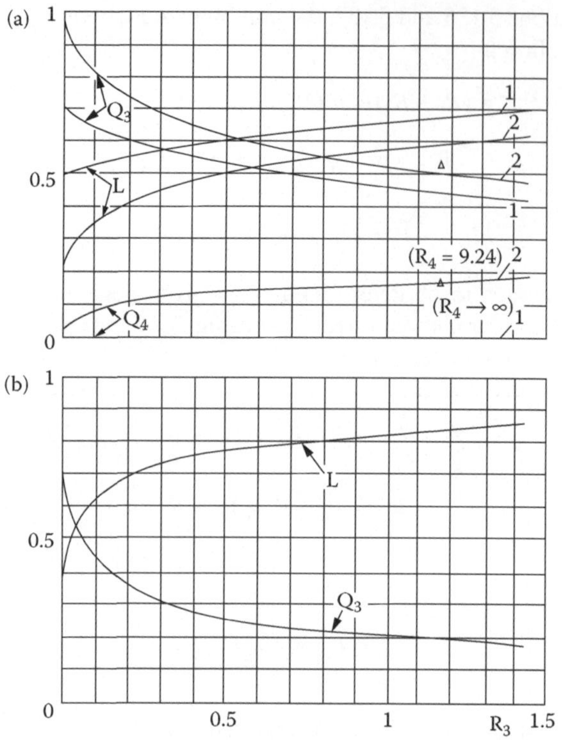

FIGURE 5.19 Changes in the amount of air injected by a DMRIE 1450*1300 crusher rotor ($n = 985$ rpm) and circulating through bypass air ducts as shown on chart **a** (1) in absence of the lower circulation ring ($R_4 \to \infty$), (2) with air circulating both in the upper (R_3 = var) and lower ($R_4 = 9.24$) rings following the chart **b**. Air flow rates have been related to $L_\infty = 3{,}03\ m^3/s$, and aerodynamic properties have been referred to $R_\infty = 100$ Pa / $(m^3/s)^2$. Δ = experimental data published by I. I. Afanasyev et al.

$$Q_3 = \sqrt{\frac{(R_1 + R_y)L^2 - \hat{P}_1}{R_3}}. \tag{5.146}$$

As noted in Figure 5.19b, in this case, the amount of injected air is lower than during operation without bypass ducts ($L < L_\infty$) but higher than with a two-ring bypass duct. It should be noted that material crushing increases R_f, thus decreasing air flow rate. It is possible that the flow rate falls so severely that the condition

$$\hat{P}_5 = R_y L^2 < P_{opt} \tag{5.147}$$

becomes true. In that case, the negative pressure in the unaspirated (lower) cowl would weaken and fall below its optimum value (P_{opt}), and dust-laden air could leak into the room. Therefore, the first case is more reliable because a dual-ring bypass design would enable evacuation of air from the lower cowl through a discharge chute and through a bypass air duct.

5.2 DUST RELEASE INTENSITY AND MITIGATION OF INITIAL DUST CONCENTRATION IN ASPIRATED AIR

5.2.1 Overview and Primary Features of Dust Release Sources

Three categories of dust release sources can be identified at ore beneficiation plants, based on the number of harmful factors and on the difficulty of responding to them.

The first category includes equipment or its separate assemblies that release dust, moisture, and heat simultaneously during operation. Steam-dust mixtures produced this way complicate the operation of ventilation systems and impede personnel operations as they infiltrate the shop floor. Most typical dust release sources at the first category of pelletization factories include drum-type fired pellet coolers, screens with water cooling for the material, and belt conveyors carrying cooled pellets. Steam-and-dust mixtures at sintering plants are formed during chilling of sintered breakage (returns) by water in drum-type coolers, initial blending of charge, and handling of cooled returns. The first category at beneficiation plants includes dust release sources from equipment used in concentrate drying houses—namely, drum-type driers and belt conveyors transporting dried concentrate.

The second category comprises sources of simultaneous dust and heat release. They are characterized by high dust concentrations in air evacuated from cowls. Typical sources in this category include filling and unloading assemblies of firing machinery, coolers, handling facilities for heated and dry materials, and facilities for their screening and transportation.

The third category includes equipment releasing relatively small quantities of dust during operation. Crushers, feeders, screens, mills, separators, and transfer facilities for transporting loose materials are representative of this category.

Deployment of high-performance machinery at pelletization plants (Poltava, Mikhailovsky, Lebedinsky, and Seversky ore beneficiation plants [OBPs]) not only enables the total number of sources to be decreased but also affects their qualitative performance. Heated wet material handling trains are eliminated from the process chain, thus eliminating or significantly reducing the first category of sources. In addition, when pellets are produced at a combined gridiron-and-furnace facility (at Poltava OBP), the number of second-category sources is reduced as well (the bed feed path—one of the most intense dust sources—is eliminated from the process equipment chain).

5.2.1.1 Dust Carryover from Aspiration Cowls

Dust escape into an aspiration network is governed by the physical and mechanical properties of the material being handled, the type and design of the equipment and cowls, and by the aerodynamic parameters of local suction units.

Table 5.5 lists average amounts of dust taken by local suction units at main machinery units of OBPs. Among all aspiration installations, the greatest specific carryover rates of material occur at agglomeration plants processing iron ore and its concentrate. Within the classification of dust emission sources, the first and second categories are responsible for the greatest carryover rates.

TABLE 5.5
Dust Carryover into Aspiration Network**

Dust-Releasing Equipment/ Assembly	Capacity (t/hr)	Average Aspiration Volume (m^3/hr)	Average Dust Concentration in Aspirated Air (mg/m^3)	Material Carryover (kg/hr)	Material Carryover Specific Rate (kg/t)
Crushing plants					
	Category 3 sources				
Conveyors (loading locations)	300	5000	2000	10.0	0.03
Cone crushers	250	1500	400	0.6	0.003
Jaw crushers	300	3000	500	1.5	0.005
Screens	200	3400	350	1.2	0.006
Dry magnetic separators	120	2000	200	0.4	0.003
Pelletization factories					
	Category 1 sources				
Screens	450	130,000	40,000	5200.0	11.5
Drum-type coolers	100	55,000	35,000	1925.0	19.3
	Category 2 sources				
Conveyors:					
(a) Loading locations	200	8000	5000	40.0	0.2
(b) Drive drums	150	4000	4500	18.0	0.12
Screens	300	100,000	20,000	2000.0	6.6
Firing machines:					
(a) Head part	400	50,000	100	50.0	0.13
(b) Tail part	400	100,000	7,000	700.0	1.75
	Category 3 sources				
Conveyors:					
Loading locations	30	1000	70	0.07	0.002
Weighing units	3	500	150	0.08	0.026
Auger mixers	30	250	120	0.03	0.001
Vibratory feeders	3	100	400	0.04	0.013
Hammer crushers	200	10,000	7000	70.0	0.35
Sintering factories					
	Category 1 sources				
Drum-type coolers	30	10,000	39,000	390.0	13.0
Conveyors:					
(a) Loading locations	30	5000	10,000	50.0	1.6
(b) Drive drums	30	4000	8000	32.0	1.1
Drum mixers (primary blending)	250	14,000	20,000	280.0	1.1
	Category 2 sources				
Sintering machines					
(a) Loading location	350	40,000	2500	100.0	0.28
(b) Unloading location	350	150,000	10,000	1500.0	4.3
Disc feeders	30	7500	20,000	150.0	5.0
Screens	100	50,000	7600	380.0	3.8

TABLE 5.5 (Continued)
Dust Carryover into Aspiration Network**

Dust-Releasing Equipment/ Assembly	Capacity (t/hr)	Average Aspiration Volume (m^3/hr)	Average Dust Concentration in Aspirated Air (mg/m^3)	Material Carryover (kg/hr)	Material Carryover Specific Rate (kg/t)
Conveyors:					
(a) Loading locations	35	8000	7000	56.0	1.6
(b) Drive drums	35	5000	5000	25.0	0.7
		Category 3 sources			
Conveyors (loading locations)	600	4000	530	2.1	0.035
Secondary mixing drum	100	4000	70	0.3	0.003
Four-roll crushers	20	5000	800	4.0	0.2
Hammer crushers	200	10,000	7000	70.0	0.35

**From primary process equipment at ore beneficiation plant

Specific dust carryover from the first category sources at agglomeration plants occurs at the drum cooler, screen, and conveyor loading cowls. For the second category, the greatest specific carryover rates commonly occur in discharge cowls of agglomeration and firing machines, in cowls of screens, and in conveyor loading facilities. For the third category, primary sources include hammer and four-roll crushers and cowls at conveyor loading locations.

5.2.1.2 Concentration and Particle Size Composition of Dust in Aspirated Air

Dust concentration and its particle size composition depend on a number of factors: material type and its particle size distribution, hardness, moisture content, structural design of handling facilities, and aerodynamic parameters of aspiration cowls (air mobility inside a cowl, air velocity at a dust receiver inlet).

The process of dust formation during processing of dust-releasing materials has proven to be extremely complex and has so far defied researchers' attempts [130–132] to come up with at least a semi-empirical computational method for determining dust concentration and particulate composition in aspirated air. To date, the only outcome from numerous studies of wet dedusting (for example, [133–135]) was the identification of increasing material moisture content as a factor reducing dust release intensity.

Therefore, we endeavored to generalize field test data summarized in previous publications [137,140]. Findings from these tests (listed in Table 5.6) indicate that dust content in evacuated air varies rather widely. Therefore, the average dust content indicated in the ninth column of Table 5.6 may be used as a guideline for dust trap selection. Dustiness is most intense at dust-releasing assemblies of agglomeration plants.

The particulate composition of dust described by distribution functions D (*d*)—and represented in a graphic form in Figures 5.20 through 5.22 using a double logarithmic coordinate grid—is more stable than dust content and depends mainly on the type of handled material and on the equipment type.

TABLE 5.6
Dust Concentration in Air Evacuated from Cowls of Dust-Generating Equipment

Material Type Aspirated Assembly	Size of Material Particles *d* (mm)	Particle Temperature t_M (° C)	Moisture Content of Material, w(%)	Material Flow Rate Gm (t/hr)	Aspiration Volume, Q (m^3/hr)	Specific Aspiration Volume (m^3/t)	Dust Concentration, C (mg/m^3) Variation Range, *C*	Average, $\overline{C}$	Percentage of Cases with Concentration $C = (0,5 \div 1,5)\overline{C}$
1	**2**	**3**	**4**	**5**	**6**	**7**	**8**	**9**	**10**
Suction from cowl at conveyor loading location									
Iron ore (transferred from conveyors/	6–70	20	1–3	50–2000	500–9000	4–70	100–600	300	67
feeders/screens)	4–10	20	3–6	150–2000	3500–4000	2–20	60–100	80	100
Iron ore (unloaded	6–100	20	1–3	70–80	2600–4200	35–60	4500–6000	5800	100
from crushers)	10–55	20	3–6	70–500	2000–9000	15–70	200–2600	1200	60
Pellets (finished stock)	10–15	70–130	0	80–500	900–20,000	4–180	400–20,000	5,000	40
Pellets (spillage, returns)	3–10	75–250	0	20–100	1000–7500	30–240	800–50,000	10,000	50
Charge (concentrate, bentonite, limestone)	0.10	20	6–8	30	1000	33	60–80	70	100
Sinter	25	150	0	35	8000	230	–	7000	–
Sinter (returns)	2–3	50–130	0–4	80–100	1500–7000	20–80	6,000–1,000	10,000	100
Crushed chalkstone	1–4	20–150	0–6	60–160	1500–3000	20–30	1500–5500	3500	100
Coke	1.5–10	20	4–10	15–20	4000–7000	250–500	450–1300	1000	100
Suction from drive drum cowl									
Pellets (finished stock)	10–15	70–350	0	80–300	2000–12,000	15–50	800–5500	3500	85

Pellets (spillage, returns)	3–10	70–150	0	20–60	1500–3000	30–90	3000–9000	6,000	100
Sinter (returns)	3	130	0–4	90	5000	60	4000–10,000	8,000	100
Suction from cone crusher cowl									
Iron ore	60–80	20	2–4	50–400	1000–2000	3–35	30–800	400	40
Suction from jaw crusher cowl									
Iron ore	80–500	20	2–5	70–500	2000–5000	10–30	150–1000	500	65
Suction from screen cowl									
Iron ore	20	20	3–5	200	3400	17	350	350	–
Pellets	10–16	70–150	0	100–500	6000–150,000	40–300	3000–40,000	20,000	40
Suction from unloading section cowl of drum-type cooler									
Sinter (returns)	3	130	0–4	30	5800	160	39,000	39,000	–
Pellets	12	150	0–3	100	55,000	550	34,000	34,000	–
Suction from unloading section cowl of secondary blending drum									
Sintering mix	1	35	8	100	4000	40	70	70	–
Suction from disc feeder cowl									
Sinter (returns)	3	500	0	30	7500	250	20,000	20,000	–
Suction from plate feeder cowl									
Pellets	16	130	0	100	3700	37	2800	2800	–
Suction from vibratory feeder cowl									
Pellets	12	150	0	350	8300	24	14,500	14,500	–
Suction from weighing unit cowl									
Crushed limestone	0.1	20	0–1	3–5	150–500	50–100	70–150	110	100
Crushed bentonite	0.1	20	0–1	2–3	300–700	150–230	60–500	170	65

continued

TABLE 5.6 (Continued)
Dust Concentration in Air Evacuated from Cowls of Dust-Generating Equipment

Material Type Aspirated Assembly	Size of Material Particles d (mm)	Particle Temperature t_M (° C)	Moisture Content of Material, w%	Material Flow Rate Gm (t/hr)	Aspiration Volume, Q (m^3/hr)	Specific Aspiration Volume (m^3/t)	Dust Concentration, C (mg/m^3): Variation Range, C	Average, $\overline{C}$	Percentage of Cases with Concentration $C = (0,5 \div 1,5)\overline{C}$
1	**2**	**3**	**4**	**5**	**6**	**7**	**8**	**9**	**10**
Suction from auger mixer cowl									
Charge (concentrate, bentonite, limestone)	0.1	20	6–8	30	170–350	6–15	70–180	120	100
Suction from vibratory feeder cowl									
Crushed bentonite	0.1	20	0–1	2–3	100	30–50	100–1000	400	65
Suction from firing machine head									
Pellets	12	100	0	100–400	10,000–50,000	100–150	200–3000	1000	40
Suction from cowl adjacent to chimney of firing machine drying zone #1									
Pellets	12	–	–	100–400	5000–10,000	25–50	200–1000	500	30
Suction from firing machine tail									
Pellets	12	200	0	100–400	20,000–100,000	200–250	2000–20,000	7,000	
Suction from firing machine cooler bin									
Pellets	12	200	0	400	100,000	250	5100	5100	–
Suction from bed bins									
Pellets	12	100	0	100–300	3000–6000	20–30	2000–14,000	8000	50
Suction from loading hoppers									
Pellets	12	100–500	0	400–600	3000–10,000	10–30	5000-20,000	10,000	40

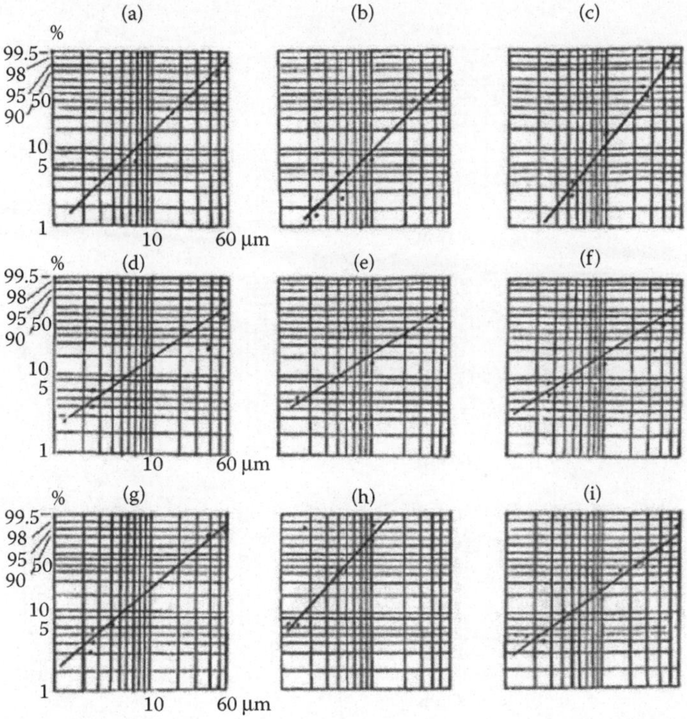

FIGURE 5.20 Particulate composition of airborne dust aspirated from conveyor cowls:

(a) Iron ore (d = 0.7…71 mm, w = 1.2…4%, u_{in} = 1.7…3.4 m/s)
(b) Chalkstone (d = 0…25 mm, w = 0…6%, u_{in} = 1.4…3.5 m/s)
(c) Sinter (d = 0…20 mm, w = 0…2%, u_{in} = 3…3.8 m/s)
(d) Sintering ore (d = 4.5…6.5 mm, w = 2.5…6%, u_{in} = 1…1.2 m/s)
(e) Coke fines and anthracite chippings (d = 0…1.1 mm, w = 4…10%, u_{in} = 1.2 m/s)
(f) Pellets (spillage: d = 9 mm, w = 0, u_{in} = 1.4…1.5 m/s)
(g) Pellets (bed: d = 12 mm, w = 0, u_{in} = 1.4…2.0 m/s)
(h) Pellets (bed: d = 10…14 mm, w = 0, u_{in} = 3.8…7.2 m/s)
(i) Pellets (finished stock: d = 10…12 mm, w = 0, u_{in} = 1.1…1.3 m/s).

5.2.2 Decrease in Dust Release Intensity

Presently, the techniques used for reducing dust content in evacuated air can be broadly identified as either passive or active.

The first (passive) group includes methods that rely on choosing a rational dust receiver layout, matching inlet velocities of air in dust receivers, and on increasing the volume of aspiration cowls. Recommendations for implementing these methods are thoroughly covered in specialized literature and are widely used in practice.

The second (active) group includes methods based on reducing the intensity of dust release (by wetting) and on using deduster cowls. Wetting the handled material is a method that is now widely used in iron ore mining and processing. There are numerous recommendations on the structure, choice, and parameters of

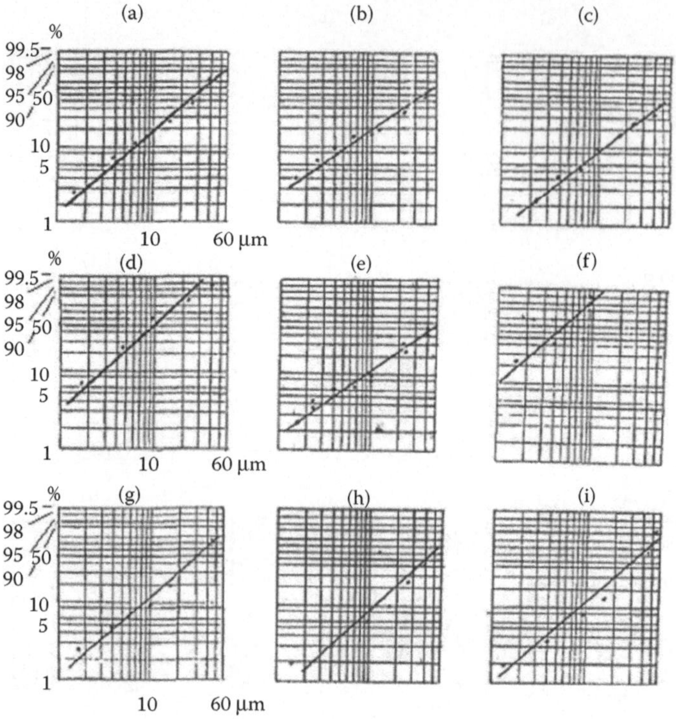

FIGURE 5.21 Particulate composition of airborne dust aspirated from process unit cowls at pelletization factories.

(a) Firing machine, loading assembly (pellets: u_{in} 6.5…7.4 m/s)
(b) Firing machine, loading assembly (pellets: d = 12…14 mm, w = 0, u_{in}= 4.0m/s)
(c) Drum-type coolers (pellets: d = 12…14 mm, w = 0, u_{in}= 8.5 m/s)
(d) Screen (pellets: d = 12…14 mm, w = 0, u_{in}= 7.1 m/s)
(e) Bin (pellets: d = 12…14 mm, w = 0, u_{in}= 1.0…2.0 m/s)
(f) Bentonite handling, semi-industrial-scale VNIIBTG installation (d = 0.1 mm, w = 0…1, u_{in}= 0.4 m/s)
(g) Vibratory feeder (bentonite: d = 0.1 mm, w = 0…1%, u_{in}= 0.4 m/s)
(h) Weighing unit (limestone: d = 0.1 mm, w = 0…1%, u_{in}= 0.4 m/s)
(i) Weighing unit (bentonite: d = 0.1 mm, w - 0…1%, u_{in}= 0.35…0.8 m/s)

wetting designs. Researchers in this area (with a focus on OBP factories) include the Laboratory of Air Dedusting for Preparation and Processing of Metallurgical Source Materials under the All-Soviet (now All-Russia) Institute for Occupational Safety in Ore Mining (VNIIBTG) [141,142].

Dust entrapment in aspirated cowls is ensured by special devices placed inside the the cowl. Work in this field was pioneered in the early 1950s by the Sverdlovsk (now Yekaterinburg) Institute for Occupational Safety (SIOT), which proposed a double-walled cowl for conveyor loading locations.

Interior vertical walls and aprons placed inside standard cowls, in addition to reducing total aspiration volumes, enabled a significant decrease of dust content in

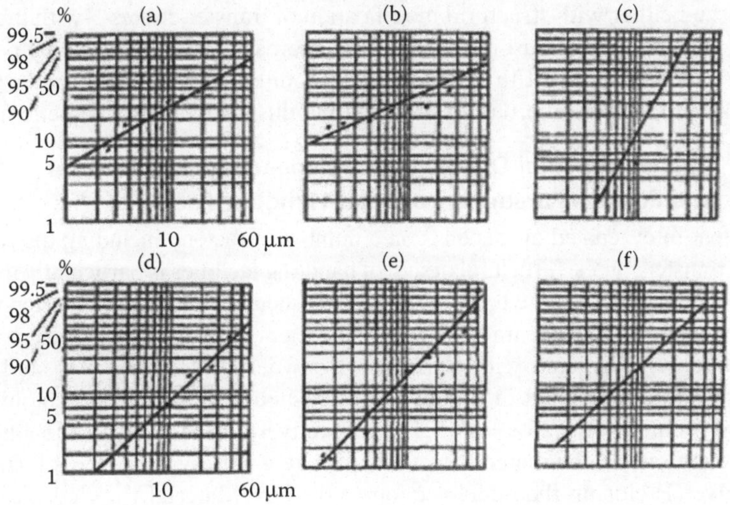

FIGURE 5.22 Particulate composition of airborne dust aspirated from process unit cowls at beneficiation and sintering factories:

(a) Cone crusher (iron ore: d = 60…100 mm, w = 3%, u_{in} = 3…3.5 m/s)
(b) Magnetic separator, $4^{\text{-drum}}$ design (concentrate: d = 0.1 mm, w = 8%, u_{in} = 2.4 m/s)
(c) Secondary blending drum (sintering mix: d = 1.0 mm, w = 8%, u_{in} − 2.7 m/s)
(d) Disc feeder (sinter return: d = 3.0 mm, w = 0, u_{in} = 2.5 m/s)
(e) Sintering machine, loading assembly (charge: u_{in} = 5.4…14.5 m/s)
(f) Drum-type cooler (return: d = 3 mm, w = 0…4%, u_{in} = 2.8 m/s).

evaluated air. These cowls were further improved by the authors of this volume in the 1960s. The recommendation was made to replace vertical walls with a chamber restricting material fall area inside the standard cowl. Dust content in evacuated air was brought down 2.5 times (from 13.4 g/m^3 to 5 g/m^3) [143].

Material carryover can be further arrested by using a foam cap [144] in the inner chamber. Dust containment efficiency with foam fed into the cowl reaches 80% [141]. Installation of cascaded grills with foam in the cowl of the hammer crusher [145] enabled a 360-fold reduction in dust content (from 9 g/m^3 down to 24 mg/m^3) [146]. An attempt was made to reduce material carryover into the aspiration network using wet dust trap cowls. To that end, nozzles with grooved drip pockets [147] or wetted vertical plates with drip pockets [148] were placed inside standard cowls. Studies [141] indicate that dust precipitation efficiency measures 37–52% with wet returns and 46–66% with ore.

Thus, using foam to reduce dust content in evacuated air is the most effective method. However, a number of factors presently limit adoption of this procedure, including process constraints on material properties, the high cost of foaming agents, and maintenance complications. Dry dust trap cowls deserve closer attention in light of the recent trend toward dry containment techniques. For example, merely installing air distribution plates in the cowl nearly doubles the reduction of dust content [141].

It would be reasonable to expect that further improvement of dry dust precipitation processes in cowls and a decrease of initial dust content in aspirated air would be concerned either with structural modification of transfer chutes—with the goal of reducing the intensity of dust release (e.g., by decreasing flow aeration by powdered materials, such as crushed limestone and bentonite), or with using other physical processes (besides pure inertia) for precipitating dust (e.g., using magnetic fields).

5.2.2.1 Changes in Total Dust Releases Depending on Structural and Process Parameters of Load-Handling Facilities

Dust content in evacuated air depends on a number of factors, including the structural design of transfer chutes. Structural design of handling facilities is particularly important when powdered materials are transferred (e.g., crushed bentonite and limestone). Chute inclination angle and height are two major parameters affecting dust release processes.

Dust release associated with transfers of powdered materials was studied at a laboratory facility depicted in Figure 5.23. Crushed limestone and bentonite, with a moisture content of 0.25% and 5.3%, respectively, were used in the study. The required flow rate of the material was ensured by a battery of washers installed on a gate valve (2) closing the outlet section of the bin. Material traveled through the chute (3) into the receiving bin (4). Dust samples were taken in a single point in all cases at a distance a = 500 mm, b = 250 mm from the dust source.

Experiments were performed with material fall height varying from 0.5 m to 1.5 m in 0.5-m steps and with chute inclination angles of 35°, 45°, 60°, 75°, and 90°. Experiment data were processed using C_0 = 1,000 mg/m^3 as the characteristic dust content value and H_0 = 3.0 m as the fall height.

The angle α was determined using the formula

$$\alpha = (\alpha_H - \alpha_e)\pi / 180, \tag{5.148}$$

where α_H is chute inclination angle (deg.); and α_e is the rest angle (deg.) ($\alpha_e = 32^\circ$ for bentonite, $\alpha_e = 34^\circ$ for limestone).

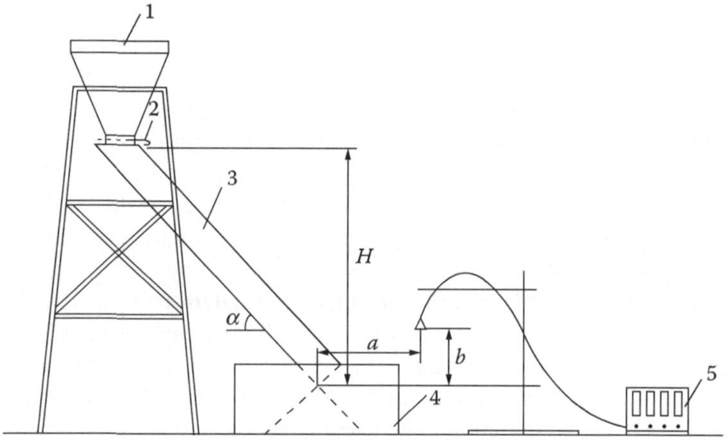

FIGURE 5.23 Laboratory facility for researching dust formation processes associated with crushed material handling: 1 = loading bin; 2 = adjustment gate; 3 = transfer chute; 4 = receiving bin; 5 = fan.

TABLE 5.7
Dust Release Intensity in Handling of Crushed Bentonite* and Limestone**

Material Fall Height (m)	Chute Inclination Angle (deg.)	Mass Flow Rate of Material (kg/s)	Average Dust Content in Air (mg/m^3)	
			For Bentonite Handling	For Limestone Handling
0.5	35	0.13	182	63
0.5	35	0.20	450	127
0.5	45	0.13	263	185
0.5	45	0.20	520	230
0.5	60	0.13	420	213
0.5	60	0.20	1160	280
0.5	75	0.13	660	270
0.5	75	0.20	1810	476
0.5	90	0.13	833	225
0.5	90	0.20	1873	580
1.0	35	0.13	160	130
1.0	35	0.20	570	150
1.0	45	0.13	425	143
1.0	45	0.20	1214	214
1.0	60	0.13	610	342
1.0	60	0.20	1900	416
1.0	75	0.13	792	463
1.0	75	0.20	2500	575
1.0	90	0.13	1150	545
1.0	90	0.20	3010	753
1.5	35	0.13	692	115
1.5	35	0.20	653	195
1.5	45	0.13	1023	220
1.5	45	0.20	1840	234
1.5	60	0.13	1276	346
1.5	60	0.20	2382	556
1.5	75	0.13	1440	743
1.5	75	0.20	2950	812
1.5	90	0.13	1554	803
1.5	90	0.20	4700	1222

*($W = 5.3\%$, $d_{cp} = 52.5$ μm, $\rho_M = 2400$ kg/m³); **($W = 0.25\%$, $d_{cp} = 52.5$ μm, $\rho_M = 2600$ kg/m³)

Experiment findings (Table 5.7) reveal that, given equal flow rates of materials, airborne dust content increases sharply with greater material fall height and chute inclination angle:

$$C / C_0 = a\alpha^n \left(H / H_0\right)^b, \tag{5.149}$$

where for bentonites $a = 2.93$; $n = 0.47$; $b = 0.85$ at a flow rate of 0.13 kg/s and $a = 5.11$; and $n = 0.57$; $b = 0.66$ at a flow rate of 0.2 kg/s; for limestone, $a = 0.95$; $n = 0.39$; $b = 0.63$ and $b = 1.14$; $n = 0.38$; and $b = 0.48$.

For example, with the inclination angle increased from 35° to 90° during bentonite handling with a flow rate of 0.2 kg/s and 0.5 m fall height, airborne dust content rose fourfold (from 450 mg/m^3 to 1873 mg/m^3); with a fall height of 1.5 m, airborne dust content increased seven times (from 653 mg/m^3 to 4700 mg/m^3).

This could be explained by aeration of the material flow similar to the phenomenon observed in shaft chutes because an increased inclination angle intensifies the aerodynamic interaction between solid particles and injected air. The unified flow of the material is split into localized jets whereby dust-like particles come off the general stream and contaminate the surrounding air. A pattern was observed when material was handled at an experimental production plant. Comparative testing of two transfer assemblies handling powdery material was performed: the first had a spiral chute, the second a prismatic chute. Crushed bentonite ($W = 0.64\%$) and limestone ($W = 3.29\%$) were used in the studies. Spiral chute dimensions were as follows: an inclination angle of the helix line $\alpha = 40°$, an outside diameter of 700 mm, a spiral width of 90 mm, a spiral border height of 50 mm, and a chute height $H = 1.5$ m. Well-known methodology was used in the performance of the studies, which included aerodynamic measurements and dust sample collection at a constant material flow rate ($G = 0.05$ kg/s).

In both cases, dust content in air evacuated by suction from the material dumping location rose with increased aspiration volumes (Table 5.8). Dependence of airborne dust content on air speed at the dust receiver inlet was noted (Figure 5.24). Given equal conditions, bentonite handling produces higher dust content than limestone because of a greater moisture content.

For the same materials, dust release intensity is lower when materials are transferred through a spiral chute than when they are transferred vertically through a prismatic chute. In the case of a spiral shute, dust content is five to ten times lower provided that optimum negative pressure ($P_y = 2$ Pa) is maintained in the cowl. In this case, dust content is lower than with an inclined chute. However, considering the unwieldiness of spiral chutes, the preferred choice for ordinary material handling (conveyor-to-conveyor transfers) would be inclined chutes with a minimum free-fall height of powdered material at inlet and outlet of the chute. Spiral chutes may be used for conveyor-to-bin transfers.

We field-tested this method in the charge shop of the Sokolovsko-Sarbaisky OBP pelletization factory [150]. At this factory, a plate diverter was used for loading crushed limestone into bins. Material free-fall height varies as the bin is loaded, reaching 4 m. Falling material forms a jet of fine particles leading to intense dust formation and to a carryover of material into the aspiration network. Individual aspiration units of uniform design are used to remove dust from limestone bins of firing machines 1 through 8. Every single unit serves two bins. Industrial-scale tests were carried out at an ATU-20-type aspiration set serving bins #20 and #21.

In order to reduce dust formation intensity and mitigate carryover of material into aspiration network, bin #21 was equipped with a loading device we proposed that was composed of a spiral chute with a helix line inclination angle of $\alpha = 40°$. Field testing indicated (Table 5.9) that material carryover into the aspiration network from

TABLE 5.8
Dust Release Intensity Survey Findings for Handling of Limestone and Bentonite

Aspiration Volume (m^3/s)	Negative Pressure inside Cowl (Pa)	Air Velocity at Dust Trap Inlet (m/s)	Velocity of Air inside Cowl (m/s)	Air Velocity in Chute (m/s)	Dust Concentration (mg/m^3)	Material Carryover (g/s)	Specific Material Carryover Rate (g/kg)
(A) Through spiral chute							
0.10	2	0.35	0.39	0.14	$\frac{70}{110}$	$\frac{0.007}{0.011}$	$\frac{0.14}{0.2}$
0.28	16	0.96	1.06	0.43	$\frac{450}{570}$	$\frac{0.13}{0.15}$	$\frac{2.6}{3.0}$
0.45	40	1.54	1.69	0.70	$\frac{1400}{2800}$	$\frac{0.61}{1.2}$	$\frac{12.2}{24.0}$
0.53	56	1.73	1.97	0.80	$\frac{2500}{3000}$	$\frac{1.32}{1.5}$	$\frac{26.4}{30.0}$
(B) Through vertical chute							
0.23	2	0.48	0.64	2.1	$\frac{300}{1100}$	$\frac{0.06}{0.26}$	$\frac{1.2}{5.2}$
0.50	10	1.06	1.40	4.5	$\frac{500}{5700}$	$\frac{0.25}{2.89}$	$\frac{5.0}{57.8}$
0.71	24	1.48	1.97	6.1	$\frac{3600}{12000}$	$\frac{2.55}{8.55}$	$\frac{51}{170}$
0.82	44	1.72	2.20	6.7	$\frac{6100}{166000}$	$\frac{26.0}{13.6}$	$\frac{100}{266}$

Numerator = limestone; denominator = bentonite.

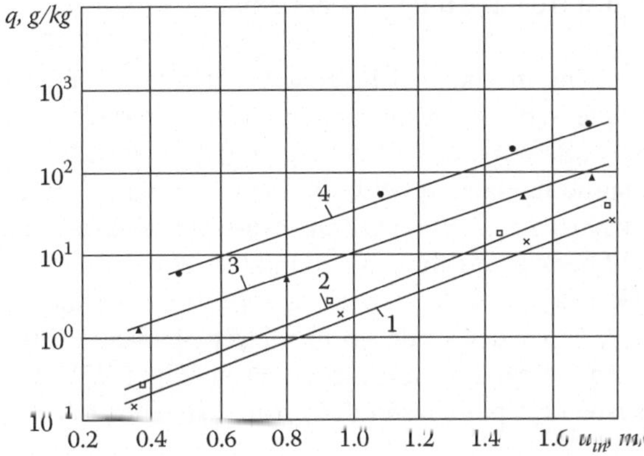

FIGURE 5.24 Plot of the relationship between air velocity at the dust receiver inlet and the specific dust carryover rate: 1(x) = transfers of limestone through a spiral chute; 2(♦) = transfers of bentonite through a spiral chute; 3(5) = transfers of limestone through a vertical chute; 4(•) = transfers of bentonite through a vertical chute.

TABLE 5.9
Results of Industrial-Scale Testing of Local Suction Units in an ATU-20 Aspiration Set

Local Suction Unit	Diameter of Suction Air Duct (mm)	Dimensions of Dust Receiver (m × m)	Air Velocity (m/s) In Air Duct	Air Velocity (m/s) In Dust Receiver	Volumetric Aspiration Flow (m^3/hr)	Dust Content in Aspirated Air (mg/m^3)	Amount of Material Carried Away (g/hr)
From bin #20	250	1.2 × 0.6	10.7	0.72	1900	–	–
From bin #21	250	1.2 × 0.6	12.0	0.81	2100	1100*/1810	2300*/3800
Downstream of dust trap	350	–	13.8	–	4800	–	–

* Data in numerator refers to a bin loaded using a spiral chute, and data in denominator refers to loading without any chute.

chalkstone bin #21 decreased more than 1.5 times after a spiral loading chute was installed.

A noticeable deviation of this effect (when compared to laboratory experiments) may be explained by the fact that the spiral chute extends only one-third of the way downward into the total depth of bin #21.

Of the process factors determining dust carryover intensity from a dumped material flow, flow rate is the most important. As an example, we will analyze this by filling bins with crushed material. Three stages can be identified with respect to dust carryover from bin-like containers:

- Dust release from material during dumping
- Precipitation of released dust in the aspirated bin
- Removal of dust from the bin by local suction

In the first stage, the mechanical breakdown of a layer of material falling off the conveyor belt, the subsequent disintegration of material flow due to collision with loading chute walls or other factors (e.g., a grill installed at the loading chute inlet), and the dynamic interaction between flowing material and air cause autohesive bonds between particles to disappear and cause particles to levitate in the stream of injected air. Held back by air flowing around the jet of a solid material, particles are unable to break away from the flow along its entire vertical course up until the final section. There, as particles pile up in a layer of material, the injected air is separated and suspended particles are carried away (Fig. 5.25). If C_e represents the concentration of suspended particles in this airflow, the dust release intensity would be expressed as

$$q = C_e\, Q_e, \tag{5.150}$$

where Q_e is the flow rate of injected air (m^3/s).

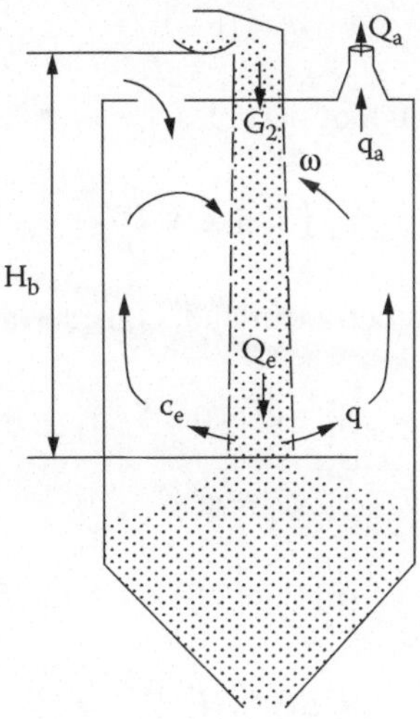

FIGURE 5.25 An illustration supporting the analysis of dust carryover from bins loaded with loose material.

It could be supposed that the concentration C_e depends on the moisture content of the material ω (%), on the volumetric concentration of material particles in the flow β, and on the relation between aerodynamic forces and the gravity force of a particle.

It is a fact that dust concentration is an exponential function of moisture content [135,151] and that a power relation of the type indicated in Equation 5.149 binds dust content in the air with free-fall height or volumetric concentration.

Therefore it can be posited that:

$$C_e = K \cdot e^{-\alpha\varpi}\beta^m \left(R / P_p\right)^n \tag{5.151}$$

where K, α are proportionality coefficients, R is the aerodynamic force equal to

$$R = \Psi \frac{\pi d^2}{4} \frac{V^2}{2} \rho, \tag{5.152}$$

P_p is particle weight, equal to

$$P_p = \frac{\pi d^3}{6} \rho_p g, \tag{5.153}$$

Ψ is the resistance coefficient of the particle; d is the equivalent particle diameter (m); V is the velocity of particle flow (m/s); ρ, ρ_p are densities of air and particle, correspondingly (kg/m^3); g is the gravity acceleration (m/s^2); and m, n are certain constants.

In the second phase, the released dust is carried away by airflows into the upper part of the bin and settles on the layer or material or on the bin walls with an efficiency of η.

In the third phase, dust is evacuated from the upper part of the bin by a local suction unit (or, absent one, through leaky joints). It is obvious that the flow rate of removed dust would be equal to

$$q_a = (1 - \eta)q. \tag{5.154}$$

Assuming a small breakthrough of particles,

$$1 - \eta = A\frac{R}{P_p}, \tag{5.155}$$

where A is a proportionality coefficient and, considering that

$$\beta = G/\left(\rho_p V S_{jet}\right),$$

$$Q_e = \varphi V S_{jet}, \tag{5.156}$$

Equation 5.154 can be rewritten (in light of Equations 5.150 through 5.153) as

$$\frac{q_a}{G} = BG^{m-1}H_b^{1.5+n-m/2}e^{-\alpha\omega}, \tag{5.157}$$

where

$$B = AKS_{jet}^{1-m}\frac{\varphi}{\rho_p^m}(2g)^{\frac{1-m}{2}}\left(\frac{3\psi\rho}{2d\rho_p}\right)^{n+1}; \tag{5.158}$$

H_b is the material dumping height (m); G is the flow rate of dumped material (kg/s); S_{jet} is the cross-sectional area of the jet formed by material (m^2); and φ is the ratio of injected air velocity to the velocity of flowing material.

Let us now review the experiment. Our findings from industrial-scale tests of local suction units in limestone and bentonite bins carried out in the pelletization factory at Sokolovsko-Sarbaisky OBP, as summarized in Table 5.10, testify to a wide variation of dust content in evacuated air and, consequently, of material losses caused by aspiration. That points to a multi-factor relationship among these variables and material properties as well as to the structural design of handling facilities.

Noticeably, limestone losses are lower than those of bentonite. Three primary factors are key determinants of losses: flow rate of material, free-fall height, and humidity.

Special surveys using bins #31 and #32 (see Tables 5.11 and 5.12) indicate that bentonite losses exceed chalkstone losses by a factor of magnitude, with flow rate of transferred material being the primary determining variable.

Processing the experimental findings indicates the possibility of evaluating losses (in%) using an empirical relationship

$$\frac{q_a}{G} = BG^{-a}H_b^c\exp(-\alpha\omega), \tag{5.159}$$

where B, a, c, and α are parameters that, in the case of bentonite, are correspondingly equal to: $B = 0.347$; $a = 0.822$; $c = 0.0977$; and $\alpha = 0.176$; and, for limestone: $B = 0.037$; $a = 0.398$; $c = 1.0859$; and $\alpha = 4.46$.

TABLE 5.10
Dust Concentration in Air Aspirated from Bins with Bentonite and Limestone*

Aspirated Assembly	Aspiration Volume (m^3/s)	Dust Concentration in Air (g/m^3)	Dust Flow Rate (g/s)	Mass Flow Rate of Material (kg/s)	Specific Dust Carryover Rate (g/kg)
Bentonite:					
Plate diverter at bin #31	0.363	1.78	0.64	0.086	7.6
Plate diverter at bin #31	0.295	2.27	0.67	0.086	7.9
Plate diverter at bin #31	0.121	0.84	0.102	0.086	1.2
Plate diverter at bin #30	0.08	1.43	0.115	0.83	1.4
Plate diverter at bin #29	0.076	4.14	0.315	0.83	3.8
Plate diverter at bin #28	0.17	2.15	0.366	0.83	0.44
Bin #31	0.976	0.63	0.615	0.14	4.3
Bin #31	0.924	0.12	0.111	0.14	0.8
Bin #30	1.36	0.38	0.517	0.83	0.62
Bin #29	1.39	2.31	3.21	0.83	3.9
Bin #28	1.3	0.87	1.13	0.83	1.4
Limestone:					
Plate diverter at bin #31	0.192	2.56	0.49	0.61	0.8
Plate diverter at bin #31	0.138	2.95	0.41	0.61	0.67
Plate diverter at bin #31	0.186	1.18	0.22	0.61	0.4
Plate diverter at bin #30	0.12	1.31	0.157	1.66	0.1
Plate diverter at bin #29	0.134	0.78	0.105	1.66	0.06
Plate diverter at bin #28	0.15	0.52	0.078	1.66	0.05
Bin #31	1.34	1.63	2.184	2.05	1.1
Bin #31	1.22	1.28	1.56	2.05	0.8
Bin #31	0.61	0.33	0.202	2.05	0.1
Bin #31	0.208	0.19	0.04	2.05	0.02
Bin #19	0.34	1.97	0.67	1.66	0.41
Bin #16	0.41	4.3	1.76	2.7	0.6
Bin #15	0.33	1.7	0.551	2.7	0.2
Bin #14	0.34	1.05	0.357	2.7	0.13
Bin #13	0.5	0.44	0.22	2.7	0.08
Bin #7	0.24	0.62	0.149	1.66	0.09

Bins located at Sokolovsko-Sarbaisky ore beneficiation plant pelletization factory.

Material losses in aspirated bins increase with greater material flow rates and falling heights as well as lower moisture content. Maximum measured losses reach 4% of flow rate for bentonite and 0.45% for limestone.

Crushed materials are normally dumped from storage bins onto a layer of wet concentrate. In this case, levitation of the dumped bentonite/limestone layer during its further transportation can be successfully prevented by using a simple technique

TABLE 5.11
Dust Release for an Aspirated Limestone Bin

Limestone Flow Rate (kg/s)	Fall Height (m)	Moisture Content of Limestone (%)	Flow Rate of Evacuated Air (m^3/s)	Airborne Dust Content (g/m^3)	Specific Dust Carryover Rate (g/kg)
1.50	2.5	0.14	0.262	5.33	0.93
1.56	3.3	0.16	0.217	2.55	0.35
0.86	3.0	0.16	0.217	2.19	0.55
0.45	2.5	0.23	0.260	0.80	0.46
0.28	2.5	0.23	0.260	1.46	1.36
0.88	2.5	0.23	0.260	1.69	0.50
1.00	3.0	0.08	0.260	2.48	0.64
1.55	2.7	0.05	0.260	1.63	0.27
0.37	2.5	0.05	0.260	1.20	0.84
0.90	2.2	0.05	0.260	0.83	0.24
1.13	2.2	0.05	0.260	0.94	0.22
0.47	1.1	0.05	0.260	0.68	0.38
0.44	4.0	0.12	0.211	1.40	0.67
0.24	3.6	0.12	0.211	2.22	1.95
1.45	4.0	0.12	0.211	4.67	0.68
1.00	1.5	0.07	0.278	1.83	0.51
0.47	1.4	0.07	0.278	2.28	1.35
1.00	2.2	0.17	0.200	2.22	0.44
2.00	1.8	0.17	0.200	4.93	0.49
1.33	1.4	0.17	0.200	3.37	0.51
0.07	3.2	0.07	0.289	1.04	4.55
0.66	2.6	0.07	0.289	1.78	0.78
2.00	2.3	0.07	0.289	4.27	0.62
4.00	2.0	0.07	0.289	4.50	0.33
0.57	1.3	0.07	0.289	3.02	1.53
0.57	2.0	0.20	0.314	0.09	0.05
1.00	1.3	0.20	0.314	0.10	0.03
0.88	2.0	0.20	0.314	0.01	0.004
2.00	2.7	0.20	0.267	7.98	1.06
0.80	2.5	0.20	0.267	4.59	1.53
0.36	2.5	0.20	0.267	3.34	2.48
1.23	3.0	0.23	0.275	2.74	0.61
2.66	2.3	0.23	0.275	3.21	0.33
0.88	2.1	0.23	0.275	1.40	0.44
0.57	2.5	0.23	0.275	0.92	0.44
0.80	2.5	0.11	0.272	2.13	0.72
2.86	1.3	0.11	0.272	4.53	0.46
0.26	1.5	0.11	0.272	0.68	0.71
0.66	2.8	0.09	0.289	3.90	1.70
3.8	2.4	0.09	0.289	5.39	0.41
2.66	2.6	0.10	0.289	10.96	1.19
2.00	2.0	0.10	0.161	4.44	0.36

TABLE 5.12
Dust Release for Aspirated Bentonite Bin

Bentonite Flow Rate (kg/s)	Fall Height (m)	Moisture Content of Bentonite (%)	Flow Rate of Evacuated Air (m^3/s)	Airborne Dust Content (g/m^3)	Specific Dust Carryover Rate (g/kg)
0.125	3.0	1.5	0.185	3.77	5.60
0.060	2.0	7.1	0.180	3.26	9.78
0.230	2.1	7.1	0.180	5.62	4.40
0.260	3.9	2.1	0.180	28.56	19.77
0.052	3.5	2.1	0.180	11.81	40.90
0.610	3.0	2.1	0.180	13.33	3.93
0.220	3.2	1.4	0.180	7.57	6.19
0.400	3.0	1.4	0.180	9.80	4.41
0.480	2.7	1.4	0.180	14.13	5.30
0.068	3.4	3.4	0.180	9.19	24.30
0.395	2.9	3.4	0.180	15.92	7.25
0.280	2.8	3.4	0.180	11.37	7.30
0.074	2.8	3.7	0.180	9.13	22.20
0.242	2.6	3.7	0.180	16.02	11.9
0.410	2.4	3.7	0.180	15.06	6.60
4.000	2.1	4.2	0.228	1.53	0.09
0.660	2.6	4.2	0.228	1.20	0.40
0.100	2.6	4.2	0.228	1.51	3.44
0.180	1.9	2.3	0.153	5.27	4.50
0.800	2.0	2.3	0.153	24.03	4.60
0.280	2.3	2.3	0.153	7.69	4.20
0.400	2.5	2.3	0.153	25.31	9.70
1.000	2.3	2.7	0.147	25.14	3.70
0.720	2.5	2.7	0.147	28.18	5.80
0.360	2.7	2.7	0.147	17.82	7.29
0.500	2.3	4.7	0.150	8.01	2.40
1.000	2.4	4.7	0.150	13.80	2.10
0.400	2.6	4.7	0.150	7.30	2.70
0.440	2.7	4.7	0.150	5.62	1.90
0.800	2.8	4.7	0.150	9.75	1.80
0.330	3.0	4.7	0.150	8.88	4.00
0.066	1.1	1.7	0.150	6.07	13.80
0.500	1.3	1.7	0.150	20.14	6.00
0.470	1.4	1.7	0.150	21.02	6.70
0.160	1.5	1.7	0.150	16.67	15.60
1.600	1.9	3.0	0.161	23.10	2.30
0.195	2.5	3.0	0.161	19.98	16.50
0.330	2.5	3.0	0.161	10.99	5.36
0.610	1.9	5.1	0.161	14.85	3.90
0.420	2.0	5.1	0.161	16.05	6.20
0.530	2.2	5.1	0.158	12.33	3.70
2.220	2.3	5.1	0.158	16.64	1.20

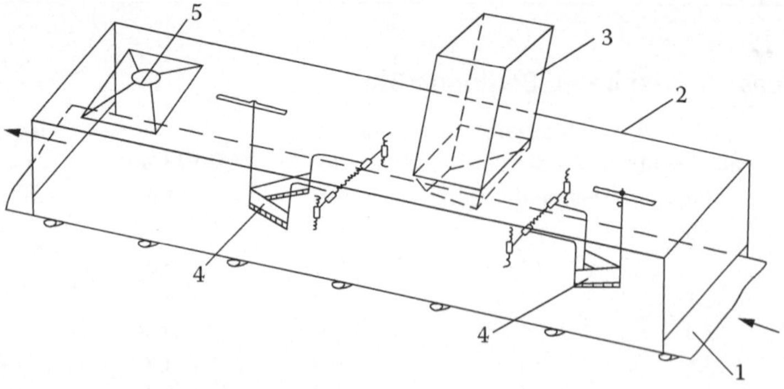

FIGURE 5.26 Layout of filtering elements inside a belt conveyor cowl: 1 = belt conveyor; 2 = aspiration cowl; 3 = loading chute; 4 = shaping element; 5 = aspiration funnel.

of placing a layer of dust-generating material inside a layer of moist concentrate. A device developed by us for that purpose [152] is placed inside an aspirated cowl at the belt conveyor loading location. It consists of two shaping elements installed upstream and downstream of the charging chute (Figure 5.26).

A shaping element (Figure 5.27) consists of two plates (6) reinforced from below with rubber-impregnated tape (7). The element is provided with nuts (14) for adjusting furrow width and submersion depth and springs (15) to maintain structural rigidity.

The designed device was tested in an industrial environment at the charge preparation house of the pelletization factory at Lebedinsky OBP. The device was housed inside a KB-7 belt conveyor cowl served by an ATU-4 aspiration unit. The latter ensures dust removal where the crushed bentonite unloaded from bins #1–#3 to the belt conveyor carrying concentrate with a moisture content of 9–12%.

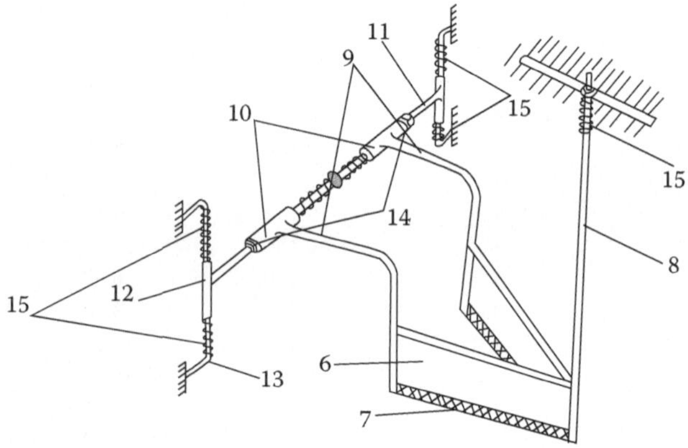

FIGURE 5.27 General structure of shaping element: 6 = plate; 7 = rubber-impregnated tape; 8 = vertical rod; 9 = bracket; 10 = horizontal hub; 11 = horizontal rod; 12 = vertical hub; 13 = vertical stand; 14 = adjusting nut; 15 = spring.

Surveys indicate that the dust removal efficiency of a TsVP #8 flushing cyclone measured 94.8% in the absence of filter elements, and the amount of material carried over into aspiration network reached 9.8 kg/hr. Airborne dust content at the KB-7 conveyor operator's workplace measured 6.7 mg/m^3.

After filter elements were installed, dust removal efficiency rose to 96.2% while the material carryover rate fell to 8.9 kg/hr. Dust content at workplaces decreased to allowed exposure limits and measured 3.2 mg/m^3.

Thus, placing a layer of powdery material atop a layer of wet concentrate prevents dust levitation from the surface of the material being transported without requiring slower conveyor belt speeds or the installation of a conveyor cowl throughout the entire length of the processing train.

5.2.3 Techniques for Intensifying Inertial Dust Precipitation in Aspirating Cowls

Based on the method used for intensifying dust precipitation, cowls can be divided into two groups. Cowls relying on water or foam to settle down dust will be referred to as wet-type deduster cowls, and those where dust precipitation is promoted by forces of inertial or electromagnetic nature without application of water will be referred to as dry deduster cowls.

Process constraints applicable to materials being handled restrict application of the former group of cowls.

Dry deduster cowls need further development in the areas of intensifying dust precipitation processes and in devising simpler and more reliable structural designs.

5.2.3.1 Inertial Trap Using a Plate Grid inside Cowl

Consider the simplest technique for dust precipitation in a cowl using a curtain of magnetized plates suspended vertically between the chute and the aspiration flange (Figure 5.28) [152]. Let us estimate the magnitude of purely inertial dust precipitation

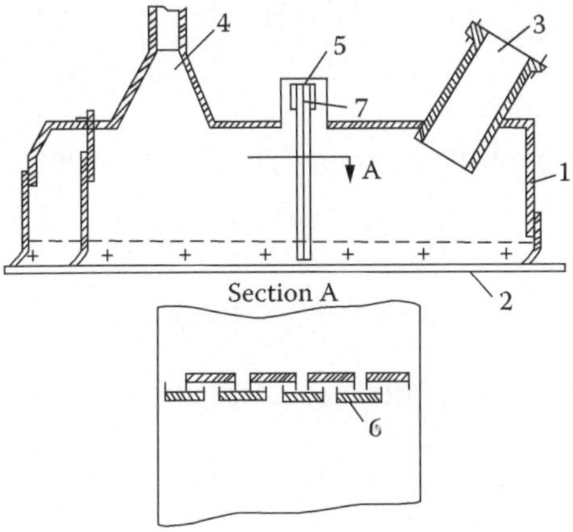

FIGURE 5.28 Aspiration cowl at conveyor loading location: 1 = cowl; 2 = conveyor belt; 3 = chute; 4 = aspiration flange; 5 = curtain; 6 = curtain plates with magnets; 7 = pivot.

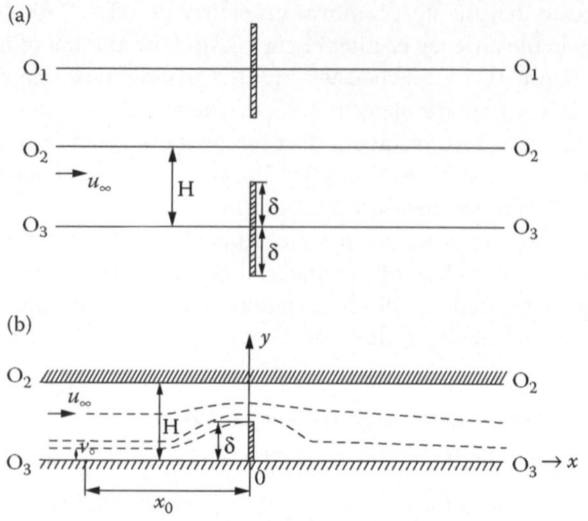

FIGURE 5.29 Velocity field of a potential flow around a grid of plates a = grid in plan; b = grid element in the XOY coordinate system.

on a single-row lattice installed at a right angle to the dust-laden flow. Then, consider a planar problem with a continuous flow pattern around particles and a potential flow rate.

We determined the velocity of the flow around the plates (Figure 5.29) using the conformal transformation method [154,155] characterized by the following combined equations

$$\bar{u}_x = \frac{\sqrt{2}}{2} \times \frac{\sqrt[4]{a^2(1-m)^2 + \Pi^2}}{\sqrt{a^2+b^2}} \sqrt{1 + \frac{\Pi}{\sqrt{\Pi^2 + a^2(1-m)^2}}}; \tag{5.160}$$

$$\bar{u}_y = -\frac{\sqrt{2}}{2} \times \frac{\sqrt[4]{a^2(1-m)^2 + \Pi^2}}{\sqrt{a^2+b^2}} \sqrt{1 - \frac{\Pi}{\sqrt{\Pi^2 + a^2(1-m)^2}}}, \tag{5.161}$$

with the following assignments for simplifying the notation:

$$\Pi = a^2 + b^2 - (1-m)b; \tag{5.162}$$

$$a = sh(\pi x/H) \times \sin(\pi y/H); \tag{5.163}$$

$$b = ch(\pi x/H) \times \cos(\pi y/H) - m; \tag{5.164}$$

$$m = \cos(\pi\delta/H); \tag{5.165}$$

$$\bar{u}_y = u_y / u_\infty; \; \bar{u}_x = u_x / u_\infty. \tag{5.166}$$

The trapping (inertial precipitation) coefficient for a grid of plates was determined using the classical method (i.e., by tracing critical paths of particles). Particle paths for a given air velocity field were traced by numerically solving combined differential equations describing a particle moving under the effect of Stokes force:

$$\left.\begin{aligned} &k\,{d\bar{v}_x}\big/{d\bar{t}} + \bar{v}_x = \bar{u}_x;\\ &k\,{d\bar{v}_y}\big/{d\bar{t}} + \bar{v}_y = \bar{u}_y;\\ &{d\bar{x}}\big/{d\bar{t}} = \bar{v}_x;\ {d\bar{y}}\big/{d\bar{t}} = \bar{v}_y, \end{aligned}\right\} \tag{5.167}$$

where $\bar{v}_x = v_x / u_\infty$; $\bar{v}_y = v_y / u_\infty$ are components of particle velocity; $\bar{u}_x = u_x / u_\infty$ $\bar{u}_y = u_y / u_\infty$ are air velocity components described by Equations 5.160 and 5.161; $\bar{t} = tu_\infty /H$ is dimensionless time; k is Stokes number, equal to

$$k = \frac{\rho_1 d^2 u_\infty}{\rho_2 18 \nu H} \tag{5.168}$$

where ρ_1, ρ_2 are particle and air densities (kg/m^3); d is equivalent particle diameter (m); ν is the kinematic viscosity factor for air (m^2/s); H is half-step of plates in grid (m); and u_∞ is the velocity of undisturbed air (m/s).

Let us determine the path of a particle passing through a point with coordinates (ε_1; $\delta - \varepsilon_2$) at initial conditions

$$\bar{x}\big|_{t=0} = 4;\ \bar{y}\big|_{t=0} = \bar{y}_0;\ \bar{v}_x\big|_{t=0} = -1 + \varepsilon_3;\ \bar{v}_y\big|_{t=0} = \varepsilon_4, \tag{5.169}$$

where $\varepsilon_1 > 0$; $\varepsilon_2 > 0$ $\varepsilon_3 > 0$; $\varepsilon_4 > 0$ are sufficiently small quantities.

The problem formulated was solved using the Kutta–Merson method with automatically chosen integration step for solving the Cauchy problem, and the boundary value problem was solved using the false position method. After the value of y_0 is determined for given k and δ, plate trap coefficient can be found as

$$\varepsilon_0 = y_0/\delta, \tag{5.170}$$

and the numerical value of y_0/H would yield the efficiency of inertial dust precipitation on the given grid. As evident from calculations (Tables 5.13 and 5.14), the trap coefficient of plates in the grid is high enough at larger Stokes numbers. However, it would be sensible to expect much less efficient dust deposition in such a grid because the majority of dust particles would ricochet against its plates. Moreover, airflow in real-world aspiration cowls would be weaker than the potential value suggests. To combat ricocheting particles and to improve dust settling in experimental conditions, plates have been magnetized (by attaching flat magnets to them).

Surveys have been carried out on a single grid element (Figure 5.30) as well as with a dual-row magnetized grid placed into the aspiration cowl of a semi-industrial

TABLE 5.13
Plate Trap Coefficient ε_0

δ/H	Stokes Number *k*								
	0.5	1	2.5	5	10	15	20	50	100
0.1	0.89	0.89	0.89	0.89	0.89	0.89	0.89	0.89	0.89
0.2	0.76	0.88	0.92	0.94	0.94	0.94	0.94	0.94	0.94
0.3	0.71	0.83	0.92	0.95	0.95	0.96	0.96	0.96	0.96
0.4	0.70	0.82	0.92	0.95	0.96	0.97	0.97	0.97	0.97
0.5	0.67	0.83	0.93	0.95	0.97	0.98	0.98	0.98	0.98
0.6	0.68	0.83	0.93	0.95	0.97	0.98	0.98	0.98	0.98
0.7	0.71	0.85	0.94	0.96	0.98	0.98	0.98	0.98	0.98
0.8	–	0.88	0.93	0.97	0.98	0.99	0.99	0.99	0.99
0.9	–	0.93	0.97	0.98	0.99	0.99	0.99	0.99	0.99

plant. Dust-laden air was passed through the magnetic grid element at a flow rate $u_\infty = 7$ m/s, with a concentration of 500 mg/m^3 (of pellet dust $\rho_1 = 4{,}000$ kg/m^3) and the following particulate composition:

Particle size (μm)	< 1.4	1.4–4.2	4.2–9.8	9.8–15	15–30	30–45	45–105
Content by mass (%)	1.2	0.7	4.4	7.4	26.3	26	34

Average Stokes number (at median diameter $d_{50} = 40$ μm) $k = \dfrac{4000 \cdot 40^2 \cdot 10^{-12} \cdot 7}{1.2 \cdot 18 \cdot 15 \cdot 10^{-6} \cdot 0.13} = 1.05$.

For a single-row grid, according to Table 5.14, the purely inertial settling factor (at $\delta/H = 0.5$) would be $\eta_1 = 0.4$; for two rows installed in a series, it would become $\eta_2 \approx 1-(1-\eta_1)^2 = 0.64$. Experiments indicate $\eta_2 = 0.5$ for an element in a dual-row

TABLE 5.14
Dust Settling Efficiency Factor for a Single-Row Grid of Plates (y_0/H)

δ/*H*	Stokes Number *k*								
	0.5	1	2.5	5	10	15	20	50	100
0.1	0.09	0.09	0.09	0.09	0.09	0.09	0.09	0.09	0.09
0.2	0.15	0.17	0.18	0.19	0.19	0.19	0.19	0.19	0.19
0.3	0.21	0.25	0.27	0.28	0.28	0.29	0.29	0.29	0.29
0.4	0.27	0.33	0.37	0.38	0.38	0.39	0.39	0.39	0.39
0.5	0.34	0.40	0.46	0.47	0.48	0.49	0.49	0.49	0.49
0.6	0.39	0.50	0.56	0.57	0.58	0.59	0.59	0.59	0.59
0.7	0.50	0.60	0.65	0.67	0.68	0.69	0.69	0.69	0.69
0.8	–	0.70	0.74	0.77	0.78	0.79	0.79	0.79	0.79
0.9	–	0.84	0.86	0.88	0.88	0.89	0.89	0.89	0.89

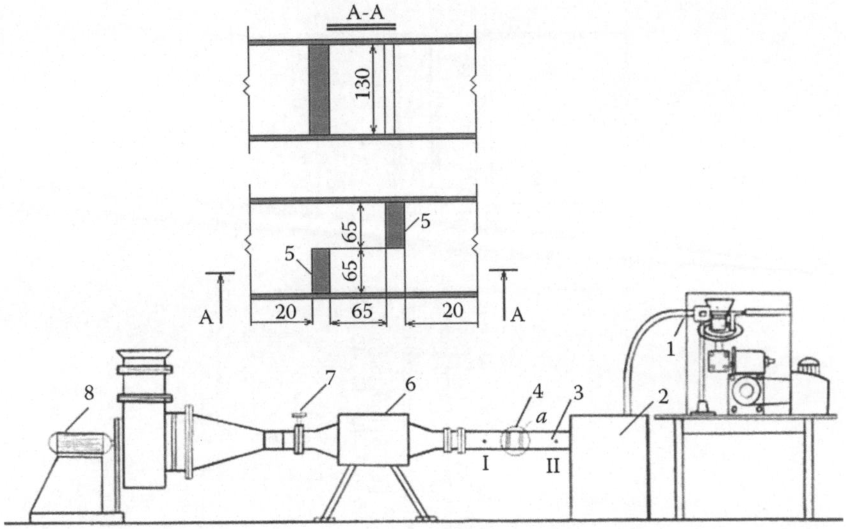

FIGURE 5.30 Layout of installation for studying dust precipitation on a magnetic grid element: 1 = dust feeder; 2 = dust chamber; 3 = duct (130 × 130); 4 = grid element; 5 = permanent magnets; 6 = static-pressure chamber; 7 = gate; 8 = fan; I, II = measurement points.

magnetic grid (with the local resistance coefficient of the grid equaling $\zeta = 10$), that is, somewhat below the theoretic value. The same result was obtained for fired pellet handling ($G = 10$ t/hr) at an experimental production plant. The grid (Figure 5.31) was installed inside the cowl, between the chute and the aspiration flange. In this case, airborne dust content inside the cowl measured 500 mg/m^3 upstream of the grid and 270 mg/m^3 downstream. The resistance coefficient equaled $\zeta_p = 14$.

Thus, rather simple devices can be used to reduce dust content by a factor of two in air evacuated from a cowl. The magnetic grid would become clogged with dust particles and fine fragments of transported material during operation. The transported material would only be able to regenerate the lower part of the grid; therefore, its resistance may rise so much that excess pressure would build at the grid inlet, causing dust-laden air to escape to the outside.

For that reason, the grid should be placed around the chute. Its dust contamination would become a boon in this layout because the amount of air injected through the chute would be reduced, and the chute would be cleaned of dust more thoroughly.

Now consider operation of this kind of device using the example of a conveyor with noticeable ferromagnetic properties filling with fired pellets. Figure 5.32 illustrates one possible layout of magnetic grids. The device comprises a cartridge-type magnetic shoe (1) made of nonmagnetic material (aluminum). Cartridges are attached to the walls of the chute (2) like pockets, extending the chute inside the aspiration cowl (3). Permanent magnet assemblies (4) made of barium ferrite are placed inside the pockets. Similar assemblies are used in magnetic separators at beneficiation plants. Assemblies are placed to form a closed magnetic system along the perimeter of the chute. There

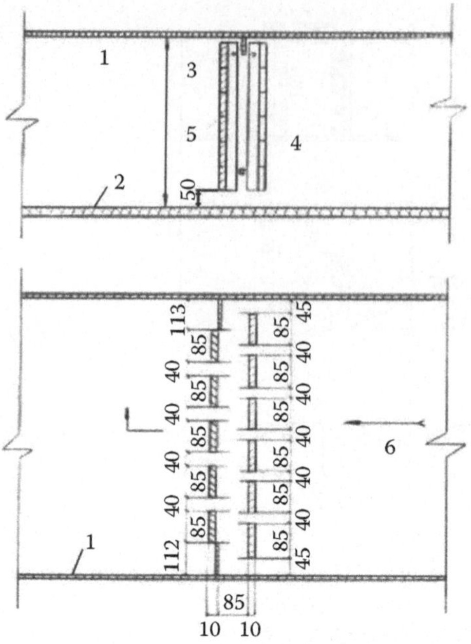

FIGURE 5.31 Magnetic grid inside aspiration cowl: 1 = cowl walls; 2 = conveyor belt; 3 = pivot; 4 = grid; 5 = magnetic element; 6 = direction of dust-laden air and conveyor belt motion.

is a 50-mm gap between the shoe and the conveyor belt. This gap was sized to enable transport of the handled material by the conveyor within the range of flow rates from 0.8 to 24 kg/s. Some material is trapped by magnets, forming a constantly renewed band. The band is 110 mm wide—enough to significantly arrest dust knockout through its own leaky spots. Field strength at 50 mm away from the belt measured 12,000 A/m.

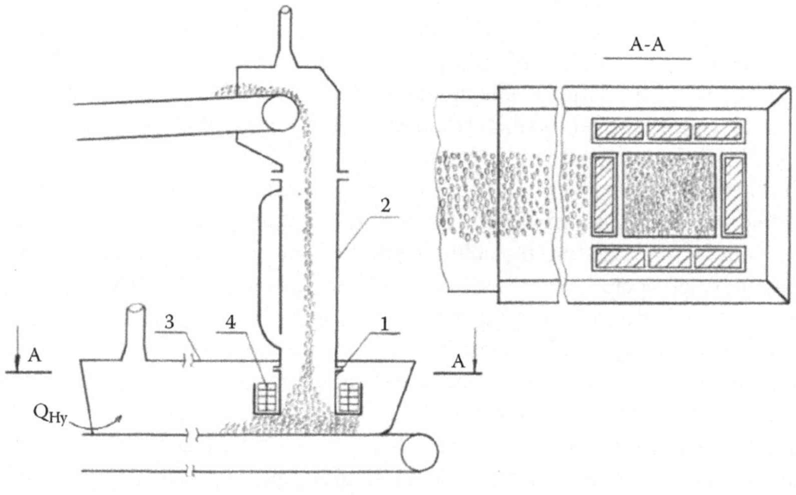

FIGURE 5.32 Magnetic shoe construction.

The design of the magnetic chute shoe described here was tested using a handling facility at a VNIIBTG production testing installation. To provide a basis for comparison, surveys without a magnetic chute (including aerodynamic surveys and dust sample collection) were previously performed at the same installation.

Survey findings are listed in Table 5.15. As the data indicates, the specific volume of aspiration (Figure 5.33) is four times lower for a handling facility equipped with a magnetic shoe. Flow rate of recirculated air in the bypass duct reaches the performance of the local suction unit. A significant reduction in evacuated air volume is attained due to the resistance that the layer of material trapped by the magnetic side of the shoe puts up to the flow of injected air. This resistance in experiments reached 120 Pa at a material flow rate of 14.6 kg/s (52.5 t/hr). Aspiration volume is also brought down due to the reduction of optimum negative pressure; for the surveyed facility, this pressure equaled 3 Pa with the magnetic shoe and 12 Pa without it (i.e., this effect is due to a reduction of the inflow of air through leaky joints in the cowl).

A reduction of the total volume of evacuated air, coupled with a significant decrease of the initial concentration, noticeably restricts dust carryover into the aspiration network (Figure 5.34).

The efficiency of the magnetic shoe for reducing the specific dust carryover rate exceeds 90%. Initial concentration of dust in aspirated air was reduced by an order of magnitude: from 3,000 mg/m^3 to 300 mg/m^3 at smaller flow rates (around 3 kg/s) and from 11 g/m^3 to 1 g/m^3 at flow rates averaging 8–10 kg/s.

Thus, the use of a magnetic shoe for belt conveyor loading has the potential of significantly reducing the required capacity of dedusting systems and of bringing down dust load on the dust trap.

TABLE 5.15
Summary of Findings from Studies of Dust Carryover in Pellet Handing

	Air Volume (m^3/s)				Dust Carryover from Cowl	
Material Flow Rate, G_M (kg/s)	Evacuated from the Cowl	Circulating in Bypass Duct	Optimum Negative Pressure inside Cowl (Pa)	Dust Content in Evacuated Air (g/m^3)	Absolute (g/s)	Specific (g/kg)
Without magnetic shoe and bypass duct						
0.8	0.37	0	5	0.4	0.148	0.18
2.67	0.51	0	34	2.5	1.275	0.477
3.9	0.54	0	24	3.0	1.62	0.415
7.35	0.74	0	28	11.0	8.14	1.11
With magnetic shoe and bypass duct						
2.44	0.14	0.08	3	0.3	0.42	0.017
2.81	0.18	Data unavail.	2	0.28	0.05	0.018
8.93	0.18	Data unavail.	2	0.96	0.173	0.020
14.56	0.28	0.21	9	1.3	0.364	0.025
23.93*	0.24	0	4	0.7	0.168	0.007

* This flow rate is manifested as a coupled mode of material flow throughout the entire chute height.

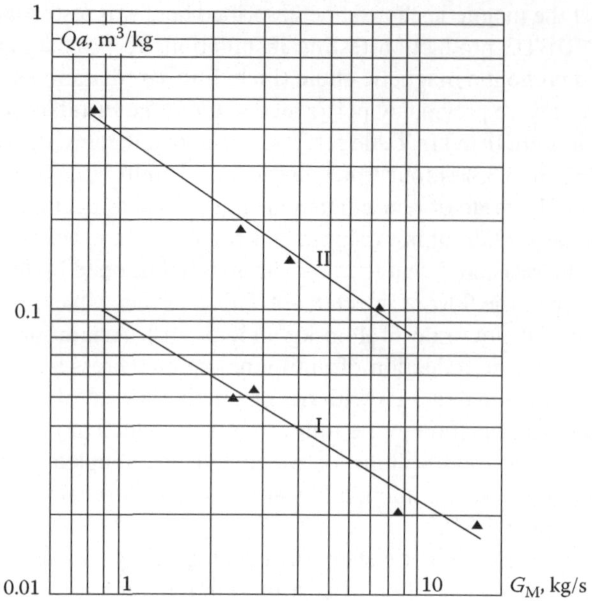

FIGURE 5.33 Specific aspiration volume as a function of material flow rate: (I) with magnetic shoe and bypass duct; (II) without magnetic shoe/bypass duct.

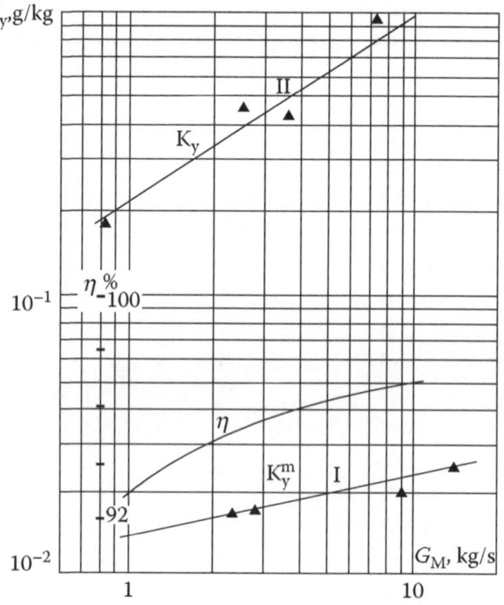

FIGURE 5.34 Changes in dust carryover rate in the local suction of a handling facility with increasing material throughput: (I) with magnetic shoe; (II) without magnetic shoe.

5.2.4. Reduction of Dust Concentration in Aspiration Funnels

5.2.4.1 Initial Dust Concentration as a Function of Air Velocity in Aspiration Funnels

Dust content in air coming into the aspiration network from the cowl is determined by many factors, including physical and mechanical properties of the material being handled, cowl design, aspiration funnel location, and air velocity in the intake section of the funnel. Many authors [59,66] recommend limiting the suction velocity. Reference standards [18,159] state particular suction velocity values depending on the coarseness of the material being handled. However, there is no substantiation for recommended suction velocities in the literature because no research has been performed so far that quantitatively estimates the relationship between suction velocities and carryover intensities.

Dust carryover upstream of the aspiration cowl was examined in a laboratory rig design (Figure 5.35). Dust was fed into the aspirated cowl (1) by means of dust feeder (2) and compressor (3). A rigid partition wall (4) was installed to ensure an even field of dustiness within the local suction area. The cowl was sized so that inside air stream velocities were as small as possible. They varied from 0.08 to 0.6 m/s in the working chamber of the cowl. Velocity at suction opening of aspiration funnel (5) was adjusted using a gate valve (6). Aerodynamic measurements and dust samplings were performed immediately in the cowl and simultaneously in the suction duct. Two materials, bentonite and pellets, were chosen for dust particle studies. Their properties are listed in Table 5.16.

Surveys involved determination of dust release source intensity J, the amount of material carried away by local suction G_y, and air velocity in the intake section of the aspiration funnel. Results indicate that the relative carryover $\bar{G}_y = G_y/J$ is intensity-independent and is constant for a given suction velocity (Figure 5.36).

Material carryover intensified as suction velocity increased, more so for pellet dust (Figure 5.37b) than for powdered bentonite (Figure 5.37a). At suction velocities of 5–5.5 m/s, virtually all pellet dust arising in the cowl was carried over into the aspiration network.

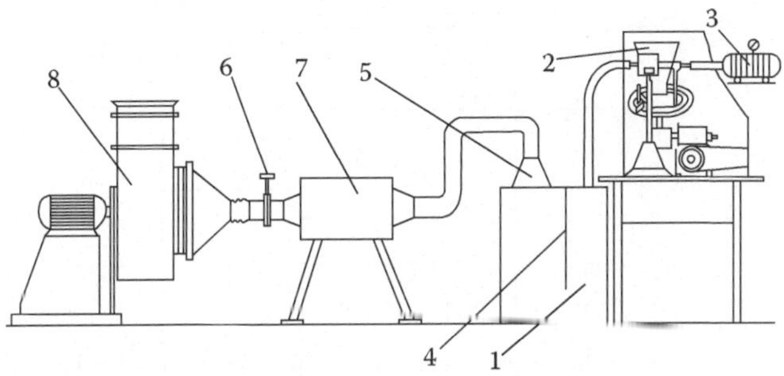

FIGURE 5.35 Laboratory rig design for examining carryover of dust-like material into aspiration network: 1 = aspirated cowl; 2 = dust feeder; 3 = compressor; 4 = rigid partition wall; 5 = aspiration funnel; 6 = gate valve; 7 = static-pressure chamber; 8 = centrifugal fan.

TABLE 5.16
Particulate Composition of Dust

Material	Density (kg/m³)	< 1.4	1.4–4.2	4.2–9.8	9.8–15	15–30	30–45	45–105	Average Particle Diameter (μm)
		Coarseness Interval (μm)							
		Fraction by Weight (%)							
Dust from bentonite	2370	1.33	1.32	5.46	16.47	31.33	2.03	42.01	42
Dust from pellets	4680	3.62	10.86	20.62	6.69	33.96	1.64	22.59	28

Less intense carryover of powdered bentonite can be explained primarily by its pronounced adhesive and autohesive properties that manifest themselves in the significant sticking of bentonite particles onto the interior surfaces of the cowl and aspiration funnel enclosures and onto the walls in the initial section of suction air duct, as well as in the coagulation of particles as they moved toward the aspiration funnel. In addition, pellet dust used in the experiment was finer than bentonite dust. In summary, an increased aspiration funnel intake opening area provides a simple yet somewhat efficient solution for reducing the carryover of material into the aspiration network.

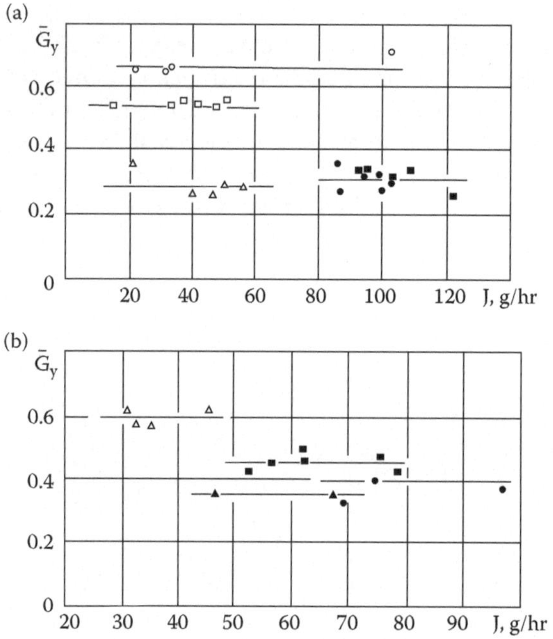

FIGURE 5.36 Dust carryover rate as a function of dust release intensity *j* (a) powdered bentonite; (b) pellet dust (at o - u_{in} = 6.63 m/s; ● - u_{in} = 3.08 m/s; □ - u_{in} = 4.96 m/s; ■ - u_{in} = 2.05 m/s; Δ - u_{in} = 4.03 m/s; ▲ - u_{in} = 0.92 m/s.

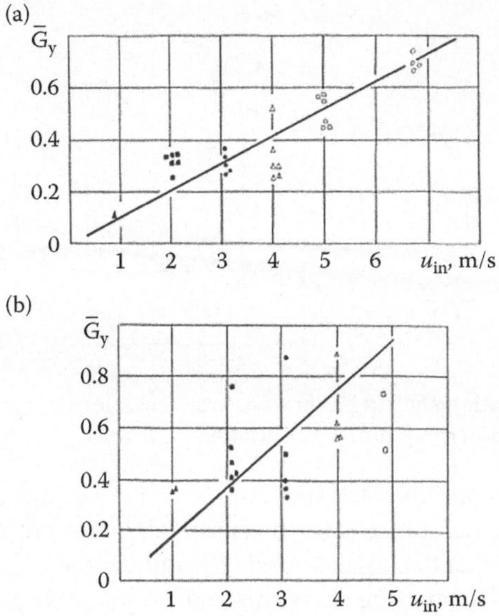

FIGURE 5.37 Dust carryover rate as a function of air velocity in the intake section of the aspiration funnel (a) powdered bentonite; (b) pellet dust (at o - u_{in} = 6.63 m/s; ● - u_{in} = 3.08 m/s; □ - u_{in} = 4.96 m/s; ■ - u_{in} = 2.05 m/s; Δ - u_{in} = 4.03 m/s; ▲ - u_{in} = 0.92 m/s).

5.2.4.2 Dust Precipitation in Dust Receivers/Separators

With regard to inertial precipitation of dust in dust receivers/separators, the following considerations guide the choice of dust receiver/separator design and optimum parameters:

- The aerodynamic layout of the separator should be simple although ensuring a constant resistance of the unit.
- Dust should be trapped dry, with sufficiently high separation efficiency.
- Means should be provided for continuously returning any precipitated material into the process.
- The use of a separator should not complicate the operation of the aspiration unit.

In accordance with these requirements, a separator was designed with a centrifugal separation area placed in a vertical plane (Figure 5.38).

The flow of dust-laden air with an operating separator comes into the contraction chamber (1), accelerating and proceeding into the expansion chamber (2) where, driven primarily by centrifugal force and gravity, dust particles are separated from the air stream. Separated dust then accumulates in the bin (3) and is returned into the operating conveyor train through a dust-release hole. The stream of air (along with any trapped dust) is removed through the aspiration air receiver (4) that is provided with a slot along the entire length of the flange confined within separator housing. A rigid partition wall (5) separates the contraction and expansion chamber.

In case of a potential rotating flow with a central outflow, streamlines are represented with logarithmic spirals [160,161]. The generatrix of the separator housing

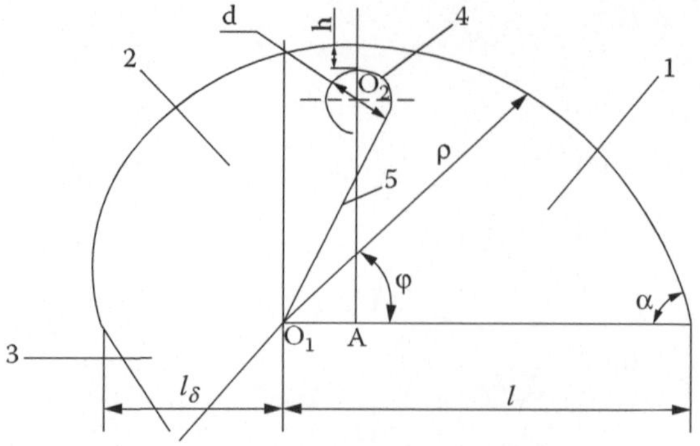

FIGURE 5.38 Illustration showing a dust receiver/separator sizing: 1 = contraction chamber; 2 = expansion chamber; 3 = bin; 4 = aspiration air receiver; 5 = rigid partition wall.

follows one of the streamlines for suppression of eddies and consequent improvement of aerodynamic resistance. This housing shape also promotes dust separation in the expansion chamber due to increased curvature of the generatrix along the flow direction of the dust-air mixture. A permanent magnetic field may be used for the same purpose in case of paramagnetic dust separation.

The following sequence is used for calculating the key variables of separator geometry:

1. Determine the required local suction capacity (per Section 5.1 of this chapter).
2. Determine the size of the separator inlet section; width b is assumed equal to the width of the aspiration cowl B_y; length l is calculated using the formula

$$l = Q_a / (B_y u_{in}),$$

   where Q_a is local suction performance (m^3/s); and u_{in} is average air velocity in a separator inlet section (m/s).
3. Plot the generatrix for the separator housing using the equation (the pole is provided by the point O_1)

$$\rho = l \cdot exp(-\varphi \cdot ctg\alpha), \tag{5.171}$$

   where ρ is the radius vector (m); φ is the polar angle (rad), ranging from 0 to π; and α is the angle between the generatrix and the separator inlet section plane (deg). Contraction chamber resistance is minimized at $\alpha < 75°$ [162].
4. Determine the height of the slot h at the dust-air mixture inlet into the expansion chamber

$$h = (u_{in} / v_s) \times l, \tag{5.172}$$

   where v_s is the average velocity of dust-air flow in the slot (accepted as equal to 15–22 m/s, similar to cyclones).

5. Determine the diameter d and width l_B of the aspiration flange

$$d = \sqrt{4Q_a / (\pi \cdot v_a)}, \tag{5.173}$$

$$l_B = 0.5d, \tag{5.174}$$

where v_a is the velocity of air inside air duct (m/s), accepted as equal to 15–20 m/s.

6. Locate the aspiration flange according to Figure 5.38. To do that, determine the distance O_1A:

$$O_1A = 0.186\ l. \tag{5.175}$$

7. Determine the dimensions of the bin inlet section; width b_δ is assumed to be equal to the width of the aspiration cowl B_y; length l_δ is calculated using the formula

$$l_\delta = 0.432\ l. \tag{5.176}$$

To judge the efficiency of centrifugal dust separation in a proposed dust receiver design, it is necessary to study the aerodynamics of the device and to determine airflow velocity fields for both separator chambers. Known equations [163] for tangential and radial velocity components (the equation describing the gyration of a free curl and the outflow equation) used to describe processes occurring in centrifugal cyclone dust traps are invalid for the proposed separator design because they were formulated for cases where air particles are moving along concentric circles with central outflow.

Due to the uniform-suction air receiver, the airflow inside the housing of a running separator is two-dimensional. Therefore, aerodynamic studies are performed using a flat separator model (Figure 5.39) with the following key parameters: $l = 0.38$ m; model width $b = 0.09$ m; $\alpha = 75°$; $h = 0.025$ m; $d = 0.065$ m; $l_B = 0.032$ m; $O_1A = 0.071$ m; and $l_\delta = 0.164$ m.

Experiments were performed with clean air (free of dust particles) with the assumption that the bin (1) for collected dust is submerged below the layer of material—its outlet was therefore plugged. Transparent plastic was used for face enclosures (2) to enable visual assessment. A gate valve was provided to adjust the volumetric flow rate of evacuated air.

A triple-channel cylindrical probe was used in experiments to determine velocity vector directions/modules and static pressures at fixed points. Measurement points were located along radius vectors. The angle between neighboring radii in all cases was equal to 15°. In addition, evacuated air volume and pressure losses in the separator were measured in every series of experiments. In addition to the total air velocity (V), tangential (V_T) and radial (V_r) velocity components were measured at each point.

Centrifugal forces present in any curved flow produce changes in the existing statistical pressure within the flow; pressure decreases from the periphery of the eddy to its axis as it happens, for example, in cyclones [164,165].

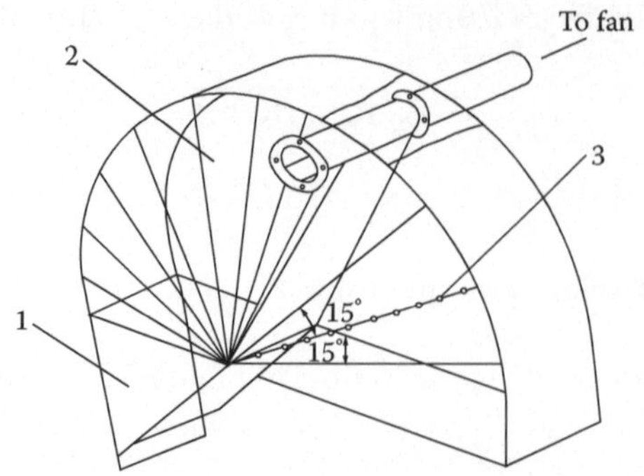

FIGURE 5.39 General layout of a dust receiver/separator: 1 = bin; 2 = face enclosure; 3 = probe insertion hole.

Obviously, large pressure gradients would mean that directions and magnitudes of velocity vectors measured using a triple-channel gauge would significantly differ from actual values.

Static pressure measurements (having negative values in our case) in every fixed point inside the separator housing are plotted in Figure 5.40. We observed (see Figure 5.40a) that negative pressure gradients are modest for every radius and tend to increase for radii with small polar angle values φ. Therefore, the greatest negative pressure gradient along the radius occurs at $\varphi = \pi/3$ with differential pressure between side probe holes (given an 8 mm diameter probe) reaching 6 Pa due to flow curvature. In all other cases, this differential pressure does not exceed 0.1–1 Pa and has virtually no impact on measurement results.

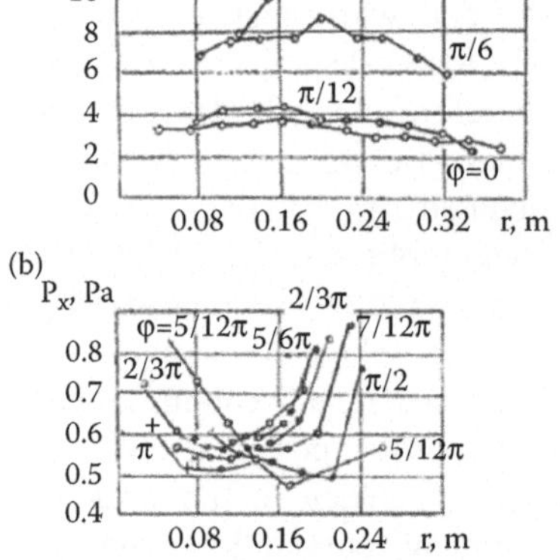

FIGURE 5.40 Distribution of negative pressure inside a dust receiver/separator enclosure: $Q_a = 0.048$ m³/s; (a) in contraction chamber; (b) in expansion chamber.

Measurements carried out beforehand indicate that the eddy core originates in the central part of the expansion chamber and has an oblong shape. A low-pressure zone coterminous with the eddy core is prominent inside this chamber (Figure 5.40b).

Pressure gradients within this zone are insignificant, but they increase markedly in peripheral areas with differential pressure between side probe holes reaching considerable values (50–70 Pa). To account for that, velocity vector directions in the expansion chamber were adjusted visually using a thread, and negative pressures in side probe holes were averaged when determining dynamic pressures.

Averaged results for multiple experiments performed at a constant air volumetric flow rate of air suction (Q_a = 0.048 m^3/s) are listed in Table 5.17. In this case, u_{in} = 1.41 m/s; v_s = 21.52 m/s; and v_a = 14.67 m/s. The value V_T is considered positive in a counterclockwise direction from the current radius and negative in the opposite direction. The value V_r is considered positive when directed away from the pole and negative when directed toward the pole.

A review of findings has revealed a variety of radius-wise changes in tangential and radial flow velocity components for contraction and expansion chambers. Therefore, experimental data for contraction and expansion chambers were treated separately for statistical processing. Most of the focus has been on the straight-flow area of the expansion chamber and process of dust precipitation from the air stream. The following empirical relations were found by processing experimental data for the contraction chamber ($0 \le \varphi \le 1.309$; $0 < S/l < 1.14$):

$$\frac{V_T}{V_T^{max}} = l^{-A((r-\rho_{max})/\rho)^2}; \tag{5.177}$$

$$V_r \Big/ V_r^{\max} = b((r/\rho)-1)+1 \tag{5.178}$$

where

$$A = 15l^{-(1.309-\varphi)}; \tag{5.179}$$

$$\rho_{max} = \rho l^{0.7(1.309-\varphi)}; \tag{5.180}$$

$$V_T^{\max} = u_{\text{in}} \frac{12+100(1.14-(S/l))^{2.44}}{1+97.5(1.14-(S/l))^{2.44}}; \tag{5.181}$$

$$b = 1.6((1.309-\varphi+0.03)/0.03)l^{-3(1.309-\varphi)}; \tag{5.182}$$

$$V_r^{\max} = -u_{\text{in}} \frac{4.1292-0.644(1.14-(S/l))}{1+6(1.14-(S/l))^{0.43}}; \tag{5.183}$$

and the straight-flow area inside the expansion chamber ($1.309 \le \varphi \le 3.14$; $1.14 < S/l < 2.2$)

$$\frac{V_T}{V_T^{\max}} = l^{-15((r/\rho)-1)^2}, \tag{5.184}$$

$$\frac{V_r}{V_r^{\max}} = 1.6((r/\rho)-1)+1, \tag{5.185}$$

TABLE 5.17
Measured Velocity Field Values in Dust Receiver/Separator

Current Radius r (mm)	Air Velocity (m/s) Total V	Tangential V_T	Radial V_r	Current Radius r (mm)	Air Velocity (m/s) Total V	Tangential V_T	Radial V_r
$\varphi = 0$ $\rho = 0.381$ m				$\varphi = 5\pi/12$ (75°) $\rho = 0.268$ m			
41	1.30	0.85	0.98	48	16.38	−0.57	16.37
71	1.42	1.31	0.55	78	14.88	−0.52	14.87
101	1.42	1.33	0.49	138	11.87	−1.65	11.75
131	1.59	1.51	0.49	168	10.03	−2.43	9.73
161	1.64	1.64	0	258	16.58	15.77	−5.12
191	1.59	1.59	0	$\varphi = \pi/2$ (90°) $\rho = 0.250$ m			
251	1.48	1.48	0	90	11.17	−4.18	10.36
281	1.30	1.16	−0.59	120	8.97	−3.36	8.32
311	1.16	0.98	−0.61	210	8.02	7.44	−3.00
341	0.92	0.72	−0.57	240	17.57	15.92	−7.43
371	0.58	0.44	−0.37	$\varphi = 7\pi/12$ (105°) $\rho = 0.233$ m			
$\varphi = \pi/12$ (15°) $\rho = 0.355$ m				73	12.31	−8.24	9.25
75	1.74	1.71	0.3	103	7.51	−5.12	5.49
105	1.74	1.54	0.82	163	6.85	4.93	−4.76
135	1.83	1.78	0.41	193	10.03	8.77	−4.86
165	1.74	1.72	0.24	64	7.60	−5.28	−5.28
196	1.69	1.69	0	94	8.35	−1.88	−8.14
255	1.59	1.54	0.4	$\varphi = 2\pi/3$ (120°) $\rho = 0.217$ m			
285	1.30	1.20	−0.51				
315	1.53	0.98	−1.17	27	14.09	−11.10	8.67
345	1.00	0.54	−0.839	57	12.58	−10.43	7.08
$\varphi = \pi/6$ (30°) $\rho = 0.331$ m				87	8.98	−6.67	6.00
81	1.64	0.84	1.41	147	7.04	4.98	−4.96
111	1.64	0.96	1.33	177	10.29	8.38	−5.87
171	1.59	1.45	0.65	207	17.57	15.06	−9.05
201	1.36	1.36	0	$\varphi = 3\pi/4$ (135°) $\rho = 0.203$ m			
231	1.53	1.53	0	133	7.14	5.99	−3.89
261	1.83	1.83	−0.10	163	11.17	8.56	−7.18
291	1.64	1.46	−0.74	193	17.57	15.92	−7.43
321	1.48	1.31	−0.69	$\varphi = 5\pi/6$ (150°) $\rho = 0.189$ m			
$\varphi = \pi/4$ (45°) $\rho = 0.308$ m				59	9.19	−9.19	0
88	1.64	0.45	1.58	89	5.55	−4.76	−2.86
118	1.83	0.83	1.63	119	4.78	1.79	−4.43
148	2.01	1.07	1.70	149	10.99	8.89	−6.46
178	2.17	1.56	1.51	179	17.57	13.65	−11.06
208	2.59	2.10	1.52	$\varphi = 11\pi/12$ (165°) $\rho = 0.176$ m			
268	2.59	2.53	−0.54	76	6.85	−5.25	−4.40
298	2.59	2.46	−0.80	106	5.79	−2.98	−4.96

TABLE 5.17 (Continued)
Measured Velocity Field Values in Dust Receiver/Separator

Current Radius r (mm)	Air Velocity (m/s) Total V	Tangential V_T	Radial V_r	Current Radius r (mm)	Air Velocity (m/s) Total V	Tangential V_T	Radial V_r
$\varphi = \pi/3$ (60°) $\rho = 0.288$ m				136	9.34	5.88	−7.26
				166	16.79	12.86	−10.79
188	3.06	0.69	2.98	$\varphi = \pi$ (180°) $\rho = 0.164$ m			
218	5.67	3.79	4.21	34	12.95	−11.10	−6.67
248	8.02	6.41	4.83	124	10.36	5.79	−8.59
278	6.85	6.77	−1.07	154	15.32	13.13	−7.89

where

$$V_T^{\max} = u_{\text{in}}[12 - 2.5((S/l) - 1.14)^2], \tag{5.186}$$

$$V_r^{\max} = -u_{\text{in}}[7.5 - 3((S/l) - 2.2)^2]. \tag{5.187}$$

The following designations are used in the accepted relations: maximum tangential and radial components of air velocity for each area, $V_T^{\max}$, $V_r^{\max}$, respectively; logarithmic spiral arc length, S, described by the formula

$$S = l\frac{\sqrt{1 + ctg^2\alpha}}{ctg\alpha}(1 - l^{-\varphi ctg\alpha}). \tag{5.188}$$

Mechanical dust particle paths (and their degrees of coarseness) were determined by numerically integrating particle dynamics in this field. Some paths are illustrated on Figure 5.41. Particles travel in spiral-shaped paths and can reach the separator generatrix, or they can become deflected from it toward the center of the expansion

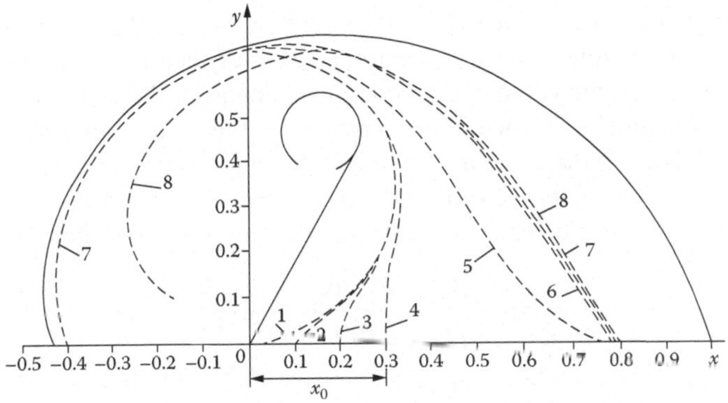

1 – κ = 10; $x_0 = 0.05$; 4 – κ = 10; $x_0 = 0.3$; 7 – κ = 100; $x_0 = 0.8$;
2 – κ = 10; $x_0 = 0.1$; 5 – κ = 10; $x_0 = 0.8$; 8 – κ = 1000; $x_0 = 0.8$;
3 – κ = 10; $x_0 = 0.2$; 6 – κ = 50; $x_0 = 0.8$;

FIGURE 5.41 Particle flow paths in a dust receiver/separator.

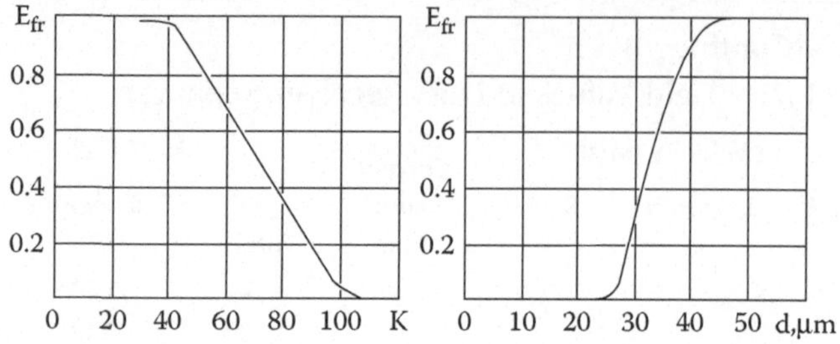

FIGURE 5.42 Calculated per-grade efficiency of a dust receiver/separator.

chamber as they near the chamber outlet. Precipitation rate E_{fr} for a given fraction of dust in the separator was determined by finding critical particle flow paths (the position of these paths was determined as x_0). This assumes that a dust particle does not return to the airflow if it reaches the separator walls.

The value E_{fr} can be determined using the expression

$$E_{\mathrm{fr}} = 1 - x_0/l. \tag{5.189}$$

Dust separator efficiency for various grades (Figure 5.42) can be determined by finding the initial position of critical paths for particles with varying diameters. The plot shows that the separator only precipitates coarse dust. Most particles less than 25 μm in diameter escape the dust trap. A rapid increase in E_{fr} occurs within a rather narrow band of 24 to 45 μm so that particles measuring 45 μm and larger are completely removed from the air stream.

Experimental surveys of dust precipitation efficiency in the separator were performed in laboratory conditions at a semi-industrial-size installation and in industrial conditions at charge preparation houses at the Lebedinsky and Sokolovsko-Sarbaisky OBPs. In both cases, tests were performed when the separator was installed on cowls of belt conveyor loading locations.

In these laboratory conditions, a dust feeder was used to supply powdered bentonite as a dust-air mixture into the cowl.

Experiments were performed at the dust-air flow inlet of the expansion chamber using varying slot heights. At every height, a series of measurements were taken to determine evacuated air volume, airborne dust content in the separator inlet opening, and aerodynamic resistance of the separator in the suction air duct. In addition, dust particles entering the separator were measured for every series of experiments. Findings from experiments are summarized in Tables 5.18 and 5.19.

The total separator efficiency was calculated using the expression [166]

$$E_0 = \frac{E_{fr1}\Phi_1}{100} + \frac{E_{fr2}\Phi_2}{100} \cdots + \frac{E_{frn}\Phi_n}{100}, \tag{5.190}$$

where E_{fr1}, E_{fr2},..., E_{frn} are calculated particle-size efficiencies of the separator (averaged for the specific fraction) determined in accordance with Figure 5.42; and Φ_1, Φ_2,...Φ_n are fractions of specific grades of particles at the separator inlet (% wt.)

Table 5.19 shows that, with h/l values ranging from 0.05 to 0.09 (that is, staying close to h/l = 0.066), the actual dust trap efficiency is always higher than the

TABLE 5.18
Particle-Size Composition of Dust Entering the Separator

Height of Slot h (mm)	<1.4	1.4–4.2	4.2–9.8	9.8–15	15–30	30–45	45–105	>105
	Fraction of material particles by weight Φ_i (%)							
30	2.8	3.2	4.4	12.1	20.1	15.1	23.1	19.2
35	1.75	2.91	6.47	24.05	22.88	18.22	23.72	0
40	1.46	3.03	3.03	14.17	17.13	4.95	20.40	31.99
50	1.8	2.2	5.1	11.4	14.9	28.2	9.7	26.7
100	1.1	1.0	3.2	4.7	17.8	11.7	16.2	44.3
Calculated per-grade efficiency	0	0	0	0	0	0.87	1.0	1.0

Coarseness Interval (μm)

calculated value. This can be explained by the fact that dust particle precipitation in the bin and on the partition wall between the separator chambers was unaccounted for in per-grade efficiency calculations.

A somewhat lower calculated efficiency prevents undersizing the dust containment performance of the separator when designing aspiration systems. If slot height is significantly increased, the empirically determined value E_0 will decrease noticeably. A decrease in the separator resistance coefficient occurs for the same reason.

The proposed dust receiver/separator design was also tested in the charge preparation house of a pelletization factory handling bentonite at the Lebedinsky OBP, together with bead-shaped directing elements installed inside cowls [167]. Measurement results indicated in Table 5.20 refer only to the dust receiver/separator unit.

TABLE 5.19
Findings of Laboratory Studies of Dust Receiver/Separator Efficiency*

Height of Slot h (mm)	h/l**	Aspiration Volume, Q_a (m³/hr)	Local Resistance Coefficient ζ	u_{in}, (m/s)	Amount of Dust (g/hr): Entering the Separator	Amount of Dust (g/hr): Evacuated by Suction	Total Dust-Trap Efficiency, E_0: Calculated	Total Dust-Trap Efficiency, E_0: Actual
30	0.054	500	3.1	0.29	23	8.7	0.55	0.62
35	0.064	500	3.1	0.29	28	13.4	0.40	0.52
40	0.073	480	3.1	0.28	33.6	11.4	0.58	0.66
50	0.091	880	2.9	0.74	460	124	0.61	0.73
100	0.182	1210	2.6	1.02	388	155	0.71	0.60

*(1 = 0.55 m, $\alpha = 75°$); **The ratio was calculated as 0.066 in the aerodynamic model of the separator.

TABLE 5.20
Findings from Industrial-Scale Tests of Dust Receiver/Separator Efficiency

Measurement Point	Measurement Section Dimensions (mm)	Air Velocity (m/s)	Volumetric Flow Rate of Evacuated Air (m^3/hr)	Dust Content in Aspirated Air (g/m^3)	Material Carryover Rate (kg/hr)	Cleaning Efficiency %
$h = 5$ mm						
Upstream of dust receiver	900 × 5	5.9	950	5.2	4.94	
Downstream of dust receiver	180	10.5	950	2.6	2.47	50.0
$h = 10$ mm						
Upstream of dust receiver	900 × 10	3.4	1100	9.75	10.7	
Downstream of dust receiver	180	12.2	1100	5.53	6.08	43.2
$h = 15$ mm						
Upstream of dust receiver	900 × 15	2.5	1200	10.0	12.0	
Downstream of dust receiver	180	13.3	1200	5.8	6.96	42.0
$h = 20$ mm						
Upstream of dust receiver	900 × 20	2.0	1300	8.96	11.6	
Downstream of dust receiver	180	14.3	1300	5.6	7.3	37.1
$h = 25$ mm						
Upstream of dust receiver	900 × 25	1.7	1380	9.12	12.6	
Downstream of dust receiver	180	15.2	1380	6.62	9.1	27.4
$h = 30$ mm						
Upstream of dust receiver	900 × 30	1.5	1420	12.1	17.1	
Downstream of dust receiver	180	15.7	1420	9.15	12.9	24.0
$h = 35$ mm						
Upstream of dust receiver	900 × 35	1.3	1500	12.45	18.6	
Downstream of dust receiver	180	16.5	1500	9.38	14.1	24.1

TABLE 5.20 (Continued)
Findings from Industrial-Scale Tests of Dust Receiver/Separator Efficiency

Measurement Point	Measurement Section Dimensions (mm)	Air Velocity (m/s)	Volumetric Flow Rate of Evacuated Air (m³/hr)	Dust Content in Aspirated Air (g/m³)	Material Carryover Rate (kg/hr)	Cleaning Efficiency %
$h = 50$ mm						
Upstream of dust receiver	900 × 50	0.98	1600	15.85	25.3	
Downstream of dust receiver	180	17.9	1600	13.1	20.9	17.3

From Lebedinsky ore beneficiation plant, charge preparation house, KB-6A bentonite conveyor, ATU-3 aspiration unit.

Industrial-scale tests of the dust receiver/separator helped determine the optimal slot height ($h = 20$ mm) from the viewpoint of material carryover into an evacuation network. Dust samplings indicate this measure resulted in a twofold reduction (from 39.7 down to 20.15 kg/hr) of the total rate of material carryover into the aspiration network. The local resistance coefficient of the dust receiver averaged $\zeta = 5$ (referred to as the dynamic pressure in the suction flange). Optimum negative pressure inside the cowl reads 4–6 Pa.

Finally, it should be noted that the dust receiver/separator can be used for processing dry loose materials in aspiration cowls of any kind of process equipment (bins, screens, crushers, etc.).

5.2.4.3 Dust Precipitation in a Local Cyclone-Type Suction Unit/Dust Trap

Aside from other techniques for mitigating carryover of coarse dust, the most commonly applied method involves maintaining minimum air velocities in receiving sections of local suction units. The effect of this technique can be significantly amplified by providing a means of coarse screening of evacuated air in the local suction unit.

This notion was explored further with the development of two designs of dust-separating local suction units based on the effects observed in inertial dust traps. The following principal requirements were taken into account: the new dust-separating suction units must retard air duct clogging, must reduce the workload on dedusting devices, and must partially return valuable dust-like material into the process cycle.

A *local suction unit with a vortex assembly* [168,169] consists of a dome (1) and a dust receiver (2) with a vortex assembly (3) and dust container (7) (Figure 5.43). The dust receiver attaches to the dome lid using four bolts (5) with nuts (6). A flange (4) is provided for removing cleaned air from the dust receiver. The local suction unit with vortex assembly was tested in the aspiration cowl at a conveyor loading location (Figure 5.44) in a semi-industrial scale VNIIBTG installation.

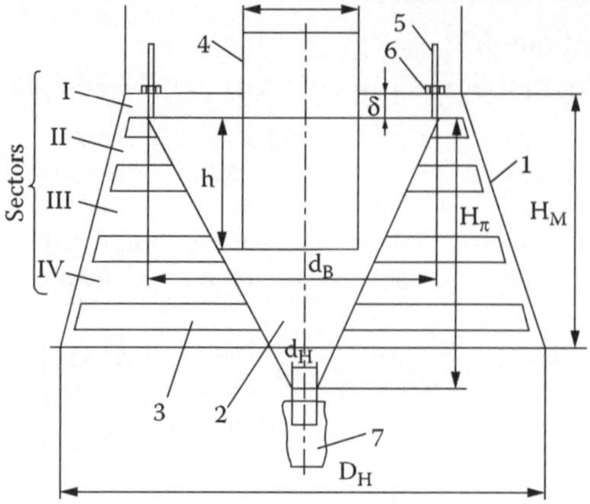

FIGURE 5.43 Local suction unit with vortex: 1 = dome; 2 = dust receiver; 3 = vortex units; 4 = flange; 5 = bolt; 6 = nut; 7 = dust container.

Dust-laden air coming through the chute (1) is forced by a fan toward the inlet section of the local suction unit (4). Inside this unit, airflow is accelerated and spun into a vortex to promote centrifugal forces that cause dust particles to precipitate on the inner surface of the dust receiver and then to fall into the dust container (or pass through a gate to fall onto the conveyor belt). Cleaned air is then routed to the flange (5) where it proceeds into the air duct system (14) of an aspiration network in the semi-industrial-scale installation.

A *cyclone-type local suction unit* [170] was tested under the same conditions as the dust-separating local suction unit with the vortex assembly. The cyclone unit (Figure 5.45) is composed of a housing (1), a tapering suction funnel (2), a vent flange (3) installed inside a conical dust receiver (4) with exhaust openings (5), guide vanes (6), and a dust container (7). A local suction unit operates as follows: Dust-laden air

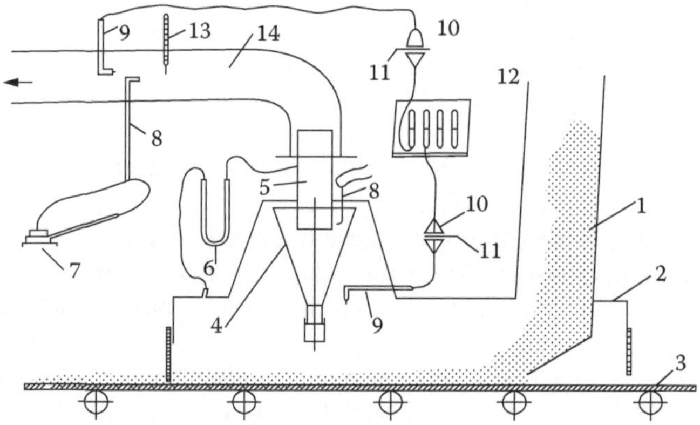

FIGURE 5.44 Test installation layout: 1 = chute; 2 = cowl; 3 = conveyor; 4 = local suction unit with vortex assembly; 5 = flange; 6 = pressure gauge; 7 = micromanometer; 8 = pitot tube; 9 = dust intake tube; 10 = cartridge; 11 = filter; 12 = blower fan; 13 = thermometer; 14 = air duct.

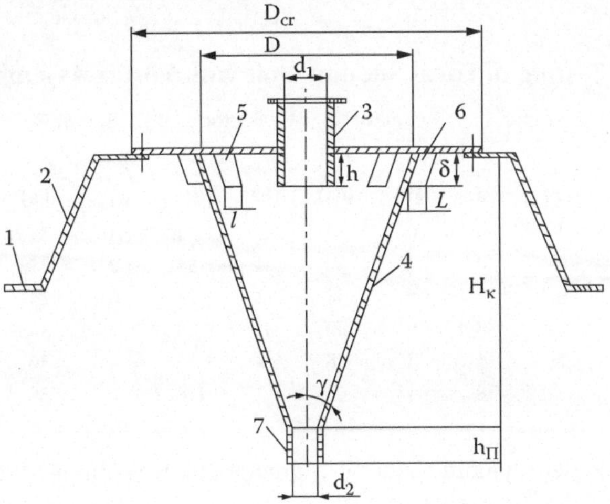

FIGURE 5.45 Cyclone-type local suction unit: 1 = housing; 2 = suction funnel; 3 = flange; 4 = dust receiver; 5 = vent hole; 6 = guide vane; 7 = dust container.

produced by the handling of loose material inside the housing (1) flows around the dust receiver (4) and then enters it through exhaust openings (5) with guide vanes (6). As air flows around the vanes, a negative-pressure zone at the vane edges causes the air stream to spread along the inner surface of the dust receiver. This results in a tangential flow of air masses inside the dust receiver (i.e., they form a vortex). The concentration of dust particles carried by the air increases due to inertial separation, causing the particles to coagulate and then to precipitate on the dust receiver walls before falling into the dust container. Cleaned air proceeds to the vent flange and enters the air duct system of the aspiration network.

Aerodynamic tests of local suction units were performed using clean air. The gap δ before the dome lid and the upper base of the dust container and the depth *h* of the flange extension inside the dust container were determined before measurements were taken. Static and dynamic air pressures were measured in sectors 1–4 (formed by the guide plates) while ascertaining static pressures in the cowl and in the evacuation flange. Findings from aerodynamic surveys are listed in Table 5.21. Findings from dust surveys are listed in Table 5.22.

As Table 5.21 indicates, increased gaps δ improve suction performance, reduce resistance, and stabilize pressure among sectors of the dust release unit. Therefore, gap size was increased for dust surveys. Data listed in Table 5.22 clearly show that, at a maximum gap width $\delta = 180$ mm, the resistance of the local suction unit with vortex assembly measured 206 Pa. With a 1314 m^3/hr airflow and initial airborne dust content of 150 mg/m^3, the dedusting efficiency of the suction unit reached 54.2% for limestone dust.

A similar procedure was used for laboratory surveys of the local suction unit. The number of variable parameters was reduced to one—the flange insertion depth *h* that was adjusted in the following range: $h = \delta, 2\delta, 3\delta$. The corresponding readings of the local resistance coefficient at the suction unit were: $\zeta = 5.5$; 11.3; 16.

TABLE 5.21
Aerodynamic Testing of Local Suction Unit with Vortex Assembly

		Sector 1		Sector 2		Sector 3		Sector 4		
δ, (mm)	Q, (m³/hr)	P_{dyn} (Pa)	P_{stat} (Pa)	P_{dyn} (Pa)	P_{stat} (Pa)	P_{dyn} (Pa)	P_{stat} (Pa)	P_{dyn} (Pa)	P_{stat} (Pa)	Resistance of Location Suction Unit (Pa)
30	850	4	30	0	34	2	24	4	32	350
		6	34	2	30	4	34	2	28	
		6	34	6	38	2	28	0	24	
60	1,090	6	30	6	30	16	28	4	34	250
		2	36	2	38	4	36	4	36	
		2	38	4	40	2	38	6	38	

The following aerodynamic suction characteristic was obtained after processing the results of the aerodynamic surveys:

- Capacity (m³/hr) 1,000–6,000
- Air velocity (m/s)
  - (a) In suction hole sections 8.8
  - (b) In the outlet flange 17.1
- Negative pressure inside cowl (Pa) 15
- Hydraulic resistance (Pa) 965
- Local resistance coefficient of the suction unit 5.5

Dust-laden air was prepared by a disc dust feeder and fed into the cowl through a chute. Table 5.23 summarizes the survey findings. The following conclusions can be made using the data at hand:

TABLE 5.22
Laboratory Field Tests of Local Suction Unit with Vortex Assembly

			Sector 1		Sector 2		Sector 3		Sector 4		Airborne Dust Content (mg/m³)			
δ, (mm)	H_n, (mm)	Q, (m³/hr)	P_{dyn} (Pa)	P_{stat} (Pa)	P_{dyn} (Pa)	P_{stat} (Pa)	P_{dyn} (Pa)	P_{stat} (Pa)	P_{dyn} (Pa)	P_{stat} (Pa)	Upstream of Suction	After Suction	Cleaning efficiency, %	Resistance of Local Suction Unit (Pa)
90	450	1240	2	20	0	20	1	20	4	20	66.6	60	9.9	220
			4	34	4	30	0	26	4	26	55.5	44.4	20	
			2	30	4	34	4	30	2	28	36.6	20	43.3	
180	550	1314	6	32	8	38	8	44	4	40	175	125	28.5	206
			2	36	10	36	4	46	4	38	56.2	31.2	44.4	
			2	36	2	44	8	48	2	40	150	68.7	54.2	

Vortex assembly at $d = 230$ mm.

TABLE 5.23
Findings from Air Cleaning Efficiency Measurement in Cyclone-Type Local Suction Unit

Range of Dust Content in Aspirated Air (mg/m^3)	Cleaning Efficiency (Percentage) with Various Insertion Depths of Inlet Flange *h* (mm)		
	350	250	125
(a) Bentonite dust ($\rho = 2420\ kg/m^3$; $d_{50} = 20\ \mu m$)			
0–100	41.1	44.4	48.7
100–200	50.0	56.2	60.5
200–350	59.2	66.4	72.8
(b) Pellet dust ($\rho = 4470\ kg/m^3$; $d_{50} = 30\ \mu m$)			
0–100	53.5	65.7	88.2
100–400	58.5	73.7	94.0
400–900	63.2	77.9	97.0

- The optimum insertion depth of the outlet flange should be equal to the height of suction holes ($h = \delta$).
- For a cyclone-type suction unit, hydraulic resistance measured 965 Pa—twice as much as for a pump with a vortex assembly.
- Cleaning efficiency measured 70% for air contaminated with bentonite dust and 97% for air contaminated with pellet dust.

Industrial testing of the local suction unit with vortex assembly was performed during chalkstone filling of bin #8 in the charge ingredient train of firing machines #1 to #8 at the Sokolovsko-Sarbaisky OBP pelletization shop. The suction air duct of bin #8 was connected to an existing ATU-5 aspiration unit. With the ATU-5 exclusively serving bin #8, air velocity in the suction air duct of the local suction unit with vortex assembly peaked at 10.4 m/s and its dedusting efficiency read 56.7%—somewhat below the expected value (60%). The local suction unit operated with an average local resistance coefficient of 4.6.

5.2.4.4 Dust Precipitation in a Local Dust-Separating Suction Unit with a Filter Element

A local suction unit/separator with a filter element (Figure 5.46) was installed in place of a dust receiver flange on the crushed material bin, comprising a dust receiver funnel (2) that houses a filter element (3) mode of polyester filter cloth, part No. 86033, Ukrainian SSR Specification (TU) 17 #3238-84.

In order to increase the filtration area, the filter element was manufactured as truncated quadrangular pyramids (4) whose top bases are attached by means of a metal frame (5) to the inner walls of the dust-receiving funnel (2); bottom bases are freely coupled via frames (6, 7) and levers (8, 9) with double-armed levers (10, 11) installed on the lid of the dust-receiving funnel. The filter element is regenerated mechanically by cams installed on the actuator shaft. Oscillating motion from the

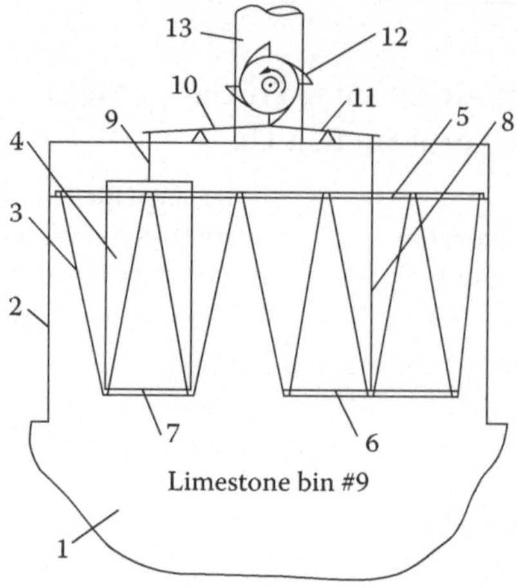

FIGURE 5.46 Dust-precipitating local suction unit with a filter element: 1 = bin; 2 = dust receiver funnel; 3 = filter element; 4 = pyramids; 5 = upper frame; 6, 7 = lower frame; 8, 9 = level; 10, 11 = double-armed lever; 12 = MEO-100/25 actuator; 13 = air duct.

actuator is transferred to the bottom base of the filter element through an intermediary frame and a lever system. The filter element regeneration mechanism operates only when the bin is loaded with loose material. It turns off when no bin loading is performed.

Aerodynamic measurements and dust sampling were performed directly inside the bin and simultaneously in the suction air duct. Surveys were performed using powdered limestone with a median diameter of 66 μm. Parameters measured during the surveys included airborne dust content inside the bin and in the suction dust, dust retention efficiency of the filter element, and resistance of the element as a function of air load and operating time. Filter cloth was continuously regenerated in laboratory conditions using a system of levers connected to the idle roller of the belt conveyor, as well as using compressed air for single-pulse or double-pulse mechanical shaking of the element.

Air load on the filter element varied from 140 to 430 $m^3/(m^2hr)$. Dust samples were taken in sequence after each regeneration of the filter element.

Dust retention efficiency of the filter element as a function of air load is plotted on Figure 5.47. Dust retention improved with decreasing air load (i.e., with increased operating time of the filter element).

After each series of experiments, the differential pressure before and after the filter element was measured, and the local resistance factor of the element was determined using the formula

$$\zeta = 2\Delta P / (\rho q^2) \tag{5.191}$$

where P is differential pressure (Pa); ρ is aspirated air density (kg/m^3); and q is air load on the filter element, $m^3/(s \cdot m^2)$. Beginning and ending times of filter element exposure to dust (t in minutes) were recorded for every series of dust samples.

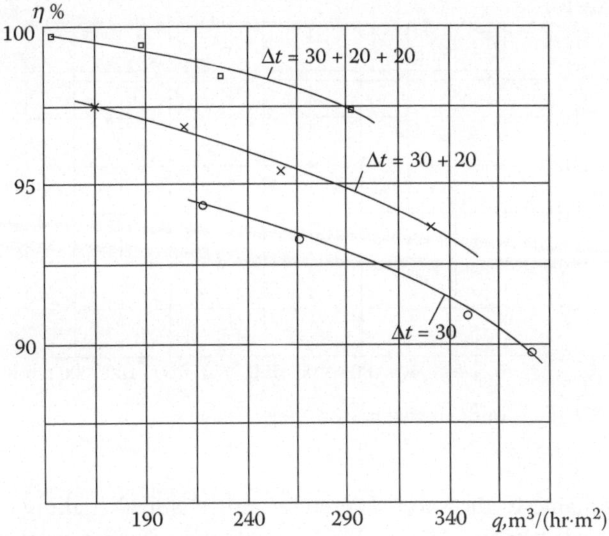

FIGURE 5.47 Dust retention efficiency of a filter element as a function of air load.

Plots of the hydraulic resistance of the filter element as a function of air load (Figure 5.48) and of filter element operating time with single-pulse and double-pulse mechanical shaking (Figure 5.49) clearly show eventual stabilization (within 220 to 270 Pa) of filter element resistance.

Regeneration of the filter element with pulses of compressed air [171,172] is the most rational and reliable method. It was not feasible to use pulse regeneration in the pelletization shop of the Sokolovsko-Sarbaisky OBP because no compressed air

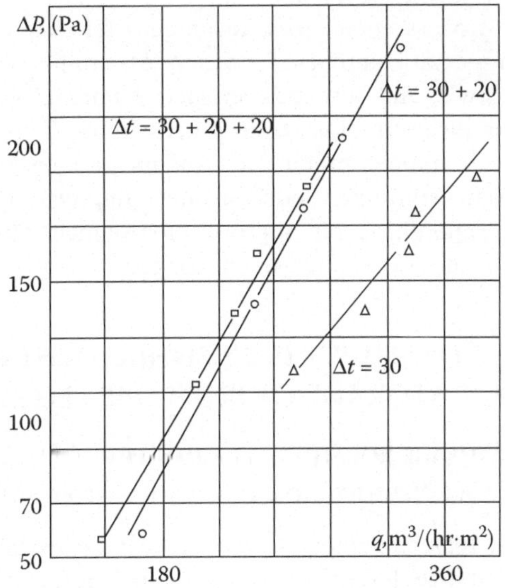

FIGURE 5.48 Plot of hydraulic resistance of filter element as a function of air load.

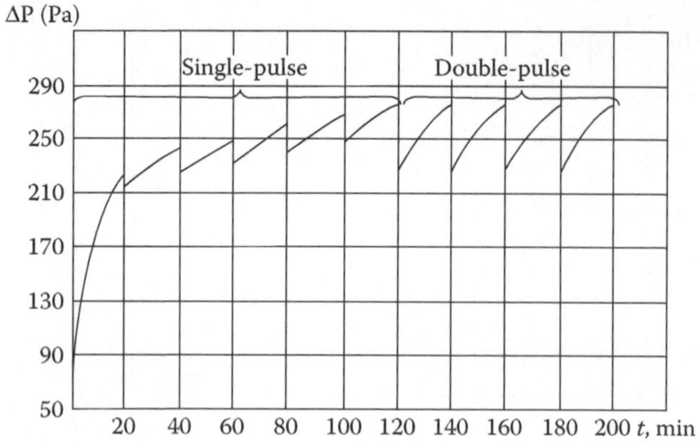

FIGURE 5.49 Changes in the resistance of filter element.

source was available in the shop. Therefore, a mechanical shaking design using a cam mechanism driven by a MEO-100/25 actuator was devised for shaking the filter element in industrial conditions. A dust-separating local suction unit with a filter element was installed on the crushed limestone bin #9 of OK-108 firing machine #3 and connected to the ATU-5 aspiration unit. Surveys (Table 5.24) indicate that dust content in air evacuated from the limestone bin #9 read 4,050 mg/m^3 and material carryover rate into the aspiration network measured 11.3 kg/hr prior to installation of the filter element. With the element installed, dust content was brought down to 391 mg/m^3, material carryover rate was reduced to 1.1 kg/hr, and dust retention efficiency of the filter element reached 90.4%. Local suction unit capacity was 2,800 m^3/hr with a hydraulic resistance of 540 Pa, and air velocity in the air duct was 15.8 m/s, with a negative pressure of 14 Pa inside the bin.

Thus, a range of process engineering techniques for shaping the flow of loose matter together with special equipment for dust release mitigation, coarse screening of air in aspiration cowls, and dust receiver suction units all have the potential of significantly reducing dust emissions from process units. This promises favorable conditions for returning trapped product into the process, enables simplification of air dedusting systems in centralized dedusting units, improves operational reliability of aspiration systems, and ultimately improves environment conditions on the shop floor and on industrial sites of ore dressing factories.

5.3 SOURCES OF FUGITIVE ATMOSPHERIC DUST RELEASES IN OUTDOOR STORAGE OF IRON ORE PELLETS

5.3.1 Outdoor Storage Locations as Atmospheric Emission Sources at Ore Beneficiation Plant Industrial Sites

OBP factory facilities include outdoor storage for blend materials and final product. They provide buffer storage, lessening the dependency of the plant on frequent supplies of raw material and on outgoing shipment of the final product.

TABLE 5.24
Findings from Industrial-Scale Testing of Local Suction Units in Bins

Sampling Location	Air Duct Diameter (mm)	Air Velocity (m/s)	Aspiration Flow Rate (m^3/hr)	Average Dust Content in Air (mg/m^3)	Material Carryover Rate (kg/hr)	Dust Retention Efficiency (%)
Before redesign						
Air duct of crushed limestone bin #8	250	10.5	1850	3370	6.25	–
Air duct of crushed limestone bin #9	250	15.8	2800	4050	11.3	–
Upstream of dust trap	570	7.6	7000	2507	17.5	–
Downstream of dust trap	350	20.8	7200	175.5	1.26	93.0
After redesign						
Air duct of crushed limestone bin #8	250	10.5	1850	1440	2.7	6.7
Air duct of crushed limestone bin #9	250	15.8	2800	391	1.1	90.4
Upstream of dust trap	570	7.5	7000	527	3.7	–
Downstream of dust trap	350	20.8	7200	35.3	0.25	93.3

Only bins #8 and #9 were loaded during measurements of dust aerodynamics. Bin #8 was loaded two-thirds full, and bin #9 was loaded half full.

Handling of loose materials is the principal process activity at outdoor storage facilities. Filling and reclaiming processes at outdoor storage locations exposed to wind involve intensive fugitive emissions of dust, resulting in dust plumes spreading far beyond the storage sites. Use of high performance equipment and a free-falling mode of material dumping cause intensive dust release; outdoor storage, therefore, is among the primary atmospheric dust emission sources at industrial sites.

5.3.1.1 Storage Process Layout

OBP factories involve outdoor storage of chalkstone, bentonite, sintering ores, and fired pellets. Chalkstone is taken in as solid stable pieces sized 5–80 mm with a moisture content of up to 5%. Bentonite is supplied in pieces sized up to 350 mm with a moisture content around 20–27% [173]. Moisture content of stockpiled sintering ore (outdoor storage at the Sokolovsko-Sarbaisky OBP) measures 6–8%. Pellets are taken in dry at temperatures ranging from 80 to 120°C.

Reviews of storage process designs [174] and industrial studies of dust release intensity carried out previously by the Laboratory for Industrial Ventilation at the All-Union (presently All-Russia) Institute for Occupational Safety in Ore Mining (VNIIBTG) and State Design Institute Kazsantekhproyekt [175] have found that most intensive dust emissions are associated with handling fired pellets at outdoor storage sites (Table 5.25). This motivated their selection as the primary survey locations.

At present, the following basic types of outdoor backup pellet storage facilities are commonly used at ore beneficiation plants:

- Storage sites equipped with high performance, continuous action machinery with dumping performed by overhead stockpiling machines and with reclaiming by rotary pickers
- Storage facilities using above-the-pile and below-the-pile conveyor galleries for dumping and reclaiming

These are low-ground storage locations where material is stockpiled in tapering or cresting piles up to 300 m long and 12–19 m high. Processing capacities enable

TABLE 5.25
Airborne Dust Content at Outdoor Storage Sites for Loose Materials

No.	Plant Name, Equipment, Material	Dumping Height (m)	Wind Speed (m/s)	Dust Content in Plume,* mg/m³ Total	≤ 10 μm
Dumping					
1	Severny OBP, pelletization factory #3, stockpiling machine, pellets	4	2	9720	–
2	Sokolovsko-Sarbaisky OBP, pellet preparation shop, tripper, pellets	21	1.43	301	270
3	Sokolovsko-Sarbaisky OBP, pellet preparation shop, grab-type handler, limestone	4	7.6	1320	–
4	Sokolovsko-Sarbaisky OBP, pellet preparation shop, grab-type handler, bentonite	4	9	80	–
5	Sokolovsko-Sarbaisky OBP, Sokolovsky ore mine, tripper, raw ore	8	6	2140	–
Reclaiming					
6	Lebedinsky OBP, pelletization factory, rotary picker, pellets	1.2	2.45	685	509
7	Severny OBP, pelletization factory #3, rotary picker, pellets	2	6.8	2370	–

OBP = ore beneficiation plant; *dust content in plume was measured 40 meters away from the source.

handling up to 500,000 metric tons of pellets per year. Maximum dumping height may reach 8–21 m, decreasing to 2–3 m as the pile is filled. Belt conveyors are used for delivering pellets to and from the storage location.

Process parameters of outdoor fired pellet storage sites are listed in Table 5.26.

Compared to indoor storage, outdoor storage locations benefit from lower capital expenditures on construction. A welcome development is the increasing use of storage facilities equipped with stockpiling machines and rotary pickers. The large capacities of these types of machinery ensure failsafe operation of pelletization factories during peak loads. This equipment enables complete mechanization of storage filling and reclaiming operations. In addition, stockpiling machines and rotary pickers are advantageous in light of industrial hygiene regulations. Containing the release of dust from running machinery of this type may be a formidable problem because the mobile working cycle is somewhat unique—yet problems can be solved by installing self-contained air dedusting solutions.

5.3.1.2 Primary Dust Emission Sources

The only means of mitigating dust emission at outdoor pellet storage sites is wet dedusting (waterspraying the material being handled). Storage equipment operating experience points to the inadequate efficiency of wet dedusting. Moreover, this method can only be applied seasonally in regions experiencing below-freezing winter temperatures, and it calls for an additional slurry pumping system. Off-season use of water results in frozen lumps of pellets that are difficult to break away from the stockpile. Using a below-the-pile conveyor gallery to reclaim pellets increases the share of manual labor. In addition, reclaiming chute packing has been observed, and waterspraying leads to thermal breakdown of hot pellets, degrading their quality as a raw material.

Handling units in storage machinery are rarely equipped with aspiration cowls. Existing cowls have low sealing quality and are not furnished with suction devices for removing dusty air. Positive pressure arising in such cowls as a result of air injection by poured pellets works to knock out dust. For that reason, the operation of virtually all machinery involved in pellet handling is accompanied by a large dust release.

In storage locations of the first type, such dust release is mainly generated by overhead stockpiling machines and rotary pickers. With overhead stockpiling machines, dust is generated at the overhead conveyor loading location and from the free-falling flow of pellets (dusty fractions are blown out with wind and are carried away by injected air from the stockpile dumping location). Dust is released during the operation of rotary pickers as the rotary wheel picks up pellets from the stockpile. Other sources include storage bin filling and unloading locations. When storage bins are loaded through an above-the-pile conveyor gallery, dust is released by the freely falling stream of pellets. All sources mentioned here are termed continuous point sources. Another classification [176] designates them as cold and ground-based sources.

Secondary dust formation as a result of windblown surface dust is insignificant in pellet stockpiles because segregation of dumped material causes fines to become buried under a layer of larger pellets.

TABLE 5.26
Specifications of Outdoor Storages Facilities for Fired Pellets

No.	Location	Type of Installation	Equipment Capacity (metric tons per hour)	Maximum Dumping Height (m)	Stockpile Volume (thousands cu. m^3)	Pellet Compression Strength (kg per pellet)	Small Fraction (5 to 0 mm) Content (%)
1	Poltava OBP, pellet preparation shop #1	Stockpiling machine,	1200	18	64	257	262
		rotary picker (Demag brand)	2200				
2	Poltava OBP, pellet preparation shop #2	Stockpiling machine,	1200	18	53.8	257	2.56
		rotary picker (Demag brand)	2200				
3	Severny OBP, pelletization factory #1	Stockpiling machine (YUMZ),	750		28		
		rotary picker (Demag brand),	500	12		219	5.08
		belt conveyor,	600		1.3		
		EKG-4,6 excavator	–		1.16		
4	Severny OBP, pelletization factory #2	Stockpiling machine,	1200	11	48.3		
		rotary picker (Lugri brand)	1500		50	227	5.50
5	Severny OBP, pelletization factory #3	Stockpiling machine,	1500	11	24.6	220	5.32
		rotary picker (Weserhutte brand)	1500		25.5		
6	Mikhailovsky OBP	Stockpiling machine,	1200	13.5	6.75	200	5.53
		rotary picker (YUMZ)	1200		29.2		
7	Sokolovsko-Sarbaisky OBP	Reversible conveyor	700	21	85	157	2.33
8	Krivyi Rih Central OBP	Dump cars, grab-type handler	500	8	140	171	6.30
9	Lebedinsky OBP	Stockpiling machine,	1200	13.5	25.4	200	3.29
		rotary picker	1200		29.2		

OBP = ore beneficiation plant.

5.3.2 Examining Dust Release Intensity at Iron Ore Pellet Storage Sites

The existing methodology for assessing the intensity of fugitive dust emission sources relies on an indirect method based on measuring dust concentration in a dust plume and on using atmospheric scattering formulas to reconstruct the dust flow rate at the origin. This calculation then serves as a quantitative estimation of dust release intensity.

Therefore, the development of fugitive dust emission estimation procedures primarily depends on choosing reasonably justified methods for calculating dust scattering. A large number of related scientific studies, conducted by Russian [177–183] as well as foreign researchers [184–187], could be tentatively classified into two groups. The first group encompasses those studies concerned with the scattering of light contaminants (gases and dust particles below 5 μm). Seminal papers were authored by A. S. Monin [177], A. M. Yaglom [178], O. G. Setton [184], and Bosanquet and Pearson [185].

The second group of studies, which deal with the dissemination of both light and heavy contaminants, include significant works by M. I. Yudin [179], M. E. Berliand [180], and J. Detry [186]. Research providing valuable practical insights into industrial site pollution by light contaminants has been conducted by V. S. Nikitin [181], I. N. Leikin [182], and V. M. Elterman [183]. With our focus on the escape of relatively large particles from the flow of dumped material, we based our calculations on known patterns of heavy contaminant distribution. These formulas provide quantitative characterization of dust release sources during the handling of fired pellets at backup storage locations using the most common online process design.

5.3.2.1 Distribution of Dust Released by a Ground-Based Source

Calculation of windborne dust concentrations from a ground-based source is based on differential equations dealing with steady-state diffusion of heavy extraneous matter in horizontally homogeneous terrain [180]:

$$u\frac{\partial q}{\partial x} + w\frac{\partial q}{\partial z} = \frac{\partial}{\partial z}k_z\frac{\partial q}{\partial z} + \frac{\partial}{\partial y}k_y\frac{\partial q}{\partial y} \tag{5.192}$$

where u is wind velocity (in meters per second, m/s); w is the vertical contaminant velocity component, taken for dust as the inverse of sedimentation speed (m/s); q is dust concentration (in grams per cubic meter, g/m^3); and k_y, k_z are horizontal and vertical components of exchange factors (m^2/s).

The coordinate origin is set at ground level under the dust release source; the OX axis is directed along median horizontal wind component; the OZ axis rises vertically; and the OY axis is perpendicular to OZ and OH axes and is directed toward the reader. For a point source located in the center $x = 0$, $y = 0$, $z = H$, the initial condition is defined as a known dust flow released to the atmosphere:

$$uq = M\delta(y)\delta(z - H) \quad \text{at} \quad x = 0, \tag{5.193}$$

where M is dust release intensity (g/s); δ is a function equal to

$$\delta(y)=0 \text{ at } y\neq 0;\quad \delta(z-H)=0 \text{ at } z\neq H;$$
$$\delta(y)=1 \text{ at } y=0;\quad \delta(z-H)=1 \text{ at } z=H.$$

Boundary conditions are based on the natural provision of diminishing concentration at an infinite distance from the source

$$q\to 0 \quad \text{at} \quad z\to\infty; \tag{5.194}$$

$$q\to 0 \quad \text{at} \quad y\to\infty \tag{5.195}$$

and on a low average turbulent flow of contaminant near the ground

$$k_z\frac{\partial q}{\partial z}=0 \quad \text{at} \quad z=0. \tag{5.196}$$

For sources located at small elevations (H measuring a few meters), a power law describes the wind speed and exchange factor

$$u=u_1\left(\frac{z}{z_1}\right)^n;\qquad k_z=k_1\left(\frac{z}{z_1}\right)^m, \tag{5.197}$$

where u_1 is the wind speed at elevation z_1 (m/s with z_1 assumed equal to 1 m); and k_1 is the exchange factor at elevation z_1 (taken as 0.2 m^2/s at z_1 = 1 m). Solution of the original Equation (5.192) led M. E. Berliand to the following expression for the ground level concentration of light contaminant (w = 0):

$$q(x,y,0)=\frac{M}{2(1+n)k_1\sqrt{\pi k_0}x^{1,5}}\exp\left[-\frac{u_1H^{1+n}}{(1+n)^2k_1x}-\frac{y^2}{4k_0x}\right]. \tag{5.198}$$

Here, the power index n = 0.15–0.2. Therefore, the power function (Equation 5.197) describing vertical wind profile in the surface air layer becomes closer to logarithmic behavior—a feature of turbulent currents—with surface unevenness of 0.01 m. Considering that z is small, linear dependence on height ($m\approx 1$) is deemed adequate for the exchange factor.

The factor k_0 correlating the exchange factor in the horizontal surface with wind speed

$$k_0=k_y/u \tag{5.199}$$

could be determined by solving the inverse problem of turbulent diffusion using contaminant distribution data or by relying on a dependent distribution of the visible outline of the smoke plume emanating from the source. This was the method used by M. E. Berliand [188] to obtain k_0 = 0.1–1 m depending on atmospheric conditions. The value k_0 increases with unstable stratification: k_0 = 0.5–1 m; for stable atmospheres: k_0 = 0.3 m.

According to M. E. Berliand and R. I. Onikul [189], ground level concentrations of heavy and light contaminants q_w and q at a distance x away from a source at elevation H are linked by the following:

$$q_w = \chi q \tag{5.200}$$

$$\chi = \frac{u_1^{\omega} H^{(1+n)\omega}}{(1+n)^{2\omega} k_1^{\omega} \Gamma(1+\omega) x^{\omega}} \tag{5.201}$$

$$\omega = \frac{w}{(1+n) k_1} \tag{5.202}$$

enabling the ground level dust concentration to be determined

$$q_w = \frac{M_w A_w}{x^{1,5+\omega}} \exp\left(-\frac{b}{x} - \left(\frac{y}{\sqrt{4k_0 x}} \right)^2 \right), \tag{5.203}$$

where, in order to simplify our notation, we assign

$$b = \frac{u_1 H^{1+n}}{(1+n)^2 k_1} \tag{5.204}$$

$$A_w = \frac{u_1^{\omega} H^{(1+n)\omega}}{2(1+n)^{1+2\omega} k_1^{1+\omega} \sqrt{\pi k_0} \Gamma(1+\omega)} = \frac{b^{\omega}}{a \Gamma(1+\omega)} \tag{5.205}$$

$$a = 2(1+n) k_1 \sqrt{\pi k_0}$$

where w is the dust particle sedimentation speed (m/s); and $\Gamma(1 + \omega)$ is a gamma function

$$\Gamma(1+\omega) = \int_0^{\infty} t^{\omega} e^{-t} dt$$

the values of which are listed in many mathematical reference manuals. A known concentration of dust along the plume center line ($y = 0$) enables us to use Equation 5.203 to determine the intensity of the source

$$M_w = \xi q_w, \quad \xi = a x^{1,5} e^{\frac{b}{x}} \frac{\Gamma(1+\omega)}{(b/x)^{\omega}}. \tag{5.206}$$

Considering that the sedimentation speed (w, m/s) depends on particle coarseness

$$w = 3,25 \cdot 10^{-5} \rho_n d_n^2, \tag{5.207}$$

where ρ_n is the density of dust material (g/cm³) and d_n is the diameter of the dust particle (μm), Equation 5.206 could be used to determine the release intensity for a narrow fraction of diameter

$$d_i = \frac{d_i^{\max} + d_i^{\min}}{2}. \tag{5.208}$$

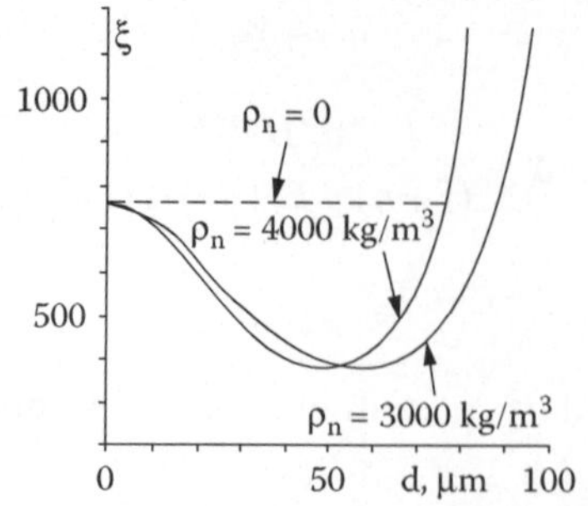

FIGURE 5.50 Behavior of ξ as a function of particle coarseness (at $k_1 = 0.2\ m^2/s$; $k_0 = 0.3$ m; n = 0.15; H = 10 m; x = 40 m; $u_1 = 1.43$ m/s).

Figure 5.50 contains a plot of changes in ξ depending on particle diameter for density $\rho_n = 3{,}000\ kg/m^3$ and $\rho_n = 4{,}000\ kg/m^3$ (in an experiment at the Sokolovsko-Sarbaisky OBP: $H = 10$ m, $u_1 = 1.43$ m/s, $x = 40$ m). Changes in ξ owe their behavior to the gamma function (Figure 5.51). As the plots illustrate, within $d < 5$ mm the ξ is virtually constant at ξ = 760.

In this area, dust distribution can be described with formulas for the light contaminant ($\omega = 0$; $\rho_n \rightarrow 0$). Intensity of polydisperse dust sources is determined by calculations for individual fractions (characterized by diameter d_i and velocity w_i). A similar approach is used to determine the ground level concentration of polydisperse dust.

Calculations could be simplified by using the concept of average polydisperse dust diameter d_{avg}. If we posit that the intensity of dust release with the coarseness d_{avg}

$$M = qe^{\frac{b}{x}} \frac{x^{1,5+\omega_{avg}}}{A_{w_{avg}}} \tag{5.209}$$

should be equal to a sum total of intensities for individual fractions (let their total number be N), then the relation

$$qe^{\frac{b}{x}} \frac{x^{1,5+\omega_{avg}}}{A_{w_{avg}}} = \sum_{i=1}^{N} qm_i e^{\frac{b}{x}} \frac{x^{1,5+\omega_i}}{A_{w_i}} \tag{5.210}$$

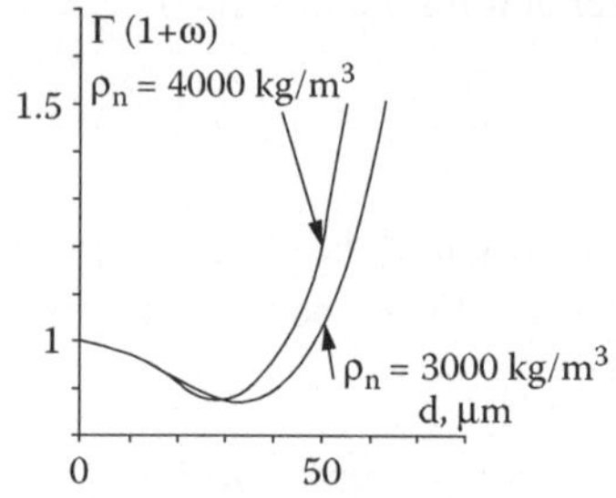

FIGURE 5.51 Behavior of the function Γ(1+ω) (at $k_1 = 0.2\ m^2/s$; n = 0.15).

could be transformed into the following equation determining the value of ω_{avg} (and, therefore, w_{avg} and d_{avg}), in view of Equations 5.202 and 5.207

$$\frac{x^{1,5+\omega_{avg}}}{A_{avg}} = \sum_{i=1}^{N} m_i \frac{x^{1,5+\omega_i}}{A_{w_i}} \tag{5.211}$$

where m_i is the proportion of particles (by mass) corresponding to fraction i, q is dust concentration (g/m^3);

$$\omega_{avg} = \frac{w_{avg}}{(1+n)k_1} \tag{5.212}$$

where w_{avg} is the sedimentation velocity for a particle of coarseness d_{avg} (m/s). In light of Equation 5.204,

$$\frac{x^{1.5+\omega_{avg}}}{\left(u_1 H^{(1+n)}\right)^{\omega_{avg}}}\left[(1+n)^{1+2\omega_{avg}} k_1^{1+\omega_{avg}} \Gamma\left(1+\omega_{avg}\right)\right] = \\ = \sum_{i=1}^{N} m_i \frac{x^{1.5+\omega_i}}{\left(u_1 H^{(1+n)}\right)^{\omega_i}}\left[(1+n)^{1+2\omega_i} k_1^{1+\omega_i} \Gamma\left(1+\omega_i\right)\right]. \tag{5.213}$$

At $\omega_i < 1$, the expressions inside the fractions can be deemed equal to $x^{1.5+\omega_{avg}} \approx x^{1.5+\omega_i}$, and then

$$\frac{1}{\left(u_1 H^{(1+n)}\right)^{\omega_{avg}}} = \sum_{i=1}^{N} m_i \frac{1}{\left(u_1 H^{(1+n)}\right)^{\omega_i}}, \tag{5.214}$$

from which we can deduce the ratio

$$\omega_{avg} = -\ln \sum_{i=1}^{N} \frac{m_i}{\left(u_1 H^{(1+n)}\right)^{\omega_i}} \Big/ \ln\left(u_1 H^{(1+n)}\right), \tag{5.215}$$

leading us to ω_{avg}.

Availability of data on ground level dust concentrations enables us to determine the intensity of its fallout on the ground surface

$$\varepsilon_w = \int_{-\infty}^{\infty} q_w w dy \tag{5.216}$$

where ε_w is the intensity of plume dust fallout per its unit length (g/s·m). In view of Equation 5.203,

$$\varepsilon_w = 2wMA_w\sqrt{\pi k_0}\,\frac{\exp(-b/x)}{x^{1+\omega}}. \tag{5.217}$$

Considering the obvious relation between the dust flow rate inside the plume and precipitation intensity

$$\frac{dG_w}{dx} = -\varepsilon_w, \tag{5.218}$$

we can determine the dust flow rate G_w (g/s) for a particular fraction carried away by the wind to a distance x from the source

$$G_w = M_w - \int_0^x \varepsilon_w dx \tag{5.219}$$

or, in view of (5.217),

$$G_w = M_w \frac{\gamma\left(\omega, \frac{b}{x}\right)}{\Gamma(\omega)}, \tag{5.220}$$

where $\gamma\left(\omega, \frac{b}{x}\right)$ is the incomplete gamma fraction

$$\gamma\left(\omega, \frac{b}{x}\right) = \int_0^{b/x} t^{\omega-1} e^{-t} dt, \tag{5.221}$$

and t is integration variable.

The following could be deduced from Equation 5.220:

$$G_w = M_w \quad \text{at } x = 0; \tag{5.222}$$

$$G_w = 0 \quad \text{at } x \to \infty. \tag{5.223}$$

The incomplete gamma function could be expressed via a tabulated *chi-square* distribution function [190]

$$\gamma\left(\omega, \frac{b}{x}\right) = \frac{\exp(-b/x)\left(\frac{b}{x}\right)^{\omega}}{\omega} + \frac{\Gamma(\omega+1)}{\omega}\left[1 - Q\left(\frac{2b}{x}|2+2\omega\right)\right], \tag{5.224}$$

where $Q\left(\frac{2b}{x}|2+2\omega\right)$ is the *chi-square* distribution function

$$Q(\chi^2|\nu) = \left[2^{\nu/2}\Gamma\left(\frac{\nu}{2}\right)\right]^{-1} \int_{\chi^2}^{\infty} t^{\nu/2-1} e^{-t/2} dt, \tag{5.225}$$

and χ, ν are independent variables, defined in our case as

$$\chi^2 = 2b/x; \quad \nu = 2 + 2\omega. \tag{5.226}$$

In view of Equation 5.224, the Equation 5.220 for the dust flow rate would assume the following computationally convenient form:

$$G_w = M_w \varphi_w, \tag{5.227}$$

where

$$\varphi_w = 1 + \frac{\left(\frac{b}{x}\right)^{\omega} \exp(-b/x)}{\Gamma(1+\omega)} - Q\left(\frac{2b}{x}\middle|2\omega+2\right) =$$

$$= 1 + \frac{\left(\frac{b}{x}\right)^{\omega} \exp(-b/x)}{\Gamma(1+\omega)}\left[1+\frac{\frac{b}{x}}{\omega+1}\right] - Q\left(\frac{2b}{x}\middle|2\omega+4\right) = \tag{5.228}$$

$$= 1 + \frac{\left(\frac{b}{x}\right)^{\omega} \exp(-b/x)}{\Gamma(1+\omega)}\left[1+\frac{\frac{b}{x}}{\omega+1}+\frac{\left(\frac{b}{x}\right)^{2}}{(\omega+2)(\omega+1)}\right] - Q\left(\frac{2b}{x}\middle|2\omega+6\right) = \ldots$$

Here, the recurrent relations for φ_w are provided in order to facilitate the use of tabular values of the function $Q\left(\chi^2 | \nu\right)$.

The value φ_w reflects relative fallout ratios of dust particles of varying coarseness. Figure 5.52 is a sample plot of φ_w for an experiment at the Sokolovsko-Sarbaisky OBP pellet storage site. As the plot illustrates, fallout of particles smaller than 15 μm can be ignored. Determining flow rates of polydisperse dust requires calculating G_w for every narrow fraction and adding together the results. The expression for average sedimentation velocity (5.215) can also be used. In this case, the average dust flow rate in the plume would be

$$G = M\left[\frac{\exp(-b/x)}{\Gamma\left(1+\omega_{avg}\right)}\left(\frac{b}{x}\right)^{\omega_{avg}} + 1 - Q\left(\frac{2b}{x}\middle|2+2\omega_{avg}\right)\right]. \tag{5.229}$$

This relies on the fact that, for smaller ω, the value of the expression inside brackets on the right side of Equation 5.228 stays virtually constant.

5.3.2.2 Field Surveys of Dust Plume Structure

Plume structure was surveyed with the goal of determining the form, geometric parameters, and motion speed of the dust plume at all stages of its formation and distribution.

Plume parameters were studied by filming it with a 16-mm Krasnogorsk-2 motion picture camera at 24 and 48 frames per second.* The filming was arranged so that a prominent object of known linear dimensions was always present within the camera's field of view on the same line with the emission source or plume segment being studied. The plume was filmed from three different perspectives: at a right angle to the plume, along the plume, and from above it.

* Postgraduate student S. I. Zadorozhny is credited with devising this method.

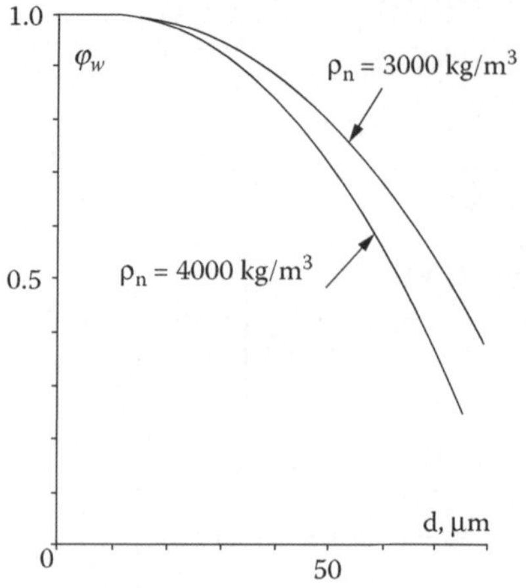

FIGURE 5.52 Changes in the ratio φ_w depending on the coarseness of particles d (at $k_1 = 0.2$ m²/s; $k_0 = 0.3$ m; $n = 0.15$; $H = 10$ m; $x = 40$ m; $u_1 = 1.43$ m/s).

The developed film was analyzed frame-by-frame using a Svityaz-M slide projector. The records were analyzed as follows: A sheet of graph paper was placed at a right angle against the projector light beam and then marked with characteristic points and plume lines, reference object position, and with dimensions using the first recorded frame. Then the film was advanced by a fixed number of frames toward the dust release progression. A series of respective marks recreated the development process in various plume locations.

In addition to providing a qualitative assessment of the factors responsible for solid phase formation and distribution in the environment, filming makes it possible to measure such factors as plume distribution velocity, plume height and width on the borders of the storage location, opening angle, and the boundaries of the zone most affected by dust.

The following handling operations were considered in studying the dynamics of the airborne dust plume:

1. Stockpile dumping from an overhead conveyor gallery
2. Reclaiming of stored pellets with a rotary picker
3. Dump car unloading

The process of stockpile dumping with dump cars was chosen for its large attendant flow of material, significant amounts of injected air, and the wide front of interaction between dumped pellets and air.

Stockpile dumping from an above-the-pile conveyor passageway was studied at a pellet storage site located at the Sokolovsko-Sarbaisky OBP. The situational layout at survey time is charted in Figure 5.53. Wind speed averaged 1.43 m/s. The product was dumped into a tapering stockpile that was, in turn, moved from the ground level of the storage site up to a height of 2.4 m. Pellets were delivered to storage from a belt

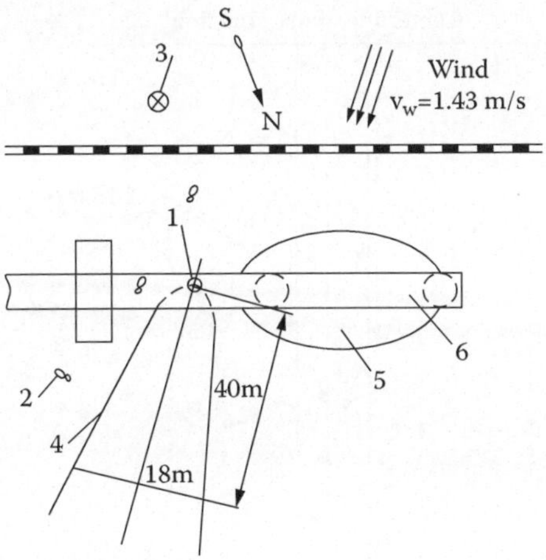

FIGURE 5.53 Situational layout chart for the field survey of a vertical linear dust emission source: 1 = location of pellet release into the stockpile; 2 = filming location; 3 = anemometer setting location; 4 = plume boundary; 5 = stockpile; 6 = above-the-pile conveyor gallery.

conveyor #28 discharge cradle handling 208.5 metric tons per hour. Pellet free-fall height decreased from 21 m to 18.6 m as the stockpile grew.

Surveys have found that stockpile dumping with a freely falling stream generates a dust plume composed of two parts: upper and lower (Figure 5.54).

The upper part of the plume is separated from the flowing material as a result of fine particles being blown out of the stream by wind. This part of the plume expands at an angle of 35°. Dust is carried away by wind, forming an angle of 19° between the flow centerline and the lower boundary of the upper part of the plume.

This angle increases with a diminishing concentration of pellets in the flow and with greater winds speeds. Increased wind speeds improve pellet stream aeration, contributing to somewhat increased dust levels in the upper part of the plume. In an extreme case, fine dust could be spread by a horizontal plume that turns sharply into the wind direction at the pellet stream mouth. The resulting velocity profile for the upper part of the plume indicates growth at the side located closer to the jet of material. Air moves faster at the lower boundary (4.57 m/s) than at the upper one (2.15 m/s).

The gravitational material flow is injected with air. When this air impinges on the stockpile, it forms a fan-shaped jet [191]. Dust particles are carried away from the pellet dumping spot by this jet and are spread along by the wind. Fan jet velocity depends on the amount of injected air, the pellet stream diameter, and on the fan jet thickness. Because the jet is displaced by wind, its thickness (in our study) measured 0.95 m on the windward side and 3.2 m on the leeward side at a diameter of 7 m. Fan jet velocity on the leeward side approached 6.8 m/s. Pellet stream diameter measured 1.93 m at a distance 1 m away from the point where it fell into the stockpile. The upper and lower airborne dust jets blend into a single plume at a distance equal to

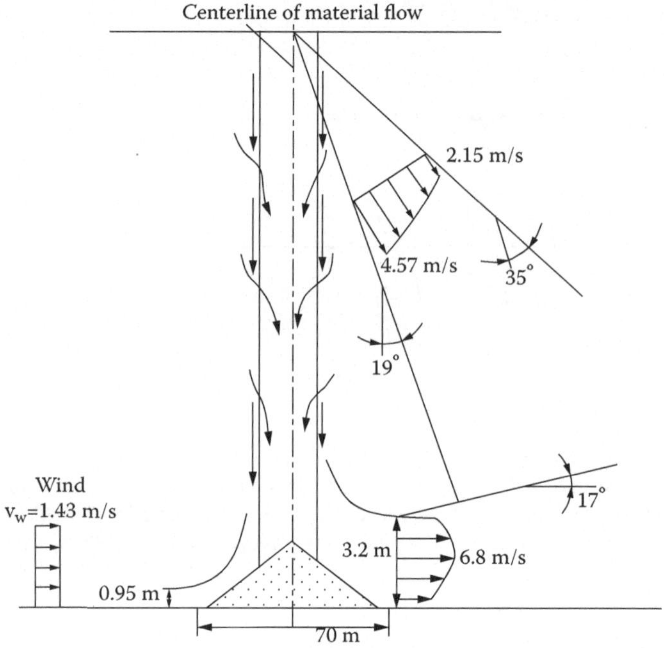

FIGURE 5.54 Reference diagram for defining the parameters of airborne dust plume and fan-shaped jet.

0.5 times the free fall height of the material at wind speeds up to 2 m/s, increasing to 3.5–4.0 times at wind speeds of 4–6 m/s.

After the single plume travels 8–10 m away from the blending point, its velocity equalizes, reaching that of the surrounding air. An aerodynamic shadow zone arises at the leeward side of the stockpile. Its length measures 2.5 times height in tapering stockpiles and 4–5 times height in cresting stockpiles. Circulation in this zone was defined by the impact range of the fan jet and by the stockpile height. Dust is spread across the entire zone, causing the plume to expand.

At sustained material flow rates of 8 to 10 tons per hour, only the upper airborne dust jet was observed. No dust emissions occurred at the location where the material was falling. In this case, the plume centerline was sharply bent, and its upper boundary became horizontal. The impact of injection forces can be illustrated with the following example. A step increase of material flow rate from 8–10 to 208.5 tons brought about intensive interaction between the flow of dumped pellets and the air, shifting the lower boundary of the plume toward the point where the material was falling into the stockpile (Figure 5.55). It is easy to see that the lower boundary moved faster as it neared its limit position, moving as fast as 4.81 m/s 2.1 m away from the jet's centerline.

Specific features of dust formation resulting from rotary picker reclamation were studied at the pellet storage site at the Lebedinsky OBP. Its cresting stockpile measured 6 m high. Pellets were taken off in the middle and lower parts of the stockpile. The rotary picker operated at a capacity of 600 tons per hour. Wind speed measured 2.45 m/s. Repetitive winnowing of a fraction of material that fell off the picker

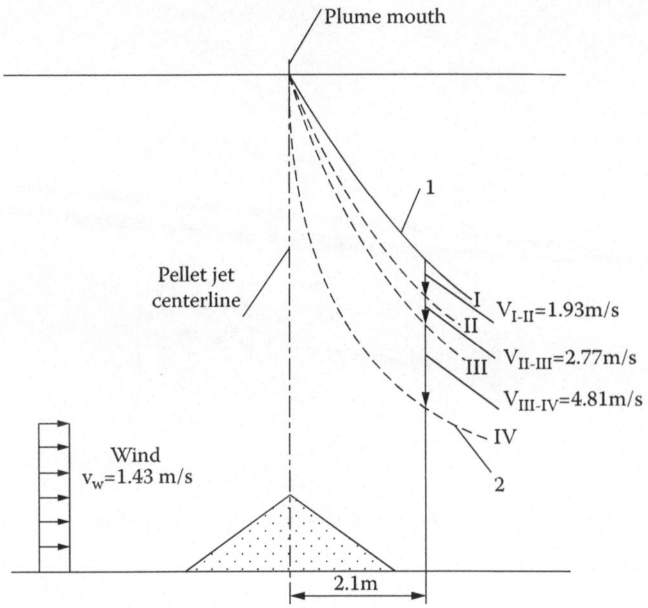

FIGURE 5.55 Relocation of the lower boundary of the upper airborne dust after step increases of material flow rate. Lower boundary location at: 1 – G_m = 8…10 t/hr; 2 – G_m = 208.5 t/hr.

buckets was responsible for dust released in the pellet reclaiming location. Another more intense instance of dust release occurred at the rotary wheel that transfers material to the conveyor.

Here, two buckets are emptied simultaneously; the new bucket brought in by the rotor begins unloading while the previous one is about to complete the cycle. The distance traveled by the material from the bucket unloading opening to the conveyor belt measures 1.5–1.8 m. At the end of the cycle, pellets fall off a slope with material flowing in a stream up to 2 m wide. Dust is blown away by the wind and is thrown up by injection airflows caused by pellets hitting the conveyor belt. The dust cloud core is formed around the upper part of the rotor wheel; its initial velocity is less than the wind speed (Figure 5.56).

Figure 5.56 shows the sequential locations of the front of the dust cloud core timed at one-second intervals. After awhile, the core disperses, swelling in volume and accelerating to the wind speed. In the main conveyor loading location, an airborne dust jet was observed spouting as far as 4 m away. It originated due to the "passing" mode of the storage bin operation—a large dumping height causes a powerful stream of injected air that lands atop the conveyor belt. Dust-generating assemblies of the rotary picker form a single plume the maximum breadth at the mouth of which is equal to the distance from the rotary wheel centerline to the point where the pellets exit the storage bin for the main conveyor. Dust release intensity during rotary picker operation is determined by the efficiency of the picker as well as the wind speed and the fines content of the pellets.

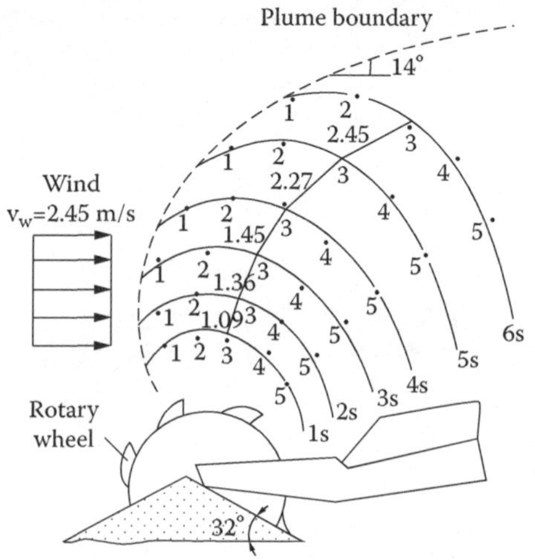

FIGURE 5.56 Initial velocity of dust plume core at the rotary picker.

The spread of the dust plume depends largely on the relative location of the dust formation source with respect to the effective wind direction. This is well illustrated by the dump car unloading example from the pellet storage site at the Zhelezorudnaya railway station in Rudny town, Kazakhstan. Here, wind speed measured 2 m/s. A single 85-ton dump car took 45 seconds to unload. Considering that unloading happened in such a short time, this source was classified as instantaneous. In this case, the dust cloud was affected by injection pressure and by external environment resistance. As unloading proceeded, the horizontal component of the dust cloud front velocity at the windward side initially decreased to zero due to counteracting wind flow (Table 5.27).

The effect of the injection pressure diminished and the cloud began to travel to the opposite site—in the wind direction, gradually accelerating to the wind speed. The vertical velocity component increased to a certain value before stabilizing in a rather narrow range. The dust cloud dissipated. When the dump car was unloaded at the leeward side, the cloud spread in the aerodynamic shadow zone. As the cloud encountered resistance from the air medium, its horizontal and vertical velocity components gradually decreased. After reaching the aerodynamic shadow boundary, the cloud gained speed equal to that of the air currents. When the extent of aerodynamic shadow was large enough, the cloud decelerated to zero and precipitated within this area. The same would happen in the absence of wind.

Therefore, it was found that the dust flare accelerates up to the velocity of a sustained plane-parallel flow 8 to 10 meters away from the confluence point of all airborne dust jets. When the plume intrudes into the aerodynamic shadow area on the leeward side of the stockpile, its velocity is equalized with the wind speed at a distance equal to 2.5 times the height of tapering stockpiles and equal to 4 to 5 times the height of cresting stockpiles.

TABLE 5.27
Changes in Dust Cloud Front Velocity during Dump Car Unloading on Windward and Leeward Sides

Windward Side			Leeward Side		
	Velocity (m/s)			Velocity (m/s)	
Time into Unloading(s)	Horizontal Component	Vertical Component	Time into Unloading(s)	Horizontal Component	Vertical Component
25	0.67	0.15	0	0	0
30	0.37	0.32	5	0	0
35	0.34	0.4	10	0.48	0.27
40	0.29	0.3	15	1.62	0.64
45	0.12	0.56	20	1.7	0.74
50	−0.34	0.6	25	0.86	0.7
55	−0.42	0.3	30	0.66	0.78
60	−0.48	0.4	35	0.28	0.52

5.3.2.3 Intensity of Primary Dust Emission Sources

The intensity of dust release depends on a number of factors (Figure 5.57), the most important being fines content in handled pellets, dumping height, wind speed, and wind direction. Dust release intensity was determined for every source emitting an independent airborne dust plume. Measurements were carried out during daylight hours except when the following factors hindered the work: wind speeds in excess of 10 m/s, choppy winds, precipitation (rain, drizzle, snow), fog, process upsets, external dust release sources, and obstacles on the path of the dust-air plume.

Measurements were carried out in the following sequence. A situational layout chart of the storage site was prepared and marked with the dust release source being surveyed. Height H of the dust release source was determined. Using the data obtained by analyzing the filmed record, the height of the source during pellet unloading with a rotary picker was assumed to be equal to 0.5 times the diameter of the rotor wheel. In the case of stockpile dumping with a freely falling flow of material, it was accepted in accordance with the formula

$$H = Lh \tag{5.230}$$

where
$L = 1/3$ at $10 \le h \le 20$; $L = 1/2$ at $5 \le h < 10$; $L = 1$ at $h < 5$, L is the factor accounting for plume shape; and h is the free-fall height of the material (m).

A streamer was used to determine wind speed and the location of the plume's longitudinal axis. A measuring station was marked on the plume axis at a fixed distance downwind from the source. It was located no closer than the point where the lower boundary of the dust plume touches the ground surface, and at least 8 to 10 meters

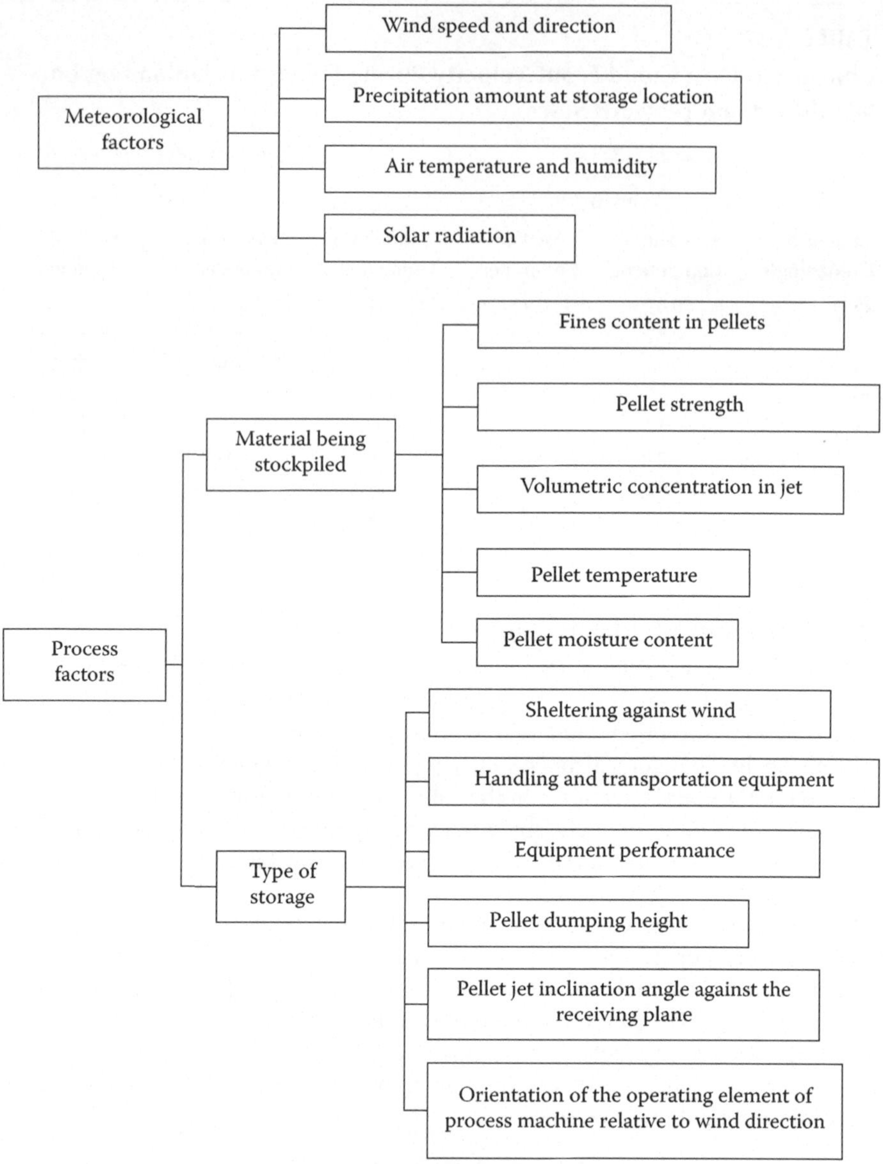

FIGURE 5.57 Factors determining dust release intensity.

away from the confluence point of all airborne dust jets emitted from the source. An anemometer was placed to measure wind speed, and air samples were taken to measure dust content at an elevation of 1 meter above ground at the measuring station. The following parameters were recorded at the same time the measurements were taken: type of handling operations, process equipment capacity, and density of pellets being handled. Samples taken were analyzed in order to determine airborne dust concentration by weight and to analyze particulate composition (by mass).

The following was used to evaluate whether data on coarse-fraction dust could be used in subsequent calculations

$$l_i = \frac{H}{w_i} u_1 > x_i, \tag{5.231}$$

where l_i is the distance traveled by a particle of fraction i in the conditions of experiment being considered. Any fractions for which this did not hold true were rejected.

Equation 5.206 was used to calculate dust release intensity for a particular fraction. In this case, the value ξ case was determined according to the formula

$$\xi = 0.447 \frac{x\sqrt{x}}{A\left(\frac{b}{\omega}, \omega\right)}, \tag{5.232}$$

where $A\left(\frac{b}{\omega}, \omega\right)$ is a parameter calculated using the formula

$$A\left(\frac{b}{x}, \omega\right) = \frac{e^{-\frac{b}{x}} \cdot \left(\frac{b}{x}\right)^{\omega}}{\Gamma(1+\omega)} = e^{-\frac{b}{x}} \cdot \left(\frac{b}{x}\right)^{\omega} \Bigg/ \int_0^{\infty} t^{\omega} e^{-t} dt, \tag{5.233}$$

with known values of b/x and ω. The following were used to calculate the latter:

$$\frac{b}{x} = 3.78 u_1 H^{1.15} / x, \tag{5.234}$$

$$\omega = 1.414 \cdot 10^{-7} \rho_n d_n^2, \tag{5.235}$$

where ρ_n is the density of dust particles (kg/m^3); and d_n is the arithmetic mean of the range of diameters for the corresponding dust fraction.

The total dust release intensity for a particular source was determined as a sum total of intensities for individual fractions.

The dust release intensity from sources was determined by measuring dust concentration in the plume at the pellet storage site at the Lebedinsky OBP (reclaiming using a rotary picker) and at the Sokolovsko-Sarbaisky OBP (stockpile dumping with a freely falling jet of material) (Table 5.28).

It is also possible to determine fractional dust release intensities using the intensity of dust settling on a horizontal plane, for example, on a substrate of known area

$$M_{w_i} = \frac{\varepsilon_{w_i}}{w_i} \xi, \tag{5.236}$$

considering that

$$\varepsilon_{w_i} = m_i \varepsilon, \tag{5.237}$$

where ε_{w_i} is the intensity of precipitation of dust fraction i on substrate (g/m^2·s); m_i is the proportion of i fraction precipitating on the substrate; ε is the total intensity of dust settling on substrate (g/m^2·s); and w_i is determined using Equation 5.207.

TABLE 5.28
Intensity of Dust Release Sources (Based on Dust Concentration in Plume)

Particulate Composition of Dust (%)/Intensity by Fraction (g/s) within Particle Size Ranges (μm)						Total Intensity (g/s)	Specific Dust Release (kg/t)
0–1.4	1.4–4.2	4.2–10	10–15	15–30	30–45		
(1) Free jet, Sokolovsko-Sarbaisky OBP; G_m = 208.5 t/h; ρ_n = 4.8 t/m^3; H = 7 m							
(a) u_1 = 1.43 m/s; x = 15 m; q_{fr} = 0.402 g/m^3							
2.51/7.66	2.19/6.64	9.48/27.40	12.91/33.24	48.72/88.05	24.18/22.70	185.59	3.21
(b) u_1 = 1.43 m/s; x = 40 m; q_{fr} = 0.28 g/m^3							
2.20/2.47	1.75/1.96	9.53/10.46	25.95/27.21	39.02/36.28	21.55/18.94	97.32	1.68
(2) Rotary picker, Lebedinsky OBP; G_m = 500 t/hr; ρ_n = 4.915 t/m^3; H = 2.5 m							
(a) u_1 = 2.45 m/s; x = 20 m; q_{fr} = 1.164 g/m^3							
1.94/3.39	1.94/3.39	9.67/16.56	26.52/43.18	38.07/54.24	22.04/29.04	149.50	1.08
(b) u_1 = 2.45 m/s; x = 60 m; q_{fr} = 0.503 g/m^3							
1.91/3.11	2.72/4.43	22.07/36.38	45.44/77.21	27.86/54.10	–	175.23	1.26

OBP = ore beneficiation plant.

Dust release intensity based on dust precipitation on a horizontal plane was determined using the research findings of the Kazsantekhproyekt Institute [175] that studied dust emissions during storage site dumping operations at the Seversky and Sokolovsko-Sarbaisky OBPs. Studies involved determining dust precipitation intensity on substrate at several measuring stations located in accordance with the aforementioned design. Calculated results are indicated in Table 5.29.

As Tables 5.28 and 5.29 indicate, the relative proportion of coarse fractions in the plume diminishes with increasing distance from the source, while that of fine fractions increases. This behavior occurs due to continued precipitation of coarse dust from the propagating plume. Dust concentration in the plume and the intensity of its precipitation on the substrate decrease with increasing distance between the source and the measuring station. This is explained by dissipation of the plume. There is a satisfactory agreement among calculated dust release intensities. Having determined the source intensity this way, we can use Equation 5.203 to calculate the surface concentration of dust along its path.

5.3.2.4 Procedure for Determining Losses of Powdered Material during Stockpiling of Fired Pellets

At outdoor storage sites, loss of powered material is defined as the amount (flow rate) of dust carried away across the storage site boundary in unit time. Losses are determined by calculating fractional dust release intensities M_i and the total intensity M.

TABLE 5.29
Intensity of Dust Release Sources (Based on Dust Precipitation on a Horizontal Plane)

Particulate Composition of Dust (% wt)/Intensity by Fraction (g/s) within Particle Size Ranges (μm)				Total Intensity (g/s)	Specific Dust Release (kg/t)
10–16	16–25	25–40	40–63		
1	2	3	4	5	6
(1) Stockpiling machine at Seversky OBP; $G_m = 600$ t/hr; $\rho_n = 4.65$ t/m³; H = 4 m					
(a) $u_1 = 1.6$ m/s; $x = 10$ m; $q_{fr} = 4.39$ g/m³; $\varepsilon = 0.3996$ g/(m²·s)					
10.38/381.78	21.86/257.01	47.10/138.76	20.66/13.65	791.20	4.75
(b) $u_1 = 2$ m/s; $x = 25$ m; $q_{fr} = 4.13$ g/m³; $\varepsilon = 0.2011$ g/(m²·s)					
33.58/598.51	34.30/219.11	32.12/68.69	–	886.31	5.32
(c) $u_1 = 2$ m/s; $x = 40$ m; $q_{fr} = 3.13$ g/m³; $\varepsilon = 0.1253$ g/(m²·s)					
45.45/615.72	34.09/178.87	20.46/43.74	–	838.33	5.03
(2) Free jet at Sokolovsko-Sarbaisky OBP; $G_m = 85$ t/hr; $\rho_n = 4.8$ t/m³; H = 4.67 m					
(a) $u_1 = 2$ m/s; $x = 10$ m; $q_{fr} = 0.47$ g/m³; $\varepsilon = 0.093$ g/(m²·s)					
1.68/57.59	5.58/56.73	34.64/73.62	58.10/18.02	205.96	8.72
(b) $u_1 = 4$ m/s; $x = 25$ m; $q_{fr} = 0.19$ g/m³; $\varepsilon = 0.0324$ g/(m²·s)					
2.56/50.92	10.58/64.84	42.33/59.71	44.53/11.67	187.14	7.93
(c) $u_1 = 7$ m/s; $x = 45$ m; $q_{fr} = 0.06$ g/m³; $\varepsilon = 0.0089$ g/(m²·s)					
2.93/35.14	13.61/50.53	44.91/38.85	38.55/6.39	130.91	5.54

OBP = ore beneficiation plant.

Fractional flow rates of dust crossing the storage site boundary at a distance x from the source are calculated using the formula

$$G_i = M_i B\left(\omega, \frac{b}{x}\right), \tag{5.238}$$

$$B\left(\omega, \frac{b}{x}\right) = \frac{\gamma\left(\omega, \frac{b}{x}\right)}{\Gamma(\omega)} = \frac{\int_0^{b/x} t^{\omega-1} e^{-t} dt}{\int_0^{\infty} t^{\omega-1} e^{-t} dt}. \tag{5.239}$$

Dust losses for the storage site as a whole are determined as a sum total of fractional rates. Calculated dust losses during free-jet storage site dumping (at the

Sokolovsko-Sarbaisky OBP) and pellet reclaiming using rotary picker (at the Lebedinsky OBP) are listed in Table 5.30.

In order to compare findings, we measured losses at the same storage sites. The following sequence of measurements was used. A measurement station was set up at a fixed distance x downwind from the source and wind speed v_B and a dust concentration field profile were determined there.

After calculating the average airborne dust content c_{avg}, the amount of dust passing through the measuring section was calculated:

$$G = \upsilon_a F c_{avg}, \tag{5.240}$$

where G is the dust flow rate (g/s); and F is the measuring section area (m^2; determined by photographing the plume at the measuring station site in longitudinal and transverse directions at a right angle to the centerline of the flare). Table 5.30 indicates that design values are close to measured values, confirming the veracity of theoretical provisions. The total amount of dust deposited from the plume on its way from the source to the measuring station could be determined—provided that conditions are equal—by subtracting the total dust flow rate from the total dust release intensity. The resulting specific dust discharge values enabled us to predict the dust emission amount for similar conditions and to calculate the atmospheric dust dispersion using the formulas indicated earlier.

TABLE 5.30
Powdered Material Losses (Based on Dust Concentration in the Plume)

Particulate Composition of Dust (% wt). Intensity by Fraction (g/s) within Particle Size Ranges (μm)						Total Losses (g/s)	
0–1.4	**1.4–4.2**	**4.2–10**	**10–15**	**15–30**	**30–45**	**Calculated**	**Actual**
(1) Free jet, Sokolovsko-Sarbaisky OBP; G_m = 208.5 t/h; ρ_n = 4.8 t/m^3; H = 7 m							
(a) u_1 = 1.43 m/s; x = 15 m; q_{fr} = 0.402 g/m^3; M = 185.69 g/s							
2.51 / 7.66	2.19 / 6.64	9.48 / 27.40	12.92 / 33.24	48.72 / 87.17	24.18 / 21.79	183.9	91.46
(b) u_1 = 1.43 m/s; x = 40 m; q_{fr} = 0.28 g/m^3; M = 97.32 g/s							
2.20 / 2.47	1.75 / 1.96	9.53 / 10.36	25.95 / 26.39	39.02 / 33.74	21.55 / 13.83	88.75	74.68
(2) Rotary picker, Lebedinsky OBP; G_m = 500 t/hr; ρ_n = 4.915 t/m^3; H = 2.5 m							
(a) u_1 = 2.45 m/s; x = 20 m; q_{fr} = 1.164 g/m^3; M = 149.50 g/s							
1.76 / 3.09	1.94 / 3.39	9.67 / 16.56	26.52 / 42.32	38.07 / 49.90	22.04 / 20.62	135.88	212.78
(b) u_1 = 2.45 m/s; x = 60 m; q_{fr} = 0.503 g/m^3; M = 175.23 g/s							
1.91 / 3.11	2.72 / 4.43	22.07 / 35.65	45.44 / 71.81	27.86 / 40.58		155.58	190.88

OBP = ore beneficiation plant.

5.3.3 Dust Release Containment in Fired Pellet Storage

Storage of iron-ore pellets is associated with intense dust release in excess of 100 g/s. Coarse particles from a moving dust plume are released early, but fine grained dust is carried away by wind far afield from the source. Calculations and field surveys carried out so far enable dust emission outside storage site boundaries to be estimated at 80% to 90% of the initial release intensity at a ground based source. As a result, containment of dust release becomes an important issue for an operating storage site.

The classification chart (Figure 5.58) shows a number of methods and means for reducing fugitive dust release. These could be implemented through organization, process design, and sanitary engineering. Efficiency could be maximized by implementing multiple means simultaneously.

5.3.3.1 Fencing Off the Flowing Material When Pellets Are Dumped into the Stockpile

Two enclosing designs—single-walled and double-walled—respectively, had been tested on the VNIIBTG process installation.

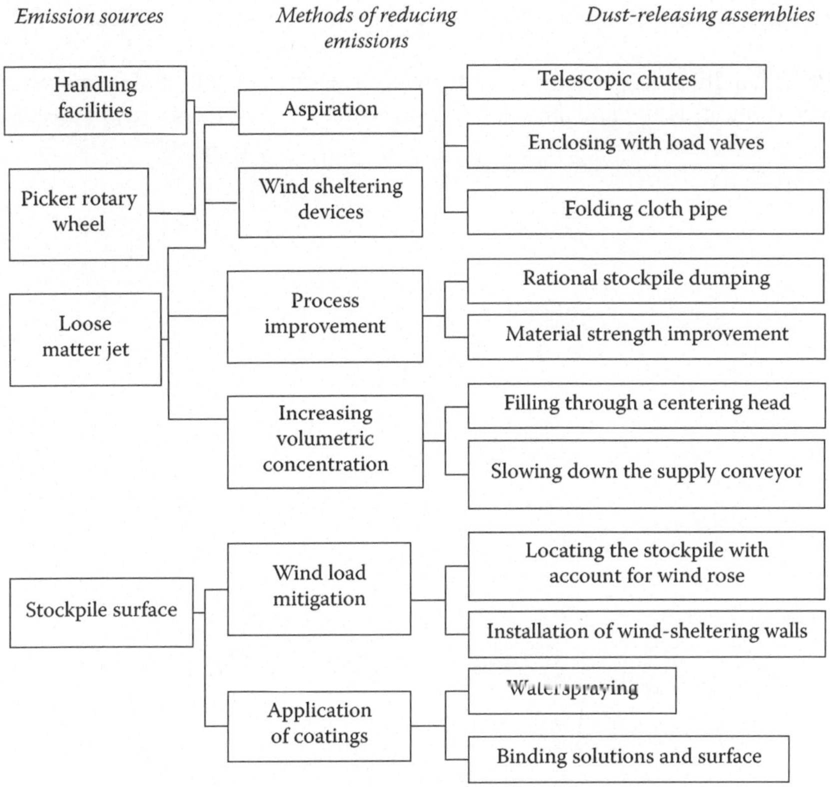

FIGURE 5.58 Means and methods of reducing fugitive dust emissions.

A single-walled enclosing structure (Figure 5.59) comprises a metalwork frame structure with hinged ports (1) in several layers over its entire surface. The conveyor (2) supplies material through a centering flange (3) and an enclosing structure (4).

A rack-and-pinion gate was used to control the capacity of the conveyor (2). Dumping height totaled 4 m. Dust-laden air was removed through an air duct (6) in the aspiration cowl (7).

A double-walled enclosing structure was installed in a small closed circuit of a semi-industrial-scale plant (Figure 5.60). Material was fed from an elevator (1) using the loading chute (2) via a connection (3) into the enclosing structure (4). By forcing the inner (5) and outer (6) ports to partially open, the material entered the receiving bin (7) and then returned to the elevator through discharge chute (8). A gate (9) was used to adjust the material flow rate. Dust-laden air was sucked away from the material dumping location via ventilation ducts (10) formed by inner and outer ports; the air then entered the aspiration manifold (11). The duct is 120 mm wide (sizing was motivated by structural reasons). Studies were carried out with the following goals:

- Verifying the expediency of a forward-flow aspiration layout
- Choosing the optimum size of the centering flange
- Verifying the design method of determining local suction efficiency

Pellet handling was accompanied by measuring the amount of air removed from the enclosing structure and the static pressures along its height. Surveys were carried

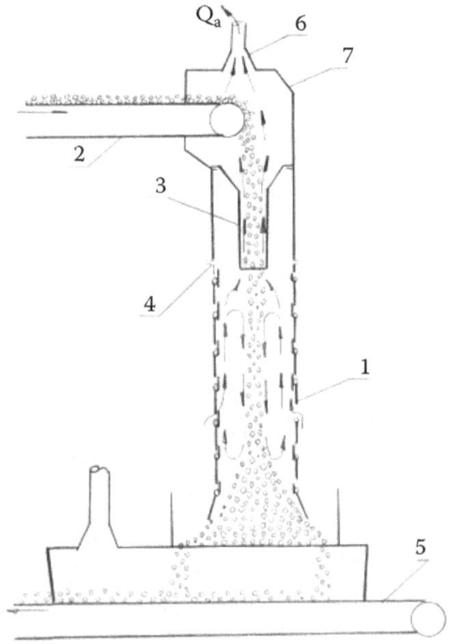

FIGURE 5.59 Enclosing structure: single-wall design.

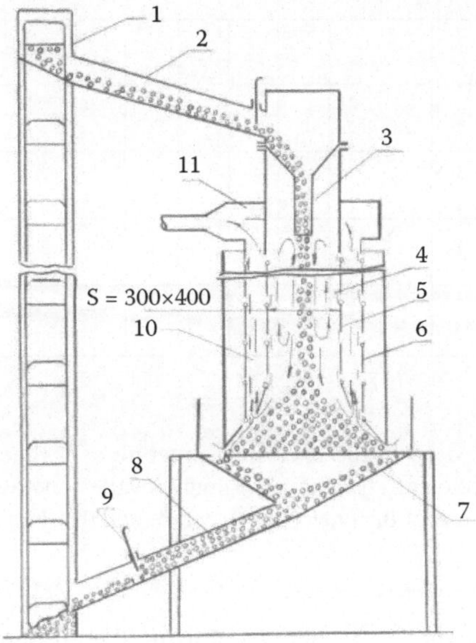

FIGURE 5.60 Enclosing structure: double-wall design.

out at different unit loads and at various ratios between cross-sectional sizes of the enclosing and the centering flange.

Benchmark tests of the aspiration systems have indicated [174] that, given equal enclosing parameters and material flow rates, the amount of evacuated air necessary for effective supply is lower in the forward-flow design than in the counterflow design. Therefore, the theoretical conclusion about expediency of an aspiration in a forward-flow design was proved by experiment.

Measurements of static pressure along the interior surface of the enclosing walls allowed for determination of the pressure gradient. It was found that, given a constant material flow rate, the pressure gradient diminishes as the cross-sectional area of the centering flange decreases (Figure 5.61). The maximum gradient, as could be expected, was observed at $k = 1$ when the flowing material occupied the entire section area of the enclosure. As flow narrowed, a vacuity formed between the flow and the enclosing walls (i.e., air recirculation area increased and the pressure drop varied sharply). At $k \geq 3$, the pressure drop virtually stabilized around a certain constant small value. Because this value is relatively minor, the pressure could be considered constantly distributed within the enclosing structure. That would significantly simplify calculation of the required evacuated air volume. Experimental surveys indicate that the optimum negative pressure between the walls is $P_2 = 12$ Pa.

Increased k not only changes the pressure gradient but also decreases the air evacuation requirement (Figure 5.62). Based on a review of experimental survey findings, 1:6 ($k = 6$) can be considered the optimum cross-section ratio between the centering flange and the enclosing structure. This value is close enough to the theoretical ratio [193].

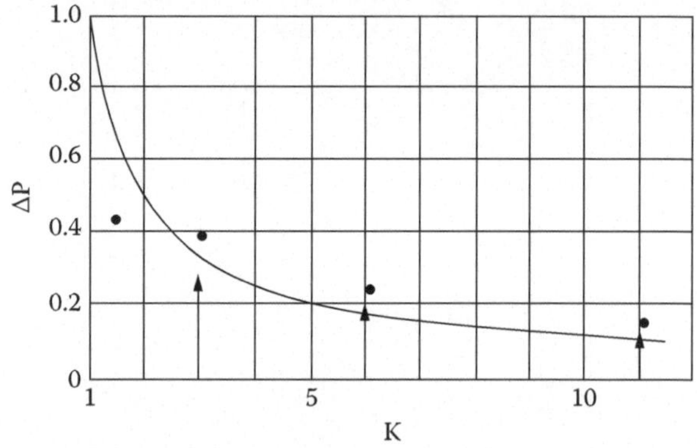

FIGURE 5.61 Pressure drop in the enclosing structure as a function of centering flange area: ● = single-walled enclosing structure; ▲ = double-walled enclosing structure ($K = S_1/S_2$; S_1, S_2—cross-sectional areas of the enclosing structure and the flange in m^2).

The volume of evacuated air is calculated using the air balance formula

$$Q_a = Q_H + Q_c. \tag{5.241}$$

The amount of air inflow through leaky locations is determined using formula

$$Q_H = 0.65 F_H \sqrt{\frac{2P_2}{\rho}}. \tag{5.242}$$

For the double-walled enclosing design described earlier, the leaking area is determined using the formula

$$F_H = N\delta\Pi, \tag{5.243}$$

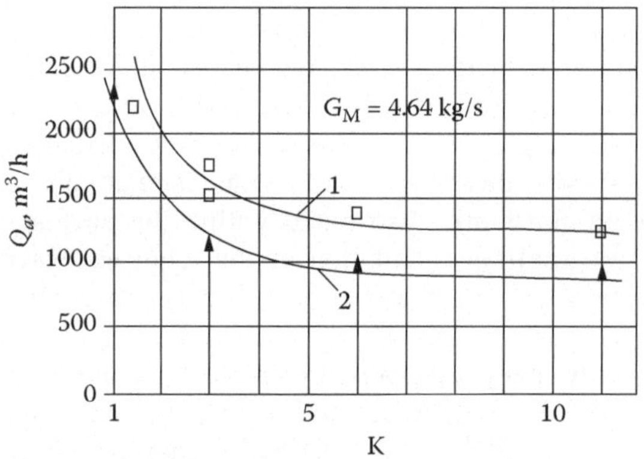

FIGURE 5.62 Aspiration volume as a function of centering flange area: 1 = single-wall enclosing structure; 2 = double-wall enclosing structure.

where N is the number of ports; Π is port perimeter (m); and δ is the conditional slot width (assumed equal to $\delta = 0.002$m). The amount of air arriving through the centering flange could be determined using known formulas deduced by us for handling facilities

$$Q_c = \varphi \upsilon_k f_c, \tag{5.244}$$

where υ_k is the speed of material flow exiting the centering flange (m/s); f_c is the cross-sectional area of the centering flange; and φ is the phase slip ratio.

Considering the distribution of leaks and the fact that the formula for Q_c produces a somewhat lower result, an adjustment factor must be introduced for the calculated relationship shown in Equation 5.241. As a result of verifying the calculation method experimentally, the value of the adjustment factor has been found to be $k_H = 1.25$.

In this case, calculated aspiration volumes would agree comfortably with measurements (Table 5.31) because relative error would not exceed the accuracy of the experiment.

Operating trials of the enclosing structure indicate that the structure is functional: ports open when the enclosure is filled with pellets and close after their release, the outflow of pellets through the ports is steady, and no dust was released outside through half-open ports and leaky spots at optimum negative pressure conditions. No dust contamination attributable to pellets moving along the stockpile surface was observed.

TABLE 5.31
Aspiration Volumes for Enclosing Structure Double-Wall Design

Material Flow Rate G_m (kg/s)	P_2, Pa	Aspiration Volumes (m³/hr) Calculated	Actual	Accuracy,%
At $f_c = 0.1$ m²				
2.12	−1.25	610	658	−7.3
4.64	−3.75	885	858	−3.1
6.15	−10.25	1294	1364	−0.5
8.54	−21.25	1766	1727	−2.3
At $f_c = 0.2$ m²				
2.12	−4.25	1021	1040	−1.8
4.64	−6.75	1244	1283	−3.0
6.15	−7.75	1450	1379	+5.1
11.27	−13.25	1725	1677	+2.9
At $f_c = 0.04$ m²				
2.12	−3.5	924	858	+7.7
4.64	−6.25	1123	1060	+5.9
6.15	−12	1405	1499	−6.3
11.27	−19	1751	1690	+3.6

The design and operating principle of an aspirated enclosing structure [192] for overhead filling of backup storage sites (Figure 5.63) rests upon using a vertical closed metal chute reinforced over its entire surface with hinged paired ports installed atop a concrete foundation and attached to the ramp steelwork at its mouth.

Structurally, this enclosure is made up of the following components: discharge chutes (1), dust-laden air manifold (2) with air receiver (3) and centering flange (4), and a closed vertical chute (5)—with ports (6) and stiffening ribs (7)—installed atop a foundation (8).

Pellets leave the dumping trolley through the discharge chute and the centering flange, entering the inner cavity of the enclosing structure. A falling stream of pellets at the stockpiling location forces open the lower row of hinged ports, passes through it, and drops into the stockpile. After the lowest level of the stockpile is filled, the next row of ports is open, and so forth, until the stockpile is completely full. When pellets are reclaimed from the stockpile, the respective ports close under their own weight.

Dust-laden air is sucked away from the material drop location through four ventilation ducts formed between paired ports and stiffening ribs.

Using paired ports and stiffening ribs reduces the occurrence of leaky spots and mitigates wind action during the aerodynamic handling processes, as well as reducing the required aspiration volumes.

Dust-laden air flows from ventilation ducts into the manifold and then passes through the air receiver and air duct system into the aspiration process unit installed in a nearby building.

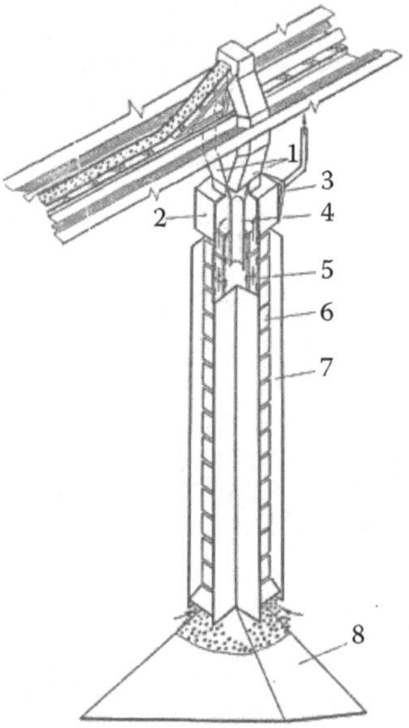

FIGURE 5.63 Aspirated enclosing structure for rack-based filling of backup storage: 1 = unloading chutes; 2 = manifold; 3 = air receiver; 4 = centering flange; 5 = a vertical chute; 6 = ports; 7 = stiffening ribs; 8 = foundation.

The cross-section of the centering device was equal to 1/6 of the inner chute section in order to improve the aerodynamic properties of the enclosing device and to eliminate the possibility of pellets hitting above the top level mark of the stockpile.

In order to ensure unobstructed operation of the loading machinery while preventing it from causing mechanical damage to the enclosing structure, the machinery must be installed on a 1.5-m-high, pyramid-shaped concrete foundation.

A prismatic enclosure model with double walls was aerodynamically tested at the VNIIBTG wind tunnel with Reynolds numbers Re = $0.52 \cdot 10^6$ (i.e., in the self-similar flow area). Parameters recorded included the temperature of circulating air, velocity pressure of free (not decelerated) flows, and the pressure exerted by the flow on the enclosure walls at flow striking angle α incremented in 15° steps at various angular protrusion lengths.

The striking angle in this case was provided by the angle between the flow velocity vector $\vec{\upsilon}$ and the inner normal n of the plane under pressure.

Test results (Figure 5.64) indicate that the wind stress factor of the enclosure is positive for faces oriented toward the flow (due to its deceleration) and turns negative on the side and back walls.

The wind stress factor was determined using the ratio

$$k = P / \left(\frac{\rho \upsilon^2}{2} \right), \tag{5.245}$$

where P is the pressure on the enclosure walls (Pa); $\rho = 1.181$ is the mass flow rate of air (kg/m^3); and $\upsilon = 8.08$ is the flow velocity (m/s).

We found that flat angular protrusions mean that more than 75% of all leaky locations are within the aerodynamic shadow zone.

Using angular protrusions in enclosing structures enables a significant reduction of the wind stream impact on the aspiration volume when the enclosure is installed at an angle $\alpha = 45°$ to the prevailing wind direction and with an angular protrusion length $h/b \geq 0.46$.

For performance studies in industrial conditions, an experimental enclosure specimen consisting of a receiving bin and three identical sections was installed under the gallery of conveyor #28 of the outdoor pelletization shop storage site at the

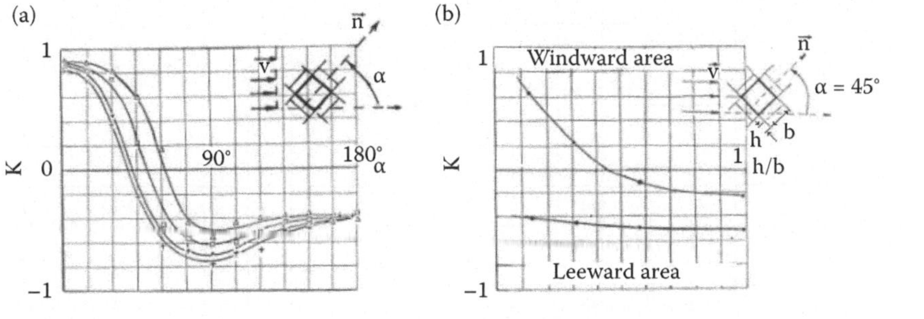

FIGURE 5.64 Behavior of the wind stress factor in a double-walled enclosure: (a) as a function of striking angle α (Δ – h/b = 0.14; □ – h/b = 0.32; • – h/b = 0.58; ⯁ - h/b = 1); (b) as a function of angular protrusion length (at $\alpha = 45°$).

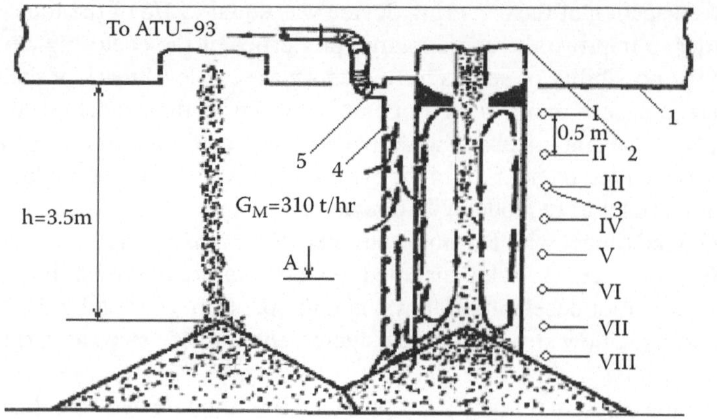

FIGURE 5.65 Industrial testing of an experimental enclosure specimen (Sokolovsko-Sarbaisky OBP): 1 = gallery of the conveyor #28; 2 = discharge chute; 3 = measurement point; 4 = loading port; 5 = local suction.

Sokolovsko-Sarbaisky OBP (Figure 5.65). The section consisted of a 1.02-m-long metal chute having three pairs of inner and outer ports on one of its sides (for releasing pellets into the stockpile). Ports were hinged on angular protrusions, providing a wind shelter. A centering washer was installed on the bottom of the bin to increase volumetric concentration of material in the flow.

Aerodynamic measurements indicated that changes in the wind stress factor (Figure 5.66) agreed with the data obtained by placing the model in a wind tunnel.

Enclosure testing was composed of three phases. Pellets were first fed into the stockpile using a free jet. Dumping height *h* varied from 3.5 to 3.2 m. Loading then proceeded through enclosures with the aspiration turned off. In the first phase, the enclosure ran with the aspiration turned on. Local suction was connected to an

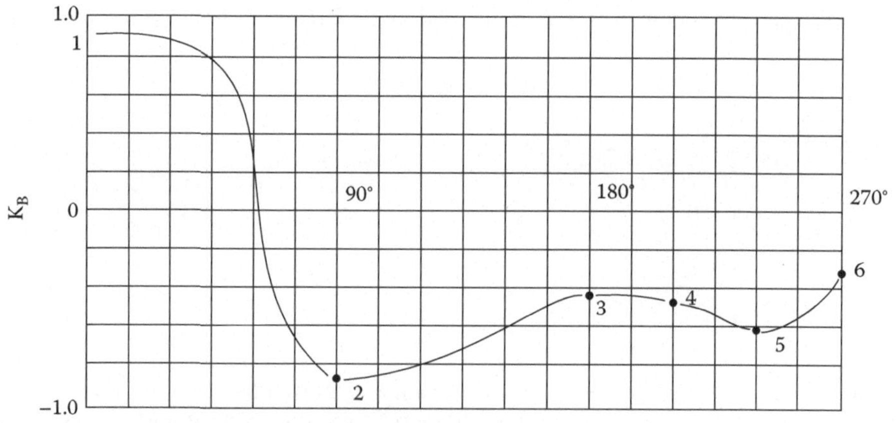

FIGURE 5.66. Behavior of the wind stress factor.

ATU-93 aspiration process unit. Conveyor #28 handled 310 t/hr. Wind speed averaged 5.96 m/s. In all three cases, air samples were taken to determine dust content at a point 40 m away from the location where the pellets fell into the stockpile, along the centerline of the plume. Tests indicated a 3.803 g/m^3 dust content in the plume in the free-jet storage unloading mode.

Dumping pellets through the enclosure with the aspiration turned off brought the airborne dust content down to 1.351 g/m^3. When the aspiration was turned on, dust content decreased further to 0.547 g/m^3.

The enclosure test specimen was designed with the following parameters:

- Inside perimeter: 2.8 m
- Height: 3.6 m
- Evacuated airflow: actual = 1350 m^3/hr; optimum (design case) = 3630 m^3/hr
- Area of leaky joints: actual = 0.535 m^2; recommended = 0.198 m^2
- Negative pressure inside duct: 13 Pa

Testing confirmed structural design choices and functional performance of the structure; therefore, it should be capable of performing dust containment functions during storage loading.

As the data illustrated here indicates, initial dust concentration in the plume was decreased by a factor of 6.9. A significant dust contamination of the air, during pellet loading through the enclosure with aspiration turned on, could be explained by insufficient performance of ATU-93. In addition, the leaky area was 2.7 times greater than what was recommended. Remedying these installation deficiencies would enable further reduction of dust released.

5.3.3.2 Local Suction Units and Aspiration Systems of Storage Facilities

The most preferable way to reduce dust released into the environment during handling operations at pellet storage sites would be to install containment cowls at handling assemblies of warehouse facilities. Such cowls would become structural parts of the machinery and would enable a manifold reduction in dust release without involving additional features.

Dedusting efficiency reaches its maximum if the dry-aspiration method is used. Foreign designers of storage equipment [194] install aspirated cowls at handling assemblies having elements that retain their spatial orientation relative to each other during operation. Wet dedusting may provide additional aid [195].

Development of integrated dedusting designs focuses on two problems: (1) validating an experimental specimen in natural conditions and (2) structurally integrating the aspiration system into a newly designed machine.

Each of these problems was solved in several phases. Experimental integrated local suction units have been installed in natural conditions on a UIK-600 overhead stockpiling machine at the Kostomukshinsky OBP outdoor pellet storage location. This machine was designed and produced by the Yuzhuralmash manufacturing group. Before the first problem could be solved, a layout chart [63,81] had to be prepared that designated dedusting equipment installation locations for the machine being served.

Although the technical solutions proposed may seem obvious, their implementation was complicated by a number of factors. This was the first practical case of using aspiration for dedusting of mobile assemblies in the design and operation of storage machinery. This unique system had its components integrated into an operating mobile machine with significant dimensional constraints, which made it difficult to select an efficient dry deduster. Thus, the following approach is used for wellhead housing construction.

The design specification for the UIK 600 stockpiling machine details the solution to these issues. After coordinating the design specification with the manufacturer, the Krivyi Rih (Ukraine) branch of the Design and Evaluation Office of the Kyiv Research Institute of Occupational Hygiene and Diseases, design documents have been developed and subsequently handed over to the Kostomukshinsky OBP staff for the fabrication of the aspiration system and its installation on stockpiling machine #1.

The aspiration system is autonomous and is made up of the following components:

1) Aspirated cowl for the boom conveyor loading location
2) Aspirated cowl for the dropper drum of the boom conveyor
3) Aspirated cowl of the stockpile dumping location
4) Air duct system with local suction units
5) Dry dedusters—an array of 4TsN-15-900P and SKTsN-34-1800 cyclone units (both types were tested)
6) DN-12,5 induced-draft fan
7) Collecting bin for trapped dust
8) Exhaust pipe

The motor of the induced-draft fan is supplied from the existing switchgear of the UIK-600 stockpiling machine. An agreement was reached with Yuzhuralmash that the power draw would not exceed the reserve of the existing 80 kW cable.

Components of the aspiration system have been structurally integrated as far as possible within the boundaries of the machine. A local suction unit for the boom conveyor loading location was installed on a chute adjoining the existing transfer assembly cowl. The distance between the aspiration funnel and the point of pellet drop onto conveyor belt measured approximately 2 m. A suction air duct was arranged 2 m above the walkway. The aspirated cowl of the dropper drum was manufactured in accordance with the design for heated loose materials [63].

The aspirated cowl of the stockpile dumping location (Figure 5.67) was furnished with a slot-type steady suction connection. The cowl has a metal duct with rectangular U-shaped aspiration chambers in front and at both side walls. At the end abutting on stockpile, the duct has a flexible seal. The boom conveyor belt cleaner was also dedusted by placing it inside the cowl at the stockpile loading location.

The air duct system (Figure 5.68) was made up of local suction units, an air duct interconnecting local suction units at the overhead floor level, dust trap piping, and an air duct connected to the induced-draft fan. Local suction units came complete with choke valves.

An array of four 4TsN-15-900P cyclone units was used for cleaning aspirated air in agreement with the Kostomukshinsky OBP (Figure 5.69). A single SKTsN-34-1800

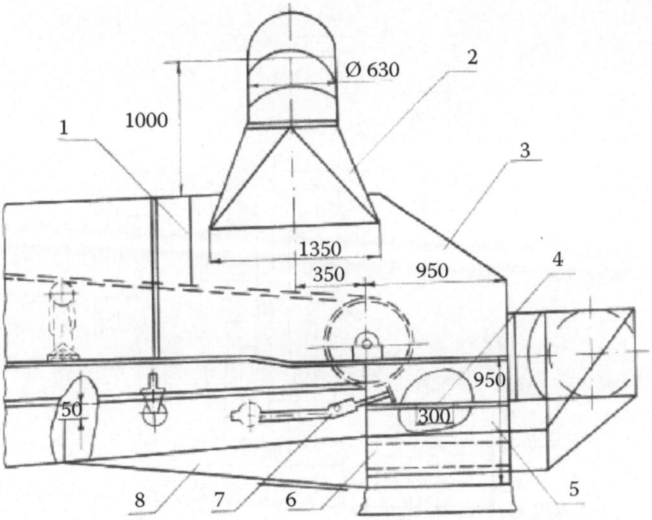

FIGURE 5.67 Aspirated cowl of the stockpile dumping location: 1= seal; 2 = local suction; 3 = dropper drum cowl; 4 = diaphragm; 5 = cowl of the stockpile dumping location; 6 = slot-hole suction; 7 = belt cleaner; 8 = belt cleaner cowl.

cyclone was added later. Cyclones were emptied into a bin under the dust trap floor for subsequent disposal of the trapped dust. The floor was located to the left of the portal at the counterweight side and was furnished with two wheeled supports. A part of the floor was attached to the dumping trolley and portal steelwork.

The induced-draft fan with its motor was located at the cyclone floor. A DN 12,5 induced-draft fan was chosen in view of operating conditions. The exhaust pipe

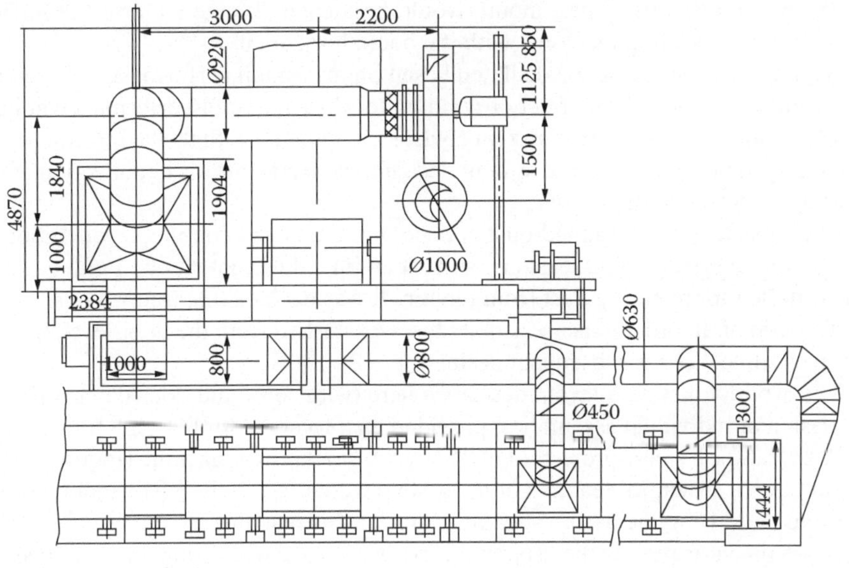

FIGURE 5.68 Air duct system of the aspiration unit in the UIK 600 stockpiling machine.

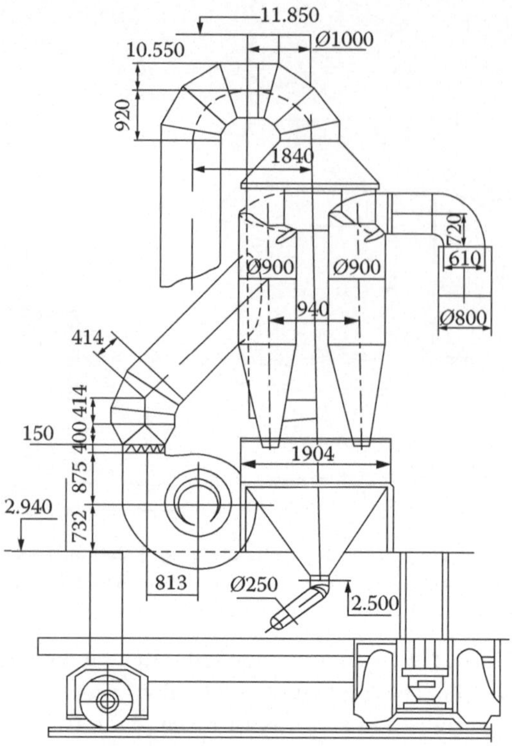

FIGURE 5.69 Aspiration cyclone array and induced-draft fan layout at the additional floor of the stockpiling machine UIK 600.

height ensured that the pipe's mouth would be located 2 meters above the highest point of the stockpiling machine, with the boom horizontal.

Aspiration volumes were calculated based on maximum performance of the process equipment and on the effect of wind on local suction unit performance at handling facilities. The aspiration system control was coupled with the laying equipment starting system. The aspiration system design included an air duct connection at the cantilever neck centerline level.

The capacity of the trapped dust collection bin enabled continuous operation of the aspiration system for at least one shift at an inlet dust content of 15 g/m^3. A gate was installed for releasing dust from the bin. It was decided that trapped dust would be disposed of. In order to achieve that, dust was flushed from the floor-level bin into the fines silo cellar for subsequent deslurring.

The aspiration system layout described here (with some additions determined by the design of individual assemblies) provided the prototype for the aspiration system in Yuzhuralmash's new project, the UISN 1200 stockpiling machine (Figure 5.70).

Additional structural features of the system were used by the Mekhanobrchermet Institute in drafting the design specifications for this machine.

A prominent feature is the aspirated cowl of the stockpile filling area, designed in cooperation with Yuzhuralmash engineers. It has a swinging suspension (Figure 5.71)

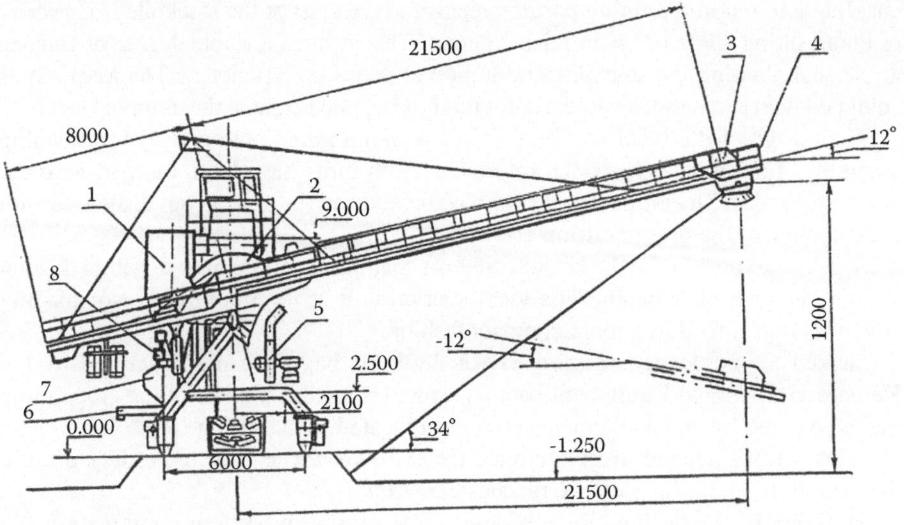

FIGURE 5.70 Aspiration diagram for the new UISN 1200 stockpiling machine: 1 = bag hose; 2 = boom conveyor loading cowl; 3 = dropper drum cowl; 4 = pile dumping cowl; 5 = induced-draft fan; 6 = auger conveyor; 7 = bin; 8 = rotary discharger.

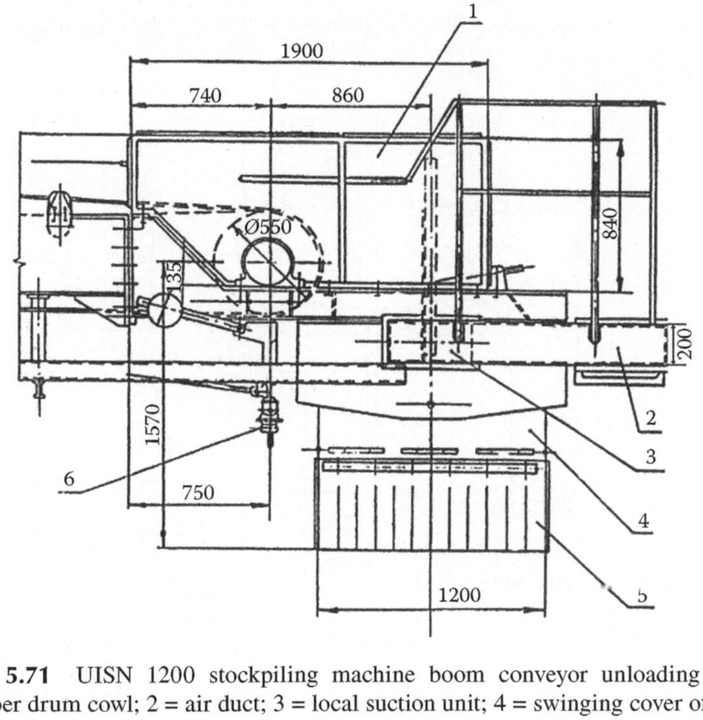

FIGURE 5.71 UISN 1200 stockpiling machine boom conveyor unloading assembly: 1 = dropper drum cowl; 2 = air duct; 3 = local suction unit; 4 = swinging cover of the stockpile loading location; 5 = elastic seals; 6 = material level sensor.

enabling it to retain a constant position relative to the crest of the stockpile irrespective of boom tilting angle or boom raising height. This ensures a stable degree of leakage at the seal abutting the stockpile and improves dedusting efficiency. The assembly is equipped with a material level sensor for monitoring pile height at the dumping location.

The design of the local suction unit connection to the boom conveyor unloading assembly (Figure 5.72) has been improved. Both units have been merged with the boom steelwork. The air duct serves as a walkway, and the new layout dispenses with local suction at the dropper drum cowl.

A new seal design for the boom conveyor loading locations was developed using spring-loaded runner bands. The local suction unit at the boom conveyor loading cowl was rearranged in a more compact fashion.

Sucked air is cleaned using a FRKN-180U-01 bag hose manufactured by the Kemerovo Chemical Equipment Factory. Dust trapped in the bin of the stockpiling machine is fed by an auger conveyor to a bin located at the storage site.

A DN-11,2U induced-draft fan (rated for 28,700 m^3/hr) is used to produce draft in the unit. Full operating pressure reaches 3.57 kPa.

In addition, the designs for positioning the aspiration system equipment in an overhead stockpiling machine, for installing cowls, and for integrated local suction units have been used by the Mekhanobrchermet Institute when designing a similar machine at the Krivyi Rih (Ukraine) Oxidized Ore Beneficiation Plant.

At the request of Yuzhuralmash Manufacturing Group and the Lebedinsky OBP, the design specifications for the new machine have been amended. The aspiration system was made a part of the machine. Following negotiations between these parties, this work was reflected in the "Supplement to Design Specifications 1-130340

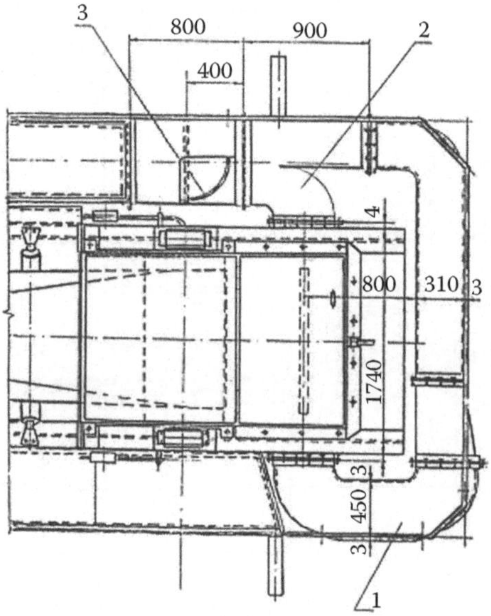

FIGURE 5.72 Local suction units connect with the discharge assembly of the UISN 1200 stockpiling machine boom conveyor: 1, 2 = local suction units; 3 = throttle valve.

TZ for the design and fabrication of an integrated aspiration system for the UISN 1400 single-boom non-rotary stockpiling machine at the final product storage site of the Lebedinsky OBP pelletization factory."

Full-scale tests pursued the following goals:

1) Measuring actual aspiration volumes and negative pressures in aspiration cowls and comparing them with optimum values
2) Testing the DN-12,5 induced-draft fan
3) Testing local suction units
4) Testing 4TsN-1-900P and TsKN-34-1800 cyclones
5) Evaluating the efficiency of the aspiration system for reducing dust emissions

Tests were carried out in line with the agreed work program at the nominal load of the process equipment. Installation defects were remedied during testing, but it should be noted that the quality of installation work was generally acceptable.

Aerodynamic measurements and dust sampling had been performed in accordance with the diagram in Figure 5.73. Values indicated correspond to design calculations. The system was tested in two modes: (I) with the choke valve at point 7 closed, and valves at points 5 and 8 half open to provide a visual sign of dust containment; and (II) with all choke valves open.

The induced-draft fan housing was manufactured at the mining and refinery equipment repair works at the Kostomukshinsky OBP. Due to engine replacement, the fan was tested at two different impeller rotation speeds: n = 985 min^{-1} and n = 1470 min^{-1} (Table. 5.32). Parameters recorded included ambient air temperature, air temperature at induced-draft fan inlet, and wind speed.

Local suction units have been tested with the goal of establishing negative pressure in all cowls so as to prevent dust knockout through process openings and leaky joints. Cowl sealing became a top priority due to the limited capacity of the

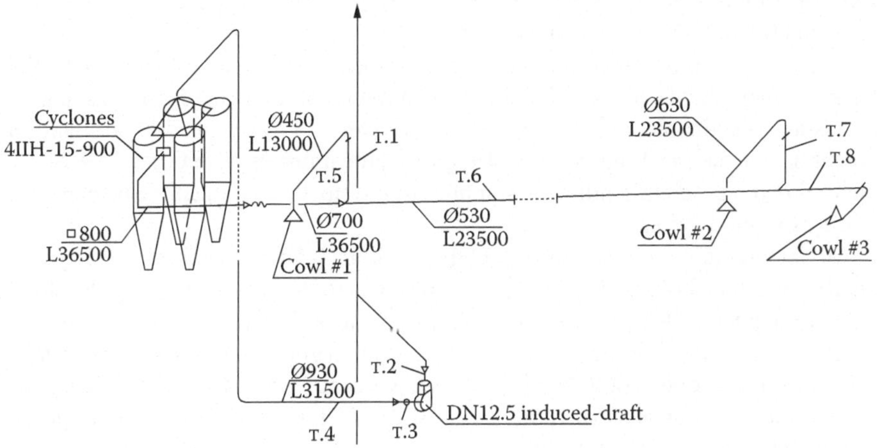

FIGURE 5.73 Aspiration system layout of the UIK 600 stockpiling machine.

TABLE 5.32
Results of Aerodynamic Tests of DN-12,5 Induced-Draft Fan

Rotation Speed, min^{-1}	Full Pressure, Pa Suction	Discharge	Total	Capacity m^3/hr
985	$\frac{-1249}{-1133}$	$\frac{\pm 263}{+295}$	$\frac{1512}{1428}$	$\frac{28,080}{30,030}$
1470	−2973	+337	3310	31,740

Numerator corresponds to Mode I, denominator corresponds to Mode II.

aspiration system and a nonexistent power margin of the supply cable. An inspection of the sealing quality of cowls revealed significant leaks, some of which had been mitigated during the commissioning of the system. For example, the total leakage in the cowl of the boom conveyor loading location reached 1.47 m^2 before tests began. Moreover, an opening with an area of 1.2 m^2 was specially created under the belt cleaner in the upper part of this cowl to enable loading from the spillage floor.

The upper part of the cowl also experienced floating leakage measuring 0.1 m^2. Depending on the boom slope angle, this occurred either in front of or behind the transfer chute. After remedying a number of installation defects, the leakage area in the boom conveyor loading location was reduced to 0.4 m^2 (i.e., by nearly 3.7 times), enabling a reduction in the volume of air evacuated by the suction units (Table 5.33).

Leaky areas at the dropper drum cowl of dumping trolley measured 0.66 m^2. Leaky spots were significantly reduced at the end side and at the bearings.

The wind-blown area under the cowl of the stockpile loading location measured 2.73 m^2. (This is the free area between the stockpile and the end of the rubber curtain along the entire cowl perimeter.) Positive dust emission containment was achieved because a reduction of the open part of the dumped pellet flow resulted in more stable operation of the local suction units in a strong wind. The pellet flow rate varied between 800 and 850 t/hr when the system was tested.

Studies indicated that actual aspiration volumes are 1.2 to 1.4 times less than design values. Nevertheless, this reduced amount of evacuated air was enough to ensure sustainable and complete containment of dust inside the handling assemblies of the machine. System design specifications required that 70% of dust released in the product-handling assemblies of the stockpiling machine be contained by the aspiration system.

Air velocity in the suction air duct at the stockpile loading location is clearly inadequate when the choke valves are open. This leads to the risk of clogging the slotted suction air duct with deposited dust. Therefore, it is desirable to operate just two suction units by abandoning suction in the dropper drum cowl of the boom conveyor. Velocities in this case would be high enough to ensure evacuation of dust-laden air.

When testing was performed at a wind speed of 2 m/s with aspiration in point 6 disconnected, a slight dynamic pressure of 3 Pa was recorded. That is, natural draft occurred with the choke valves open. This should be taken into account when the wet

TABLE 5.33
Results of Aerodynamic Testing of Local Suctions and Cowls

	Cowl Location		
Measured Parameter	**Boom Loading**	**Dropper Drum**	**Stockpile Dumping**
Air duct diameter (mm)	450	630	630
Air pressure (Pa)	112	0	125
–dynamic	166	76	26
–static	−157	0	−494
	−269	−107	−92
Air velocity in air duct	13.1	0	13.9
of the local suction unit (m/s)	15.9	10.7	6.3
Temperature of evacuated air (°C)	5.5	5	4
Volumetric flow rate of evacuated air (m^3/hr)	7520	0	15,610
	9080	12,010	7020
Total volumetric flow rate of evacuated air (m^3/hr)		23,110	
		28,110	
Dust content in evacuated air (mg/m^3)	22,810	0	8330
	9960	25,020	2100
Dust carryover into aspiration network (kg/hr)	171.53	0	130.03
	90.44	300.49	14.74
Dust density (kg/m^3)		4760	
Pressure losses at suction (Pa)	45	–	369
	103	31	66
Local resistance factor of suction unit	0.57	–	2.95
	0.62	0.41	2.54
Air velocity in process opening of suction unit (m/s)	13.1	0	15.77
	15.9	3.92	7.09
Negative pressure in cowl behind chute (Pa)	−2	−8	−10
	−6	−30	−12
Total area of leaky joints in cowl (m^2)	0.4	0.4	2.25

Numerator corresponds to Mode I, denominator corresponds to Mode II.

dedusting system is operated in seasons with above-freezing temperatures: choke valves should be closed to avoid clogging the network.

The 4TsN-15-900P cyclone array was tested at modes similar to those defined earlier, and the TsKN-34-1800 cyclone was tested at fully open choke valves (Table 5.34). Cyclones installed on the suction line are airtight. A dust removal gate enables removal of trapped dust from the bin. Cleaning efficiency was determined in view of the data obtained during tests of local suction units and cowls.

Significant pressure losses in the SKTsN-34-1800 cyclone necessitated increasing the speed of the induced-draft fan impeller to 1470 min^{-1}, and the fan motor was replaced.

TABLE 5.34
Results of Dust Trap Testing

	Test Results	
Indicators	n_1 = 985 min⁻¹	n_2 = 1470 min⁻¹
Dust traps		
(1) Dust trap type	4TsN-15-900P	SKTsN-34-1800
(2) Dynamic pressure upstream of cyclone (Pa)	116/135	146
(3) Pressure losses in cyclone (Pa)	840/930	2,560
(4) Air velocity at cyclone inlet (m/s)	13.3/16.16	25.05
(5) Air velocity inside cyclone (m/s)	2.64/3.21	3.47
(6) Local resistance factor	190/142	360
(7) Air leaks into cyclones (%)	14.81/7.62	4.9
(8) Average airborne dust content (mg/m³)		
—Upstream of cyclones	13,038/14,432	8860
—Downstream of cyclones	690 1183.3	778
(9) Dust balance (kg/hr)		
—Entered cyclones	301.66/405.67	281.22
—Trapped by cyclones	282.83/369.66	256.3
—Escaped to atmosphere	18.73/36.01	24.69
(10) Cleaning efficiency (%)	94.71/91.8	91.22

Numerator corresponds to Mode I, denominator corresponds to Mode II.

While the SKTsN-34-1800 cyclone was tested, leaky areas in the cowls increased somewhat (up to 30%) because seals lost their tightness.

As Table 5.34 shows, large amounts of dust were carried away into the aspiration network: 301.56 kg/hr in Mode I and 405.67 kg/hr in Mode II. That made it possible to separate (isolate) dust in the path upstream from stockpiling machine or to introduce measures aimed at reducing dust concentrations in evacuated air. Specially designed local dust separator suction units can be used for that purpose.

Cyclones operate at their maximum efficiency, which is almost equal for all types of dust traps tested: When using a TsKN-34-1800 cyclone with a significant hydraulic resistance at the first cleaning stage, the full pressure of the induced-draft fan must be increased.

A cowl, even without the suction of dust-laden air, significantly reduces fugitive dust emissions into the atmosphere. Dust samples were taken at four measuring stations (Figure 5.74) along the wind in the breathing zone to determine specific dust content values. Samples were taken at various boom positions and with aspiration alternatively turned on and off (Table 5.35).

Surveys were performed before and after repairs on the stockpiling machine that included improving the cowl sealing and relocating the belt cleaner into the cowl at the stockpile dumping location. As part of test sampling, background dust content and wind speed were recorded. Pellet flow rate varied between 800 and 850 t/hr.

As is seen in Table 5.35, cowl installation in the stockpile loading location together with minimum dumping height permitted by the process design (without aspiration) reduced the airborne dust content by a factor varying from 6.4 to 10.8 compared to the worst case. With aspiration, this reading was 16.1 times lower; dust containment efficiency measured 93.3%.

Survey findings indicate that dust is carried away beyond the storage site boundaries. Dust content in the air with aspiration turned off rises as pellet dumping height increases, reaching 437.5 mg/m^3.

Various local integrated suction units removing dust from aspiration air can be used to mitigate dust emissions during operation of storage-site machinery:

- For stockpiling machines, local suction units should be installed as follows—in the overhead conveyor loading location (ensuring cowl sealing when the cantilever tilt angle changes) and where the pellets drop into the stockpile (providing a centering flange and a swinging chute with flexible sealing shields in the upper part and ensuring steady removal of contaminated air from the cavity of this chute)
- Rotary pickup machines require local suction units at the boom conveyor loading assembly (with sealing provided in loading and unloading locations of rotary wheel buckets) and at the boom conveyor unloading assembly. A discharge bin should be designed for a "buried" operation.

Equipping the U1K 600 stockpiling machine at the Kostomukshinsky OBP pelletization factory with an integrated set of local suction units and aspiration air cleaners led to more than a tenfold decrease in dust emissions during pellet stockpiling.

5.4 FUGITIVE EMISSIONS AND CONTAINMENT OF DUST DURING LOADING OF IRON-ORE PELLETS IN RAILWAY CARS

5.4.1 Dust Emission Containment Designs for Loading Bins

A process widely used today for loading pellets and sinter into railway cars is based on an intermediate bin. This loading design is currently in place at a number of ore beneficiation plants, including Kachkanarsky, Sokolovsko-Sarbaisky, Lebedinsky, Mikhailovsky, and Kostomukshinsky OBP's.

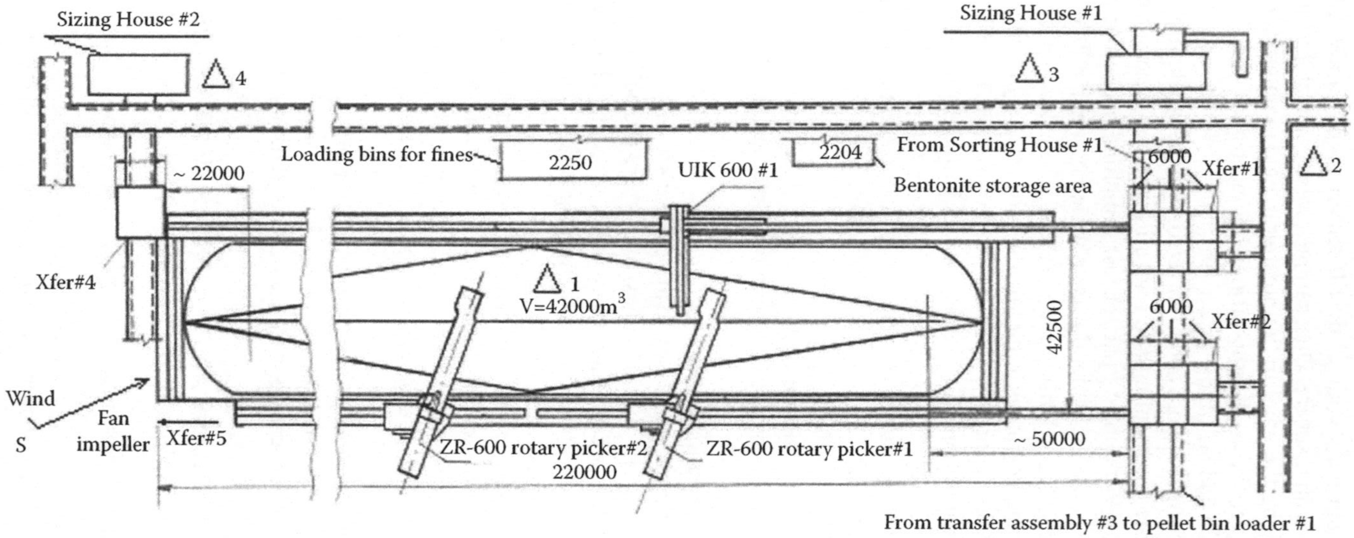

FIGURE 5.74 Situational layout diagram of the outdoor pellet storage site at Kostomukshinsky OBP (Δ 1, Δ2, Δ3, Δ4 – measuring stations, Xfer – transfer assemblies).

TABLE 5.35
Results of Efficiency Studies Carried Out for Aspiration System and Cowls

No.	Measuring Station/ Distance to machine (m)	Condition of Aspiration	Position of Cowl at Loading Location	Dust Content in Air (mg/m³)	Efficiency of Containment	
					Factor	%
Prior to repairs (northerly wind)						
1	1/30	On	Raised by 3.5 m	1744.6	5.0	79.8
2	1/30	Off	Raised by 3.5 m	8637.5	–	–
3	1/30	Off	Lowered onto stockpile	870.8	9.9	89.92
4	1/30	On	Lowered onto stockpile	212.5	41.1	97.54
After repairs (southerly wind)						
5	2/150	Off	Lowered onto taper	11.54	6.4	80.83
6	2/150	Off	Raised by 3.5 m	62.3	–	–
7	2/150	On C_{out} = 1183.3 mg/m³	Lowered onto taper	7.68	10.8	87.24
8	3/180	On C_{out} = 1183.3 mg/m³	Lowered onto taper	8.85	–	–
9	4/100	Off	Raised by 5 m	437.5	–	–
10	4/100	On C_{out} = 1183.3 mg/m³	Lowered onto taper	29.17	16.1	93.3

When loading bins are present, apron conveyors supplying pellets to and from the bins are commonly provided with local air suction. When cars are loaded inside a tunnel housing, the housing is equipped with an air suction system.

Figures 5.75 and 5.76 illustrate loading the bin aspiration layout currently in place at pelletization factories of the Kachkanarsky and Sokolovsko-Sarbaisky OBPs, as well as loading tunnel dedusting layout at the Krivyi Rih Central OBP, the Seversky OBP sintering factory #1 at the Novokrivorozhsky (presently ArcelorMittal Krivyi Rih) OBP, and sintering factories at the Zhdanovsky (Mariupol') and Novolipetsky steel works.

Industrial-scale tests of loading bin aspiration systems at pelletization factories at the Kachkanarsky OBP, the Sokolovsko-Sarbaisky OBP, and the loading tunnel housings of the Krivyi Rih Central OBP, the Novokrivorozhsky OBP, and Ilyich Iron and Steel Works (Mariupol', Ukraine) have been performed by the All-Union Institute for Occupational Safety in Ore Mining [196].

State Design Institute Santekhproyekt performed tests at dedusting systems of loading bin array #1 and at loading tunnel housings at the Seversky OBP and Novolipetsky Steel Works. The loading bin array #2 at the Seversky OBP was inspected by the Seversky OBP Laboratory for Dust Measurements. Findings from industrial-scale tests are listed in Table 5.36.

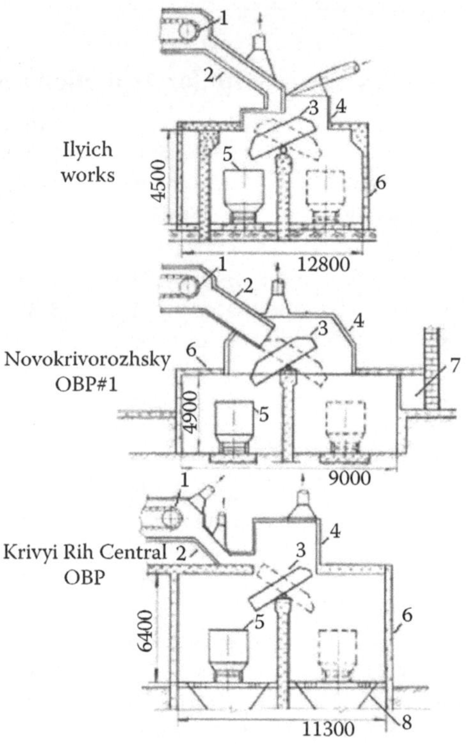

FIGURE 5.75 Dedusting layouts for loading tunnels in multi-point loading mode: 1 = sintering (firing) unit; 2 = discharge chute; 3 = fly chute; 4 = fly chute cowl; 5 = car; 6 = tunnel; 7 = gallery; 8 = spillage bin.

A significant dust release occurs at railway car loading stations. The resulting dust cloud (with dust concentrations reaching 4 to 10 g/m^3) contaminates the industrial site atmosphere as it spreads around, aided by convectional flows and by wind. In the majority of cases, existing aspiration facilities designed to trap dust at car loading stations are unable to completely contain dust emissions. Aspirated air volumes per loaded car range from 25,000 m^3/hr at the Krivyi Rih Central OBP to 230,000 m^3/hr at Novolipetsky Iron and Steel Works. A centralized aspiration dedusting system in the loading tunnel of sintering factory #2 at Ilyich works is by far the most efficient [198]; there is virtually no release of contaminated air into the industrial site. An insignificant dust release occurs through the end openings of the housing only with the most unfavorable wind direction.

Installation of telescopic chutes at loading bins #2 and #3 of the Sokolovsko-Sarbaisky OBP pelletization factory enabled significant reduction of industrial site air pollution from dust. Yet, with significant leakage areas and low volumes of evacuated air, noticeable dust release still occurs during the initial startup and topping-up of the hopper car [199]. Airborne dust content at working sites in the loading bin structure measures 20 to 30 mg/m^3.

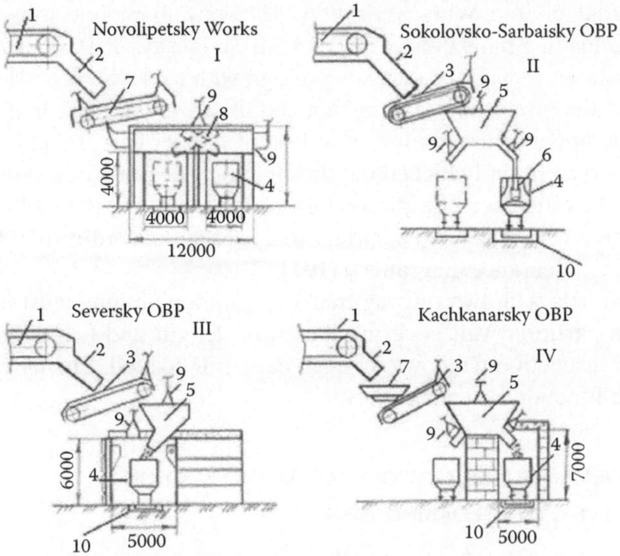

FIGURE 5.76 Dedusting layouts for transfer bins in single-point loading mode: 1 = sintering (firing) machine ; 2 = discharge chute; 3 = apron or belt-type conveyor; 4 = hopper car; 5 = intermediate bin; 6 = telescopic chute; 7 = cup-type or linear cooler; 8 = fly chute or mobile tray; 9 = local suction unit; 10 = scale car.

TABLE 5.36
Characteristics of Loading Tunnels at Pelletization and Sintering Factories

Plant Name	Number of Loading Points	Tunnel Length (m)	End Opening Area (m^2)	Volumetric Flow Rate of Evacuated Air (m^3/hr)	Air Velocity in Dust Receiver (m/s)	Average Dust Concentration in Evacuated Air (g/m^3)
Pelletization factories						
Seversky OBP: Loading bin #1	1	36	36	3 × 27,000	3.6	2.4
Loading bin #2	1	18	36	3 × 26,600	7.4	1.2
Kachkanarsky OBP	1	48	46	2 × 50,000	3.0	2.0
Krivyi Rih Central OBP	8	216	68	8 × 25,000	4.1	1.5
Sintering factories						
Novolipetsky Steel Works	1	18	32	1 × 230,000	3.3	2.0
Novokrivorozhsky OBP #1	6	84	47	6 × 23,800	2.8	4.6
Ilyich Iron and Steel Works	6	90	51	6 × 114,000	3.6	9.0

OBP = ore beneficiation plant; GOST 9238-73 mandates a minimum area of end opening of 26 m^2 [197].

With the goal of improving aspiration efficiency in mobile tray operations at the railway car loading bin #3 of the Sokolovsko-Sarbaisky OBP pelletization factory, Mekhanobr Institute proposed a dedusting design with a telescopic cowl (Figure 5.77). A drawback of the proposed design, given conditions at the OBP, is that it makes it difficult for the operator to monitor the filling of hopper cars. In addition, the cowl fails to contain dust initially kicked out through the side ports of a railway car.

Improving the efficiency of aspiration units designed for existing loading tunnels requires reducing the area of end openings to 2 × 17 m^2 (according to GOST 9238-73 "Construction … clearance diagrams") [197].

In filling tunnels with two railway tracks, aspiration volume must be reduced by constructing a partition wall covering the entire length and height of the housing structure with inspection windows for each car being loaded. The fly chute must be sealed so as to function as a choke valve.

5.4.2 Performance Calculations for Local Suction Units of Pellet Loading Bins

Apron conveyors (at the Sokolovsko-Sarbaisky OBP) or belt conveyors (at the Seversky and Kachkanarsky OBPs) are used to load pellets into bins at pelletization factories. The bins are adequately sealed, and pellets are transferred through closed chutes. Therefore, the procedure used to calculate the performance of local suction from conveyor driving drum cowls and from bins is completely identical to that used for ordinary process facilities handling hot materials.

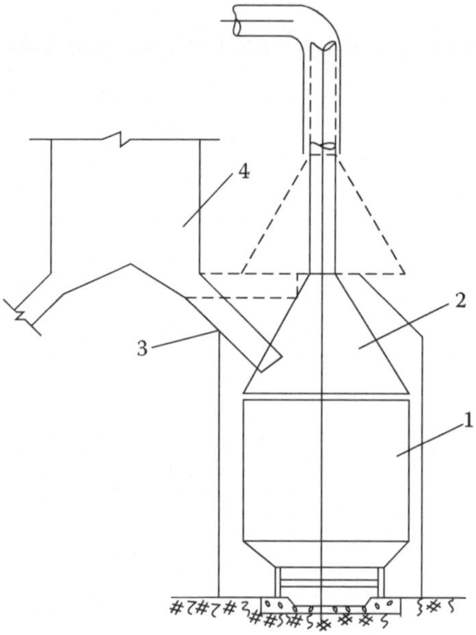

FIGURE 5.77 Dedusting system layout railway car loading bin #3 at the pelletization factory of Sokolovsko-Sarbaisky OBP (preliminary design): 1 = hopper car; 2 = telescopic cowl; 3 = mobile tray; 4 = loading bin.

When pellets are loaded through telescopic chutes, dust emission must be reduced by keeping the dust receiver installed in the upper part (Figure 5.78) under a negative pressure sufficient for inducing upward air circulation in chutes (opposite to falling material). It is important to cover the very beginning of the loading process when both the vertical distance traveled by material and the clearance between the chute end and hopper bottom are the greatest.

The following design calculation method [200,201] is used for similar aspiration processes with telescopic chutes.

The velocity of air drawn through the clearance between the chute and the hopper bottom is selected so that it does not preclude the knocking out of particulate dust when the material starts flowing.

Then, in succession, beginning from the lowest mobile section of the chute, negative pressure in joints must be calculated that would ensure the counterflow of the required amount of air in view of the resistance created by falling material and the inevitable suction of outside air leaking in through imperfect seals in joints.

The volume of air to be evacuated is determined using the formula

$$L_a = 0,65 F_{HK} \sqrt{2P_K / \rho_0} + L_K, \tag{5.246}$$

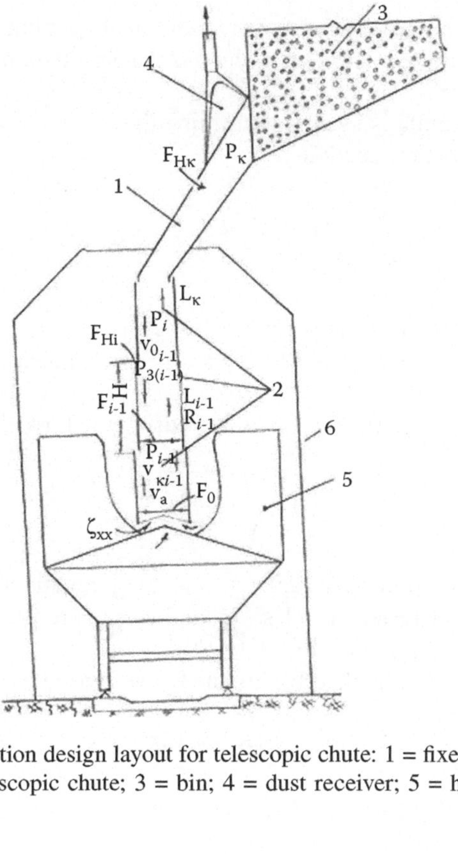

FIGURE 5.78 Aspiration design layout for telescopic chute: 1 = fixed immobile chute; 2 = mobile sections of telescopic chute; 3 = bin; 4 = dust receiver; 5 = hopper car; 6 = tunnel housing.

where L_k is the amount of air received from the last mobile section of the telescopic chute (m^3/s); P_k is the negative pressure in the dust receiver (Pa); ρ_0 is the density of ambient air (kg/m^3); and F_{HK} is the leakage area in the dust receiver (m^2).

The amount of air inflow through chute section i is determined using the formula

$$L_i = L_{i-1} + 0,65\sqrt{2P_i / \rho_0}\, F_{Hi}, \tag{5.247}$$

and the negative pressure at the inlet of this chute section is found using the formula

$$P_i = P_{i-1} + P_{ei-1} - P_{Ti-1} + R_{i-1}L_{i-1}^2, \tag{5.248}$$

where L_{i-1} is the amount of air flowing through the upstream section of the chute (m^3/s); F_{Hi} is the total leakage area in the joint between ith and i-1-th sections of the chute (m^2); P_{i-1} is the negative pressure at the inlet of the previous chute section (Pa); P_{ei-1} is the resistance to airflow created by falling material in the previous chute section (Pa); P_{Ti-1} is the thermal head in the previous chute section (Pa); R_{i-1} is the hydraulic variable of the previous chute section, equal to

$$R_{i-1} = \zeta_{i-1}\rho / \left(2F_{i-1}^2\right), \tag{5.249}$$

where ζ_{i-1} is the resistance factor of the previous chute section; F_{i-1} is the live section area of the previous chute section (m^2); and ρ is the density of the air inside the chute (kg/m^3).

The following formula is used to determine the resistance of falling material to air stream in the first chute section

$$P_{ei-1} = \frac{G_M}{3F_{i-1}\omega^2}\left[\left(v_{ki-1} + \frac{L_{i-1}}{F_{i-1}}\right)^3 - \left(v_{Hi-1} + \frac{L_{i-1}}{F_{i-1}}\right)^3\right], \tag{5.250}$$

where G_M is the material flow rate (kg/s); ω is the terminal velocity of particles (m/s); and v_{Hi-1}, v_{ki-1} are material velocities at the inlet and outlet of i-1-th chute section (m/s).

Terminal velocity of particles is calculated using the formula

$$\omega = \sqrt{\frac{4}{3}\frac{\rho_M g d_r}{\rho\psi}}, \tag{5.251}$$

where ρ_M is the material density (kg/m^3); g is the gravitational acceleration (m/s^2); d_r is the diameter of material particles (m); and ψ is the drag coefficient of material particles.

Thermal head can be calculated using the known formula

$$P_{Ti-1} = (\rho_0 - \rho)\, gH_{i-1}, \tag{5.252}$$

where H_{i-1} is the height of i–1-th chute section (m).

Calculation begins with the first chute section where the flow rate of evacuated air is defined as

$$L_i = F_s u_s, \tag{5.253}$$

where F_s is the live section area of the gap between this chute section and the car bottom (m²); and u_s is the velocity of air in this section preventing the knocking-out of dust particles (m/s).

Negative pressure at the inlet of the first section is assumed equal to

$$P_1 = \zeta_{in} u_s^2 \rho_0 / 2, \tag{5.254}$$

where ζ_{in} is the local resistance factor for air ingress through the opening.

With reference to the conditions at the Sokolovsko-Sarbaisky OBP pelletization factory, the method described here was used to calculate optimum aspiration volumes for pellet loading into railroad cars through telescopic chutes with the following input parameters: G_M = 260 kg/s; ρ_M = 3500 kg/m³; u_{in} = 5 m/s. Calculations indicate that maintaining air counterflow in the chute requires a suction capacity of 13,500 m³/hr at every telescopic chute. This would ensure a negative pressure of 750 Pa, enough to maintain a countercurrent of air.

In order to prevent dust carryover out of the loading tunnel, aspiration volume must be adjusted for an air velocity in the leeward end opening $\vartheta_n < 0$ (Figure 5.79). A simplified physical model of aerodynamic processes occurring in the tunnel may be described as follows. The oncoming air stream (at a velocity of w m/s) produces an excess pressure at the windward end opening (in section 1-1);

$$p_1 = P_1 - P_{atm}, \tag{5.255}$$

and in the leeward end opening

$$p_2 = P_2 - P_{atm}. \tag{5.256}$$

Here, P_1, P_2, P_{atm} are, respectively, absolute pressures in the leeward and windward end openings and in undisturbed flow (atmospheric pressure).

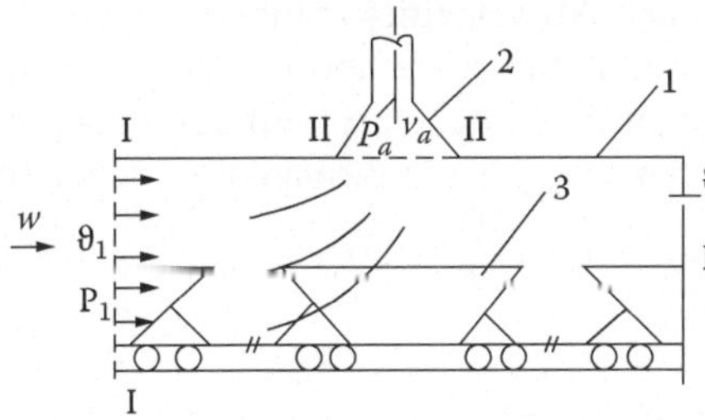

FIGURE 5.79. Aspiration design layout for loading tunnel: 1 = tunnel housing; 2 = local suction unit; 3 = hopper car.

The effect of this pressure and local suction unit operation is to force air at a velocity ϑ_1 through the windward end opening into the tunnel. As air travels above heated material, it becomes warmer and passes into the local suction unit.

A certain negative pressure must be maintained in the local suction cross-section (section 2-2) to compensate for the thermal head causing dust-laden air to escape to the outside through leaky areas in the tunnel. In order to simplify subsequent analysis, we will posit the pressure in this section to be equal to p_2

$$p_a = P_a - P_{atm} \approx P_2 - P_{atm} = P_2. \tag{5.257}$$

The first use of Bernoulli's equation for quantitative analysis of wind blowing through a tunnel is described elsewhere [196]. However, the authors of that work were unable to correctly account for mass forces (Archimedean forces) arising due to heat exchange.

In the present study, we shall try to account for the effect of Archimedean forces on evacuated air moving above the heated surfaces of pellets being handled.

The equation of total energy conservation law can be expressed in its integral form for air in the tunnel as

$$\frac{\partial}{\partial \tau}\int_w \rho\left(U+\frac{\vartheta^2}{2}\right)d\omega = -\int_s \rho\left(U+\frac{\vartheta^2}{2}\right)\vec{\vartheta}\cdot d\vec{S} + \int_w \rho\vec{M}\cdot\vec{\vartheta}d\omega + \int_s \vec{P}_n\cdot\vec{\vartheta}dS + \int_w \rho j_m d\omega, \tag{5.258}$$

where U is the internal energy, equal to $U = cT$ in our case; c is the specific heat capacity (J/kg K); T is air temperature (K); ϑ is air velocity (m/s); and $\vec{M}$ is mass force vector, defined through the Archimedean force in our case as the equation

$$\rho\vec{M} = \vec{g}(\rho - \rho_0),$$

where ρ, ρ_0 are air densities inside and outside the housing (kg/m^3); $\vec{g}$ is the gravitational acceleration vector (m/s^2); $\vec{P}_n$ is the surface force vector (Pa); W is the volume of air inside the tunnel (m^3); S is the area of tunnel surface confining the air (m^2); and j_m is the intensity of heat exchange with heated material J/(kg · s).

The following simplifying assumptions will be used by us for a quantitative evaluation of integrals comprising Equation 5.258.

The process is stationary. Air velocities at the open ends of the tunnel and in the inlet section of the aspiration flange are normal and are directed at normal to the cross-sections. Effective surface forces are limited to friction forces and to local resistance forces. They can be expressed through the friction ratio

$$\int_{S_{w.t.}} \vec{P}_n\cdot\vec{\vartheta}dS = \zeta_m Q_a \vartheta_1^2 \rho/2, \tag{5.259}$$

where $S_{w.t.}$ is the area of windward tunnel opening (m^2); ϑ_1 is the air velocity inside the 1–1 section (m/s); and Q_a is the volumetric flow rate of aspirated air (m^3/s; air velocity in the leeward opening was assumed to be zero).

Work performed by mass forces will be determined using the following assumptions. The flow inside the housing will be divided into two zones. In the first zone (at the air inlet into the tunnel), airflow will be considered horizontal $\left(\vec{\vartheta}\cdot\vec{M}=0\right)$. Although for the flow in the second zone (with the volume $Q_a h/\vartheta$), there is a uniformly rising flow $\left(\vec{\vartheta}\cdot\vec{M}=\vartheta g\left(\rho_0-\rho\right)=const\right)$. Then,

$$\int_w \rho\vec{M}\cdot\vec{\vartheta}d\omega = g\left(\rho_0-\rho\right)hQ_a. \tag{5.260}$$

In view of these assumptions, Equation 5.258 would become

$$\begin{aligned}&\rho_0 cT_0\vartheta_1F_1+\rho_0\vartheta_1F_1\vartheta_1^2/2-\rho_{cp}\vartheta_a cTF_a-\rho\vartheta_aF_a\vartheta_a^2/2=\\&=g\left(\rho_0-\rho\right)hQ_a+P_1F_1\vartheta_1-P_aF_a\vartheta_a-\zeta_m\rho Q_a\vartheta_1^2/2+q_m,\end{aligned} \tag{5.261}$$

where F_1 is the area of the windward open end (m²); F_a is the dust receiver inlet area (m²); and q_m is the amount of heat released by heated pellets in the car.

It is possible to split Equation 5.261 into two equations, balancing thermal and kinetic energy, respectively, that, after some transformations, would assume the form

$$q_m = cG_a\left(T-T_0\right), \tag{5.262}$$

$$\frac{\vartheta_1^2}{2}-\frac{\vartheta_a^2}{2}=\frac{g\left(\rho_0-\rho\right)h}{\rho}+\frac{P_1}{\rho_0}-\frac{P_a}{\rho}-\zeta_m\frac{\vartheta_1^2}{2}. \tag{5.263}$$

Considering that

$$\frac{\rho_0}{\rho}\approx\frac{T}{T_0}, \tag{5.264}$$

and expressing excess pressures p_1 and p_2 through wind pressure ratios k_1 and k_2

$$p_1 = k_1\rho_0 w^2/2; \qquad p_2 = k_2\rho_0 w^2/2, \tag{5.265}$$

after some transformations, in view of Equations 5.255 through 5.257, a joint solution of Equations 5.262 and 5.263 (assuming $\rho_0/\rho\approx 1$) would yield

$$G_a=\rho_0F_1w\sqrt{\frac{k_1-k_2}{\sum\zeta}+2gh\frac{q_m}{cG_aT_0w^2\sum\zeta}}, \tag{5.266}$$

where

$$\sum\zeta = 1-\left(\frac{F_1}{F_a}\right)^2+\zeta_m. \tag{5.267}$$

Therefore, determining suction performance not only requires a knowledge of meteorological conditions at the loading tunnel site and of the tunnel's aerodynamic properties but also the amount of heat released by material. The value of q_m could be determined using the formula

$$q_m = \alpha F_c\left(t_M - t_o\right), \tag{5.268}$$

where α is the heat exchange coefficient (kW/m^2 K); F_c is the area of heat-releasing surface assumed to be equal to the plan area of the car being loaded (m^2); and t_M, t_0 are material and air temperatures (°C).

The heat exchange coefficient is dependent on a number of factors, including the specific loading process in use, structural features of the car and its loading facility, air mobility, and so on. For a car loading with telescopic chutes, the heat exchange factor was determined using the air flow/heat balance (Equation 5.262) during industrial-scale aspiration testing in the filling tunnel of the Sokolovsko-Sarbaisky OBP. Testing results are summarized in Table 5.37.

Here G_a', G_a'' is the performance of the first and second local suctions; and t_a', t_a'' is the temperature of the evacuated air, °C. As an illustration, we shall calculate local suction capacity for the filling tunnel given conditions at the Sokolovsko-Sarbaisky OBP. Inputs: $F_1 = 26.5$ m^2; $w = 6.8$ m/s; $\rho_0 = 1.2$ kg/m^2; $k_1 - k_2 = 1$; $F_a = F_c = 29$ m^2; $c = 1.08$ kJ/kg K; $T_0 = 293$ K; $h = 4$ m; and $q_m = 228$ kW.

Air drag in the tunnel will be determined using the formula

$$\zeta_m = \zeta_0\left(F_1 / F_0\right)^2,$$

where ζ_0 is the local resistance coefficient referred to a dynamic head in the tunnel opening section (adjusted for the car inside the tunnel and additional curtains covering the tunnel opening at the Sokolovsko-Sarbaisky OBP, $F_0 = 6$ m^2).

At $F_0/F_1 = 6/26.5 = 0.226$, $\zeta_0 = 1.63$ [202], then

$$\zeta_m = 1,63(26,5/6)^2 = 32.$$

Using the formula (5.267), we can find that

$$\sum\zeta = 1 - (26,5/29)^2 + 32 = 32,2.$$

A substitution of initial data and the calculated $\sum\zeta$ into Equation 5.266 gives us

TABLE 5.37
Results of Local Suction Tests in Filling Tunnel of Sokolovsko-Sarbaisky Ore Beneficiation Plant

G_a', kg/s	t_a', °C	G_α'', kg/s	t_a'', °C	q_m, kW	α, kW/m^2 K
Mode I (t_0 = 1.4 °C)					
10.1	15	16.1	7	226	0.02
Mode II (t_0 = 15 °C)					
6.12	31	22.2	19	180	0.015

$$G_a = 1,2 \cdot 26,5 \cdot 6,8\sqrt{\frac{1}{32,2} + \frac{2 \cdot 9,8 \cdot 4 \cdot 228}{1 \cdot G_a \cdot 293 \cdot 6,8^2 \cdot 32,2}},$$

from which we could find $G_a = 38$ kg/s ($L_a = 115000$ m^3/hr).

5.4.3 Improving Aspiration Efficiency for Pellet Handling in Transfer Bin Housings

Following industrial-scale testing at the pelletization factory of Sokolovsko-Sarbaisky OBP, it was proposed the aspiration efficiency of transfer bin array #2 be improved by implementing the following measures at the car loading station handling pellets: constructing a tunnel housing with an air suction capacity of 115,000 m^3/hr, increasing aspiration flow rates in process chutes up to 13,500 m/hr, and installing spacious cowls at apron conveyor drums with an air suction capacity of 8,000 m^3/hr [199].

The engineering design department of the plant followed the issued recommendations and prepared shop drawings for aspiration system redesign in loading bin array #2. Dedusting systems have been retrofitted as provided by the prepared technical documents.

Dust release is reliably contained by two operating aspiration units: ATU-1a and ATU-1. The former enables air removal from drive drum cowls at apron conveyors #51 and #52 and from the bin. Dust is removed from aspiration air by the SIOT flushing cyclone with an angular coagulator designed by VNIIBTG at its inlet.

The other unit (ATU-1) maintains operation of local suction units at loading tunnels and telescopic chutes (Figures 5.80 through 5.82). A dust trap designed by VNIIBTG composed of an angular coagulator and a plate-type drip pan is used to clean the air.

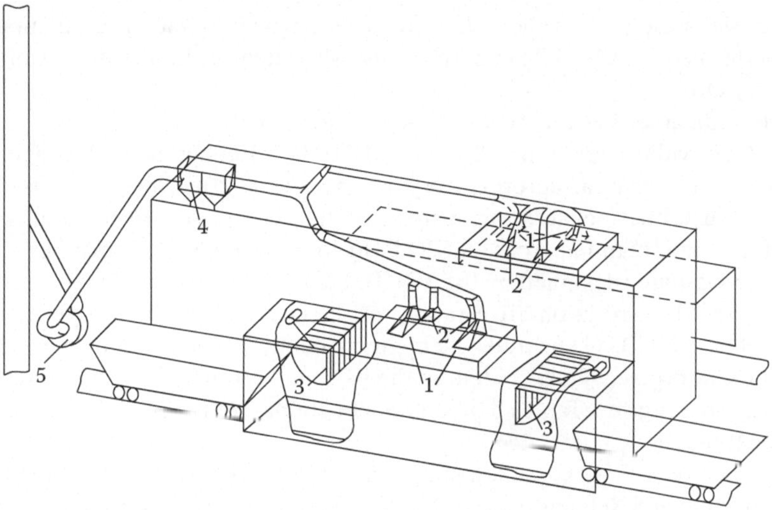

FIGURE 5.80 Improved aspiration system layout for the filling tunnel at Sokolovsko-Sarbaisky OBP: 1 = suction from tunnel; 2 = suction from telescopic chutes; 3 = sealing curtains; 4 = angular coagulator with a drip pan; 5 = induced-draft fan.

(a)

(b)

FIGURE 5.81 Dust release during pellet loading into cars at the Sokolovsko-Sarbaisky OBP loading bin #2: (a) unmitigated, (b) with tunnel housing.

Industrial-scale tests of the efficiency of redesigned aspiration installations had been performed by VNIIBTG and the Dust Measurement Lab at the Sokolovsko-Sarbaisky OBP.

Tests indicated that the performance of local suction units is close to volumetric flow values recommended by VNIIBTG. Reduction of leakage areas in drive drum cowls at apron conveyors #51 and #52 had made it possible to maintain sustained negative pressures and had completely eliminated knock-outs of dust into manned areas. A SIOT flushing cyclone, used together with an angular coagulator designed by the VNIIBTG Institute, resulted in a significant improvement in aspiration air cleaning efficiency. The ATU-1a aspiration unit designed by VNIIBTG ensured safe hygienic conditions at the conveyor #51/#52 operator's workplace. Dust content in the workplace atmosphere decreased from 42 mg/m^3 to 2 mg/m^3. Tables 5.38 and 5.39 summarize the results of industrial-scale testing.

As a part of ATU-1 commissioning activities, the Laboratory for Small-Scale Mechanization and Reliability at the Sokolovsko-Sarbaisky OBP installed operator-controlled rubber curtains in end openings of the loading tunnel (Figure 5.83). The Dust Measurements Laboratory of the OBP carried out industrial-scale testing of angular coagulator efficiency as summarized in Table 5.40.

FIGURE 5.82 General view of aspirated tunnel for pellet loading into railway cars at the Sokolovsko-Sarbaisky OBP.

These tests revealed a low degree of dust-laden air cleaning and a significant carryover of moisture in droplet form. The engineering design department of the OBP in cooperation with VNIIBTG prepared engineering documentation for redesigning the drip pan. Upon completion of retrofitting work on the drip pan, the VNIIBTG Institute carried out industrial-scale efficiency tests at the ATU-1 aspiration unit.

Industrial-scale test results are summarized in Tables 5.40 and 5.41. As the data indicates, the total capacity of telescopic chute suction units was virtually equal to the recommended value. The total volume of aspiration out of the tunnel housing equaled 76,000 m^3/hr, well below the recommended value of 113,000 m^3/hr. A short knock-out of dust from the car was observed during initial startup of material flow, as well as during final topping-up of the hopper. Airborne dust content measured 4.3 g/m^3 at discharge ports of the car and 2.8 g/m^3 at the exit from the telescopic chute skid. Total upstream dust content of the dust trap measured 6.8 g/m^3. Residual dust content in air discharged to the atmosphere measured 0.12 to 0.16 g/m^3. Dust trap efficiency reached 97.7%.

The recommended aspiration layout for the filling tunnel and telescopic chutes together with increased ATU-1 performance will ensure complete containment of dust release caused by filling cars with pellets.

TABLE 5.38
Summary of Industrial-Scale Test Results at ATU-1a Local Suction Units

Suction Unit Designation	Flow Rate of Evacuated Air (m^3/hr)	Flow Rate of Evacuated Air Prior to Retrofitting (m^3/hr)	Negative Pressure in the Cowl (Pa)	Dust Concentration in Evacuated Air (g/m^3)		Material Carryover (kg/hr)	
				Before Retrofitting	After Retrofitting	Before Retrofitting	After Retrofitting
From cowl of conveyor #51 drive drum	$\frac{8000}{12,000}$	6000	$\frac{6}{9}$	5.6	1.1	34	13.0
From pellet storage bin	$\frac{10,000}{10,800}$	1700	$\frac{4}{20}$	47	6.7	80	72
From cowl of conveyor #52 drive drum	$\frac{8000}{12,900}$	8300	$\frac{6}{13}$	41	1.6	340	21
From pellet storage bin	$\frac{10,000}{7700}$	2100	$\frac{4}{13}$	54	7.0	113	54

Note: In fractions, the numerator indicates recommended levels, whereas the denominator indicates actual levels.

TABLE 5.39
Summary of Test Results of Two-Stage Cleaning System Complete with Angular Coagulator and SIOT Flushing Cyclone

(1) Angular Coagulator

Main dimensions (mm)
Length = 1380
Width = 520
Height = 1400
Capacity (m^3/hr) = 46,000
Hydraulic resistance (Pa) = 950

(2) SIOT Flushing Cyclone

Main dimensions (mm)
Cyclone diameter = 2256
Outlet flange diameter = 940
Capacity (m^3/hr) = 46,000
Hydraulic resistance (Pa) = 620

(3) Airborne Dust Content (mg/m³)

Before cleaning = 3800
After cleaning = 100

(4) Particulate Dust Composition (%)

Prior to cleaning							
Fraction (μm)	1.4	1.4–4.2	4.2–9.8	9.8–15	15–30	30–45	45–105
%	2.6	2.6	9.2	30.2	23	18	14.4
After cleaning							
Fraction (μm)	1.4	1.4–4.2	4.2–9.8	9.8–15	15–30	30–45	45–105
%	9.7	4.7	23	0.8	48.4	13.4	–

(5) Cleaning efficiency (97.5%)

SIOT = Sverdlovsk (now Yekaterinburg) Institute for Occupational Safety.

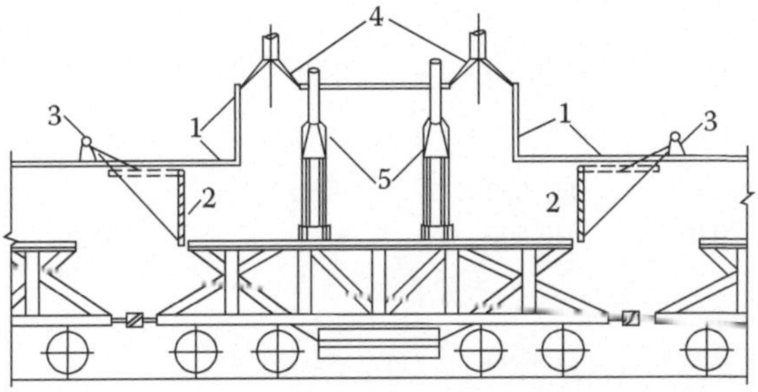

FIGURE 5.83 Layout drawing for sealing curtain installation inside loading tunnel: 1 = tunnel-type housing; 2 = sealing curtain; 3 = drive motor; 4 = local suction units in tunnel; 5 = local suction units at telescopic chutes.

TABLE 5.40
Test Results for Angular Coagulator with Design Capacity of 120,000 m³/hr

Air Evacuation Capacity (m³/hr)	Average Dust Concentration in Aspirated Air (g/m³): Upstream of Dust Trap	Average Dust Concentration in Aspirated Air (g/m³): Downstream of Dust Trap	Material Carryover (kg/hr): Entering Dust Trap	Material Carryover (kg/hr): Released to Atmosphere	Cleaning Efficiency (%)
$\frac{136,600}{102,000}$	$\frac{88.8}{6.82}$	$\frac{2.25}{0.16}$	$\frac{1213}{696}$	$\frac{308}{19}$	$\frac{75}{97.7}$

Main dimensions (mm): length = 5200; width = 2400; height = 2200; inlet flange diameter = 1430; outlet flange diameter = 1430. Numerators reflect results of tests carried out before drip pan retrofitting, and denominators refer to tests carried out after retrofitting.

It can be concluded that fired pellet car loading units are the most powerful sources of fugitive dust emissions at pelletization factory sites. Dust release intensity depends on fine content (5 mm class), pellet strength properties, pellet temperature, and on falling height. Telescopic chutes may be used to reduce the free-fall height of pellets, considering their good flowing behavior.

Dust release while loading a railway car with pellets should be localized either by using an aspirated tunnel with cars loaded inside it or by using aspirated cars equipped with sealing lids and telescopic air suction flanges [216,217].

Prevention of dust release from the loading tunnel requires installing a counter-flow aspiration system for chutes and evacuating air from the upper part of the tunnel

TABLE 5.41
Performance of Local Suction Units in Filling Tunnel

Local Suction Unit Designation	Unit Capacity (m³/hr): Actual	Unit Capacity (m³/hr): Recommended	Air Velocity in Air Duct (m/s)	Average Dust Concentration in Aspirated Air (g/m³)	Material Carryover (kg/hr)
Suction from left-hand side of filling tunnel	29,900	56,500	12.9	0.59	17.6
Suction from right-hand side of filling tunnel	46,200	56,500	20.0	4.3	198.0
Total:	76,100	113,000			215.6
Suction from left-hand telescopic chute	20,000	13,500	15.6	10.6	212
Suction from right-hand telescopic chute	10,500	13,500	8.1	12.0	126
Total:	30,500	27,000			338

by means of suction. Calculations of the required air volume to be evacuated from telescopic chutes must reflect injection properties of the pellet stream. In addition, air evacuation from the tunnel must be designed with regard to heat release from warm pellets and the effects of wind. Flow rates in aspiration units may be reduced by sealing end openings with movable barriers and by choosing tunnel length so that cars would occupy these openings.

Conclusion

The studies we conducted are aimed at improving the theoretical basis for design of ventilated suction hoods used for bulk material transfer groups (i.e., the most typical and common sources of dust emissions in dust-forming material reprocessing). The basic findings and resulting conclusions are as follows:

1. Aerodynamically, bulk material streams belong to a class of two-component flows with a discrete carrying medium of solid particles falling at a growing rate while also serving as a pseudo-continuous medium of induced air. The accelerated motion of the carrying medium determines the asymptotic nature of air induction aerodynamic processes (Figures 3.11 and 4.11), which depend on the geometry of transport channels; the stream kinematics and the intensity of the dynamic interaction of components, distribution, fineness and composition of particles in a stream; and the material's temperature and humidity (see Section 1.3).
2. Due to its physical and mechanical properties, a bulk material stream in a chute features an accelerated motion with an exponential distribution of particles affected by the channel cross-section depth (Figure 2.5), which significantly changes conditions of the dynamic intercomponent interaction—the heat and mass exchange—as compared to one-dimensional models. Particle distribution streams may be divided into two classes (Figure 2.6): streams featuring pseudo-uniform particle distribution—where a saltating motion of particles with a light concentration gradient prevails, and streams featuring laminar motion—where most particles collide at the chute bottom, with a small portion of saltating particles above those particles with a heavy concentration gradient.
3. Solid particles of bulk materials produced by mechanical crushing are geometrically close to compacted regular-shaped bodies (Figure 2.11) for which geometric form factor values for the midsection surface and area virtually coincide. These materials should be divided into two groups, based on their aerodynamic resistance: the first group is made up of sharp-grained particles (iron ore, chalkstone, agglomerate, granite) that features a range of geometric form and resistance factors; the second one is the group of round-shaped particles (iron-ore pellets) having a geometric form factor close to one and a resistance coefficient that is close to the s sphere drag (Table 2.1).
4. A simple method for experimentally determining the aerodynamic resistance of bombarding particles in a stream without interfering with the latter's structure has been substantiated and developed (Equations 2.149 and 3.19). Based on measurements of closed chute wall static pressure at the transfer of solid particles (Figure 3.2), this method enabled us to obtain the basic characteristics of bulk material suction properties, that is, an averaged aerodynamic drag factor for particles in a stream. It was theoretically

proven and experimentally demonstrated that a transverse particle concentration gradient results in an averaged aerodynamic drag factor decrease with an increase in the bulk concentration (Equation 3.29).

5. By means of averaging velocities, bulk concentrations, and intercomponent interaction forces in the channel section, it is possible to formulate a one-dimensional equation illustrating the dynamics of a particle stream (Equation 3.37) and the induced air (Equation 3.38) in closed chutes. The aerodynamic processes for short chutes ($h < 0.3$) are precisely described by Equation 3.99, depicting a uniformly accelerated particle stream and the uniform motion of induced air (Figure 3.8). The induced air volume is limited at the chute inlet due to the braking effect of a particle stream. As the Butakov–Heykov number increases, the slip ratio of components (Equation 3.121) also increases, reaching an asymptotic value equal to $\varphi = (1+n)/2$.
6. Due to non-uniformity of particles distribution, the intensity of intercomponent heat exchange (Equation 3.182) in a chute is lower (by an order) than the intensity of heat exchange in a vertical channel having uniform particle distribution (Figure 3.17). Counteracting forces resulting from the heat exchange of buoyancy forces contribute to a decrease in the suction properties of a bulk material stream (Equation 3.214). The direction and value of the air velocity in a chute at the transfer of a heated material is determined by the relation between heat and induction pressures. When transferring wet heated materials, the amount of induced air increases due to water vapors that result from condensation and from auxiliary interphase pressure forces (Equations 3.229 and 3.232).
7. At instant chuting of a heated bulk material, an increase in induction pressure is strongly related to a variation in transferred material flow (Equation 3.286); due to considerable inertness of the intercomponent heat exchange, temperature variations are far behind variations in transferred material quantity (Figure 3.20). A gradual change in particle flow rate at the beginning and at the end of a material transfer is associated with excessive (peak) pressures against steady-state values (Figure 3.21). This is explained by the maximum induction head at a certain material flow rate (Equations 3.299 and 3.300). There are no pressure surges if the material flow rate does not reach the necessary value.
8. The dynamics of air flows in a stream of freely-falling particles may be described by the boundary-layer equation. Due to a great mass of particles, the carrying (solid) component dynamics do not depend on the hydrodynamical field, which distinguishes these streams from gas jets carrying solid impurities. The main forces that form air jet flows in a stream of freely falling particles are those of intercomponent interaction and turbulent viscosity (Equations 4.67 and 4.68). The intercomponent interaction that forces increase the quantity of the inducing jet motion (Equation 4.74) which distinguishes these jets from free gas jets.
9. Regularities of air flows in an axisymmetric stream of freely falling particles may be described by Equation 4.326, the solutions of which are quantitatively and qualitatively consistent with the experimental data

(Figures 4.19 and 4.20). If a jet is short ($h < 0.5$), turbulent viscosity forces are insignificant and the induced air volume may be determined using design ratios of a one-dimensional problem. If a jet is long ($h > 1$), turbulent viscosity forces noticeably increase the induced air volume due to the air medium being engaged in the axial motion (Figure 4.22).

10. When a free stream of particles is enclosed with impenetrable walls, air inflow is hindered; hence, closed circulation flows occur in $1 < r < 2$. As the distance between the walls and the stream surface becomes shorter ($r \rightarrow 1$), the length of these whirls and the velocity variation amplitude in the external stream is reduced to zero (Figures 4.25 and 4.27) although flow velocity in a particle stream tends to maintain a constant value equal to the initial velocity. As the distance to the channel walls becomes longer, the whirls also become longer. With $r >$ there is only an external reverse flow, which area is decreased inversely as r.
11. When the channel section is partially filled with a particle stream, averaged integral equations for a boundary layer may be used as a basis for making one-dimensional equations (Equations 4.469 and 4.470) that describe the motion of a two-component stream (internal flow) and the air flow in a cavity limited to the stream surface and the channel walls (external flow). The general solutions of these equations may be used to derive formulas for one-dimensional problems regarding a chute with pseudo-uniform distribution of particles as well as a free jet of freely falling particles. This creates the base for developing of a universal computation methodology for induced air volumes.
12. The following procedures should be used when determining the optimum output of dust control systems and the efficient arrangement of suction nozzles is required:
    a. With conveyer transfers, a combined air induction model should be applied. For a chute receiving funnel, air induction with a free material stream should be considered; for the straight portion of a chute, it is the operation of serial steady-flow injectors.
    b. For complex groups—for instance, with a cascade arrangement of process equipment—the aerodynamic link of hoods should be considered. When reprocessing wet heated materials, it is necessary to consider the temperature head value, the intensity of steam emissions, and interphase pressure.
    c. With high-speed equipment such as hammer crushers, it is necessary to consider the ventilating capability of actuating elements and bypass areas with extreme excessive pressure values.
13. Evaluation of computation methods at industrial enterprises engaged in reprocessing a wide range of bulk materials (iron and complex ore, granite, chalkstone, magnesite, charred coal, charge, agglomerate, pellets) showed that design capacities of suction hoods that satisfactorily match actual test data. There were normative documents developed on this basis to become widely used in the design and arrangement of suction systems for dressing, sintering, and pellet plants. Operation of ventilation plants built in

accordance with this data at new (Mikhailovskiy and Kostomukshskiy mining and concentration complexes) and reconstructed (Sokolovsko-Sarbayskiy, Lebedinskiy, and Yuzhny mining and concentration complexes) plants has shown that the optimum operating modes of dust exhaustion systems make it possible to reduce air dustiness to maximum permissible concentration levels.

14. In order to reduce gross emissions, it is necessary to apply the technical dust control means at all stages of dust formation (see Section 5.2):
    a. In chutes—by arranging an indiscrete mass of transferred material or by stacking dried powders in a wet concentrate bed (the dust concentration reduction degree is 10–80%, the suction volume reduction degree is 40–60%)
    b. In hoods—by filtering the induced air through a granular bed, a grate of magnetized plates, or a foam blanket (the dust concentration reduction degree is 60–90%, the suction volume reduction degree is 50–70%)
    c. In dust-collecting bags—by intensifying the gravity sedimentation in cyclone dust separators or by filtering through that fabric adapters (the aspiration air treatment degree is 50–95%) which decreases the initial dust concentration by one or two orders and the volume of purified air by one order and also enables an advance treatment of suction emissions to MPC in more compact and highly efficient central dusters
15. The following technical means may be used to reduce the most major fugitive dust emissions:
    a. For open pellet storage—ventilated shafts, when material is free poured from the above-stack gallery (an 80% gross emission reduction level was achieved at the SSGOK) and built-in suction hoods and suction systems for stackers (a 90% gross emission reduction level was achieved at the Kostomukshskiy mining and concentration complex)
    b. For facilities loading burnt pellets in rail cars—a ventilated tunnel with a system of telescopic chutes (a 90% dust emission reduction level was achieved at the SSGOK).
16. Professional experience gained and generalizations made suggest that further directions for the development of the theory and practice of air containment and dedusting in bulk materials reprocessing are as follows:
    a. Studying the dynamics of a bulk material stream and developing technical means of bulk material transfer management in a bound mode of particle motion (i.e., introduction of dust-free transfers)
    b. Studying transportation modes of dust-laden flows in air ducts and arranging the controlled processes of dust particle sedimentation therein with subsequent development of dust sedimentation suction systems for vapor and dust mixtures
    c. Developing closed suction systems to enable dust-laden air recirculation and to reduce the quantity of energy consumed by supply and exhaust ventilation
    d. Developing automated suction systems as subsystems of bulk material reprocessing flows

Appendix

Initial Aerodynamic Equations for a Bulk Material Stream

The basic equations of heterogeneous media dynamics can be used to explain the mechanism of interaction between a bulk material stream and air, as well as to formulate the basic provisions of the dynamic approach to the solution of problems concerning air suction.

The fundamental efforts of domestic [26,27,33,36,41,47,70] and foreign scientists [12,20,57,66] devoted to multicomponent flow mechanics explain the interaction of components, provide a mathematical description of such interaction regularities, and analyze a wide range of practical tasks. The most theoretically and experimentally elaborate tasks are those where the carrying medium is continuous (liquid or gas) even though the carried or stationary medium is discrete (solid particles, fluid drops, gas bubbles). These are flows of aerosols [19,22,23,70], suspensions [12,15,16,17,18,35,66], gas suspensions [21,24,25,26,39,71] and liquid-gas mixtures [27,31,36,43,44,45,46,47,49,58,61,62,63,65,72], fluidization [6,7,8,14,34], ground filtration [47,48,52,53], hydro- and pneumo-transport [13,51,55,56,64], jets with impurities [1,2,3,4,5,11,37,38], sediments [9,10,28,29,30,67,68,69], and storms [32,33].

There are two methodical approaches used to describe the mechanics of multicomponent streams: the phenomenological method—where heterogeneous media streams are regarded as a motion of interpenetrating multispeed continua—and the method of averaging accounting equations for the classical mechanics in space and time microscales.

The streams of bulk material and air in question are two-component flows containing solid particles (for clarity, we will designate them as the first component) and a gaseous pseudo-continuous medium (the second component, i.e., air). Representation of these streams as multicomponent flows is reasonable in cases with a non-homogeneous solid stream, for instance, when there is a particle stream of different size. So as not to complicate the task, for now, we will consider two-component solid particle streams having average characteristics (in size, shape, and mass of particles).

A.1 PHENOMENOLOGICAL METHOD OF DYNAMIC EQUATION DEVELOPMENT FOR TWO-COMPONENT STREAM

The phenomenological method is based on the same provisions as the mechanics of a heterogeneous continuous medium where it is assumed that, despite their smallness, the elementary volume of the mixture as well as the elementary volumes of components contain a sufficient number of particles to enable a static ensemble averaging of physical parameters of these particles in a valid approximation. There are two ways of averaging the parameters that describe the i-component: either in the full volume or in a portion of the volume filled with the i-component. For instance, for density, the first case is

$$\rho_{ip} = n_i m_i, \tag{A.1}$$

and the second is

$$\rho_i = n_i m_i / \beta_i, \tag{A.2}$$

where n_i is a number of particles of the i-component in the unit volume; m_i is the weight of particles of the i-component; and β_i is a fraction of the mixture volume filled with particles of the i-component.

Thus, in the first case, we have a partial density, and in the second case, we have the true density of particles of the i-component. There is an apparent relation between them:

$$\rho_{ip} = \beta_i \rho_i. \tag{A.3}$$

The obvious relation available for a two-component mixture (for convenience, we assigned the symbol Σ to adding from $i = 1$ to $i = 2$) is

$$\rho = \rho_{1p} + \rho_{2p} = \beta_1\rho_1 + \beta_2\rho_2 = \sum \beta_i\rho_i;\ \beta_1 + \beta_2 = \sum \beta_i = 1, \tag{A.4}$$

and for a mass-averaged (barycentric) velocity of this mixture

$$\rho\vec{\vartheta} = \sum \beta_i \rho_i \vec{\vartheta}_i. \tag{A.5}$$

For the sake of convenience, in some cases, we use diffusion velocities ω_i that are general mixture mass center velocities of components:

$$\vec{\omega}_1 = \vec{\vartheta}_1 - \vec{\vartheta};\quad \vec{\omega}_2 = \vec{\vartheta}_2 - \vec{\vartheta};\quad \left(\sum \beta_i \rho_i \vec{\omega}_i = 0\right). \tag{A.6}$$

A.1.1 Inter-Component Interaction

External surface and bulk forces of the mixture are determined using the similar relations:

$$\vec{\Pi} = \sum \beta_i \vec{\Pi}_i;\quad \rho\vec{M} = \sum \beta_i \rho_i \vec{M}_i. \tag{A.7}$$

Due to a small bulk concentration of solid particles in the streams in question ($\beta_1 << 1$) and an ignorable gas pressure effect on particle deformation, the surface force equation may be significantly reduced:

$$\vec{\Pi} \approx \vec{\Pi}_2;\ (\beta_1 \vec{\Pi}_1 = 0;\ \beta_2 \vec{\Pi}_2 \approx \vec{\Pi}_2). \tag{A.8}$$

A bulk force in the Earth's gravitational field may be represented as accounting for the buoyancy force using the obvious relations:

$$\beta_1\rho_1\vec{M}_1 = \beta_1(\rho_1 - \rho_2)\vec{g} \approx \beta_1\rho_1\vec{g}; \quad (\rho_1 >> \rho_2), \tag{A.9}$$

$$\beta_2\rho_2\vec{M}_2 = \beta_2(\rho_2 - \rho_0)\vec{g}, \tag{A.10}$$

where ρ_0 is a density of air outside the stream of particles.

The specific mixture energy E per the medium mass unit is considered as a sum of internal U and kinematic K energy:

$$E = K + U. \tag{A.11}$$

Therefore, the internal energy is assumed to be additive by mass

$$\rho U = \sum \beta_i\rho_i U_i, \tag{A.12}$$

although the kinematic energy defines only the microscopic movement of components*

$$\rho K = \sum \beta_i\rho_i \frac{\vartheta_i^2}{2}. \tag{A.13}$$

Then, the specific mixture energy may be presented as

$$\rho E = \sum \beta_i\rho_i E_i, \tag{A.14}$$

$$E_i = U_i + \frac{\vartheta_1^2}{2}. \tag{A.15}$$

Note that

$$\rho K \neq \frac{1}{2}\rho\vartheta^2, \tag{A.16}$$

because, in view of Equations A.5 and A.6, we have:

$$\rho K = \rho\frac{\vartheta^2}{2} + \sum \beta_i\rho_i \frac{\omega_i^2}{2}, \tag{A.17}$$

that is, the kinematic energy of a multispeed medium is determined by a motion as an integral with speed ϑ and by velocities of relative motion of components.
The internal energy of components is related to the enthalpy as follows:

$$e_i = U_i + \frac{P_i}{\rho_i}. \tag{A.18}$$

* In fact, there are also small-scale flows (having the specific linear dimension of inclusions). For instance, reverse carrying fluid flows close to solid particles due to their relative motion and the random motion of inclusions. The phenomenological theory of mixture mechanics does not account for the kinematic energy of such motion.

Because solid particles cannot be compressed with relation to air:

$$P_1 = P_2 = P. \tag{A.19}$$

The main problem in describing the mechanics of two-component streams is a corrective accounting of interaction between the components. In general, the inter-component interaction in phase transitions is characterized by mass, pulse, and energy. Phase transformations may, for example, occur as a result of moisture evaporation from the surface of solid particles of a heated wet material. With neglect of cross effects as of the initial approximation, the mass transfer will be characterized by the bulk evaporation rate J that is equal to:

$$J = S_p n_p \alpha_m (T_1 - T_s), \tag{A.20}$$

where T_1 is the particle surface temperature; T_s is the point of saturation; S_p is a particle surface; α_m is the mass transfer coefficient; and n_p is the particle concentration. Due to the mass conservation law,

$$J_{12} = -J_{21} = J. \tag{A.21}$$

Here, indices 12 and 21 denote the phase transition direction (1⟶2 is for evaporation; 2⟶1 is for condensation).

Mass exchange in phase transitions results in the inter-component exchange of pulse and energy. The primary role in pulse exchange belongs to the force interaction:

$$\vec{P}_{12} = -\vec{P}_{21} = \vec{F}_{12} + J\vec{\vartheta}_{12}. \tag{A.22}$$

Here, F_{12}, the interaction force applied to air from solid particles,

$$\vec{F}_{12} = n_p \vec{R}_{12} = \beta_1 \frac{\rho_1}{m_p} \vec{R}_{12} \tag{A.23}$$

is a bulk force resulting from aerodynamic resistance of particles due to the relative velocity of components $\vec{\vartheta}_1 - \vec{\vartheta}_2$, ($\vec{R}_{12}$ is the aerodynamic action of a particle on air, m_p is the particle mass). The velocity of the evaporating portion of the gaseous component at a particle's surface is assumed as

$$\vec{\vartheta}_{12} = \vec{\vartheta}_1. \tag{A.24}$$

The inter-component energy exchange ($E_{12} = -E_{21}$) is associated with the transfer of the mechanical work of interaction forces $\vec{F}_{12}$, heat exchange, and of phase transitions. The heat exchange, like the mass exchange, may be evaluated using the difference of the component temperatures

$$Q_{12} = -Q_{21} = S_p n_p \alpha_m (T_1 - T_2). \tag{A.25}$$

A change in the energy of components during phase transformations is due to the fact that, during evaporation, the second component receives the first component mass J that carries the total energy

$$U_{12}+\frac{1}{2}\vartheta_1^2.$$

Let us analyze the interaction of components specifically with regard to a stream of wet particles that entrain unheated air ($T_1 > T_2$). First, let us establish the aerodynamic characteristics of the stream. We will let moisture be physically and mechanically related with the material of particles (according to Rehbinder classification), and air is a mixture of non-condensable gas and water vapor (condensable gas). Thus, we have a stream of four components, the characteristics of which can be denoted with the corresponding inferior indices: *m* is for the material of particles, *B* is for particle moisture, g is for non-condensable portion of air, and *n* is for airborne water vapors. The moisture and vapor content is determined by moisture (*m*) that is obviously related to moisture content (*d*) widely used in the engineering practice as follows:

$$m_1=\frac{d_1}{1+d_1};\ m_2=\frac{d_2}{1+d_2}; \tag{A.26}$$

where

$$d_1=\frac{\beta_B\rho_B}{\beta_m\rho_m};\ d_2=\frac{\beta_n\rho_n}{\beta_g\rho_g}. \tag{A.27}$$

Therefore, the density, internal energy, and enthalpy of components will be determined by the additive relations

$$\left.\begin{aligned}&\beta_1\rho_1=\beta_B\rho_B+\beta_m\rho_m;\quad \beta_B+\beta_m=\beta_1;\\ &\beta_2\rho_2=\beta_n\rho_n+\beta_g\rho_g;\quad \beta_n+\beta_g=\beta_2;\\ &\beta_1\rho_1U_1=\beta_B\rho_BU_B+\beta_m\rho_mU_m;\quad \beta_1\rho_1e_1=\beta_B\rho_Be_B+\beta_m\rho_me_m;\\ &\beta_2\rho_2U_2=\beta_n\rho_nU_n+\beta_g\rho_gU_g;\quad \beta_2\rho_2e_2=\beta_n\rho_ne_n+\beta_g\rho_ge_g.\end{aligned}\right\} \tag{A.28}$$

A.1.2 Accounting Equations

Let us write down accounting equations for mass, pulse, and energy for the *i*-component of a mixture filling volume *V* that is limited by surface *S* with outer normal line $\vec{f}$. Bear in mind that, here, the interaction of *j*-component with *i*-component is an external effect ($i = 1.2; j = 1.2; i \neq j$).

For the mass of components, we have

$$\left.\begin{aligned}&\frac{\partial}{\partial t}\int_V\beta_m\rho_m dV=-\int_S\beta_m\rho_m\vartheta_{1f}dS;\quad \frac{\partial}{\partial t}\int_V\beta_B\rho_B dV=-\int_S\beta_B\rho_B\vartheta_{1f}dS-\int_V JdV;\\ &\frac{\partial}{\partial t}\int_V\beta_g\rho_g dV=-\int_S\beta_g\rho_g\vartheta_{2f}dS;\quad \frac{\partial}{\partial t}\int_V\beta_n\rho_n dV=-\int_S\beta_n\rho_n\vartheta_{2f}dS+\int_V JdV.\end{aligned}\right\} \tag{A.29}$$

from which, accounting for Equation A.28, we obtain

$$\frac{\partial}{\partial t}\int_V \beta_1\rho_1\, dV = -\int_S \beta_1\rho_1\vartheta_{1f}\, dS - \int_V J\, dV; \tag{A.30}$$

$$\frac{\partial}{\partial t}\int_V \beta_2\rho_2\, dV = -\int_S \beta_2\rho_2\vartheta_{2f}\, dS + \int_V J\, dV, \tag{A.31}$$

where ϑ_{1f} and ϑ_{2f} are hereinafter the projections of the components' velocity vectors on the outer normal line $\vec{f}$ of surface dS.

For the pulse, accounting for Equations A.8, A.22, and A.24, we have

$$\frac{\partial}{\partial t}\int_V \beta_1\rho_1\vec{\vartheta}_1\, dV = -\int_S \beta_1\rho_1\vec{\vartheta}_1\vartheta_{1f}\, dS + \int_V \beta_1\rho_1\vec{M}_1\, dV - \int_V \vec{F}_{12}\, dV - \int_V J\vec{\vartheta}_1\, dV; \tag{A.32}$$

$$\frac{\partial}{\partial t}\int_V \beta_2\rho_2\vec{\vartheta}_2\, dV = -\int_S \beta_2\rho_2\vec{\vartheta}_2\vartheta_{2f}\, dS + \int_V \beta_2\rho_2\vec{M}_2\, dV + \int_S \vec{\Pi}_{2f}\, dS + \int_V \vec{F}_{12}\, dV + \int_V J\vec{\vartheta}_1\, dV, \tag{A.33}$$

where $\vec{\Pi}_{2f}$ is a surface vector for the force applied to the surface with outer normal line $\vec{f}$.

When setting up the energy relations, we use the energy conservation laws of classical mechanics formulated for continuous media by L. I. Sedov [59]. For instance, according to the work-kinetic energy theorem, a change in the kinetic energy of the continuous media final volume is equal to a summarized elementary work of external mass and surface forces and internal mass and surface forces:

$$\frac{\partial}{\partial t}\int_V \beta_1\rho_1\frac{\vartheta_1^2}{2}\, dV = -\int_S \beta_1\rho_1\frac{\vartheta_1^2}{2}\vartheta_{1f}\, dS - \int_V J\frac{\vartheta_1^2}{2}\, dV + \int_V \beta_1\rho_1\vec{M}_1\cdot\vec{\vartheta}_1\, dV - \int_V \vec{F}_{12}\cdot\vec{\vartheta}_1\, dV; \tag{A.34}$$

$$\frac{\partial}{\partial t}\int_V \beta_2\rho_2\frac{\vartheta_1^2}{2}\, dV = -\int_S \beta_2\rho_2\frac{\vartheta_2^2}{2}\vartheta_{2f}\, dS + \int_V J\frac{\vartheta_1^2}{2}\, dV + \int_V \beta_2\rho_2\vec{M}_2\cdot\vec{\vartheta}_2\, dV + \int_V \vec{F}_{12}\cdot\vec{\vartheta}_2\, dV + \int_S \vec{\Pi}_{2f}\cdot\vec{\vartheta}_2\, dS + \int_V N\, dV. \tag{A.35}$$

Here, the second integrals of the right-hand sides of the equations represent the inter-component kinetic energy exchange resulting from evaporation; the third integrals represent the work of external bulk forces; the fourth integrals stand for the work of inter-component interaction forces; the fifth integral of the right-hand side of Equation A.35 is for the work of external surface forces; and the sixth integral denotes the work of internal surface forces. N is also called the power of internal forces per volume unit [41]. The explicit expression for N is obtained by comparing differential equations for kinetic energy on the one side that are written based on the work-kinetic energy theorem with the equation obtained by the scalar multiplication of the differential momentum conservation equation by speed on the other side.

The following can be obtained for the internal momentum conservation equation, subject to the common thermodynamical law of internal energy variation resulting from heat inflow (with the deduction of the work of internal forces):

$$\frac{\partial}{\partial t}\int_V \beta_1\rho_1 U_1\, dV = -\int_S \beta_1\rho_1 U_1 \vartheta_{1f}\, dS - \int_V JU_{ns}\, dV - \int_V Q_{12}\, dV - \int_V JP\left(\frac{1}{\rho_{ns}} - \frac{1}{\rho_B}\right) dV, \quad \text{(A.36)}$$

$$\frac{\partial}{\partial t}\int_V \beta_2\rho_2 U_2\, dV = -\int_S \beta_2\rho_2 U_2 \vartheta_{2f}\, dS + \int_V JU_{ns}\, dV + \int_V Q_{12}\, dV +$$

$$+\int_V JP\left(\frac{1}{\rho_{ns}} - \frac{1}{\rho_B}\right) dV + \int_S q_f\, dS + \int_{V} W\, dV. \quad \text{(A.37)}$$

Here, the second integrals characterize the internal energy exchange resulting from evaporation; the third integrals represent the inter-component heat exchange; the fourth integrals stand for the evaporating moisture expansion work; the fifth term of the right-hand side of Equation A.37 represents the exogenous inflow of heat due to the thermal conductivity; and the sixth term denotes the work of internal forces resulting from the force interaction of components. The explicit expression for W is obtained by comparison with the accounting equation for the full mixture in general:

$$\frac{\partial}{\partial t}\int_V (\beta_1\rho_1 E_1 + \beta_2\rho_2 E_2) dV = -\int_S (\beta_1\rho_1 E_1 \vartheta_{1f} + \beta_2\rho_2 E_2 \vartheta_{2f}) dS + \int_S \vec{\Pi}_{2f}\cdot\vec{\vartheta}_2 dS +$$

$$+\int_V (\beta_1\rho_1 \vec{M}_1\cdot\vec{\vartheta}_1 + \beta_2\rho_2\vec{M}_2\cdot\vec{\vartheta}_2) dV + \int_S q_{2f} dS. \quad \text{(A.38)}$$

Integral relations following the application of the well-known Gauss–Ostrogradskiy formula

$$\int_S A\vartheta_f\, dS = \int_V \frac{\partial A\vartheta_k}{\partial x_k} dV \quad \text{(A.39)}$$

are converted into the corresponding differential equations. Repeated inferior index k hereinafter means the summation over coordinate axes ox_k ($k = 1, 2, 3$). Because volume V is selected randomly, the accounting mass equations (A.29, A.30, and A.31) are transformed into the differential equations of continuity:

$$\left.\begin{aligned}
&\frac{\partial}{\partial t}\beta_m\rho_m + \frac{\partial}{\partial x_k}\beta_m\rho_m\vartheta_{1k} = 0; \quad \frac{\partial}{\partial t}\beta_B\rho_B + \frac{\partial}{\partial x_k}\beta_B\rho_B\vartheta_{1k} = -J;\\
&\frac{\partial}{\partial t}\beta_g\rho_g + \frac{\partial}{\partial x_k}\beta_g\rho_g\vartheta_{2k} = 0; \quad \frac{\partial}{\partial t}\beta_n\rho_n + \frac{\partial}{\partial x_k}\beta_n\rho_n\vartheta_{2k} = J;\\
&\frac{\partial}{\partial t}\beta_1\rho_1 + \frac{\partial}{\partial x_k}\beta_1\rho_1\vartheta_{1k} = -J; \quad \frac{\partial}{\partial t}\beta_2\rho_2 + \frac{\partial}{\partial x_k}\beta_2\rho_2\vartheta_{2k} = J.
\end{aligned}\right\} \quad \text{(A.40)}$$

Accordingly, for pulse

$$\left.\begin{aligned}&\frac{\partial}{\partial t}\beta_1\rho_1\vec{\vartheta}_1+\frac{\partial}{\partial x_k}\beta_1\rho_1\vec{\vartheta}_1\vartheta_{1k}=\beta_1\rho_1\vec{M}_1-\vec{F}_{12}-J\vec{\vartheta}_1;\\&\frac{\partial}{\partial t}\beta_2\rho_2\vec{\vartheta}_2+\frac{\partial}{\partial x_k}\beta_2\rho_2\vec{\vartheta}_2\vartheta_{2k}=\beta_2\rho_2\vec{M}_2+\frac{\partial\vec{\Pi}_{2k}}{\partial x_k}+\vec{F}_{12}+J\vec{\vartheta}_1.\end{aligned}\right\}\qquad\text{(A.41)}$$

Or for accounting (A.40)

$$\left.\begin{aligned}&\beta_1\rho_1\frac{d\vec{\vartheta}_1}{dt}=\beta_1\rho_1\vec{M}_1-\vec{F}_{12};\\&\beta_2\rho_2\frac{d\vec{\vartheta}_2}{dt}=\beta_2\rho_2\vec{M}_2+\frac{\partial\vec{\Pi}_{2k}}{\partial x_k}+\vec{F}_{12}+J(\vec{\vartheta}_1-\vec{\vartheta}_2).\end{aligned}\right\}\qquad\text{(A.42)}$$

Hereinafter,

$$\frac{d}{dt}=\frac{\partial}{\partial t}+\vartheta_k\frac{\partial}{\partial x_k}$$

is a substantial derivative functional, and $\vec{\Pi}_{2k}$ is the force applied to a surface perpendicular x_k axis.

The explicit expression for $\dfrac{\partial\vec{\Pi}_{2k}}{\partial x_k}$ is obtained with the assumption that the gaseous component is a Newtonian fluid having the dynamic-viscosity coefficient μ^* that accounts for properties of the carrying medium and properties of inclusions [60]. In order to exclude further triple indexing, assume that $\vec{\Pi}_{2k}\equiv\vec{P}_k;\vec{\vartheta}_2\equiv\vec{u}$.

The surface force projections on coordinate axes based on the generalized Newton's law for viscous fluid are related with the medium velocity projections by the known equation (A.41)

$$P_{ij}=\begin{cases}\mu^*\left(\dfrac{\partial u_i}{\partial x_i}+\dfrac{\partial u_j}{\partial x_j}\right) & \text{at}\quad j\neq i,\ j=1,2,3;\\[2ex] -P+2\mu^*\dfrac{\partial u_i}{\partial x_i} & \text{at}\quad j=i.\end{cases}$$

Thereby

$$\frac{\partial\vec{\Pi}_{2k}}{\partial x_k}\equiv\frac{\partial\vec{P}_k}{\partial x_k}=-gradP+\mu^*\vec{l}_k\frac{\partial}{\partial x_k}\left(\frac{\partial u_k}{\partial x_k}\right)+\mu^*\nabla^2\vec{u}.\qquad\text{(A.43)}$$

Here $\vec{l}_k$ is the unit vector of x_k axis, and the repeated index k denotes the double summation over coordinate axes.

Scalar multiplication of both sides of Equation A.42 by the velocity vector of components results in the following relations for kinematic energy

$$\left.\begin{aligned} \beta_1\rho_1 \frac{d}{dt}\frac{\vartheta_1^2}{2} &= \beta_1\rho_1\vec{M}_1\cdot\vec{\vartheta}_1 - \vec{F}_{12}\cdot\vec{\vartheta}_1; \\ \beta_2\rho_2 \frac{d}{dt}\frac{\vartheta_2^2}{2} &= \beta_2\rho_2\vec{M}_2\cdot\vec{\vartheta}_2 + \vec{\vartheta}_2\cdot\frac{\partial\vec{\Pi}_{2k}}{\partial x_k} + \vec{F}_{12}\cdot\vec{\vartheta}_2 + J(\vec{\vartheta}_1 - \vec{\vartheta}_2)\cdot\vec{\vartheta}_2 \end{aligned}\right\}. \tag{A.44}$$

From Equations A.34 and A.35:

$$\left.\begin{aligned} \frac{\partial}{\partial t}\beta_1\rho_1\frac{\vartheta_1^2}{2} + \frac{\partial}{\partial x_k}\beta_1\rho_1\frac{\vartheta_1^2}{2}\vartheta_{1k} &= -J\frac{\vartheta_1^2}{2} + \beta_1\rho_1\vec{M}_1\cdot\vec{\vartheta}_1 - \vec{F}_{12}\cdot\vec{\vartheta}_1; \\ \frac{\partial}{\partial t}\beta_2\rho_2\frac{\vartheta_2^2}{2} + \frac{\partial}{\partial x_k}\beta_2\rho_2\frac{\vartheta_2^2}{2}\vartheta_{2k} &= J\frac{\vartheta_1^2}{2} + \beta_2\rho_2\vec{M}_2\cdot\vec{\vartheta}_2 + \vec{F}_{12}\cdot\vec{\vartheta}_2 + \frac{\partial}{\partial x_k}\vec{\Pi}_{2k}\cdot\vec{\vartheta}_2 + N. \end{aligned}\right\} \tag{A.45}$$

or, Equation A.40

$$\left.\begin{aligned} \beta_1\rho_1\frac{d}{dt}\frac{\vartheta_1^2}{2} &= \beta_1\rho_1\vec{M}_1\cdot\vec{\vartheta}_1 - \vec{F}_{12}\cdot\vec{\vartheta}_1, \\ \beta_2\rho_2\frac{d}{dt}\frac{\vartheta_2^2}{2} &= \beta_2\rho_2\vec{M}_2\cdot\vec{\vartheta}_2 + \frac{\partial}{\partial x_k}\vec{\Pi}_{2k}\cdot\vec{\vartheta}_2 + \vec{F}_{12}\cdot\vec{\vartheta}_2 + J\frac{\vartheta_1^2 - \vartheta_2^2}{2} + N. \end{aligned}\right\} \tag{A.46}$$

By comparing Equations A.44 and A.46, we determine

$$N = -\vec{\Pi}_{2k}\cdot\frac{\partial\vec{\vartheta}_2}{\partial x_k} - J\frac{(\vartheta_1 - \vartheta_2)^2}{2}. \tag{A.47}$$

The following equations are obtained for internal energy of components and full energy of the mixture based on Equations A.36, A.37, and A.38:

$$\frac{\partial}{\partial t}\beta_1\rho_1 U_1 + \frac{\partial}{\partial x_k}\beta_1\rho_1 U_1\vartheta_{1k} = -JU_{ns} - Q_{12} - JP\left(\frac{1}{\rho_{ns}} - \frac{1}{\rho_B}\right); \tag{A.48}$$

$$\frac{\partial}{\partial t}\beta_2\rho_2 U_2 + \frac{\partial}{\partial x_k}\beta_2\rho_2 U_2\vartheta_{2k} = JU_{ns} + Q_{12} + JP\left(\frac{1}{\rho_{ns}} - \frac{1}{\rho_n}\right) + \frac{\partial q_k}{\partial x_k} + W; \tag{A.49}$$

$$\begin{aligned} &\frac{\partial}{\partial t}(\beta_1\rho_1 E_1 + \beta_2\rho_2 E_2) + \frac{\partial}{\partial x_k}(\beta_1\rho_1 E_1\vartheta_{1k} - \beta_2\rho_2 E_2\vartheta_{2k}) = \\ &= \frac{\partial}{\partial x_k}(\vec{\Pi}_{2k}\cdot\vec{\vartheta}_2) + \beta_1\rho_1\vec{M}_1\cdot\vec{\vartheta}_1 + \beta_2\rho_2\vec{M}_2\cdot\vec{\vartheta}_2 + \frac{\partial q_k}{\partial x_k}. \end{aligned} \tag{A.50}$$

On the other hand, adding Equations A.45, A.48, and A.49 term by term, and keeping in mind Equation A.15, we obtain the following equation for full mixture energy:

$$\frac{\partial}{\partial t}(\beta_1\rho_1 E_1+\beta_2\rho_2 E_2)+\frac{\partial}{\partial x_k}(\beta_1\rho_1 E_1\vartheta_{1k}-\beta_2\rho_2 E_2\vartheta_{2k})=\frac{\partial}{\partial x_k}(\vec{\Pi}_{2k}\cdot\vec{\vartheta}_2)+\beta_1\rho_1\vec{M}_1\cdot\vec{\vartheta}_1+$$

$$+\beta_2\rho_2\vec{M}_2\cdot\vec{\vartheta}_2-\vec{F}_{12}\cdot(\vec{\vartheta}_1-\vec{\vartheta}_2)+\frac{\partial q_k}{\partial x_k}+N-JP\left(\frac{1}{\rho_n}-\frac{1}{\rho_B}\right)+W. \quad \text{(A.51)}$$

Comparing the result with Equation A.50, we obtain

$$W=\vec{F}_{12}\cdot(\vec{\vartheta}_1-\vec{\vartheta}_2)-N+JP\left(\frac{1}{\rho_n}-\frac{1}{\rho_B}\right). \quad \text{(A.52)}$$

In view of the resulting equation for W, additive relations (A.28), definition (A.18), and equations of continuity (A.40), Equations A.48 and A.49 can be rewritten as

$$\beta_B\rho_B\frac{dU_B}{dt}+\beta_m\rho_m\frac{dU_m}{dt}=-J(e_{ns}-e_B)-Q_{12}; \quad \text{(A.53)}$$

$$\beta_n\rho_n\frac{dU_n}{dt}+\beta_g\rho_g\frac{dU_g}{dt}=J(e_{ns}-e_n)+Q_{12}+\frac{\partial q_k}{\partial x_k}-N+\vec{F}_{12}\cdot(\vec{\vartheta}_1-\vec{\vartheta}_2)+JP\left(\frac{1}{\rho_n}-\frac{1}{\rho_B}\right). \quad \text{(A.54)}$$

or

$$\beta_1\rho_1 c_1\frac{dT_1}{dt}=-J[c_n(T_1-T_s)+r]-Q_{12}, \quad \text{(A.55)}$$

$$\beta_2\rho_2 c_2\frac{dT_2}{dt}=Jc_n(T_s-T_2)+Q_{12}+\frac{\partial q_k}{\partial x_k}-N+\vec{F}_{12}\cdot(\vec{\vartheta}_1-\vec{\vartheta}_2)-$$

$$-J\frac{P}{\rho_B}+\frac{\partial}{\partial t}\beta_2 P+\frac{\partial}{\partial x_k}\beta_2 P\vartheta_{2k}. \quad \text{(A.56)}$$

In most of the practical cases, dissipative terms are neglected due to their smallness, and the enthalpy chamber for a gaseous component is described as follows:

$$\beta_2\rho_2 c_2\frac{dT_2}{dt}=Jc_n(T_s-T_2)+Q_{12}+\frac{\partial q_k}{\partial x_k}. \quad \text{(A.57)}$$

Due to homogeneity of the second component, the heat inflow equation is

$$\vec{q}=-\lambda gradT, \quad \text{(A.58)}$$

$$\frac{\partial q_k}{\partial x_k}=-\lambda\nabla^2 T. \quad \text{(A.59)}$$

The following result should be noted to finish the description of energy interactions among the components. As is clear from Equation A.44, the mixture momentum

$$\beta_1\rho_1\frac{d}{dt}\frac{\vartheta_1^2}{2}+\beta_2\rho_2\frac{d}{dt}\frac{\vartheta_2^2}{2}=\beta_1\rho_1\vec{g}\cdot\vec{\vartheta}_1+\beta_2\cdot(\rho_2-\rho_0)\vec{g}\cdot\vec{\vartheta}_2+\vec{\vartheta}_2\cdot\frac{\partial\vec{\Pi}_k}{\partial x_k}- \\ -\vec{F}_{12}\cdot(\vec{\vartheta}_1-\vec{\vartheta}_2)-J(\vec{\vartheta}_2-\vec{\vartheta}_1)\cdot\vec{\vartheta}_2 \quad (A.60)$$

is increased due to the work of surface and bulk forces and is decreased due to the work of inter-component forces and of interphase transformations. Momentum losses resulting from inter-component interactions are applied to the increase of the internal mixture energy and the kinematic energy of small-scale flows within or around inclusions. This fact, as already stated, was not considered in theoretical efforts, which regarded the motion of air entrained by a solid stream in terms of the work-kinetic energy theorem.

A.2 SPACE–TIME METHOD OF AVERAGING ACCOUNTING EQUATIONS

Without the relevant corrections, the quoted results of the phenomenological theory of heterogeneous media mechanics cannot be used to analyze the motion of streams containing large inclusions, which are the streams of a bulk material and air in question. This condition is associated with the averaging method that requires a sufficiently large number of particles in the volume unit. In our case, preserving this condition would make it necessary to choose volume dimensions proportional to or larger than the typical stream dimensions (dimensions of the channel wherein a bulk material is flowing). The only possible way out of this situation is probably the application of the space and time averaging method that was first used by F. I. Frankl [67,68] in a study of pumps and by A. K. Dyunin [32,33] in a study of storms, after which we used it to study the bulk material stream in chutes [40,42]. Variables should be averaged in macrovolume V (limited by surface S) for which typical dimensions L_V, L_S are larger than the dimensions of inclusions a (diameter of particles, distance between the particles, etc.) but at the same time are smaller than the typical macroscopic dimension of L (the channel length and diameter):

$$L >> L_V(L_S) >> a >> dx. \quad (A.61)$$

Because a particle volume is larger than the unit of volume

$$V_p >> dV, \quad (A.62)$$

there is no point in discussing bulk particle concentration in the microvolume. Instead, we introduce a signpost to the existence of the γ_i^I component, which is a unit value if volume dV only contains i-component, or $\gamma_i^I = 0$, if the volume lacks this component.

Let us write down mass, pulse, and energy transfer equations for a two-component stream without phase transformations.

A.2.1 Mass Transfer Equation

Here, instantaneous velocity vector projections as well as signposts to the existence of components are not necessarily continuous functions of spatial coordinates and time:

$$\frac{\partial}{\partial t}\int_V \gamma_i' \rho_i' \, dV = -\int_S \gamma_i' \rho_i' \vartheta_{if}' \, dS. \tag{A.63}$$

A differential mass transfer equation may be obtained from integral Equation A.63. For this purpose, in order to ensure the continuity of scalar and vector values included in the equation and derivatives of the same with spatial coordinates and time, we need to apply the operation of averaging and transition to an infinitesimal volume.

All scalar and vector values are averaged by space and time cylinder 2ωΔt (i.e., in a constant area ω and in a constant range 2Δτ), using the following formula:

$$\bar{A}_i' \equiv A_i = \frac{1}{2\omega \Delta t}\int_{2\Delta\tau}\int_{\omega} A_i' \, d\omega \, d\tau. \tag{A.64}$$

A hyphen above the physical value A_i hereinafter means an averaging operation.

Having been averaged as described here, a signpost to existence of components shall represent nothing but a bulk concentration of these components. Other physical values (density, velocity, etc.) should not be averaged by the full volume ω but by the portion of this volume that is filled with the component in question. To this effect, we use the following equation that is a pre-image of dependencies suggested for these purposes by F. I. Frankl:

$$A_i = \left(\frac{\overline{\gamma_i' A_i'}}{\beta_{i\omega}'}\right) = \frac{1}{2\Delta\tau}\int_{2\Delta\tau}\left(\frac{1}{\beta_{i\omega}'\omega}\int_{\omega} A_i' \, d\omega_i'\right) d\tau, \tag{A.65}$$

where*

$$\beta_{i\omega}' = \frac{1}{\omega}\int_{\omega} \gamma_i' \, d\omega; (\beta_{i\omega}' \cdot \omega = \omega_i')$$

$$d\omega_i' = \gamma_i' \, d\omega.$$

Having been averaged as described earlier, the values and derivatives of the same time and coordinate are continuous. Therefore, Gauss–Ostrogradskiy transformations may be applied to the surface integrals. By integrating both sides of Equation A.63 in volume 2Δτω

* Averaging by Frankl appears as follows: $A_i = \frac{\overline{\gamma_i \cdot A_i}}{\beta_i}$ which $\neq \left(\frac{\overline{\gamma_i \cdot A_i}}{\beta_{i\omega}'}\right)$. This inaccuracy was studied for the first time in 1965 [33].

$$\frac{\partial}{\partial t}\int_V\left(\int_{2\Delta\tau}\int_{\omega}\gamma_i'\,\rho_i\omega\,d\tau\right)dV=-\int_S\left(\int_{2\Delta\tau}\int_{\omega}\gamma_i'\,\rho_i'\,\vartheta_{if}'\,d\omega\,d\tau\right)dS \tag{A.66}$$

and dividing both sides by $2\Delta\tau\omega$, we obtain

$$\frac{\partial}{\partial t}\int_V\overline{\gamma_i'\,\rho_i'}\,dV=-\int_S\overline{\gamma_i'\,\rho_i'\,\vartheta_{if}'}\,dS \tag{A.67}$$

or

$$\int_V\left(\frac{\partial}{\partial t}\overline{\gamma_i\rho_i}+\frac{\partial}{\partial x_k}\overline{\gamma_i\rho_i\vartheta_{ik}}\right)dV=0, \tag{A.68}$$

from which, in view of the randomness of volume V,

$$\frac{\partial}{\partial t}\overline{\gamma_i'\,\rho_i'}+\frac{\partial}{\partial x_k}\overline{\gamma_i'\,\rho_i'\,\vartheta_{\iota\kappa}'}=0. \tag{A.69}$$

Remove the averaging signs using the method of successive averages, first in volume ω and then in time interval $2\Delta\tau$. Keep in mind that the averaged values and their instantaneous and pulsating values for i-component are related by the following apparent equations:

$$A_i'=A_{i\omega}'+A_{i\omega}'';\int_{\omega}A_{i\omega}''\,d\omega_i'=0; \tag{A.70}$$

$$A_{i\omega}'=A_i+A_{i\tau}'';\int_{2\Delta\tau}A_{i\tau}''\,d\tau=0. \tag{A.71}$$

From which the volume average (at given instant)

$$A_{i\omega}'=\frac{1}{\beta_{i\omega}'\omega}\int_{\omega}A_i'\,d\omega_i'. \tag{A.72}$$

As an example, we will consider the averaged product of instantaneous values $\gamma_i'\,\rho_i'$:

$$\overline{\gamma_i'\,\rho_i'}=\frac{1}{2\Delta\tau\omega}\int_{2\Delta\tau}\int_{\omega}\gamma_i'\,\rho_i'\,d\omega d\tau=\frac{1}{2\Delta\tau}\int_{2\Delta\tau}\left(\frac{\beta_{i\omega}'}{\omega\beta_{i\omega}'}\int_{\omega}\rho_i'\,d\omega_i'\right)d\tau=\frac{1}{2\Delta\tau}\int_{2\Delta\tau}\beta_{i\omega}'\rho_{i\omega}'d\tau$$

or

$$\overline{\gamma_i'\,\rho_i'}=\beta_i\rho_i+\frac{1}{2\Delta\tau}\int_{2\Delta\tau}\beta_{i\tau}''\,\rho_{i\tau}''\,d\tau-\beta_i\rho_i+\overline{\beta_{i\tau}''\,\rho_{i\tau}''}. \tag{A.73}$$

Thus,

$$\overline{\gamma_i'\,A_i'}=\beta_iA_i+\psi''(\gamma_iA_i), \tag{A.74}$$

where

$$\psi''(\gamma_i A_i) = \overline{\beta''_{i\tau} A''_{i\tau}}. \tag{A.75}$$

Accordingly,

$$\overline{\gamma'_i A'_i B'_i} = \beta_i A_i B_i + \psi''(\gamma_i A_i B_i), \tag{A.76}$$

where

$$\psi''(\gamma_i A_i B_i) = \overline{\beta_i A''_{i\tau} B''_{i\tau}} + \overline{\beta''_{i\tau} A_i B''_{i\tau}} + \overline{\beta''_{i\tau} A''_{i\tau} B_i} + \overline{\beta''_{i\tau} A''_{i\tau} B''_{i\tau}} + \overline{\gamma'_i A''_{i\omega} B''_{i\omega}}. \tag{A.77}$$

In view of the resulting relations, the averaged mass transfer equation (A.69) will appear as

$$\frac{\partial}{\partial t}\beta_i\rho_i + \frac{\partial}{\partial x_k}\beta_i\rho_i\vartheta_{1k} = -\frac{\partial}{\partial t}\psi''(\gamma_i\rho_i) - \frac{\partial}{\partial x_k}\psi''(\beta_i\rho_i\vartheta_{1k}). \tag{A.78}$$

It is apparent that, with a constant density of components,

$$\frac{\partial}{\partial t}\beta_i + \frac{\partial}{\partial x_k}\beta_i\vartheta_{1k} = -\frac{\partial}{\partial x_k}\overline{\beta''_{i\tau}\vartheta''_{ik\tau}} \tag{A.79}$$

or

$$\frac{\partial}{\partial t}\beta_i + \frac{\partial}{\partial x_k}K_{\vartheta_k}\beta_i\vartheta_{1k} = 0, \tag{A.80}$$

where K_{ϑ_k} is a correction factor accounting for pulsations of a bulk concentration and for a velocity of components:

$$K_{\vartheta_k} = 1 + \overline{\beta''_{i\tau}\vartheta''_{ik\tau}} / (\beta_i\vartheta_{ik}). \tag{A.81}$$

A.2.2 Pulse Transfer Equation

The integral pulse balance equation for i-component is

$$\frac{\partial}{\partial t}\int_V \gamma'_i\rho'_i\vec{\vartheta}'_i\,dV = -\int_S \gamma'_i\rho'_i\vec{\vartheta}'_i\vartheta'_{if}\,dS + \int_V \gamma'_i\rho'_i\vec{M}'_i\,dV + \int_S \gamma'_i\vec{\Pi}'_{if}\,dS + \int_V \gamma'_i\vec{r}'_{ji}f'_\phi\,dV. \tag{A.82}$$

Here, the last term of the sum in the right-hand side of the equation represents the force of interaction between components; $\vec{r}'_{ji}$ is a vector of the media impact force applied to a particle surface unit; and f'_ϕ is a particle surface in the mixture volume unit:

$$\vec{r}'_{ji} = -\vec{r}'_{ji} = \frac{\vec{R}_{ji}}{S_p} = \frac{1}{S_p}\int_{S_p}\vec{\Pi}_{jis}\,dS_p;\; f_p = \frac{dS_p}{dV}, \tag{A.83}$$

where $\vec{\Pi}_{jis}$ is a vector of force on a particle surface with outer normal line $\vec{f}_p$. By averaging integral relation Equation A.82 and making some transformations, we obtain

$$\frac{\partial}{\partial t}\overline{\gamma_i' \rho_i' \vec{\vartheta}_i'} + \frac{\partial}{\partial x_k}\overline{\gamma_i' \rho_i' \vec{\vartheta}_i' \vartheta_{ik}} = \overline{\gamma_i' \rho_i' \vec{M}'_i} + \frac{\partial}{\partial x_k}\overline{\gamma_i' \vec{\Pi}_{ik}} + \overline{\gamma_i' r_{ji}' f_\phi'} \qquad \text{(A.84)}$$

or, by removing the averaging signs,

$$\frac{\partial}{\partial t}[\beta_i\rho_i\vec{\vartheta}_i + \psi''(\gamma_i\rho_i\vec{\vartheta}_i)] + \frac{\partial}{\partial x_k}[\beta_i\rho_i\vec{\vartheta}_i\vartheta_{ik} + \rho_i\psi''(\gamma_i\vec{\vartheta}_i\vartheta_{ik}) + \rho_{i\omega}''\overline{\gamma_i'\vec{\vartheta}_i'\vartheta_{ik}'} + \rho_{i\tau}''\overline{\gamma_i'\vec{\vartheta}_i'\vartheta_{ik}'}] =$$

$$= \beta_i\rho_i\vec{M}_i + \psi''(\gamma_i\rho_i\vec{M}_i) + \frac{\partial}{\partial x_k}[\beta_i\vec{\Pi}_{ik} + \psi''(\gamma_i\vec{\Pi}_{ik})] + \vec{r}_{ji}f_\phi + \phi''(\vec{r}_{ji}f_\phi), \qquad \text{(A.85)}$$

where

$$\phi''(\vec{r}_{ji}f_\phi) = \overline{\vec{r}_{ji\omega}''f_{\phi\omega}''} + \overline{\vec{r}_{ji\omega}''f_{\phi\tau}''} + \overline{\vec{r}_{ji\tau}''f_{\phi\omega}''} + \overline{\vec{r}_{ji\tau}''f_{\phi\tau}''}. \qquad \text{(A.86)}$$

Let us introduce correction factors accounting for pulse instants

$$K_{\vec{\vartheta}} = 1 + \psi''(\gamma_i\rho_i\vec{\vartheta}_i)/(\beta_i\rho_i\vec{\vartheta}_i);\ K_R = 1 + \phi''(\vec{r}_{ji}f_\phi)/(\vec{r}_{ji}f_\phi); \qquad \text{(A.87)}$$

$$\left.\begin{aligned} K_{\vartheta\vartheta_k} &= 1 + \frac{\rho_i\psi''(\gamma_i\vartheta_i\vartheta_{ik})}{\beta_i\rho_i\vartheta_i\vartheta_{ik}} + \frac{\rho_{i\omega}''\overline{\gamma_i'\vartheta_i'\vartheta_{ik}'} + \rho_{i\tau}''\overline{\gamma_i'\vartheta_i'\vartheta_{ik}'}}{\rho_i\beta_i\vartheta_i\vartheta_{ik}}; \\ K_M &= 1 + \frac{\psi''(\gamma_i\rho_i\vec{M}_i)}{\beta_i\rho_i\vec{M}_i};\quad K_\Pi = 1 + \frac{\psi''(\gamma_i\vec{\Pi}_{ik})}{\beta_i\vec{\Pi}_{ik}}; \end{aligned}\right\} \qquad \text{(A.88)}$$

and express the averaged force of the components' interaction in terms of a particle drag force and the number of particles per volume unit

$$n_p = \beta_1/V_p;\ f_\phi = n_pS_p = \beta_1S_p/V_p;$$

$$\vec{r}_{21}f_\phi = \frac{\vec{R}_{21}}{S_p}\frac{\beta_1}{V_p}S_p = \frac{\beta_1}{V_p}\vec{R}_{21}. \qquad \text{(A.89)}$$

Then, the pulse transfer equation will appear as follows:

$$\frac{\partial}{\partial t}K_{\vec{\vartheta}}\beta_i\rho_i\vartheta_i + \frac{\partial}{\partial x_k}K_{\vec{\vartheta}\vartheta_k}\beta_i\rho_i\vec{\vartheta}_i\vartheta_{ik} = K_M\beta_i\rho_i\vec{M}_i + \frac{\partial}{\partial x_k}K_\Pi\beta_i\vec{\Pi}_{ik} + K_R\frac{\beta_1}{V_p}\vec{R}_{ji},$$

(A.90)

which differs from the dual-speed continuous medium equation only by the presence of correction factors.

A.2.3 Energy Transfer Equation

The integral momentum balance equation (with considering internal forces) appears as follows:

$$\frac{\partial}{\partial t}\int_V \gamma_i' \rho_i' \frac{(\vartheta_i^2)'}{2} dV = -\int_S \gamma_i' \rho_i' \frac{(\vartheta_i^2)'}{2}\vartheta_{if}' \, dS + \int_V \gamma_i' \rho_i' \vec{M}_i' \cdot \vec{\vartheta}_i' \, dV + + \int_S \gamma_i' \vec{\Pi}_{if}' \cdot \vec{\vartheta}_\iota' \, dS + \int_V \vec{r}_{ji} \cdot \vec{\vartheta}_i' f_\phi' \, dV. \tag{A.91}$$

By averaging and undertaking some transformations similar to those we performed for mass and pulse transfer equations, we obtain

$$\frac{\partial}{\partial t} K_{\vartheta^2}\beta_i\rho_i \frac{\vartheta_i^2}{2} + \frac{\partial}{\partial x_k} K_{\vartheta^2\vartheta_k}\beta_i\rho_i \frac{\vartheta_i^2}{2}\vartheta_{ik} = K_{\vec{M}\cdot\vec{\vartheta}}\beta_i\rho_i\vec{M}_i\cdot\vec{\vartheta}_i + \frac{\partial}{\partial x_k} K_{\vec{\Pi}\cdot\vec{\vartheta}}\beta_i\vec{\Pi}_{ik}\cdot\vec{\vartheta}_i + K_{\vec{R}\cdot\vec{\vartheta}}\frac{\beta_1}{V_x}\vec{R}_{ji}\cdot\vec{\vartheta}_i, \tag{A.92}$$

where

$$\left.\begin{aligned}
&K_{\vartheta^2} = 1 + \frac{\psi''\left(\gamma_i\rho_i\frac{\vartheta_i^2}{2}\right)}{\beta_i\rho_i\frac{\vartheta_i^2}{2}}; \quad K_{\vec{\Pi}\cdot\vec{\vartheta}} = 1 + \frac{\psi''\left(\gamma_i\vec{\Pi}_{ik}\cdot\vec{\vartheta}_i\right)}{\beta_i\vec{\Pi}_{ik}\cdot\vec{\vartheta}_i};\\
&K_{\vartheta^2\vartheta_k} = 1 + \frac{\rho_i\psi''\left(\gamma_i\frac{\vartheta_i^2}{2}\vartheta_{ik}\right) + \overline{\rho_{i\omega}''\gamma_i'\left(\frac{\vartheta_i^2}{2}\right)'\vartheta_{ik}'} + \overline{\rho_{i\tau}''\gamma_i'\left(\frac{\vartheta_i^2}{2}\right)'\vartheta_{ik}'}}{\beta_i\rho_i\vartheta_i^2\vartheta_{ik}/2};\\
&K_{\vartheta^2\vartheta_k} = 1 + \frac{\rho_i\psi''\left(\gamma_i\frac{\vartheta_i^2}{2}\vartheta_{ik}\right) + \overline{\rho_{i\omega}''\gamma_i'\left(\frac{\vartheta_i^2}{2}\right)'\vartheta_{ik}'} + \overline{\rho_{i\tau}''\gamma_i'\left(\frac{\vartheta_i^2}{2}\right)'\vartheta_{\iota\kappa}'}}{\beta_i\rho_i\vartheta_i^2\vartheta_{ik}/2}.
\end{aligned}\right\} \tag{A.93}$$

Let us write down the following equation for heat transfer (ignoring the mechanical work dissipation and heat inflow due to heat conductivity):

$$\frac{\partial}{\partial t}\int_V \gamma_i' \rho_i' c_i' T_i' dV = -\int_S \gamma_i' \rho_i' c_i' T' \vartheta_{if}' \, dS + \int_V q_{ji}' f_\phi' dV, \tag{A.94}$$

from which

$$\frac{\partial}{\partial t} K_T\beta_i c_i\rho_i T_i + \frac{\partial}{\partial x_k} K_{T\vartheta_k}\beta_i c_i\rho_i T_i\vartheta_{ik} = K_q\frac{\beta_1}{V_p}Q_{ji} \tag{A.95}$$

where

$$\left.\begin{aligned}
&K_T = 1 + \frac{\psi''(\gamma_i \rho_i T_i)}{\beta_i \rho_i T_i}; \quad K_q = 1 + \frac{\phi''(q_{ji} f_\phi)}{q_{ji} f_\phi};\\
&K_{T\vartheta_k} = 1 + \frac{\rho_i' \, \psi''(\gamma_i T_i \vartheta_{ik}) + \overline{\rho_{i\omega}'' \gamma_i' \, T_i' \, \vartheta_{ik}'} + \overline{\rho_{i\tau}'' \gamma_i' \, T_i' \, \vartheta_{ik}'}}{\beta_i \rho_i T_i \vartheta_{ik}};\\
&q_{ji} f_\phi = \beta_1 Q_{ji} / V_p.
\end{aligned}\right\} \qquad \text{(A.96)}$$

Thus, a material flow and air it entrains is characterized, on the one hand, by a macroscale process of the induced air translational movement and, on the other hand, by a microscale pulsation of components—a mechanism of inter-component interaction expanded in a theoretical description. An alternative method consists of developing approximations. However, a quite efficient theory of aerodynamic processes in a bulk material gravity flow exists that, in a sense, would combine a comparative usability with an ample exactitude and similarity. This precise theory was suggested herein. Conclusions and results drawn from this theory are in qualitative and quantitative line with the experimental data.

REFERENCES

1. G. N. Abramovich. On the influence of solid particles or drops on the structure of an eddy gas jet. *Reports of the Academy of Sciences of the USSR*, 1970. V. 190. No. 5. pp. 1052–1055.
2. G. N. Abramovich, V. I. Bazhanov, T. A. Girshovich. Eddy stream with solid impurities. *News of the Academy of Sciences of the USSR*, 1972. No. 6. pp. 41–49.
3. G. N. Abramovich, T. A. Girshovich. In dispersion of solid particles in eddy streams. *Reports of the Academy of Sciences of the USSR*, 1973. V. 212. No. 33. pp. 573–576.
4. G. N. Abramovich, S. Yu. Krashennikov, A. N. Sekundov, I. P. Smirnov. Turbulent mixing of gas jets. Moscow: Nauka, 1974. p. 272.
5. G. N. Abramovich, T. A. Girshovich. Eddy streams carrying solid or liquid impurities. Steam-and-fluid streams. Minsk, 1977. pp. 155–175.
6. G. L. Babukha, A. A. Shreiber. Impact of the collision of polydisperse material particles on motion and interphase heat exchange in a vertical two-phase stream. *PMTF*, 1966. No. 4. pp. 23–29.
7. G. L. Babukha, A. A. Shreiber. On motion of polydisperse material particles in a vertical stream. *PMTF*, 1967. No. 2. pp. 50–57.
8. G. L. Babukha, A. A. Shreiber. Interaction of polydisperse material particles in two-phase streams. Kiev: Naukova Dumka, 1972. p. 173.
9. G. I. Barenblatt. On motion of suspended particles in a turbulent flow filling a half-space or a two-dimensional open channel of a finite depth. *PMM*, 1955. V. 19. No. 1. pp. 61–68.
10. G. I. Barenblatt. On motion of suspended particles in a turbulent flow. *PMM*, 1953. V. 17. No. 3. pp. 53–61.
11. Yu. G. Borschevskiy, I. M. Fedotin, A. M. Kolodin. Two-phase turbulent flow streams. Kiev: Texnika ,1972. p. 112.
12. G. Brenner. Rheology of two-phase systems. Rheology of suspensions: Coll. Moscow: Mir, 1975. pp. 11–67.
13. B. I. Bronstein, O. M. Todes. The theory of pneumatic transport. *ZhTF*, 1953. V. 23. No. 1. pp. 110–126.

14. Yu. A. Buevich. Two-fluid hydrodynamics of a suspended bed. *News of the Academy of Sciences of the USSR*, 1966. No. 4. pp. 94–100.
15. Yu. A. Buevich, I. N. Schelchkova. Rheological properties of homogeneous fine suspensions. Steady flows. *IFZh*, 1977. V. 33. No. 5. pp. 872–879.
16. Yu. A. Buevich, V. G. Markov. Rheological properties of homogeneous fine suspensions. Unsteady flows. *IFZh*, 1978. V. 34. No. 6. pp. 1007–1013.
17. Yu. A. Buevich. Patterns of laminar steady flows of two-component dispersion systems. Vertical streams. *News of the Academy of Sciences of the USSR*, 1968. No. 3. pp. 53–62.
18. Yu. A. Buevich. Successive approximations in hydromechanics dispersion systems. *PMM*, 1971. V. 35. No. 3. pp. 464–481.
19. A. S. Burchakov, E. M. Moskalenko. Dynamics of aerosols in mine roadways. Moscow: Nauka, 1965. p. 67.
20. R. Busroid. Gas flow with suspended particles. Moscow: Mir, 1975. p. 378.
21. V. Ya. Vanatoa. Lateral motion of particles in tubes. Two-phase eddy streams: Coll. Under the editorship of M. K. Laats. Tallin: Academy of Sciences of Estonian SSR, 1979. pp. 72–77.
22. V. M. Voloschuk. Introduction to hydrodynamics of coarsely dispersed aerosols. Moscow: Hydrometeoizdat, 1971. p. 208.
23. V. M. Voloschuk. Approximate solution for Fokker-Planck equation for aerosol particles. *News of the Academy of Sciences of the USSR*, 1970. No. 2. pp. 155–162.
24. Z. R. Gorbis. Criterion equations of convective heat exchange in two-phase gas suspension flows. *News of the Academy of Sciences of the USSR*, 1958. No. 9.
25. Z. R. Gorbis. Differential equations of heat exchange in dispersed fluidized bed streams. *IFZh*, 1962. No. 10.
26. Z. R. Gorbis. Heat exchange and hydromechanics of dispersed through flows. Moscow: Energiya, 1970. p. 424.
27. M. L. Deich, G. A. Philippov. Gas dynamics of two-phase media. Moscow: Energiya, 1968. p. 423.
28. M. A. Dementiev. On group hydraulic size. *VNIIG News*, 1962. V. 71.
29. M. A. Dementiev. General equations and dynamic similarity of suspension-carrying streams. *VNIIG News*, 1963. V. 73.
30. M. A. Dementiev. Dynamic equations and relative phase velocities of two-phase suspended streams. *VNIIG News*, 1978. V. 123. pp. 3–11.
31. I. N. Dorokhov, V. V. Kafarov, R. I. Nigmatulin. Continuum mechanics methods for description of multiphase multicomponent mixtures with chemical reactions and heat- and mass transfer processes. *PMM*, 1975. V. 39. No. 3. pp. 485–495.
32. A. K. Dyunin. Mechanics of storms. Novosibirsk: Siberian Department of the Academy of Sciences of the USSR, 1963.
33. A. K. Dyunin, Yu. T. Borschevskiy, N. A. Yakovlev. Mechanics of multicomponent streams. Novosibirsk: Siberian Department of the Academy of Sciences of the USSR, 1965. p. 74.
34. S. S. Zabrodskiy. Hydrodynamics and heat exchange in a fluidized (boiling) bed. Moscow: Gosenergoizdat, 1963.
35. A. N. Krayko, L. E. Sternin. On the theory of dual-velocity continuous medium streams with solid or liquid particles. *PMM*, 1965. V. 29. No. 3. pp. 418–429.
36. S. S. Kutateladze, M. A. Styrikovich. Hydrodynamics of liquid-gas systems. Moscow: Energiya, 1976.
37. M. K. Laats, F. A. Frishman. On assumptions applied in design of a two-phase stream. *News of the Academy of Sciences of the USSR*, 1970. No. 2. pp. 186–191.
38. M. K. Laats, F. A. Frishman. Eddy transfer processes in a two-phase stream. Eddy transfer processes in shifted eddy streams. *Coll. Tallin*, 1973. pp. 104–199.

39. A. M. Levshakov. Phenomenological equation for suspensions of matter in gas. Physics of air dispersion systems. Kiev: Odessa, 1978. No. 17. pp. 46–49.
40. I. N. Logachev. Fluid equations of a bulk material stream in chutes. Ventilation and air treatment (Tr. NIIrudventilyatsiya). Moscow: Nedra, 1970. pp. 159–162.
41. L . G. Loytsyanskiy. Fluid mechanics. Moscow: Nauka, 1973. p. 849.
42. O. D. Neikov, I. N. Logachev. Aspiration in production of powder materials. Moscow: Metallurgy, 1973. p. 224.
43. R. I. Nigmatulin. Equations of hydromechanics and compression wave in a dual-velocity and two-component continuous medium given phase transformations. *News of the Academy of Sciences of the USSR*, 1967. No. 5. pp. 33–47.
44. R. I. Nigmatulin. Some questions of hydrodynamics of two-phase polydisperse media. *News of the Academy of Sciences of the USSR*, 1968. No. 3. pp. 63–67.
45. R. I. Nigmatulin. Methods of continuous medium mechanics for the description of multiphase mixtures. *PMM*, 1970. V. 34. No. 6. pp. 1097–1112.
46. R. I. Nigmatulin. Small-scale streams and surface effects in hydromechanics of multiphase media. *PMM*, 1971. V. 35. No. 3. pp. 451–463.
47. R. I. Nigmatulin. Mechanics of heterogeneous media. Moscow: Nauka, 1978. p. 336.
48. V. N. Nikolaevskiy, K. S. Basnev, A. T. Gorbunov, G. A. Zotov. Mechanics of saturated porous media. Moscow: Nedra, 1970.
49. Kh. A. Rakhmatulin. Gas dynamics of interpenetrating motions of compressible media. *PMM*, 1956. V. 20. No. 2.
50. A. E. Smoldyrev, A. V. Taylevskiy. Pneumatic transport of unit loads. Moscow: Mashinostroeniye, 1979. p. 158.
51. N. A. Slezkin. Differential equations of fine material motion. *Reports of the Academy of Sciences of the USSR*, 1952. V. 86. No. 2.
52. N. A. Slezkin. Basic motion equations for a deformable medium of particles with a variable mass. *Reports of the Academy of Sciences of the USSR*, 1951. V. 79. No. 1.
53. N. A. Slezkin. On differential filtration equations. *Reports of the Academy of Sciences of the USSR*, 1951. V. 79. No. 5.
54. A. E. Smoldyrev. Hydrodynamic and kinematic characteristics of non-homogeneous dispersion system motion in tubes. *Reports of the Academy of Sciences of the USSR*, 1959. V. 129. No. 5.
55. A. E. Smoldyrev. On basic motion regularities of hydraulic fluids and air-and-coal mixtures in tubes. *Reports of the Academy of Sciences of the USSR*, 1961. V. 137. No. 6.
56. A. E. Smoldyrev. Pipeline transport (theoretical elements and design basis). Moscow: Nedra, 1970. p. 271.
57. S. Sou. Hydrodynamics of multiphase systems. Moscow: Mir, 1971. p. 536.
58. L. E. Sternin. Gas dynamics of two-phase nozzle flows. Moscow: Mashinostroenie, 1974. p. 212.
59. L. I. Sedov. Continuous medium mechanics. Moscow: Nauka, 1976. p. 535.
60. V. M. Safray. On application of cell models for calculation of dispersion systems viscosity. *PMTF*, 1970. No. 1.
61. S. G. Teletov. Issues concerning hydrodynamics of two-phase systems. *Moscow University Bulletin: Ser. Math., Mech., Astron., Phys., Chem.*, 1958. V. 2.
62. S. G. Teletov. Hydrodynamic equations of two-phase liquids. *Reports of the Academy of Sciences of the USSR*, 1945. V. 50. pp 99–102.
63. G. Wallis. One-dimensional two-phase streams. Moscow: Mir, 1972.
64. V. Uspenskiy. Pneumatic transportation of suspended materials. Moscow: Metallurgizdat, 1952.
65. D. F. Fayzullaev. Laminar flow of multiphase media in pipelines. Tashkent: Academy of Sciences of the USSR, 1966.
66. A. Fortie. Mechanics of suspensions. Moscow: Mir, 1971. p. 264.

67. F. I. Frankl. On the theory of suspended sediments motion. *Reports of the Academy of Sciences of the USSR*, 1953. V. 92. No. 2. pp. 247–250.
68. F. I. Frankl. Energy equation for motion of fluids containing suspended sediments. *Reports of the Academy of Sciences of the USSR*, 1955. V. 102. No. 5. pp. 903–906.
69. F. I. Frankl. On the set of equations for motion of suspended sediments. Study of the study maximum flow, wave action and motion of sediments: Coll. Moscow: Academy of Sciences of the USSR, 1960.
70. N. A. Fux. Mechanics of aerosols. Moscow: Academy of Sciences of the USSR, 1955. p. 352.
71. G. N. Khudyakov. On motion of solid particles in a suspension of matter in gas. *News of the Academy of Sciences of the USSR*, 1953. No. 7. pp. 1022–1034.
72. G. V. Tsiklauri, V. S. Danilin, I. L. Seleznev. Adiabatic two-phase flows. Moscow: Atomizdat, 1973. p. 447.

References

1. I. G. Aramanovich, V. I. Levin. Equations of mathematical physics. Moscow: Nauka, 1964. p. 303.
2. I. I. Afanasiev, F. I. Danchenko, Yu. I. Pirogov. Dedusting at crushing and dressing plants. Moscow: Nedra, 1989. p. 198.
3. I. I. Afanasiev, I. N. Logachev et al. Air dedusting at plants of mining and concentration complexes. Moscow: Nedra, 1972. p. 184.
4. L. I. Baron. Mine technical rock science. Subject and method of research. Moscow: Nauka, 1977. p. 324.
5. L. I. Baron. Lumpiness and its measurement methods. Moscow: Academy of Sciences of the USSR, 1960. p. 124.
6. L. I. Baron, N. F. Andrianov, G. Cherepanov. Single-piece measurement of lump in open-cut mines. Moscow: Rotopr. Skochinsky Institute of Mining Affairs, 1962. p. 47.
7. R. Bernstein, V. V. Pomerantsev, L. Shagalova. Generalized method for computation of the aerodynamic resistance of filled sections: Topics of aerodynamics and heat transmission in boiler and furnace processes. Under the editorship of M. Knorre. Moscow: Gosenergoizdat, 1958. pp. 267–289.
8. O. A. Bogaevskiy, U. Kh. Bakirov. Air entrainment with a falling material: Optimization of ventilation systems in the metal mining industry. Sverdlovsk: TSNIIP, 1971. V. 3. pp. 169–175.
9. L. Boltsman. Gas theory readings. Moscow: Gostekhizdat, 1956.
10. E. N. Boshnyakov. Ventilation in non-ferrous metallurgy primary production shops. Moscow: Metallurgy, 1985. p. 160.
11. E. N. Boshnyakov. Method for calculation of suction air exchanges. *Water Supply and Sanitary Engineering*, 1965. No. 11. pp. 14–20.
12. Yu. A. Buevich. On motion resistance of a particle suspended in a turbulent medium. *Journal of the Academy of Sciences of the USSR*, 1966. No. 6. pp. 182–183.
13. Yu. A. Buevich. Phase interaction in concentrated dispersion systems. *PMTF*, 1966. No. 3. pp. 115–117.
14. R. Busroid. Gas flow with suspended particles. Moscow: Mir, 1975. p. 378.
15. S. E. Butakov. Aerodynamics of plant air systems. Moscow: Profizdat, 1949. p. 268.
16. I. A Vakhrushev, A. I. Skoblo. Study of thermal losses in an ascendant stream of gas suspension. VNIIP Studies. Moscow: Gostoptekhizdat, 1959. V. 8.
17. I. P. Vereschagin, V. I. Levitov, G. Z. Mirzabelyan, M. M. Pashin. Fundamentals of electrogasdynamics of dispersion systems. Moscow: Energiya, 1974. p. 480.
18. A. Ya. Sharipov, I. N. Logachev et al. Temporary guidelines for calculation of volumes of air induced from hoods of crushing and wet material processing equipment. Alma-Ata: AO GPI Santekhproject, 1973. p. 96.
19. O. D. Neikov, I. N. Logachev, V. A. Minko et al. Temporary guidelines for calculation of volumes of air induced from hoods in transfer points of dust-forming materials. A3-611. Moscow: GPI Santekhproject, 1973. p. 31.
20. A. M. Gervasiev, V. D. Olifer. Some findings related with chuting of bulk materials: Dedusting ventilation. Sverdlovsk. VNIOT, 1973. pp. 3–9.
21. A. M. Gervasiev. Plant and process ventilation. Sverdlovsk: Metallurgizdat, 1960. V. 9. pp. 13–20.
22. L. A. Grushkov. Ventilation of crushing and milling plants. Sverdlovsk: Metallurgizdat, 1956. p. 95.

23. A. M. Golyshev. "Study of local exhaust ventilation in burning and sieving of iron-ore pellets." Thesis, Cand.Sc. (Engineering), Krivoy Rog, 1981. p. 278.
24. Z. R. Gorbis. Heat exchange and hydromechanics of dispersed through flows. 2nd ed. Moscow: Energiya, 1970. p. 424.
25. Z. R. Gorbis, R. A. Bakhtiozin. On the aerodynamic characteristic of graphite particles. *Izv. vuzov. Energetics*, 1961. No. 11. pp. 101–104.
26. Z. R. Gorbis. Heat exchange of dispersed through flows. Moscow: Energiya, 1964. p. 296.
27. N. F. Graschenkov, V. Kharkovskiy, B. Tsoy. Determining volumes of air induced from hoods in points of bulk materials transfer to the conveyer. *Izv. vuzov. Mining Journal*, 1977. No. 10. pp. 78–80.
28. A. Ya. Dobromyslov. Regarding the theory of vertical pipelines design in building-drainage systems. Sanitary engineering. Moscow: Stroyizdat, 1970. V. 33. pp. 64–78.
29. V. P. Zhuravlev. "Improvement of hydraulic dust control systems for working and development coal faces (by the example of Karaganda basin)": Thesis, Doctor of Engineering, Karaganda, 1973. p. 412.
30. V. P. Zhuravlev, E. F. Demisheva, L. A. Spirin. Aerodynamic methods of coal dust control. Rostov: Rostov University Publishing, 1988. p. 142.
31. Ya. Zayonchkovski. Dust control in industry (translated from Polish). Moscow: Stroyizdat, 1969. p. 350.
32. Ya. I. Zilberberg. "Study of suction in dedusting of continuous transfer groups." Thesis abstract, Cand.Sc. (Engineering), Karaganda, 1980. p. 22.
33. R. L. Zenkov. Bulk cargo mechanics. Moscow: Mashgiz, 1952.
34. I. E. Idelchik. Hydraulic resistance guide. Moscow: Mashinostroenie, 1975. p. 559.
35. V. M. Pankov, I. N. Logachev, P. P. Mamkin et al. Guidelines for comprehensive improvement of working conditions at iron and steel industry dressing plants. Leningrad: Mekhanobr, 1984. p. 166.
36. A. V. Shelketin, I. N. Logachev et al. Study and research of efficient methods of dedusting hot material transfers. Krivoy Rog: NIImetallurgventilyatsiya, 1965. p. 132.
37. M. T. Kamyshenko. Dedusting of bulk material transfer points in crushing and transport shops. Moscow: Profizdat, 1955. p. 99.
38. K. A. Karpov. Tables of $\omega(z)=e^{-z^2}\int_0^z e^{x^2}\,dx$ in a complex domain. Moscow: Academy of Sciences of the USSR, 1954. p. 535.
39. P. I. Kilin. "Dedusting of conveyer feed ends at ore mining enterprises." Thesis abstract, Cand.Sc. (Engineering), Sverdlovsk, 1977.
40. P. I. Kilin, V. I. Shapotaylo. Studies of labor protection institutes of All-Union Central Council of Trade-Unions. Moscow: Profizdat, 1975. V. 96. pp. 3–8.
41. A. I. Kuprin, Yu. V. Glinyaniy, G. V. Kleshnina, G. Pirogov, A. M. Tikhontsev. On estimation criteria of shape of particles moving in a stream of fluid: Aggregates. Collected papers of VNIInerud. Moscow: Tolyatti, 1970. V. 28. pp. 17–25.
42. I. N. Logachev. On circulation of air in chutes when transferring unheated bulk materials: Occupational safety in the ore mining industry. Moscow: Nedra, 1987. pp. 39–45.
43. I. N. Logachev, L. M. Chernenko. Peculiarities of air induction in a free jet of coalescent powder: Improvement of occupational safety at ore mining enterprises. Moscow: Nedra, 1989. pp. 65–69.
44. I. N. Logachev, S. I. Zadorozhniy et al. Research and development of dust control methods for open pellet storages at pellet plants of mining and concentration complexes. Krivoy Rog: VNIIBTG, 1978. p. 149.
45. I. N. Logachev, S. I. Zadorozhniy, L. M. Chernenko et al. Research and development of fundamentals of containment and elimination of dust in reprocessing of ores and concentrates. Krivoy Rog: VNIIBTG, 1980. p. 240.

46. I. N. Logachev. Thermal head in hot material transfers. Air ventilation and treatment. No. 4. Moscow: Nedra, 1970. pp. 120–124.
47. I. N. Logachev. *Intercomponent heat exchange in a stream of a bulk material in closed chute transfers*. Proceedings of the 4th Republic Interuniversity Conference for Evaporation, Combustion and Gas Dynamics of Dispersion Systems. Odessa: OGU, 1969. p. 30.
48. I. N. Logachev, A. M. Golyshev. Peculiarities of the dynamic interaction of a powdered material and air in chutes: Dust control of process flows in the construction materials industry. Moscow: MISI; BTISM, 1984. pp. 10–19.
49. I. N. Logachev. Suction in bulk material transfers at sintering plants: Local exhaust ventilation. Moscow: MDNTI, 1969. pp. 93–106.
50. I. N. Logachev, R. N. Shumilov. Movement of bulk materials in chutes. Air ventilation and treatment: Collected papers of NIImetallurgventilyatsiya. No. 4. Moscow: Nedra, 1970. pp. 124–129.
51. I. N. Logachev. *Air mechanics of a one-dimensional flow of a granular medium in chutes*. Proceeding of the All-Union Interuniversity Scientific Conference for Processes in Dispersed through Flows. Odessa: Lomonosov OTI, 1967. p. 23.
52. I. N. Logachev. "Study of suction in hot material transfers." Thesis, Cand.Sc. (Engineering), Krivoy Rog, 1969.
53. I. N. Logachev. One-dimensional flow of a bulk material in chutes. Air ventilation and treatment. Moscow: Nedra (NIIrudventilyatsiya), 1970. V. 6. pp. 121–128.
54. I. N. Logachev. "Study of suction in hot material transfers." Thesis abstract, Cand.Sc. (Engineering), Krivoy Rog, 1971. p. 20.
55. I. N. Logachev. *Reduction in power of dust emissions in bulk material transfers at ore preparation plants*. Collective proceedings of the international conference. Belgorod: BelGTASM, 1997. V. 9. pp. 30–36.
56. A. P. Lyubimova. "Analysis of suction dedusting parameters for coal transfer and reprocessing groups at dressing plants." Thesis abstract, Cand.Sc. (Engineering), Kemerovo, 1977.
57. P. V. Lyaschenko. Gravity dressing methods. Moscow: Gostoptekhizdat, 1940.
58. V. A. Minko. Fundamentals of industrial ventilation and pneumotransport. Moscow: MISI, BTISM, 1975. p. 129.
59. V. A. Minko et al. Dedusting in cast shops and mechanical engineering enterprises. Moscow: Mashinostroenie, 1987. p. 224.
60. V. A. Minko. Some topics of aerodynamics concerning fine bulk material gravity flows. *IFZH. T.* XVI, No. 6, 1969. pp. 1045–1051.
61. V. A. Minko. Dust control of construction materials production processes. Voronezh: VGU, 1981. p. 175.
62. D. M. Mints. The theory of water treatment technology. Moscow: Stroyizdat, 1964. p. 156.
63. I. N. Logachev, A. M. Golyshev, L. M. Chernenko et al. Suction exhausts and hoods of process equipment of ore preparation plants [album]. Krivoy Rog: VNIIBTG, 1986.
64. Yu. I. Morozov. "Study of heat exchangers with a falling layer." Thesis, Cand.Sc. (Engineering), Kiev, 1967.
65. R. I. Nigmatulin. Mechanics of heterogeneous media. Moscow: Nauka, 1978. p. 336.
66. V. V. Nedin, O. D. Neikov. Dust control at mines. Moscow: Nedra, 1965. p. 200.
67. O. D. Neikov, Ya. I. Zilberberg. Study of air mechanics of closed chutes with powder materials gravity flow: Prevention of a sudden combustion of powders and explosions of gas-dispersion systems. Kiev: Naukova Dumka, 1975. pp. 196–203.
68. O. D. Neikov, I. N. Logachev, R. N. Shumilov. Aspiration of vapor and dust mixtures in dedusting of process equipment. Kiev: Naukova Dumka, 1974. p. 127.

69. O. D. Neikov, I. N. Logachev. Air suction and dedusting in production of powders. Moscow: Metallurgy, 1981. p. 192.
70. O. D. Neikov, I. N. Logachev. Ventilation in production of powder materials. Moscow: Metallurgy, 1973. p. 224.
71. V. D. Olifer. "Study of ventilation in chuting of bulk materials." Thesis abstract, Cand.Sc. (Engineering), Sverdlovsk, 1974.
72. V. Ordinants. The efficiency of exhaustion of dust-laden air from hoods: Ferrous metals [translation from German]. 1975. No. 4. pp. 24–25.
73. OST 14-17-98-83. SSBT. Metallurgic raw materials preparation: Ventilation—A method for calculation of capacity of local exhausts of bulk material transfer hoods. Moscow: MCHM USSR, 1984. p. 31.
74. V. P. Pavlov. On hydrodynamics and heat exchange in a stream of bombarding particles: Heat and mass transfer. V. IX. Minsk. pp. 278–296.
75. P. N. Platonov. "Motion peculiarities of grain streams." Thesis, Doctor of Engineering, Moscow, 1969.
76. P. N. Platonov. Motion peculiarities of granular materials in chutes. *Studies of the Odessa Institute of Flour-Milling Industry and Elevating Equipment Engineers*, 1948. V. 2. pp. 126–131.
77. V. A. Popov. Movement of dust materials on descend trough LIIVT studies, 1971. pp. 87–98.
78. V. A. Popov. "Dust control of processes of dusty materials motion in port transport plant transfer groups." Thesis abstract, Cand.Sc. (Engineering), Leningrad, 1972.
79. A. M. Gervasiev, I. N. Logachev, R. N. Shumilov et al. Design guidelines for heating and ventilation systems for sintering plants in the iron and steel industry. Design guidance materials. A3-500. Moscow: GPI Santekhproject, 1971. p. 31.
80. P. Roach. Computational fluid dynamics. Moscow: Mir, 1980. p. 616.
81. I. N. Logachev, A. M. Golychev, S. I. Zadorozhniy et al. Design guidelines for heating and ventilation systems for pellet plants in the iron and steel industry. Krivoy Rog; Alma-Ata: VNIIBTG; GPI Santekhproject; Mechanobrchermet, 1985. p. 114.
82. L. I. Sedov. Similarity and dimensional methods in mechanics. Moscow: Nauka, 1965. p. 388.
83. A. Semenov. "Study of ventilation in chuting of hot materials." Thesis abstract, Cand.Sc. (Engineering), Sverdlovsk, 1975. p. 18.
84. A. N. Serbin, N. I. Krasikov et al. Research and development of methods and means of dedusting in regular bulk transfers. Krivoy Rog: VNIIBTG, 1973. p. 193.
85. A. S. Serenko. Air dedusting in the refractory industry. Moscow: Metallurgizdat, 1953. p. 144.
86. A. S. Serenko, G. A. Protsenko, A. V. Sheleketin. Air dedusting at iron ore crushing-and-sorting and dressing plants. Moscow: Metallurgizdat, 1957. p. 164.
87. A. S. Serenko. "Dust control of the main flows of solid material reprocessing." Thesis abstract, Cand.Sc. (Engineering), Kiev, 1952.
88. V. I. Smirnov. A course in higher mathematics. V. II. Moscow: Technical and Theoretical Literature Publishing, 1956. p. 328.
89. S. Sou. Dynamics of charged suspensions: Rheology of suspensions. Moscow: Mir, 1975. pp. 140–284.
90. S. Sou. Flow dynamics of multiphase systems. Moscow: Mir, 1971. p. 536.
91. O. S. Bogdanov, ed. Ore processing: Preparative processes. Moscow: Nedra, 1972. V. 3. p. 448.
92. A. Stezhko et al. Improvement of working conditions at mining and concentration complexes. Moscow: Nedra, 1990. p. 172.

93. I. N. Logachev, A. M. Golychev et al. Guidelines for calculation of volumes of air induced from sieve hoods and the adjacent equipment when sieving dusty materials. Krivoy Rog: VNIIBTG, 1980. p. 47.
94. O. D. Neikov et al. Guidelines for calculation and arrangement of ventilation in the metal powder industry. Dusty material transfer groups. Moscow: USSR Minchermet, 1979. p. 105.
95. G. M. Fichtengolz. Calculus course. Moscow: Fizmatgiz, 1963. V. 3. p. 382.
96. A. Fortie. Mechanics of suspensions. Moscow: Mir, 1971. p. 264.
97. N. A. Fux. Mechanics of aerosols. Moscow: Academy of Sciences of the USSR, 1955. p. 352.
98. J. Happel, G. Brener. Hydrodynamics at low Reynolds numbers. Moscow: Mir, 1976. p. 630.
99. G. N. Khudyakov. On motion of solid particles in suspension of matter in gas. *Journal of the Academy of Sciences of the USSR*, 1953. No. 7. pp. 1022–1034.
100. M. L. Chertousov. Hydraulics. Moscow: Gosenergoizdat, 1962.
101. P. Ch. Chulakov, N. N. Karabekov, K. Salimzhanov. *Izv. vuzov. Mining Journal*, 1972. No. 9. pp. 46–50.
102. A. V. Shelketin. Determining volume of air for ventilation of equipment of crushing-screening plants: Silicosis control. Moscow: Academy of Sciences of the USSR, 1959. pp. 135–140.
103. V. P. Shidlovskiy. Introduction to rarefied gas dynamics. Moscow: Nauka, 1965.
104. G. Schlichting. Boundary-layer theory. Moscow: Foreign Literature Publishing, 1956. p. 526.
105. I. F. Yarembash. Mine air treatment after blasting operations. Moscow: Nedra, 1979. p. 191.
106. D. M. Anderson. Dust control design by the air induction technique. *Industr. Medicine and Surgery*, 1964. No. 2. p. 68.
107. B. Degner. *Bergbautechnic*, 1969. No. 7. p. 369.
108. T. Hath. *Pact Huitiene Annee*, 1954. No. 6. p. 425.
109. W. C. L. Hemeon. Plant and process ventilation. New York, NY: Industrial Press, 1955.
110. C. W. Kruse, W. O. A. Bianconi. Air prom induced in enclosed inclined chutes of materials handling systems. *Industrial Hygiene Association*, 1966. No. 3. pp. 220–227.
111. S. Larson. "Air induction by falling material basis for exhaust hood design." Thesis, M.Sc., University of Pittsburgh, 1952. p. 16.
112. I. N. Morrison. *Rock Prod*, 1970. No. 11. pp. 67–71.
113. R. T. Pring, I. F. Knudson, R. Dennis. *Industrial and Engin. Chem.*, 1949. V. 41. p. 2242.
114. M. W. Reeks, G. Skyrme. The dependance of particle deposition velocity on particle inertia in turbulent pipe flow. I. *Aerosol Science*, 1976. V. 7. No. 6. pp. 485–495.
115. H. E. Voegeli. *Air Conditioning, Heating, and Ventilation*, 1963. No. 8. pp. 77–80.
116. I. N. Logachev, I. I. Afanasiev. The theory of design of dust-forming equipment local exhausts. Air ventilation and treatment. Moscow: Nedra, 1972. V. 7. pp. 196–206.
117. I. N. Logachev, I. I. Afanasiev. The theory of process equipment ventilation design: Silicosis control. Moscow: Nauka, 1974. V. 9. pp. 136–139.
118. I. N. Logachev, V. I. Stukanov. Dust control in dressing and sintering of mineral products. In A. Kuzmich, ed., Guide book on dust control in the mining industry. Moscow: Nedra, 1982. p. 240.
119. N. P. Popova. "Development of design methods and means of ventilation for mixture bins at iron and steel works." Thesis abstract, Cand.Sc. (Engineering), Moscow, 1983. p. 23.

120. E. A. Dmitruk. Dust control at feed-milling plants. Moscow: Agropromizdat, 1987. p. 85.
121. O. D. Neykov, E. N. Boshnyakov. Laboratory study of ventilation of material transfer points: Collective papers of the Krivoy Rog branch of the Institute of Mining Affairs of the Academy of Sciences of the Ukrainian SSR. Moscow: Gosgortekhizdat, 1962. V. 1. pp. 145–154.
122. A. V. Kalmykov, L. F. Zhurbinskiy. Dust and noise control at dressing plants. Moscow: Nedra, 1984. p. 222.
123. K. G. Rudenko, A. V. Kalmykov. Dust elimination and collection in reprocessing of mineral products. Moscow: Nedra, 1987. p. 264.
124. E. N. Boshnyakov, A. N. Serbin. Study of ventilated hoods in ore conveyer loading points. Air ventilation and treatment. Moscow: Nedra, 1968. V. 2. pp. 122–135.
125. R. L. Zenkov, I. I. Ivashkov, L. N. Kolobov. Continuous transport machines. Moscow: Mashinostroenie, 1980. p. 304.
126. A. D. Brook et al. Centrifugal blowers. Moscow, 1975. p. 416.
127. V. P. Osokin. Hammer crushers. Moscow: Energiya, 1980. p. 176.
128. T. Solomakhova, K. V. Chebysheva. Centrifugal blowers. Aerodynamic configurations and characteristics. Moscow: Mashinostroenie, 1980. p. 176.
129. I. I. Afanasiev, A. P. Kolesnik, I. N. Logachev. Calculation of ventilated volumes for desintegrators and reverse hammer crushers. *Water Supply and Sanitary Engineering*, 1969. No. 6. pp. 24–26.
130. I. N. Logachev, V. A. Minko, A. P. Kolesnik. Study of aerodynamics and design engineering of suction hoods for bulk material transfers. Krivoy Rog: NIIrudventilyatsiya, 1970. p. 131.
131. V. A. Minko, N. G. Abramkin et al. Research of dedusting of process equipment and workstations in production of lime-sand brick and glass products using a foam and dust suppression technique. Section 1. Suction systems research. Research memorandum of Belgorod Technical Institute of Construction Materials. Belgorod, 1976. p. 290.
132. B. A. Buzunov. Study of the intensity of dust emissions in crushing of ore at processing plants. TsNIIPP studies. V. 15. Sverdlovsk, 1976.
133. I. I. Afanasiev et al. Research and development of methods and means of dedusting ore delivery conveyers. Krivoy Rog: VNIIBTG, 1977. p. 154.
134. A. V. Kalmykov. Dedusting of crushing departments. Moscow: Nedra, 1976. p. 207.
135. A. V. Shelketin, I. I. Afanasiev, R. N. Shumilov. Humidity effect of wet charge materials and recycling on air dust content. Air ventilation and treatment: Collective papers. No. 3. Moscow: Nedra, 1967. pp. 21–26.
136. V. A. Silantiev, Yu. V. Timofeev, F. B. Usmanov et al. Technical report on development of normative documents on sieve ventilation design. Alma-Ata: AO GPI Santekhproject, 1974. p. 132.
137. V. A. Silantiev, Yu. V. Timofeev, F. B. Usmanov et al. Field tests and development of guidelines for dedusting of process equipment: Section 1, crushing and mincing equipment. Alma-Ata: AO GPI Santekhproject, 1972. p. 221.
138. I. I. Afanasiev, I. N. Logachev et al. Performance analysis of dust control equipment used at SSGOK pellet plants. Krivoy Rog: VNIIBTG, 1973. p. 236.
139. I. I. Afanasiev, I. N. Logachev, L. M. Chernenko. Tables of particle size distribution in the air induced from transfer group hoods. Krivoy Rog: VNIIBTG, 1973. p. 236.
140. V. A. Silantiev, E. D. Nesterov, A. Ya. Sharipov, N. Tsitsorin et al. Field tests for determination of particle size distribution and dust content in the air of bulk material transfer groups: Section 2. Alma-Ata: AO GPI Santekhproject, 1972. p. 99.
141. I. I. Afanasiev et al. Research and development of methods and means of dedusting ore delivery conveyers. Krivoy Rog: VNIIBTG, 1977. p. 154.

142. I. I. Afanasiev, V. M. Marinchenko, L. K. Saplinov et al. Research and development of methods and means of dust control in crushing and classification of iron. Krivoy Rog: VNIIBTG, 1976. p. 146.
143. I. I. Afanasiev, I. N. Logachev, R. N. Shumilov et al. Ventilation and dedusting of raw material recovery and preparation departments of iron ore sintering plants. Krivoy Rog: NIImetallurgventilyatsiya, 1967. p. 185.
144. I. N. Logachev, L. K. Saplinov, T. N. Nikolina, I. I. Afanasiev. Suction hoods of belt loading points. USSR Certificate of Authorship No. 559021. Bulletin of Inventions, 1977. No. 19.
145. I. I. Afanasiev, L. K. Saplinov, Yu. I. Pirogov. Dust control device for ventilated hoods. USSR Certificate of Authorship No. 591213. Bulletin of Inventions, 1978. No. 5.
146. A. V. Arkhipov, I. N. Zavertkin, I. I. Afanasiev et al. Foam installation for dust control of hammer crushers. TsNIIchermet Bulletin, 1975. No. 15. pp. 57–58.
147. I. I. Afanasiev, R. N. Shumilov, V. M. Marinchenko, L. D. Korennoy, L. M. Chernenko. Ventilated hoods of belt conveyers. USSR Certificate of Authorship No. 477073. Bulletin of Inventions, 1975. No. 26.
148. V. M. Marinchenko, V. L. Petukhov, R. N. Shumilov. Conveyer loading group dust exhaust device. USSR Certificate of Authorship No. 588161. Bulletin of Inventions, 1978. No. 2.
149. R. N. Shumilov, V. M. Marinchenko, I. I. Afanasiev et al. Ventilated hoods of belt conveyers. USSR Certificate of Authorship No. 595517. Bulletin of Inventions, 1978. No. 8.
150. I. N. Lagachev, A. M. Golyshev, L. M. Chernenko et al. Research and development of optimizing techniques for ventilated hoods of bulk material transfer groups. Krivoy Rog: VNIIBTG, 1982. p. 90.
151. I. N. Logachev, R. N. Shumilov. *Ventilation of hot and steamy material channels at sintering plants*. Collective proceedings of the 11th Coordination Meeting on Industrial Ventilation. Moscow: NIIOT, 1969. pp. 29–32.
152. I. N. Logachev, L. M. Chernenko, A. M. Golyshev et al. Device intended for formatting a bulk material stream on a conveyer belt. USSR Certificate of Authorship No. 1105419. MKI V65 G 47/22/Discoveries, Inventions, 1984. No. 28. p. 54.
153. R. K. Beletskiy, N. N. Grichina. Measuring parameters of dust and gas streams in the iron and steel industry. Moscow: Metallurgy, 1979. p. 80.
154. I. N. Lagachev, A. M. Golyshev, L. M. Chernenko et al. Integrating a set of arrangements to minimize the amount of material entrained by suction. Krivoy Rog: VNIIBTG, 1982. p. 51.
155. I. A. Zavertaylo. "Research for means of reduction of the aspiration air dustiness in bulk material transfers." Thesis,. . . Cand.Sc. (Engineering), Krivoy Rog, 1993. p. 202.
156. L. A. Vasiliev. Shadow methods. Moscow: Nauka, 1968. p. 400.
157. L. Holder, R. Nort. Shadow methods in aerodynamics. Moscow: Mir, 1966. p. 179.
158. L. N. Lyakhovskiy, N. Syrkin. Air mechanics of a blast flowing into a jet of different density. *Zh. T. F.*, 1939. No. 9. pp. 54–68.
159. O. D. Neikov et al. Guidelines for design and arrangement of dust exhaustion systems in the metal powder industry. Dusty material drop groups. Chelyabinsk, 1979. p. 84.
160. S. G. Ushakov, N. I. Zverev. Inertial dust separation. Moscow: Energiya, 1974. p. 168.
161. A. Bogatykh. Cyclone foaming equipment. Leningrad: Mashinostroyeniye, 1978. p. 224.
162. M. P. Kalinushkin. Fan installations Moscow: Higher School, 1967. p. 259.
163. L. G. Loytsyanskiy. Fluid mechanics. Moscow: Nauka, 1973. p. 847.
164. V. N. Uzhov, A. Yu. Valdberg. Pre-treatment of industrial gases. Moscow: Chemistry, 1975. p. 216.
165. A. I. Pirumov. Air dedusting. Moscow: Stroyizdat, 1981. p. 296.

166. V. M. Aleshin, A. Yu. Valdberg, G. M. Gordon et al. Dust collection in metallurgy. Moscow: Metallurgy, 1984. p. 336.
167. I. N. Logachev, A. M. Golyshev, L. M. Chernenko. Reduction in collector dust losses in ventilation at iron-ore pellet plants. *Mining Journal*, 1985. No. 3. pp. 57–59.
168. Certificate of Authorship 1458297 USSR. B65 G21/00.
169. I. N. Logachev, L. M. Chernenko et al. Development of efficient local dust exhausts for mixture bins of bentonite and chalkstone at SSGPO TsPO. Krivoy Rog: VNIIBTG, 1989. p. 90.
170. I. N. Logachev, S. I. Zadorozhniy, A. Ya. Shvets, A. D. Ratoshnyuk. Suction hoods of the belt conveyer loading points. USSR Certificate of Authorship 1686185, 1991. Bull. No. 39.
171. Reclaimer at iron ore pellet terminal. *Mining Mag*, 1969. V. 120. No. 3. pp. 191–193.
172. I. V. Sobolev. Equipment for transportation, storage and preparation of iron ores and iron ore pellets: Iron and steel industry. Chermetinformatsiya Institute Bulletin, 1966. No. 20. pp. 58–59.
173. N. N. Berezhnoy, V. V. Bulychev, A. A. Kostin. Production of iron ore pellets. Moscow: Nedra, 1977. p. 240.
174. I. N. Logachev, A. M. Golychev, S. I. Zadorozhniy. Research and development of methods and means of dust control of open pellet storages at pellet plants of mining and concentration complexes. Krivoy Rog: VNIIBTG, 1978. p. 149.
175. A. Ya. Sharipov, V. A. Silantiev et al. Designed means of dust emissions in open storage of bulk materials. Alma-Ata: AO GPI Santekhproject, 1982. p. 93.
176. Guidelines for design of dispersion of harmful substances contained in emission of enterprises, SN369-74. Moscow: Stroyizdat, 1975. p. 41.
177. A. Monin. On diffusion at the terminal velocity. *Journal of the Academy of Sciences of the USSR*, 1955. No. 3. pp. 234–248.
178. A. M. Yaglom. On eddy diffusion in the ground layer. *Journal of the Academy of Sciences of the USSR: Atmospheric and Oceanic Physics*, 1972. No. 6. pp. 579–594.
179. M. I. Yudin. Regarding dispersion of solid particles in turbulent stream. *Meteorology and Hydrology*, 1946. No. 5. pp. 12–20.
180. M. E. Berlyand. Present troubles of atmospheric diffusion and air pollution. Leningrad: Hydrometeoizdat, 1975. p. 448.
181. V. Nikitin, N. G. Maksimkina, V. T. Samsonova et al. Ventilation of industrial sites and adjacent areas. Moscow: Stroyizdat, 1980. p. 200.
182. I. N. Leykin. Dispersion of ventilation emission of chemical enterprises (design and calculation). Moscow: Chemistry, 1982. p. 224.
183. V. M. Elterman. Chemical production ventilation. Moscow: Chemistry, 1971.
184. O. G. Setton. Micrometeorology. Leningrad: Hydrometeoizdat, 1958. p. 120.
185. C. Bosanquet, J. Pearson. The spread of smoke and gases from chimneys. *Trans. Faraday Soc.*, 1936. V. 32. pp. 1249–1263.
186. G. Detri. Atmosphere should be clean. Moscow: Progress, 1973. p. 378.
187. K. Wark, C. F. Warner, W. T. Davis. Air pollution: Sources and control. Moscow: Mir, 1980. p. 544.
188. M. E. Berlyand. Eddy exchange factor determination by the contours of smoke emitted from industrial pipes. *Meteorology and Hydrology*, 1961. No. 6. pp. 13–20.
189. A. E. Berlyand, I. Onikul. Principal physics of industrial emissions dispersion in the air. *Tr. GGO*, 1968. V. 234. pp. 3–27.
190. M. Abramovich and I. Stigan, eds. Reference book for special functions with formulas, graphs, and tables. Moscow: Nauka, 1979. p. 832.
191. O. D. Neikov, I. N. Logachev, G. I. Vasilieva. Regarding a gas suspension formation in motion of free gravitational flows of bulk materials: Prevention of sudden combustions of powders and explosions of gas-dispersion systems. Kiev: Naukova Dumka, 1975. pp. 28–32.

192. I. N. Logachev, V. V. Kachanov, L. D. Korennoy et al. Ventilated shaft for stacking of bulk materials USSR Certificate of Authorship 901213. Bulletin of Inventions, 1982. No. 4.
193. I. N. Logachev, S. I. Zadorozhniy et al. Elaboration of methodical guidelines for determination and design of unorganized sources of dust emissions in open storage of dust materials at plants of mining and concentration complexes. Krivoy Rog: VNIIBTG, 1982. p. 101.
194. R. H. Wohlbier, ed. Stacking, blending, and reclaiming of bulk materials. Series on bulk material handling, 1975–1977. V. 1. No. 5.
195. *International Journal of Storing, Handling, and Transporting Bulk*, 1989. V. 9. No. 3.
196. I. N. Logachev, A. M. Golyshev, V. V. Kachanov et al. Research for methods of air-hardening in agglomerating departments of iron ore sintering plants. Krivoy Rog: VNIIBTG, 1970. p. 119.
197. Rail tracks of iron and steel works. Moscow: Metallurgy, 1975. p. 271.
198. I. N. Logachev, V. V. Kachanov et al. Dedusting of pellet and agglomerate discharge areas. Krivoy Rog: VNIIBTG, 1973. p. 75.
199. Research for methods of air dedusting and occupational safety improvement at iron ore pellet plants. Experimental industrial check of efficiency of ventilation of the main sources of dust emissions in departments of SSGOK pellet plants. Krivoy Rog: VNIIBTG, 1974. p. 236.
200. Research for methods of air dedusting and occupational safety improvement at iron ore pellet plants. Krivoy Rog: VNIIBTG, 1975. p. 148.
201. I. N. Logachev, V. V. Kachanov, G. N. Maltsev. Dynamic interaction of a gravitational material flow and air in telescopic chutes: Occupational safety in the metal mining industry. Moscow: Nedra, 1978. pp. 64–68.
202. I. E. Idelchik. Flow resistances (physical and mechanical basis). Moscow: Gosenergoizdat, 1954. p. 316.
203. P. V. Beresnevich, V. A. Mikhaylov, S. S. Filatov. Aerology of open-cut mines. Moscow: Nedra, 1990. p. 280.
204. V. A. Minko, N. G. Abramkin. Determination of particle size distribution in ventilated air ducts of transfer groups. *Refractory Products*, 1974. No. 8. pp. 23–28.
205. V. A. Minko, V. I. Ilyin, N. G. Abramkin. Study on calculation of the maximum size of dust particles in the induced air of bulk material transfer groups. Moscow: Moscow Civil Engineering Institute, 1974. V. 6.
206. I. N. Logachev, A. M. Golyshev et al. Efficient methods of containment of dust emissions in transfer of bulk materials at ore mining enterprises. Moscow: Chermetinformatsiya Institute, 1986. V. 2. p. 15.
207. I. N. Logachev, A. M. Golychev, S. I. Zadorozhniy et al. Research for new means of containment of dust emissions in transfer of bulk materials at plants of mining and concentration complexes. Krivoy Rog: VNIIBTG, 1986. p. 87.
208. I. N. Logachev, A. M. Golyshev et al. Local exhausts and hoods of process equipment of ore preparation plants (album). Krivoy Rog; Alma-Ata, 1986. p. 87.
209. Yu. V. Krasovitskiy, A. V. Malinov, V. V. Durov. Dedusting of artificial gases in the pottery industry. Moscow: Chemistry, 1994. p. 272.
210. V. N. Uzhov, A. Yu. Valdberg et al. Dedusting of industrial gases. Moscow: Chemistry, 1981. 322.
211. P. A. Kouzov, A. D. Malgin, G. M. Skryabin. Dedusting of gases and air in the chemical industry. Leningrad: Chemistry, 1982. p. 256.
212. A. A. Tsytsura. "Air dedusting management on the basis of the physical and chemical approach to mechanisms of interaction of dust aerosols and dispersed fluids." Thesis, Doctor of Engineering, Karaganda. p. 359.

213. A. Kuzmich, ed. Guide book on dust control in the mining industry. Moscow: Nedra, 1982. p. 240.
214. A. Ya. Dobromyslov. Design and construction of building drainage systems. Moscow: Stroyizdat, 1978. p. 120.
215. N. M. Anzheurov, I. N. Logachev. Optimum aspirating volumes of coalescent powder transfers. Shukhov BSTU Bulletin. Belgorod, 2003. No. 6.
216. V. V. Kachanov, F. I. Danchenko, I. N. Logachev et al. Arrangement for bulk material transfers. USSR Certificate of Authorship 1504188. Bulletin of Inventions, 1989. No. 32.
217. V. S. Vaschenko, V. V. Kachanov, I. N. Logachev et al. Arrangement for covering open wagons in bulk material transfers. USSR Certificate of Authorship 544595. Bulletin of Inventions, 1977. No. 4.

Index

A

Accounting equations
phenomenological method, 367–373
space–time method, 25, 373–379
Active component force interaction, 23, 30–32, 83, 102
Aerodynamic characteristics of particles, air injection, 78–84
monofractional streams, 78–81
polyfractional streams, 81, 83–84
Aerodynamic characteristics of particles, gravitational flow, 33–74
experimental arrangement for study of, 34
method for evaluating, 65–74
peculiarities of motion in chutes, 33–45
sedimentation of particles, 55–65
single particles, 45–55
Aerodynamic characteristics of solid-particle jets, 133
axially symmetric jets of freely falling particles, 184–196
in channels, 196–211
flat jets of loose matter, 144–184
initial equations, 134–144
Aerodynamic characteristics of unsteady particle flow, 120–131
smooth changes in material flow, 129–131
sudden changes in material flow, 120–129
Aerodynamic process research, 24–29
experimental studies, 13, 26–28
industrial evaluation, 28–29
mathematical modeling, 13–14, 25–26
Aerodynamic resistance of bombarding particles (ARBP), 10
Agglomerate
aerodynamic characteristics, 82
dust emission intensity, 1–4
geometric shape, 49
induced air flow, 99
particle concentration and size composition in aspirated air, 253
particle-size distribution, 85
pressure variation, 88, 94
specific dust emissions from, 5–6
Air dedusting. *See* Dust release containment and air dedusting solutions
Air drag (motion drag), 14, 45
of particles in bulk material streams, 79–80
of unsteady particle streams, 65
Air energy (power), 14
Air injection in chutes, 75
aerodynamics of unsteady particle flow, 120–131
heat and mass exchange, 110–120
isothermal flow, 75–110
Air injection in jets of freely falling particles, 134–196
axially symmetric jets, 184–196
flat jets of loose matter, 144–184
initial equations, 134–144
Air injection research
Butakov–Hemeon model, 14–22
semiempirical models, 22–24
Air suction
periods of study, 12–14
theoretical models of, 12–29
Air suction dust control process (dry method), 8–11. *See also* Local suction units
Air velocity, 19
in aspiration funnels, 279–280
effect of heat and mass exchange, 116–118
monofractional stream conditions, 79
Air viscosity forces, 47, 158–160
All-Russia Institute for Occupational Safety in Ore Mining (VNIIBTG), 26, 258, 277, 300, 321, 327, 341, 351–353
Altmark, M. K., 12–13
Anderson, D. M., 14, 21, 99
Anti-suction (reverse air flow), 12–13, 24, 88–89, 211
ARBP (aerodynamic resistance of bombarding particles), 10
Archimedean forces, 113, 348
Aspiration layouts, 222–242
aspiration volumes, 246–250
cascade installations, 236–237
conveyor–crusher–conveyor transfers, 229–232
conveyor–screen–conveyor transfers, 232–234
conveyor-to-conveyor transfers, 224–229
dry magnetic separation assembly, 234–236
dust concentration reduction in aspiration funnels, 279–298
injection-driven air discharge calculations for complex chute configurations, 237–242
loading bins, 351–357
outdoor storage, 329–339
Axially symmetric particle jets, 137, 139, 157, 210
air injection in, 184–196
fluid dynamics equations, 141–144

B

Bakirov, U. Kh., 18, 99
Ballistic method of particle stream velocity measurement, 41
Basset–Boussinesq–Oseen equation, 60
Basset force, 57
Bentonite, 1
  dust composition in evacuated air, 258
  dust concentration in evacuated air, 254–256
  dust emission containment designs for loading bins, 340
  dynamic interaction of air and bulk material streams in laminar flow, 101
  fugitive dust releases in outdoor storage, 299–300
  reduction of dust concentration in aspiration funnels, 279–281, 288–289, 291, 295
  structural and process parameters of load-handling facilities, 260–263, 266–267, 269–270
Berliand, M. E., 303–304
Bernoulli's equation, 348
Bernstein, R., 107
Bianconi, W. O. A., 14, 21
Bin-type (pyramid-shaped) chutes, 9, 33, 104–107, 224–226, 233
Blasius method, 152, 154, 156
Bogaevskiy, O. A., 18, 99
Bosanquet, C., 303
Boshnyakov, E. N., 13–14, 24, 99, 217
Boundary-layer equations, 14, 143–144, 175, 360
Bulk material transfer, 1–32
  aerodynamic characteristics of chuted particle gravitational flow, 33–74
  aspiration layouts, 224–234
  classification of bulk material streams, 29–32
  phenomenological method of dynamic equation development, 363–373
  prevalence of, 1
  space–time method of averaging accounting equations, 373–379
  theoretical models of air suction, 12–29
  transfer groups as air pollution sources, 1–12
Bunker-type chutes, 11
Buoyancy force, 45–46
Butakov, S. E., 12–13, 18–19, 26, 97–99
Butakov–Neykov model, 14–22, 96, 360

C

Cascade installations, 236–237
Cauchy problem, 152*n*, 273
Chalkstone, 33, 52, 102
  aerodynamic characteristics, 54, 82
  drag factor, 54
  dust composition in evacuated air, 257
  dust concentration in evacuated air, 254
  fugitive dust releases in outdoor storage, 299
  induced air flow, 97
  particle-size distribution, 85
  reduction of dust concentration in aspiration funnels, 295
  structural and process parameters of load-handling facilities, 266
Channels
  aerodynamics of solid-particle jets in, 196–211
  exponentially distributed particles and height of, 39
  geometry of, 29–30
  pressure variation in, 66–72
Chargers, 100–101
Chen equation, 60
Christiansen, E., 52–53
Chulakov, P. Ch., 16–17, 99
Chute pressure variation and distribution, 66–72
  across vertical channel length, 69
  air injection of particle streams in prismatic chutes, 86–95
  in vertical tube at various initial velocities and bulk concentrations, 71
Chutes, 11. *See also specific types of chutes*
  aerodynamic characteristics of particle gravitational flow in, 33–74
  aerodynamic process research, 25–27
  air injection in, 75–131
  Butakov–Hemeon model of air suction, 14–22
  classification of bulk material streams, 29, 31–32
  defined, 33
  design parameters of, 7–8
  inclination angle, 35
  reverse air flow (anti-suction), 12
  semiempirical models of air suction, 22–24
  shape of, 33
  transfer subgroups, 33–34
Classification of bulk material streams, 29–32
  dynamic interaction intensity, 29–31
  geometry of channels, 29–30
  material temperature and humidity, 29–30, 33
  particle bulk concentration distribution, 29–32
  particle fineness and composition, 29–31
  stream kinematics, 29–31
Coal, 1, 22–24
  aerodynamic characteristics, 54
  drag factor, 54
  induced air flow, 99
  particle-size distribution, 85
Compressive effect, 219–221
Cone crushers, 12, 28, 230–231, 252, 255, 259
Constrained motion, 35, 37, 41–42

Conventional airborne velocity, 92, 95–96
Conveyor–crusher–conveyor transfers, 229–232
Conveyors
aerodynamic process research, 27
Butakov–Hemeon model of air suction, 19, 21
hydraulic dust control, 8
intensity of dust emissions of, 3–4
semiempirical models of air suction, 22, 24
specific dust emissions of, 2, 5
transfer subgroups, 33–34
Conveyor–screen–conveyor transfers, 232–234
Conveyor-to-conveyor loading, 33–34, 224–229
Conveyor-to-equipment loading, 33–34
Cooled materials, 34
Coolers, 115, 251–253, 258–259, 343
Cowls, 213
air replacement calculations, 242, 247–250
decrease in dust release intensity, 257–259, 262–263, 270–271
dust carryover from, 251–253
dust release and containment in loading, 342, 344, 351–352, 354
dust release and containment in outdoor storage, 301, 322, 329–337, 339
inertial traps using plate grids inside, 271, 276–277
initial equations, 214–215
intensifying inertial dust precipitation in, 271–278
layout of filtering elements inside, 270
local suction units with vortex assembly, 292
minimum negative pressure, 215–222
particle concentration and size composition in aspirated air, 254–257
performance calculation, 222–237, 241–242
reduction of dust concentration in aspiration funnels, 279–280, 282–283, 288–289, 291, 293–294, 298
thermal pressure in, 221
Crushed materials, 1, 33
conveyor–crusher–conveyor transfer, 229–231
dust concentration in evacuated air, 254–256
heat exchange, 111–112
industrial-scale testing of local suction units in bins, 299
modes of motion, 35, 37
particle distribution, 38–39
particle-size distribution, 85
structural and process parameters of load-handling facilities, 260, 262
Crushers, 2, 4, 133, 250–251
bypass-equipped, 248–249
cascade installations, 236
cone, 12, 28, 230–231, 252, 255, 259
conveyor–crusher–conveyor transfers, 229–232
dimensionless aerodynamic parameters of, 245
dust concentration in evacuated air, 254
four-roll, 230–231, 253
hammer, 242, 245–249, 252–253, 259, 361
jaw, 229–230, 252, 255
materials transfer, 3, 34, 229–232
rotary, 214, 242–244, 246
Cyclone-type suction units and dust traps, 271, 283, 291–295, 330–332, 335, 337–338, 351–352, 362
Cylindrical chutes, 33, 209

D

Dampers, 28, 51, 77–78
Degner, B., 13, 24
Dennis, R., 12, 21, 99
Detry, J., 303
Differential momentum conservation equation, 76, 369
Dirichlet formula, 63
Dobromyslov, A. N., 87
Double-walled enclosing structures, 321–325, 327
Dressing process, 2, 4
Dry magnetic separation assemblies, 234–236
Dry method (air suction) dust control process, 8–11. *See also* Local suction units
Dust carryover
aerodynamic testing of local suctions and cowls, 337
in air aspirated from bins, 267–269
air velocity in aspiration funnels, 279–281
from aspiration cowls, 251–253
inertial traps using plate grids inside cowls, 278
structural and process parameters of load-handling facilities, 263–265
Dust dilution, 8
Dust emissions
danger of, 1
during iron ore pellet loading in railway cars, 339–357
in outdoor iron ore pellet storage, 298–339
primary means of control, 8–12
Dust emissions from bulk transfer
alternating stages of, 5
causes of immediate dusting of ground level air, 2
intensity of, 1–8
transfer groups, 1–2
Dust fractions
origin of, 2
weight content of, 7
Dust plume structure, 309–315
Dust precipitation
in dust receivers/separators, 281–291

intensifying inertial in aspirating cowls, 271–278
in local cyclone-type suction units/dust traps, 291–295
in local dust-separating suction units with filter elements, 295–298
Dust receivers/separators, 213–214, 257
dust precipitation in, 281–293, 295–296
fugitive dust releases during loading, 343, 345–346
fugitive dust releases in storage, 298
industrial-scale tests of efficiency, 290–291
measured velocity field values in, 286–287
structural and process parameters of load-handling facilities, 262–264
Dust release containment and air dedusting solutions, 213–357
decrease in dust release intensity, 257–271
dust concentration reduction in aspiration funnels, 279–298
dust release sources, 251–257
fugitive dust emissions and containment during iron ore pellet loading, 339–357
fugitive dust releases in outdoor iron ore pellet storage, 298–339
intensifying inertial dust precipitation in aspirating cowls, 271–278
local suction capacity calculation, 214–250
Dust release intensity, 251–298
decrease in, 257–271
dust concentration reduction in aspiration funnels, 279–298
dust release sources, 251–257
intensifying inertial dust precipitation in aspirating cowls, 271–278
releases during iron ore pellet loading in railway cars, 339–357
releases in outdoor iron ore pellet storage, 303–320
Dust traps, 259
air velocity at, 263
cyclone type, 283, 291–295, 337–338
efficiency of, 288–289, 353
selecting, 253–256
Dynamic shape of particles, 50–53
Dyunin, A. K., 373

E

Eddy run length, 204–205, 208
Elterman, V. M., 303
Energy theory model, 12
Energy transfer equation, 14–15, 378–379
Equipment-to-conveyor loading, 33–34
Equipment-to-equipment loading, 33–34
Euler criterion, 100, 104, 117
Experimental semicommercial suction plant, 27–28
Exponentially distributed particles, 359
channel height and, 39
jets of, 134, 137, 144, 169–175, 185, 198–199
External (outdoor) transfer groups, 1–2

F

Fencing off of flowing material, 321–329
Flat particle jets, 144–184, 187, 243
changes in airflow, 161
effect of pressure gradient, 175–184
exponentially distributed particles, 169–175
fluid dynamics equations, 141, 143
one-dimensional problem, 157–168
self-similar motion equations, 144–154
structure of air streams in, 144–184
uniformly distributed particles, 154–157
volumetric concentration, 135
Flow continuity law, 39–40
Flow strength, 35–36
Flow-to-wall contact condition, 43
Fluid dynamics equations, 139–144
Foam, 259, 271, 362
Fortie, A., 60, 62*n*
Fourier method, 126, 177
Four-roll crushers, 230–231, 253
Frankl, F. I., 373–374, 374*n*
Free-falling material stream, 5–6, 11, 41
Free particle jets, 11, 29, 133, 141, 207, 241, 361
air injection in, 134–196
air stream structure in flat jets, 144–184
axially symmetric jets, 184–196
dust release intensity, 318–320, 328–329
Froude number, 35, 35*n*, 37–38, 40
Fugitive atmospheric dust releases
during iron ore pellet loading in railway cars, 339–357
in outdoor iron ore pellet storage, 298–339
Funnels
aspiration, 215–216, 218, 270, 279–298, 330
receiving, 2–3, 11, 34, 43, 295–296, 361
suction, 292–293

G

Gaiduk, V. P., 194–195
Galloping particle motion (saltation), 37–40
Gauss–Ostrogradskiy transformations, 369, 374
Geometric shape of particles, 47–50
Gervasiev, A. M., 22
Golyshev, A.M., 194–195
Gorbis, Z. R., 35*n*, 45, 52, 54–55, 111–112
Granite, 33, 118, 167
aerodynamic characteristics, 54, 82
drag factor, 54
geometric shape, 49–50
heat exchange, 112

induced air flow, 97, 99
modes of motion, 35, 37–38
particle distribution, 39
pressure variation, 94, 130
Granular materials, 30–31, 33, 211, 362
Graschenkov, N. F., 16, 99
Grashof number, 117
Grid elements, 271–278

H

Hammer crushers, 252–253, 259, 361
aspiration volumes, 246–249
as fans, 242–246
Hath, T., 14, 20, 99
Heat and mass exchange, effect of, 110–120
air velocity in chutes, 116–118
heat exchange in inclined chutes, 111–113
influence of mass exchange on volumes of induced air, 118–120
thermal head, 113–116
Heated materials, 30, 111, 116–118, 211, 221–223, 226*n*, 251, 348–349, 360–361, 366
Hemeon, W. C. L., 14–22, 26, 99
High-performance dedusting of suction emissions, 8
Hoods
aerodynamic process research, 25–28
Butakov–Hemeon model of air suction, 21
classification of bulk material streams, 31
design parameters of, 7–8
semiempirical models of air suction, 22–24
Hydraulic dust control (watering materials), 3–4, 8–9, 253, 301, 329

I

Ilyich Iron and Steel Works, 341–343
Impulse transfer equation, 139
Inclined chutes, 35, 105, 111–113, 131, 225, 228–229, 233, 262
Induced air, 8, 11
air mechanics of particle stream with high bulk concentrations, 107
Butakov–Hemeon model of air suction, 14–22
influence of mass exchange on volumes of, 118–120
interaction of air and chuted material streams in laminar flow, 104–106
particle streams in prismatic chutes, 95–101
Induction, defined, 12
Induction head, 13, 95–97, 100, 105, 116–117, 360
Induction pressure, 25, 89–91, 93–95, 102–103, 109, 121, 127, 129–131, 360
Indurating machines, 2, 5, 28
Inguletsky complex (InGOK), 1–2, 4
Intermediate motion, 35
Internal transfer groups, 2
Iron ore, 1–2, 28, 33, 115, 167, 257, 259
aerodynamic characteristics, 54, 82
drag factor, 54
dust concentration in evacuated air, 254–255
dust emission intensity, 3–4
fugitive dust emissions and containment during loading, 339–357
fugitive dust releases in outdoor storage, 298–339
geometric shape, 49
heat exchange, 112
induced air flow, 99
particle-size distribution, 85
specific dust emissions from, 5–6
Ishchuk, I. G., 8
Isothermal flow, 75–110
air injection in bin chutes with uniform distribution of particles, 104–107
air injection with particle streams in prismatic chutes, 84, 86–101
air mechanics of particle streams with high bulk concentrations, 107–110
averaged aerodynamic characteristic of particles, 78–84
peculiarities of dynamic interaction of air and bulk material streams in laminar flow in chutes, 101–104

J

Jaw crushers, 229–230, 252, 255
Jet axis equation, 43
Jet discontinuity, 37

K

Kachkanarsky ore beneficiation plant, 1, 339, 341, 343–344
Kamyshenko, M. T., 13, 22, 24, 99
Kazsantekhproyekt Institute, 300, 318
Kemerovo Chemical Equipment Factory, 334
Kharkovskiy, V. S., 16
Khudyakova, G. N., 54
Kilin, P. I., 18, 99
Kinked chutes, 33, 42
Knudson, I. F., 12
Korabekov, N. N., 16–17
Kostomukshinsky ore beneficiation plant (KostGOK), 1, 6, 329–330, 335, 339–340, 362
Kovdorskiy complex, southwestern Russia, 1
Krivbass mining and concentration complexes, 1–2, 4
Krivyi Rih Central ore beneficiation plant, 341–343
Kruse, C. W., 14, 21
Kutta–Merson method, 273

L

Laminar motion, 40, 47, 101–104, 359
Lateral force, 45–46
Lebedinsky ore beneficiation plant, 1, 28, 251, 270, 362
  loading, 339
  outdoor storage, 300, 312, 317–318, 320, 335
  reduction of dust concentration in aspiration funnels, 288–291
Leikin, I. N., 303
Limestone, 1
  dust composition in evacuated air, 258
  dust concentration in evacuated air, 254–256
  fugitive dust releases in outdoor storage, 299–300
  reduction of dust concentration in aspiration funnels, 293, 296, 298
  structural and process parameters of load-handling facilities, 260, 262–263, 266–268
Loading, 339–357
  conveyor to conveyor, 33–34, 224–229
  conveyor to equipment, 33–34
  dust emission containment designs, 339–344
  equipment to conveyor, 33–34
  equipment to equipment, 33–34
  improving aspiration efficiency, 351–357
  performance calculations for local suction units, 344–351
Loading tunnels, 341–342, 344, 347, 349, 351–352, 355–356
Local suction units, 214–250
  air replacement calculations, 242–250
  aspiration layouts and performance calculation, 222–242
  dust precipitation in cyclone-type units, 291–295
  dust precipitation in dust-separating units with filter elements, 295–298
  initial equations, 214–215
  loading bins, 344–351
  minimum negative pressure, 215–222
  outdoor storage, 329–339
Logachev, I. N., 13–14, 162
Lump materials, 30–31, 102, 211, 228
Lyashchenko, P. V., 47, 67, 69, 108

M

Maclaurin's series, 169
Magnetic grid elements, 273–276
Magnetic shoe, 276–278
Mass transfer equation, 374–376
Mekhanobrchermet Institute, 332, 334, 344
Mikhailovsky ore beneficiation plant, 1, 251, 339, 362
Minimum negative pressure, 215–222
  air injection and suction spectrum of unit, 215–219
  compressive effect, 219–221
  optimizing choice of, 221–222
  thermal pressure in cowl, 221
Minko, V. A., 7, 13–14, 17, 84, 84*n*, 99
Mixed component force interaction, 30–32
Momentum conservation equation, 75–76, 98, 158, 369
Monin, A. S., 303
Monofractional stream conditions, 30, 78–82, 228
Morozov, Yu. I., 112
Morrison, N. I., 21
Motion drag (air drag), 14, 45
  of particles in bulk material streams, 79–80
  of unsteady particle streams, 65

N

Navier–Stokes equation, 133
Neykov, O. D., 13–14, 103, 162, 217
Nikitin, V. S., 303
Nonuniformly accelerated particle streams, 30
Nonuniform particle distribution, 30, 47, 360
Novo-Krivorozhsky complex (NKGOK), 1–2, 4, 341–343
Novolipetsky Iron and Steel Works, 341–343
Nusselt number, 113

O

Obstacle surface equation, 44
Olenegorsky complex, southwestern Russia, 1
Olifer, V. D., 23, 74*n*, 99
One-dimensional problem and equations, 25, 195, 199, 359–361
  channel pressure variation, 66
  isothermal flow, 75–78, 98–99, 104
  jets of freely falling particles, air injection in, 157–168
  jets of solid particles in channels, 200–211
  thermal head, 114
Onikul, R. I., 304
Oseen-type relations, 53
Outdoor (external) transfer groups, 1–2
Outdoor storage, fugitive dust releases from, 298–339
  dust release containment, 321–339
  dust release intensity, 303–320
  primary dust emission sources, 301–302
  storage process layout, 299–301

P

Particle distribution (concentration)
  air mechanics of particle streams with high bulk concentrations, 107–110

changes in volumetric particle concentration in jets of material, 134–137
in chutes, 38–40
exponentially distributed particles, 169–175
nonuniform, 30, 47, 360
pseudo-uniform, 30–32, 40, 77, 104, 224, 229, 359, 361
uniform, 30–31, 73, 77, 79–80, 104–107, 154, 173, 190, 360
Particle relative velocity vector, 45
Passive component force interaction, 30–31, 101–104
Pavlov, V. P., 24
Pearson, J., 303
Pellets, 2, 5, 99, 115, 274–275, 295
aerodynamic characteristics, 54, 82
drag factor, 54
dust carryover, 277, 280–281
dust composition in evacuated air, 257–258
dust concentration in evacuated air, 254–256
dust emission intensity, 1–4
dust exhaust system design basis development, 28
dust release intensity and mitigation, 251–252
fugitive dust releases during loading in railway cars, 339–357
fugitive dust releases in outdoor storage, 298–339
particle-size distribution, 85
pressure variation, 88, 91, 94
reduction of dust concentration in aspiration funnels, 279–280
specific dust emissions from, 5–6
suction properties of hot particle streams, 12
Pettyjohn, E., 52–53
Phenomenological method of dynamic equation development, 25–26, 363–373
accounting equations, 367–373
inter-component interaction, 364–367
Photographic method of particle stream velocity measurement, 41
Plane-parallel flow
axially symmetric jets of freely falling particles, 184, 187, 189
changes in volumetric particle concentration in jets of material, 135
dust plume structure, 314
flat jets of loose matter, 146, 148, 155–156, 169
solid-particle jets in channels, 197–200
Plate grids, 271–278
Platonov, P. N., 35
Poltavsky complex, 2, 5, 251
Polydisperse dust, 306, 309
Polyfractional stream conditions, 30, 228
aerodynamic characteristics of particles, air injection, 81, 83–84
particle-size distribution, 85
Pomerantsev, V. V., 107
Popov, V. A., 19
Powdered materials, 30–31, 101–103, 102*n*, 211, 260, 262, 279–281, 318–320
Prandtl, L., 133, 140
Pressure gradient, 86, 93, 222, 323
dust precipitation in dust receivers/separators, 284–285
effect on air stream structure in flat particle jets, 175–184
Pring, R. T., 12
Prismatic chutes, 29, 33, 107–108, 113, 262
air injection with stream of particles in, 84, 86–101
conveyor-to-conveyor transfers, 224–225, 228
Pseudo-uniform particle distribution, 30–32, 40, 77, 104, 224, 229, 359, 361
Pulse transfer equation, 376–377
Pyramid-shaped (bin-type) chutes, 9, 33, 104–107, 224–226, 233

Q

Quadratic law, 105, 189

R

Rayleigh curve, 53, 74
Reduced particle resistance coefficient, 10
Resistance coefficient, 53–55
Reverse air flow (anti-suction), 12–13, 24, 88–89, 211
Reversed direction of particle drag force, 16
Reynolds number, 46, 54, 60, 74, 81, 107, 113, 117, 142, 327
Rock coarseness (rock diameter), 48
Rotary crushers, 214, 242–244, 246
Rotary pickers, 300–302, 312–315, 317–318, 320, 340
Rounded particles, 53–54, 108

S

Salimzhanov, K. S., 16–17
Saltation (galloping particle motion), 37–40
Santekhproyekt Institute, 341
Schlichting, G., 45, 74
Screening
cascade installations, 236–237
conveyor–screen–conveyor transfers, 232–234
dust carryover, 252–253
local cyclone-type suction units/dust traps, 291
particle concentration and size composition in aspirated air, 254–255, 258
as source of dust emission, 251

Sedimentation, 55–65
aerodynamic drag of accelerating particles, 57, 60–65
formulas for calculating in vertical air streams, 58–60
particle motion in air stream, 55–57
Sedov, L. I., 45, 368
Self-similar motion equations
axially symmetric jets of freely falling particles, 185–196
flat jets of loose matter, 144–154
Semenov, A. S., 111–112
Semiempirical models of air suction, 22–24
Serenko, A. S., 13, 24, 210–211
Setton, O. G., 303
Seversky ore beneficiation plant (Northern complex; SevGOK), 1–2, 4, 251, 300, 319, 341, 343–344
Shagalova, L., 107
Shaping elements, 270
Sharp-grained particles, 48, 53–54, 108
Shelketin, A. V., 13, 22, 24
Sieves
air velocity variance, 104
chutes adjacent to, 11
dedusting, 28
equipment to equipment loading pattern, 34
specific dust emissions of, 5
Single-particle aerodynamic characteristics, 45–55
dynamic shape, 50–53
geometric shape, 48–50
resistance coefficient, 53–55
Single-walled enclosing structures, 321–322, 324
Sintering process and equipment, 2, 4, 213
dust carryover, 252
dust concentration in air evacuated from, 254–255, 259
dust release and mitigation in aspirated air, 251
particle concentration and size composition in aspirated air, 257
suction properties of hot particle streams, 12
SIOT (Sverdlovsk Institute for Occupational Safety), 258, 355
Skoblo, A. I., 54
Slip ratio, 104, 161, 167, 223, 325, 360
Slyusarenko, G. V., 235
Sokolovsko-Sarbaysky ore beneficiation plant (SSGOK), 1, 6, 28, 262, 264–267, 288, 295, 297–298, 362
fugitive dust releases during loading, 339, 341–344, 347, 350–353
fugitive dust releases in outdoor storage, 299–300, 306, 309–310, 317–320, 328
Solid-particle jets, aerodynamics of in channels, 196–211
one-dimensional flow, 200–211
plane-parallel flow, 197–200
Solid-particle jets, air injection in, 133–196
axially symmetric jets, 184–196
flat jets of loose matter, 144–184
initial equations, 134–144
Solid stream suction properties, study of, 10–14
Space-time method of averaging accounting equations, 25, 373–379
energy transfer equation, 378–379
mass transfer equation, 374–376
pulse transfer equation, 376–377
Spiral chutes, 9, 262–264
Splicing method, 153, 169–170, 172, 192
Steady settling velocity measurement, 65
Steel balls
aerodynamic characteristics, 52–54
drag factor, 73–74, 74
Stokes flow-round area, 53, 55–56, 62, 64–66
Stokes law, 46, 57
Stokes number, 273–274
Stoylensky complex, 1
Strouhal number, 62
Suction volume, 8
Sverdlovsk (Yekaterinburg) Institute for Occupational Safety (SIOT), 258, 355

T

Tapered chutes, 9, 26, 50–51
Telescopic chutes, 3, 9, 321, 342–343, 345–347, 350–351, 353, 355–357, 362
Thermal head, 12, 113–116, 211, 214, 221–222, 346, 348
Thermal pressure in cowls, 221
Tip chutes, 33, 35, 40–42, 47, 76
Transfer groups, 1–12
intensity of dust emissions, 1–8
primary means of dust emission control, 8–12
subgroups, 33–34
Tsoy, B., 16
Tsytsura, A. A., 8
Turbulent friction stress tensor, 140
Two-way chutes, 9

U

Unconstrained motion, 35, 37–38, 41–42
Uniformly accelerated particle streams, 12, 30, 41, 78, 90, 109, 360
Uniform motion, 30–31, 108, 198, 208
Uniform particle distribution, 30–31, 73, 77, 79–80, 173, 190, 360
air injection in bin chutes with, 104–107
air streams in flat jets of loose matter, 154–157
Unit particle resistance coefficient, 10
Ural Promstroy NII project, 29

V

Vacuum-gauge pressure, 7, 23–25, 87, 99–100, 104
Vakhrusheva, I. A., 54
Vertical chutes, 12, 16, 19, 33, 35*n*, 73, 74*n*, 84*n*, 107, 111–112, 225, 263, 326
Viscous flow around particles, 45, 137, 139, 146, 148, 168, 187
Viscous stress tensor, 139
VNIIBTG (All-Russia Institute for Occupational Safety in Ore Mining), 26, 258, 277, 300, 321, 327, 341, 351–353
Voegeli, H. E., 87
Volumetric concentration of particles
  in axially symmetric jets of freely falling particles, 184
  dust release intensity and, 316
  exponentially distributed particles, 169
  fluid dynamics equations, 139
  fugitive dust emission reduction methods, 321
  in jets of loose matter, 134–136
  moisture content and, 265
  self-similar motion equations, 144

W

Water drops, 72–74
Watering materials (hydraulic dust control), 3–4, 8–9, 253, 301, 329
Weber number, 73
Wet material, 30, 102, 118, 120, 251, 366

Y

Yaglom, A. M., 303
Yudin, M. I., 303
Yuzhny complex (Southern complex; YuGOK), 1–2, 4, 362
Yuzhuralmash Manufacturing Group, 329–330, 332–334

Z

Zadorozhny, S. I., 309*n*
Zenkov, R. L., 44
Zhdanovsky steel works, 341
Zhelezorudnaya railway station, Rudny, Kazakhstan, 314
Zhuravlev, V. P., 8
Zilberberg, Ya. I., 103, 211